THE FUNGI

VOLUME IVB
A Taxonomic Review with Keys:
Basidiomycetes and Lower Fungi

Contributors to This Volume

G. C. AINSWORTH

CONSTANTINE J. ALEXOPOULOS

M. W. DICK

D. M. DRING

C. L. DUDDINGTON

RUBÉN DURÁN

J. J. ELLIS

KENNETH A. HARRISON

C. W. HESSELTINE

G. F. LAUNDON

ROBERT W. LICHTWARDT

R. F. R. McNABB

D. N. PEGLER

RONALD H. PETERSEN

KENNETH B. RAPER

ALEXANDER H. SMITH

F. K. SPARROW

P. H. B. TALBOT

GRACE M. WATERHOUSE

THE FUNGI

An Advanced Treatise

Edited by

G. C. AINSWORTH

FORMERLY OF THE COMMONWEALTH MYCOLOGICAL INSTITUTE
KEW, SURREY, ENGLAND

FREDERICK K. SPARROW

DEPARTMENT OF BOTANY
UNIVERSITY OF MICHIGAN
ANN ARBOR, MICHIGAN

ALFRED S. SUSSMAN

DEPARTMENT OF BOTANY
UNIVERSITY OF MICHIGAN
ANN ARBOR, MICHIGAN

VOLUME IVB
A Taxonomic Review with Keys:
Basidiomycetes and Lower Fungi

1973

ACADEMIC PRESS New York and London
A Subsidiary of Harcourt Brace Jovanovich, Publishers

ACADEMIC PRESS, INC.
111 Fifth Avenue, New York, New York 10003

United Kingdom Edition published by
ACADEMIC PRESS, INC. (LONDON) LTD.
24/28 Oval Road, London NW1

Library of Congress Cataloging in Publication Data

Ainsworth, Geoffrey Clough, DATE   ed.
      The fungi; an advanced treatise.

      Vol. 4B edited by G.C. Ainsworth, F. K. Sparrow, and
A. S. Sussman.
      Includes bibliographies.
      CONTENTS: v. 1. The fungal cell–v. 2. The fungal
organism–v. 3. The fungal population.   [etc.]
      1.   Fungi.   I.   Sussman, Alfred S.   II.   Sparrow,
Frederick Kroeber, DATE   [DNLM: 1.   Fungi.
QK 603 A297f]
QK603.A52   589.2   65–15769
ISBN 0–12–045644–3 (v. 4B)

To the memory of

GEORGE WILLARD MARTIN

1886—1971

Contents

EUMYCOTA

MASTIGOMYCOTINA

MASTIGOMYCOTINA, PLASMODIOPHOROMYCETES

MASTIGOMYCOTINA, CHYTRIDIOMYCETES, AND HYPHOCHYTRIDIOMYCETES

MASTIGOMYCOTINA, OOMYCETES

8. Leptomitales

M. W. Dick

9. Lagenidiales

F. K. Sparrow

10. Peronosporales

Grace M. Waterhouse

ZYGOMYCOTINA, ZYGOMYCETES

11. Mucorales

C. W. Hesseltine and J. J. Ellis

12. Entomophthorales

Grace M. Waterhouse

13. Zoopagales

C. L. Duddington

ZYGOMYCOTINA, TRICHOMYCETES

14. Trichomycetes

Robert W. Lichtwardt

BASIDIOMYCOTINA, TELIOMYCETES

15. Uredinales

G. F. Laundon

20. Aphyllophorales II: The Clavarioid and Cantharelloid Basidiomycetes

Ronald H. Petersen

21. Aphyllophorales III: Hydnaceae and Echinodontiaceae

Kenneth A. Harrison

22. Aphyllophorales IV: Poroid Families

D. N. Pegler

23. Agaricales and Related Secotioid Gasteromycetes

Alexander H. Smith

BASIDIOMYCOTINA, GASTEROMYCETES

24. Gasteromycetes

D. M. Dring

List of Contributors

Numbers in parentheses indicate the page on which the authors' contributions begin.

G. C. Ainsworth (1), Formerly of The Commonwealth Mycological Institute, Kew, Surrey, England

Constantine J. Alexopoulos (39), Department of Botany, University of Texas at Austin, Austin, Texas

M. W. Dick (113, 145), Department of Botany, University of Reading, Reading, England

D. M. Dring (451), The Herbarium, Royal Botanic Gardens, Kew, Surrey, England

C. L. Duddington (231), Formerly of the Department of Life Sciences, The Polytechnic of Central London, London, England

Rubén Durán (281), Department of Plant Pathology, Washington State University, Pullman, Washington

J. J. Ellis (187), Northern Regional Research Laboratory, Agricultural Research Service, U. S. Department of Agriculture, Peoria, Illinois

Kenneth A. Harrison (369), University of Michigan Herbarium, Ann Arbor, Michigan

C. W. Hesseltine (187), Northern Regional Research Laboratory, Agricultural Research Service, U. S. Department of Agriculture, Peoria, Illinois

G. F. Laundon (247), Department of Agriculture, Horticultural Research Centre, Levine, New Zealand

Robert W. Lichtwardt (237), Department of Botany, University of Kansas, Lawrence, Kansas

R. F. R. McNabb[1] (303, 317), Microbiology Department, Lincoln College, Canterbury, New Zealand

D. N. Pegler (397), The Herbarium, Royal Botanic Gardens, Kew, Surrey, England

[1] Deceased.

Ronald H. Petersen (351), Department of Botany, University of Tennessee, Knoxville, Tennessee

Kenneth B. Raper (9), Departments of Bacteriology and Botany, University of Wisconsin, Madison, Wisconsin

Alexander H. Smith (421), Department of Botany and University Herbarium, University of Michigan, Ann Arbor, Michigan

F. K. Sparrow (61, 85, 159), Department of Botany, University of Michigan, Ann Arbor, Michigan

P. H. B. Talbot (317, 327), Waite Agricultural Research Institute, Glen Osmond, South Australia

Grace M. Waterhouse (75, 165, 219), Formerly of the Commonwealth Mycological Institute, Kew, Surrey, England

Preface

Volume IV of this treatise in which an attempt is made to provide generic keys for all the major groups of fungi is unique among multiauthor works. Never before in modern times has such a distinguished group of world specialists on the taxonomy of fungi provided for the professional mycologist such a comprehensive survey of fungal classification at the generic level.

It may be noted that although contributors were asked to conform to a standard pattern, the need for differences in the approaches to some groups— and also the well-known individualism of taxonomists which we were loath to suppress—led to contributions which show a rather wide and, at times, radical variation in treatment. In addition, although contributors to allied groups consulted with one another, a number of genera, and even families, are duplicated by different authors—that is, they appear in the keys of more than one of the higher taxa. Such duplication, as any student of the fungi knows, is, indeed, a reflection of currently unresolved taxonomic problems. These differences and duplications should, however, help rather than hinder the many who will have occasion to consult this volume which is comprised of two parts, IVA and IVB, each part being published separately.

The two parts of Volume IV complete the treatise as planned, and although there is no intention of issuing new editions of any of the volumes the possibility of updating the treatise by one or more supplementary volumes is being considered.

We would like to take this opportunity to thank all the contributors to this treatise for their hard work and for their forbearance with editorial pedantry, and, in conclusion, on behalf of ourselves and every contributor, to thank the staff of Academic Press for their patient and unobtrusive work which has done so much to lighten our labors.

G. C. Ainsworth
F. K. Sparrow
A. S. Sussman

Contents of
Previous Volumes

VOLUME I

THE FUNGAL CELL

VOLUME II
THE FUNGAL ORGANISM

VOLUME III

THE FUNGAL POPULATION

VOLUME IVA

ASCOMYCETES AND FUNGI IMPERFECTI

CHAPTER 1

Introduction and Keys to Higher Taxa

G. C. AINSWORTH

Formerly of the
Commonwealth Mycological Institute
Kew, Surrey, England

During recent years, the traditional approaches to the status, circumscription, and taxonomic arrangement of fungi have been seriously questioned. It is customary to classify fungi as plants, if only on the basis that they are not animals, but as Stafleu (1969) writes, "The times in which we recognized two kingdoms, *Regnum vegetabile* and *Regnum animale*, are perhaps surviving only in the subdivision of biology in some of our universities . . ."; the "seemingly fundamental division in two has proved to be untenable." Where should fungi be classified today?

I. STATUS OF FUNGI

Dissatisfaction with the division of organisms into two kingdoms is not a new phenomenon. A century ago Haeckel proposed the taxon "Moneres" and subsequently developed a four-kingdom system. Since then there have been a number of proposals for the reorganization of the major taxa. Representative modern examples are the four-kingdom systems of Copeland (1956) and Barkley (1968) and the five-kingdom system of Whittaker (1969). The writings of these authors may be consulted for details of their own and other systems.

There is general agreement that taxonomy should endeavor to reflect phylogeny and that in the virtual absence of a fossil record much of the detailed phylogeny of microorganisms is speculative. However, by supplementing the morphological approach to taxonomy with the results of nutritional and biochemical studies, it is possible to arrive at arrangements which, in general, correspond with current beliefs on possible evolutionary sequences. There is, for example, widespread agreement that the bacteria and

1

blue-green algae, which comprise the kingdom Mychota of Copeland and the kingdom Monera of Barkley and Whittaker, are distinct from other organisms and show primitive characters. Copeland classifies the fungi, with the algal and protozoan groups, in his kingdom Protoctist, mainly in the phylum Inophyta; Barkley's treatment, in which fungi are assigned to the kingdom Protista, mostly as the phylum Mycophyta, is very similar. Whittaker, on the other hand, in an attractive and well-argued arrangement, raises the fungi to the status of a kingdom.

Whittaker's two primitive kingdoms are the prokaryotic Monera and the eukaryotic Protista, and he suggests that it was from Protista-like ancestors that three nutritionally distinct lines have developed: (1) the *plant kingdom*, characterized by photosynthesis, (2) the *animal kingdom*, characterized by ingestive nutrition, and (3) *fungi*, in which nutrition is absorptive.

Recent additional evidence in support of the distinctness of fungi has been derived from the comparative studies by Nolan and Margoliash (1968) on cytochrome c in different organisms. Cytochrome c, a component of the terminal respiratory chain of enzymes in aerobic organisms (see Lindenmayer, 1965), is widely distributed in both plants and animals. It is concluded that the cytochrome c enzymes in all organisms are homologous, that they arise from homologous gene loci, and that each possesses an evolutionary history. From these premises, the interpretation of the experimental findings suggests that the fungi studied form a phylogenetic line distinct from the animal and plant kingdoms. Wheat, the representative of the plant kingdom investigated, was found to be more closely related to man than to fungi from the point of view of its cytochrome c. Whatever the final outcome of such studies, fungi are clearly organisms which, as Martin (1968) concluded in Volume III, "may reasonably be treated as a discrete major taxonomic unit."

II. CIRCUMSCRIPTION OF THE FUNGI

The circumscription of Fungi is not always well defined. The best-known example of a group of uncertain affinity is that of the slime molds, which has long been claimed by both mycologists (as the Myxomycetes) and zoologists (as the Mycetozoa). Other groups, if less familiar, are of equally uncertain status: the cellular slime molds (Acrasiomycetes), for example, are claimed as Protozoa, a taxon to which the Hydromyxomycetes (including the Labyrinthulales) have been referred.

At the turn of the century, it was a widely held belief that fungi of the class Phycomycetes were derived from algae, a belief reflected in the nomenclature adopted by Clements and Shear (1931) in their "Genera of Fungi" for the orders of Phycomycetes. This hypothesis, which fell into disrepute, has recently been revived in modified form as seen by the classification of the Oomycetes (the only fungal group exhibiting true cellulose in the cell wall) with the algae by both Copeland (1956) and Barkley (1968) and by the German mycologist Kreisel (1969), while the Hyphochytriales (or Hypho-chytridiomycetes) are also associated with biflagellated algal groups (Heterokontae) by Barkley, Copeland, and Kreisel. In this treatise, following tradition, the Acrasiomycetes and Myxomycetes are included as fungal groups, as are the Oomycetes, but the Hydromyxomycetes are excluded. The recent discovery of biflagellate zoospores (see Mastigomycotina, Chapter 4) in the latter group, however, will necessitate a reconsideration of their status.

III. TAXONOMIC ARRANGEMENTS

For the past fifty years, most taxonomic arrangements of the fungi have differentiated the plasmodial Myxomycota) from the mycelial (Eumycota) forms, which comprise the bulk of fungi. Three classes of the latter have been universally recognized—Phycomycetes, Ascomycetes, and Basidiomy-cetes—based on the types of sexually produced spores. These are grouped with the Deuteromycetes, or Fungi Imperfecti, which are characterized by the possession of asexual spores only. Because the class Phycomycetes was patently a miscellaneous assemblage, following the lead of Sparrow (1959), its constituents are with increasing frequency being accommodated in a series of classes: the Chytridiomycetes, Hyphochytridiomycetes, Oomy-cetes, Zygomycetes, and Trichomycetes. Although it is still held that the Ascomycetes are monophyletic, there have been major adjustments within the group, particularly as a result of the significance now given to uni- and bitunicate asci as taxonomic criteria. For a hundred years, treatment of the basidiomycetes, which include so many of the larger fungi, has been dominated by the Friesian approach which had—and still has—the attrac-tion of differentiating fungi into groups mainly on field characters. More critical microscopic studies of basidiocarp structure and morphogenesis and the recognition that there has been much evolutionary convergence are leading to an abandonment of the Friesian system and the development of more natural groupings. For many basidiomycetes, as for many ascomy-cetes, the older descriptions are inadequate, and new descriptions have yet to be prepared, so that in many areas the current classification of these fungi is in a state of transition. Some of the other classes recognized in the present account, for example, the Pyrenomycetes and Gasteromycetes, are also clearly heterogeneous, but they are still useful pigeonholes for ordering many common fungi.

In these volumes, the taxonomic arrangement is based on that proposed by Ainsworth (1966) for use in the current edition of the "Dictionary of Fungi" (Ainsworth, 1971). Fungi are treated here as either a separate kingdom or, for the more conservative, as a subkingdom of the plant kingdom, with two divisions, the Myxomycota, for plasmodial forms, and the Eumycota, for nonplasmodial forms which are frequently mycelial. Five subdivisions of the latter are recognized, including the Ascomycotina (for ascomycetes) and the Basidiomycotina (for basidiomycetes), while the imperfect fungi are classified for convenience as the Deuteromycotina, although in a hierarchical classification it is incorrect to equate these fungi with the ascomycetes and the basidiomycetes to which they are subsidiary both taxonomically and by nomenclature. Most lichenized fungi (lichens) have been omitted, as they are to be the subject of a separate book.

IV. KEYS TO THE HIGHER TAXA

FUNGI

It is difficult to give a concise diagnostic definition of fungi. The main characteristics of the group are

Nutrition: heterotrophic (photosynthesis lacking) and absorptive (ingestion rare).

Thallus: on or in the substratum and plasmodial amoeboid or pseudoplasmodial; or in the substratum and unicellular or filamentous (mycelial), the last, septate or nonseptate; typically nonmotile (with protoplasmic flow through the mycelium) but motile states (e.g., zoospores) may occur.

Cell wall: well-defined, typically chitinized (cellulose in Oomycetes).

Nuclear status: eukaryotic, multinucleate, the mycelium being homo- or heterokaryotic, haploid, dikaryotic, or diploid, the last being usually of limited duration.

Life cycle: simple to complex.

Sexuality: asexual or sexual and homo- or heterothallic.

Sporocarps: microscopic or macroscopic and showing limited tissue differentiation.

Habitat: ubiquitous as saprobes, symbionts, parasites, or hyperparasites.

Distribution: cosmopolitan.

KEY TO DIVISIONS OF FUNGI

1. Plasmodium or pseudoplasmodium present . . . . **Myxomycota** I

1'. Plasmodium or pseudoplasmodium absent, assimilative phase
  typically filamentous . . . . **Eumycota** II

I. Myxomycota

KEY TO CLASSES OF MYXOMYCOTA

1. Assimilative phase a plasmodium 2

1′. Assimilative phase free-living amoebae which unite as a pseudoplasmodium
before reproduction . . . . **Acrasiomycetes**[1] p. 12

   2(1) Plasmodium forming a network ("net plasmodium") . . . . . . **Labyrinthulales**[2]

   2′(1) Plasmodium not forming a network 3

3(2′) Plasmodium saprobic, free-living . . . . **Myxomycetes** p. 53

3′(2′) Plasmodium parasitic within cells of the
host plant . . . . **Plasmodiophoromycetes**[3] p. 83

II. Eumycota

KEY TO SUBDIVISIONS OF EUMYCOTA

1. Motile cells (zoospores) present; perfect-state
spores typically oospores . . . . **Mastigomycotina** (p. 65) III

1′. Motile cells absent 2

   2(1′) Perfect state present 3

   2(1′) Perfect state absent . . . . **Deuteromycotina** VII

3(2) Perfect-state spores zygospores . . . . **Zygomycotina** IV

3′(2) Zygospores absent 4

   4(3′) Perfect-state spores ascospores . . . . **Ascomycotina** V

   4′(3′) Perfect-state spores basidiospores . . . . **Basidiomycotina** VI

III. Mastigomycotina

KEY TO CLASSES OF MASTIGOMYCOTINA

1. Zoospores posteriorly uniflagellate
(flagella whiplash-type) . . . . **Chytridiomycetes** p. 101

1′. Zoospores not posteriorly uniflagellate 2

   2(1′) Zoospores anteriorly uniflagellate
   (flagella tinsel-type) . . . . **Hyphochytridiomycetes** p. 112

   2′(1′) Zoospores biflagellate (posterior flagellum whiplash-type;
   anterior tinsel-type); cell wall cellulosic . . . . **Oomycetes** p. 75

[1]Excluded from the Myxomycota by Martin and Alexopoulos (1969).
[2]Excluded from this treatment.
[3]Treated as a class of the Mastigomycotina.

IV. Zygomycotina

KEY TO CLASSES OF ZYGOMYCOTINA

1. Saprobic or, if parasitic or predacious, having mycelium immersed in
host tissue . . . . **Zygomycetes**　p. 191
1'. Associated with arthropods and attached to the cuticle or digestive tract by a holdfast
and not immersed in the host tissue 　. . . . . **Trichomycetes**　p. 244

V. Ascomycotina

KEY TO CLASSES OF ASCOMYCOTINA

1. Ascocarps and ascogenous hyphae lacking; thallus
mycelial or yeastlike 　. . . . . **Hemiascomycetes**　Vol. IVA
1'. Ascocarps and ascogenous hyphae present; thallus mycelial 　. 2
　　2(1') Asci bitunicate; ascocarp an
　　ascostroma 　. . . . . **Loculoascomycetes**　Vol. IVA
　　2'(1') Asci typically unitunicate; if bitunicate,
　　ascocarp an apothecium 　. 3
3(2') Asci evanescent, scattered within the astomous ascocarp which
is typically a cleistothecium; ascospores aseptate 　. . . . . . **Plectomycetes**　Vol. IVA
3'(2') Asci regularly arranged within the ascocarp as a
basal or peripheral layer 　. 4
　　4(3') Exoparasites of arthropods; thallus reduced; ascocarp a
　　perithecium; asci inoperculate 　. . . . . **Laboulbeniomycetes**　Vol. IVA
　　4'(3') Not exoparasites of arthropods 　. 5
5(4') Ascocarp typically a perithecium which is usually ostiolate (if astamous, asci not
evanescent); asci inoperculate with an apical pore or
slit 　. . . . . **Pyrenomycetes**　Vol. IVA
5'(4') Ascocarp an apothecium or a modified apothecium, frequently macrocarpic,
epigean or hypogean; asci inoperculate or
operculate 　. . . . . **Discomycetes**　Vol. IVA

VI. Basidiomycotina

KEY TO CLASSES OF BASIDIOMYCOTINA[4]

1. Basidiocarp lacking and replaced by teliospores
(encysted probasidia) grouped in sori or scattered within the
host tissue; parasitic on vascular plants 　. . . . . **Teliomycetes**　p. 251
1'. Basidiocarp usually well-developed; basidia typically
organized as a hymenium; saprobic or rarely parasitic 　. 2
　　2(1') Basidiocarp typically gymnocarpous or semiangiocarpous;
　　basidia phragmobasidia (**Phragmobasidiomycetidae**, p. 310) or

　　[4]If yeastlike, see p. 11.

holobasidia (**Holobasidiomycetidae**, p. 323); basidiospores
ballistospores . . . . **Hymenomycetes** p. 307

2'(1') Basidiocarp typically angiocarpous; basidia holobasidia;
basidiospores not ballistospores . . . . **Gasteromycetes** p. 451

VII. Deuteromycotina

KEY TO CLASSES OF DEUTEROMYCOTINA

1. Budding (yeast or yeastlike) cells with or without pseudomycelium
characteristic; true mycelium lacking or not
well-developed . . . . **Blastomycetes** Vol. IVA

1'. Mycelium well-developed, assimilative budding cells absent 2

2(1') Mycelium sterile or bearing spores directly or on
special branches (sporophores) which may be variously
aggregated but not in pycnidia or acervuli . . . . . . **Hyphomycetes** Vol. IVA

2'(1') Spores in pycnidia or acervuli . . . . **Coelomycetes** Vol. IVA

REFERENCES

Ainsworth, G. C. (1966). A general purpose classification for fungi. *Bibl. Syst. Mycol.* No. 1:1–4.

Ainsworth, G. C. (1971). "Ainsworth and Bisby's Dictionary of the Fungi," 6th ed. Commonwealth Mycol. Inst., Kew, Surrey, England.

Barkley, F. A. (1968). "Outline Classification of Organisms," 2nd ed. Hopkins Press, Providence, Massachusetts.

Clements, F. E., and C. L. Shear. (1931). "The Genera of Fungi." Wilson, New York.

Copeland, H. F. (1956). "The Classification of Lower Organisms." Pacific Books, Palo Alto, California.

Kreisel, H. (1969). "Grundzüge eines natürlichen Systems der Pilze." Cramer, Lehre.

Lindenmayer, A. (1965). Carbohydrate metabolism. 3. Terminal oxidation and electron transport. *In* "The Fungi" (G. C. Ainsworth and A. S. Sussman, eds.). Vol. 1, pp. 301–348. Academic Press, New York.

Martin, G. W. (1968). The origin and status of fungi. *In* "The Fungi" (G. C. Ainsworth and A. S. Sussman, eds.), Vol. 3, pp. 635–648. Academic Press, New York.

Martin, G. W., and C. J. Alexopoulos. (1969). "The Myxomycetes," p. 30. Univ. of Iowa Press, Iowa City.

Nolan, C., and E. Margoliash. (1968). Comparative aspects of primary structures of proteins. *Annu. Rev. Biochem.* **37**:727–790.

Sparrow, F. K. (1959). Interrelationships and taxonomy of the aquatic phycomycetes. *Mycologia* **50**:797–813.

Stafleu, F. A. (1969). Biosystematic pathways anno 1969. *Taxon* **18**:485–500

Whittaker, R. H. (1969). New concepts of kingdoms of organisms. *Science* **163**: 150–160.

Myxomycota
Acrasiomycetes

CHAPTER 2

Acrasiomycetes[1]

KENNETH B. RAPER

Departments of Bacteriology and Botany
University of Wisconsin
Madison, Wisconsin

I. INTRODUCTION

The microorganisms presently considered as constituting the class Acrasiomycetes may or may not represent a natural group, but they all have in common a trophic stage consisting of myxamoebae or, in a few cases, minute plasmodia, and all show some level of cellular differentiation incident to fructification. For these reasons they are brought together here.

Whether the Acrasiomycetes should be included with the fungi, or should be considered more properly as protozoa that exhibit varying levels of cellular integration and differentiation, poses a question of long standing which I will not attempt to resolve. Instead, I will assume that since these organisms, along with the Myxomycetes, have been isolated and studied primarily by mycologists, historical precedent if not demonstrable kinship with the fungi warrants their inclusion in this volume. Furthermore, since mycologists generally adhere to the International Code of Botanical Nomenclature, I will do my best to conform the names of families, orders, and subclasses used in this chapter to botanical rather than zoological usage.

No single type of fructification can be cited as characterizing the Acrasiomycetes since these vary markedly in structure, form, and dimensions. What can be said is that fructifications arise from small amoeboid protoplasts that undergo substantial differentiation, singly in some forms and collectively in others—cellular differentiation in the latter being clearly governed by different levels of intercellular coordination and controls.

In genera of the former type, comprising the Protostelidae (Protostelids), the vegetative stage may consist of uninucleate or plurinucleate myxamoebae,

[1]This summary report is based in substantial part upon studies supported by research grants from the National Science Foundation (No. GB-8624) and the National Institutes of Health, U. S. Public Health Service (AI ₦ 04915–17).

or of myxamoebae and small plasmodia that later fragment into smaller protoplasts before fruiting occurs. In either case the resulting fructifications, or *sporocarps*,[2] consist of delicate, upright tubular stalks bearing one to few spores. Some genera among these minute and presumably primitive forms may represent progenitors of the Myxomycetes while others may approximate ancestral types of such genera as *Dictyostelium* and *Polysphondylium* (see L. S. Olive, 1970).

In other and more highly developed genera, the trophic stage consists *only* of myxamoebae that are typically uninucleate; and these subsequently aggregate to form closely coordinated cell associations, or *pseudoplasmodia*, from which later develop multicellular fruiting structures, or *sorocarps*.[2] Among such aggregating types, now generally known as the "Cellular Slime Molds," varying degrees of intercellular organization characterize the pseudoplasmodia and differing levels of cellular differentiation are achieved by the associated myxamoebae during fructification. Two general subgroups, or subclasses, can be distinguished: (1) the Acrasidae wherein the myxamoebae aggregate without forming definite streams, the cells instead converging as individuals toward common centers from which subsequently develop fruiting structures that may range from simple or branched columns of encysted cells in some forms, to sorocarps with globose sori borne on definite stalks composed of modified but not highly differentiated cells in others; (2) the Dictyostelidae, wherein the myxamoebae collect into definite streams during aggregation and subsequently produce sorocarps which, except in the genus *Acytostelium*, consist of stalks, or *sorophores*, composed of vacuolate, parenchymalike cells, bearing terminal dropletlike spore masses, or *sori*.

Basic differences occur also in the amoeboid or trophic stages of the three subclasses: the myxamoebae (and small plasmodia) of the Protostelidae are characterized by nuclei with single, centrally positioned nucleoli and filose pseudopodia; the myxamoebae of the Acrasidae have nuclei of a similar pattern, apparently, but show lobose pseudopodia; whereas the myxamoebae of the Dictyostelidae contain nuclei with two or more peripheral nucleoli and have filose pseudopodia.

KEY TO SUBCLASSES OF ACRASIOMYCETES

1. Sporulation not preceded by aggregation of myxamoebae; sporocarps 1- to few-spored; trophic stage consisting of myxamoebae or minute plasmodia, both with filose pseudopodia

---

[2]The term *sporocarp* is used to designate small, stalked fructifications bearing one to few spores that arise from single uni- or plurinucleate protoplasts (L. S. Olive, 1967). In contrast, the term *sorocarp* is applied to fructifications that arise from communities of cells and consist of a definite spore mass, or sorus, borne upon a supporting stalk (Harper, 1926).

and nuclei with single centrally positioned nucleoli; flagellate stage present in some genera, lacking in others . . . . **Protostelidae** p. 11

1′. Sporulation preceded by aggregation of myxamoebae to form pseudoplasmodia; sorocarps multispored; trophic stage consisting of uninucleate myxamoebae; pseudopodia filose in some genera and lobose in others; flagellate cells absent 2

2(1′) Aggregating myxamoebae do not form streams in developing pseudoplasmodia; fructifications may or may not show definite sori and sorophores; myxamoebae with lobose pseudopodia and nuclei with single centrally positioned nucleoli
. . . . **Acrasidae** p. 17

2′(1′) Aggregating myxamoebae form convergent streams in developing pseudoplasmodia; sorocarps with well-defined sori and sorophores: myxamoebae with filose pseudopodia and nuclei with 2 or more peripheral nucleoli . . . . **Dictyostelidae** p. 25

II. SUBCLASS PROTOSTELIDAE

The vegetative stage of this subclass ranges from uninucleate or pluri-nucleate myxamoebae to small, multinucleate, and often reticulate plasmodia, the latter lacking the characteristic rhythmic ebb and flow of protoplasm seen in the plasmodia of the Myxomycetes. The pseudopodia are typically filose, infrequently appearing lobose in some genera and under some conditions. Flagellate cells are characteristic of some genera, not of others. The fruiting bodies, or sporocarps, arise from uninucleate or multi-nucleate protoplasts and consist of tubular stalks of varying proportions bearing one to few spores. Contractile vacuoles are present in the active protoplasts of all genera. The nuclei possess a single, usually centrally positioned nucleolus which may arise by the fusion of several nucleolar bodies following nuclear division.

Of widespread occurrence throughout the world, protostelids have been isolated by L. S. Olive and Stoianovitch (1960, 1966a,b, 1971a,b) from dead wood, tree bark, humus, soil, dung, and especially from dead plant structures such as florets, pods, capsules, berries, etc. Information concerning the isolation and cultivation of these microorganisms may be found in the several papers by Olive and by Olive and Stoianovitch, who not only discovered the protostelids but have described all known members of this singular group.

The process of sporocarp formation in the Protostelidae is described by L. S. Olive (1970): "The prespore cell at first becomes hemispherical, then hat-shaped as its protoplasm concentrates in the center, leaving a thinner margin. A thin protective sheath, which begins to develop over the protoplast at this time, surrounds the sporocarp throughout its ontogeny. A steliogen appears in the lower part of the protoplast and begins to secrete the stalk. A narrow portion of the steliogen extends into the upper part of the stalk during its development, keeping the stalk tube hollow. At the end of the process of sporogenesis, which may require little more than half an

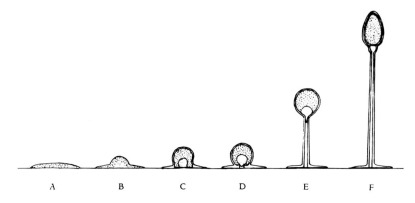

FIG. 1. Diagrammatic illustration of sporogenesis in *Nematostelium ovatum* Olive. A, Prespore cell; B, hat-shaped stage; C, delimination of steliogen and sheath; D, beginning of stalk development; E, intermediate stage in stalk development, with steliogen extending into stalk tube; F, mature sporocarp. (Courtesy of L. S. Olive, 1970.)

hour, the protoplast secretes a wall and develops into a spore at the tip of the stalk.... The spores may be round, oval pyriform, or elongate." The development of such a sporocarp is illustrated in Fig. 1.

A.   Order Protosteliales

The order possess the characteristics of the subclass.

KEY TO FAMILIES OF PROTOSTELIALES

1. Sporocarps produced from prespore cells directly on the substrate, with or without a flagellate stage   2

1'. Sporocarps produced on gelatinous spore horns . . . . **Ceratiomyxaceae**   p. 16

    2(1) Flagellate cells lacking   . . . . **Protosteliaceae**   p. 12

    2'(1) Flagellate cells present   . . . . **Cavosteliaceae**   p. 15

1.   Protosteliaceae (L. S. Olive, 1962)

These forms do not show flagellate cells at any stage; the vegetative phase consists of either uninucleate, or both uni- and plurinucleate myx-amoebae, or in some species of small reticulate plasmodia. They have sporo-carps with stalks ranging from short and subulate to long and slender, usually bearing a single spore; in some species the spores are apophysate and deciduous, in others they are nonapophysate and nondeciduous.

KEY TO GENERA OF PROTOSTELIACEAE

1. Trophic stage uninucleate or plurinucleate protoplasts   2

1'. Trophic stage a small reticulate plasmodium   3

Protostelium (L. S. Olive and Stoianovitch, 1960). The sporocarps in this genus typically bear single spores (sometimes two-spored in one species) on slender stalks with a reduced apophysis at the base of the spore. The spores are variable in form, globose (in two species), bell-shaped to pyriform (in two species), or oval to peanut-shaped (one species). The myxamoebae are typically uninucleate but plurinucleate cells are not uncommon in some species, and are generally hyaline but showing an orange pigmentation in one species. The genus includes five species, of which *P. mycophaga* may be considered representative (Figs. 2–4).

Protosteliopsis (L. S. Olive and Stoianovitch, 1966a). The sporocarps resemble those of *Protostelium* but their stalks are less slender and somewhat irregular, nonapophysate and tending to deliquesce in water; the spores are globose, typically uninucleate, and nondeciduous. The myxamoebae are typically uninucleate or less commonly multinucleate, with filose or sometimes lobose pseudopodia. A single species, *P. fimicola*, has been described.

Schizoplasmodium (L. S. Olive and Stoianovitch, 1966b). In this genus, the sporocarps are short stalked, bearing single multinucleate spores with subtending apophysis; the spores are globose to subglobose, and are forcibly discharged by a gas bubble mechanism. The germinating spore gives rise to a multinucleate protoplast which develops into a small reticulate plasmodium that subsequently fragments to yield prespore cells from which sporocarps develop. The genus presently consists of a single species, *S. cavostelioides*.

Nematostelium (L. S. Olive, 1970). The sporocarps consist of slender stalks that bear single multinucleate spores subtended by an apophysis; the spores are oval or globose with a distinct hilum, deciduous but not forcibly discharged, germinating to produce a thin reticulate plasmodium with filopodia or reticulopodia, subsequently segmenting into multinucleate prespore cells from which the sporocarps develop. Two species, *N. ovatum* and *N. gracile,* comprise the genus, both of which were previously described as *Schizoplasmodium* (Olive and Stoianovitch, 1966b) and subsequently reassigned to this genus.

Schizoplasmodiopsis (L. S. Olive, 1967). Here the sporocarps consist of

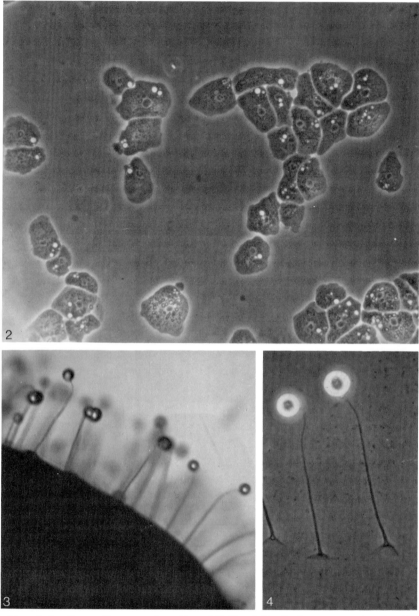

FIGS. 2–4. *Protostelium mycophaga* O. and S. Fig. 2. Postfeeding myxamoebae showing contractile vacuoles (white) and nuclei with centrally placed nucleoli, × 410. Fig. 3. Sporocarps as seen in side view showing spores and supporting stalks, × 300. Fig. 4. Sorocarps photographed on agar beneath coverglass showing spores and delicate stalks with expanded disclike bases, × 600. Figs. 2 and 4: phase microscopy.

relatively short, upright stalks bearing single globose spores, which are nonapophysate and nondeciduous. The spores are uninucleate, germinating to release protoplasts that develop into reticulate plasmodia by growth and/or fusion of adjoining protoplasts; the plasmodia subsequently fragment to produce prespore cells from which the sporocarps arise, or form irregular cysts that frequently contain enucleate sporoid bodies. The genus contains a single species, *S. pseudoendospora*.

2. Cavosteliaceae (L. S. Olive, 1964)

These forms have flagellate cells and the amoeboid stage ranges from uninucleate cells to small reticulate plasmodia, producing stalked sporocarps bearing one or two spores.

KEY TO GENERA OF CAVOSTELIACEAE

1. Sporocarps very short-stalked, bearing one or two uninucleate spores
. . . . **Cavostelium** p. 15

1'. Sporocarps with comparatively long thin stalks.
  2(1') Sporocarps typically bearing single and mostly uninucleate spores, nondeciduous
. . . . **Planoprotostelium** p. 15

  2'(1') Sporocarps bearing single multinucleate spores, deciduous
. . . . **Ceratiomyxella** p. 15

Cavostelium (L. S. Olive, 1964). The vegetative stage of this genus consists mainly of uninucleate, but occasionally plurinucleate amoeboid cells. The myxamoebae become flagellate in water; they are typically uniflagellate but often show two or more flagella. The sporocarps are short-stalked with an apophysis at the stalk apex; they are one to two spored in one species (*C. apophysatum*) and consistently two-spored in a second species (*C. bisporum*).

Planoprotostelium (L. S. Olive and Stoianovitch, 1971b). The trophic stage typically consists of uninucleate, less commonly bi- or even trinucleate amoeboid cells with filose pseudopodia, one or more contractile vacuoles and orange pigmented cytoplasm. The myxamoebae become flagellate in water, commonly with from two to five flagella. The sporocarps have stalks which are comparatively long and slender, usually bearing single globose spores; aberrant structures are often short-stalked with bilobed spores. The single pigmented species, *Planoprotostelium aurantium,* superficially resembles *Protostelium mycophaga* but differs from it in possessing non-deciduous spores and myxamoebae that become flagellate in water (Olive and Stoianovitch, 1971b).

Ceratiomyxella (L. S. Olive and Stoianovitch, 1971a). In this genus, the trophic stage is amoeboid, typically taking the form of a small, reticulate plasmodium which subdivides to yield multinucleate prespore cells, each of which develops into a sporocarp bearing a single deciduous multinucleate spore. The stalks are comparatively long and slender with apical expansions (apophyses). Upon germination, zoocysts develop from the emergent protoplasts; these subsequently yield up to eight uninucleate cells that become flagellate in water. One species, *C. tahitiensis,* with two varieties has been described. The genus is believed by Olive and Stoianovitch (1971a) to be transitional between *Cavostelium* and *Ceratiomyxa.* Stages in the life cycle are illustrated in Fig. 5.

3.   Ceratiomyxaceae (Schroeter, 1897)

This family, represented by the single exosporous genus *Ceratiomyxa* (Schroeter, 1897) was included with the Protostelids in L. S. Olive's revised classification of the Mycetozoa (1970), hence its inclusion in the group key (p. 12). However, the singular slime molds that comprise the genus and family are generally considered to represent a subgroup of the Myxomycetes and they are so treated in this volume. See p. 51.

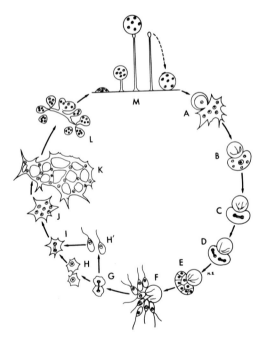

FIG. 5.   *Ceratiomyxella tahitiensis* O. and S. Diagram of life cycle. A, Spore germination; B, cyst formation; C, D, two successive divisions of the single surviving nucleus; E, delimitation of 8 cells following third mitosis; F, release of flagellate cells; G, cell division; H, amoeboid cells formed; H', flagellate cells formed; I, J, early stages in plasmodial development; K, mature reticulate plasmodium; L, prespore cell formation; M, sporogenesis. (Courtesy of L. S. Olive, 1970.) Figure originally entitled "Life cycle of protostelid isolate Ta-67-7."

III. SUBCLASS ACRASIDAE

The vegetative stage of this subclass consists of small, independently feeding and dividing myxamoebae that later aggregate as individual responding cells to form loose, multicellular associations from which develop fructifications wherein the constituent myxamoebae may or may not show divergent patterns of differentiation. Cell aggregation occurs without stream formation and in response to chemotactic stimuli of unknown character. Fructifications range from simple unbranched of branched columns of encysted cells in some forms, to others in which simple supportive stalks, usually composed of somewhat modified cells, bear spores either in chains or in definite sori. The myxamoebae are hyaline or pigmented in pink shades, with lobose pseudopodia (often developing as sudden hyaloplasmic extensions), usually with one or sometimes more contractile vacuoles, and typically a single nucleus with a conspicuous, centrally positioned nucleolus. Actively moving myxamoebae commonly assume a limax-form and in some cases show constrictions near the posterior ends, delimiting uroid regions that may bear small filose pseudopodia. Myxamoebae that do not enter aggregations, hence they do not contribute to fructifications, normally encyst as individual cells to form microcysts. Feeding is accomplished by the engulfment and subsequent digestion of bacterial cells, and in some genera yeasts and fungus spores as well.

Some slime molds belonging to this subclass appear to be coprophilous in habit, having been found only on animal dungs. Others have been isolated from decaying plant parts, from rotting wood and the bark of living trees, from decaying mushrooms, from a paste of brewer's yeast, and from the surface soil and overlying leaf litter of deciduous forests. Some genera can be cultivated quite successfully in the laboratory, some are more refractory, while others are known only from original descriptions and collections made in the field.

The degree of relationship between the Acrasidae and the Dictyostelidae is open to question, as is indeed the relationship among the several forms presently included in the former subclass. This stems is substantial part from the inadequacy of most original descriptions (Cienkowsky, 1873; van Tieghem, 1880; E. W. Olive, 1901, 1902), in part from a dearth of papers relating to these slime molds, and in part from the recurrent problems experienced in attempts to reisolate them or, when found, to successfully cultivate and confidently identify them under comparable laboratory conditions. Faced with such obstacles, it becomes very difficult to draw up a satisfactory classification of the slime molds under consideration, and the arrangement here proposed must be regarded as tentative, incomplete, and subject to revision as additional cultures are isolated and more definitive information is obtained concerning those now in hand. For the present we believe it prudent to envision the subclass Acrasidae as consisting of a single order, the Acrasiales, and two families, the Acrasiaceae and the Guttulinaceae, realizing that one or more additional families may need to be included later.

In his treatment of the Guttulinaceae, E. W. Olive (1902) distinguished between spores and pseudospores, the latter representing resting cells that returned to the trophic state by simple rehydration whereas the former germinated to release their protoplasts. We have been unable to substantiate this distinction, especially in forms where the entire structure seemingly consists of similarly differentiated cells which may, when young, return to the trophic state without casting off a containing wall, whereas the same cells when older would undergo true germination. Instead of using the term pseudospore for cells where behavior is age dependent, I will refer to these simply as encysted cells. The term spore will still be applied to cells especially differentiated for propagative functions, even though these may seemingly differ little from other cells that form structural components of the supportive stalks upon which the spores are borne.

A.   Order Acrasiales

The order possesses the characteristics of the subclass.

KEY TO FAMILIES AND GENERA OF ACRASIALES

1. Fructifications consisting of simple or branched chains of spores borne on ensheathed stalks composed of cells exhibiting limited differentiation. Spores in simple or branched chains stalk cells somewhat elongate or polygonal
. . . . **(Acrasiaceae) Acrasis** p. 19

1'. Fructifications consisting of more or less well-defined sori borne on stalks composed of wedge-shaped or compacted cells showing limited differentiation . . .   **(Guttulinaceae)** 2

1''. Fructifications consisting of simple or branched columns lacking any evidence of supporting stalks, the entire structures consisting of similarly encysted myxamoebae
. . . . **Unassigned Isolates**   p. 23

   2(1') Sori globose, clearly defined and borne on stalks composed of superimposed wedge-shaped cells, or cells less strongly differentiated and surrounded by a thin hyaline sheath
. . . . **Guttulina**   p. 21

   2'(1') Sori globose to somewhat clavate, less clearly defined and often appearing confluent with the supporting stalks composed of cells that may or may not appear clearly different from the spores . . . . **Guttulinopsis**   p. 22

1.   Acrasiaceae (Acrasieae) (van Tieghem, 1880)

The spores in this family occur in simple or branched chains borne on definite to poorly formed stalks, the cellular elements of which may differ but little from the spores that are borne terminally.

Acrasis van Tieghem, 1880. The fructifications consist of simple or branched chains of spores borne on stalks that vary from simple uniseriate columns of somewhat elongate cells to irregular columns of compacted cells showing limited differentiation.

Two species have been described, of which the type *A. granulata* van Tieghem (1880), isolated from a paste of brewer's yeast used as a substrate for *Dictyostelium roseum,* is known only from the original publication. As reported, this was characterized by simple stalks consisting of single tiers of somewhat elongate cells, bearing a simple chain of globose spores, often unequal in size, violaceous brown in color and with surfaces covered with minute roughenings. The lower cell of the stalk was said to be dilated, forming a palmlike holdfast and often surrounded by additional supporting cells. In rich cultures on yeast, several structures, each composed of a stalk and terminal spore chain, would develop simultaneously and in juxtaposition to produce coremiumlike structures. Microscopic characteristics of the myxamoebae were not given, precluding comparison with those of other species and genera of the cellular slime molds. It was in this species that van Tieghem first observed cell aggregation without cell fusion preparatory to fructification, and it is from this lack of cell fusion that the name *Acrasis* is derived.

The second species, *A. rosea* (Olive and Stoianovitch 1960), possesses some of the same basic characteristics, e.g., the formation of spores in chains borne upon stalks which in small structures consist of a single tier of somewhat elongate cells. It differs from van Tieghem's species in other particulars, especially in the formation of branched chains of spores, in the formation of larger stalks that show no evidence of being composed of multiple and adherent tiers of cells, and in the absence of basal holdfasts whether the stalks are uniseriate or many cells in thickness. The myxamoebae of *A. rosea,* faintly pinkish in color, are characterized by lobose pseudopodia, a single nucleus with a centrally positioned nucleolus and one or two contractile vacuoles. Walls of the globose spores are smooth and give a positive test for cellulose when stained with chloroiodide of zinc. *Acrasis rosea* is common on dead plant parts, being first isolated from dry florets of the marsh grass *Phragmites*. Stages in the development of *A. rosea* are shown in Figs. 6 – 9.

One surmises that the fructifications of *A. granulata* showed a considerably greater degree of cellular differentiation than do those of *A. rosea*; and if this is confirmed, when and if the former species is rediscovered and studied in laboratory cultures, it may prove desirable to place the two species in different genera, possibly returning *A. granulata* to the Dictyosteliales where it was placed by E. W. Olive in his "Monograph of the Acrasieae" (1902).

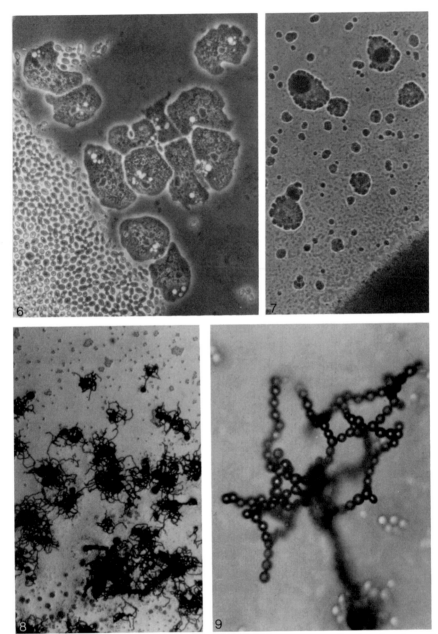

FIGS. 6–9. *Acrasis rosea* O. and S. Fig. 6. Vegetative myxamoebae at the edge of a colony of *Rhodotorula* showing contractile and food vacuoles, lobose pseudopodia, and nuclei with centrally placed nucleoli, × 450. Fig. 7. Myxamoebae forming simple aggregations preparatory to fruiting, × 38. Fig. 8. Mature fructifications, × 16. Fig. 9. A single fruiting structure showing characteristic chains of spores, × 300. Fig. 6: phase microscopy.

2. Guttulinaceae (Guttulineen) (Zopf, 1885)

The spores occur in sori of variable form and dimensions, borne on definite to ill-defined stalks in which the cellular elements may or may not show appreciable modification.

Guttulina (Cienkowsky, 1873). The fructifications consist of dropletlike sori borne on short stalks composed of superimposed wedge-shaped cells, or somewhat angular but less highly differentiated cells compacted together and surrounded by a thin hyaline slime sheath.

The type species, *G. rosea* (Cienkowsky, 1873), first observed on rotting wood, is known from the original description and from more recent collections from the bark of trees made by N. E. Nannenga-Bremekamp of Doorwerth, Netherlands. As described by Cienkowsky, the sorocarps consisted of microscopic droplets of rose color (sori) 70 μm in size, not ensheathed, and borne on stalks of the same length. The cells of the heads (spores) were spherical whereas those of the stalk were pressed together and wedge-shaped in appearance. The myxamoebae contained a red plasm and a nucleus, as did the spores and apparently the stalk cells. In water, the amoebae were described as resembling *Amoeba limax* Dujardin. Writing about *G. rosea* some years later, Zopf (1885) recorded that the membranes surrounding the spores bore fine tubercles or papillae, as did those in the specimens collected by Mrs. Nannenga-Bremerkamp when observed by her and subsequently by me. Unfortunately, neither of us has yet succeeded in reisolating and cultivating the slime mold. The presence of rose-colored cytoplasm may indicate a relationship to *Acrasis rosea* as L. S. Olive (1965) has suggested, and perhaps to *Protostelium mycophaga* as well. Fructification of *Guttulina rosea* is illustrated in Figs. 10–13.

Two additional species, *Guttulina aurea* and *G. sessilis*, were described by van Tieghem in 1880, the first said to resemble *G. rosea* but to differ in color, and the second reported to consist of simple milk-white droplets resting directly on the substrate. Neither species was seen by E. W. Olive, nor have we encountered any isolates that fit van Tieghem's brief descriptions, unless perchance the latter was redescribed as *Guttulinopsis vulgaris* by E. W. Olive (1901). A third species described as *Guttulina protea* by Fayod in 1883 was reassigned two years later by Zopf (1885) to a new genus *Copromyxa* as *C. protea* upon the basis that the constituent myxamoebae underwent no differentiation into stalk and head cells as was true of Cienkowsky's *Guttulina*.

A slime mold, "Pan-35," possessing some of the characteristics of *G. rosea* was isolated several years ago in our laboratory by Dr. Dietrich Kessler, and later discussed by the writer as a species of *Guttulina* (Raper, 1960, pp. 589–591, Figs. 31–38). This is characterized by globose, unwalled, milk-white sori up to 250 μm in diameter borne on tapering stalks that may reach 500 μm or more in length, the stalks being constructed of somewhat

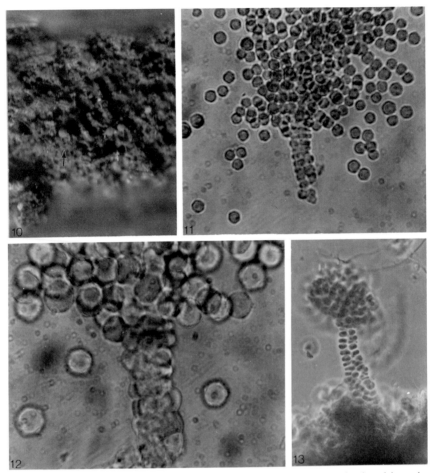

Figs. 10–13. *Guttulina rosea* Cienk. Fig. 10. Three minute sorocarps developed in moist chamber upon a piece of bark encrusted with algae and lichens, × 20. Fig. 11. A single sorocarp mounted in water with sorus broken and spores dispersed, × 304. Fig. 12. A portion of the preceding enlarged, × 648. Fig. 13. A stained preparation showing wedge-shaped character of the stalk cells, × 186. Specimens and stained preparation contributed by Mrs. N. E. Nannenga-Bremekamp.

modified myxamoebae compacted within a surrounding sheath. Whether or not this slime mold, of which we now have four isolates, should be included in a broadened concept of Cienkowsky's genus is still open to question and no formal description has been published.

Guttulinopsis (E. W. Olive, 1901). As delineated by E. W. Olive, fructifications consist of sessile or stalked sori, the sori composed of pseudospores and the cells of the stalk usually slightly elongated. Myxamoebae were reported to have lobose pseudopods but were not otherwise characterized in his text; however, they were illustrated showing one or more contractile vacuoles and single nuclei with centrally positioned nucleoli. Three species were described: *G. vulgaris*, *G. clavata* and *G. stipitata* (E. W. Olive, 1901, 1902). The first of these was said to be common on the dungs of various animals with fructifications usually short stalked, sometimes sessile, 150 to 500 μm in height and almost equally as wide. Its frequent occurrence on dung has been confirmed in this laboratory and also in that of Professor L. S. Olive, who in 1965 published a full account of its development and structure to which the reader is referred for further information. The other two species, considerably larger than *G. vulgaris*, are known only from E. W. Olive's original meagre descriptions (1901, 1902) and from the not too informative illustrations included in his latter paper.

Some years ago, I discussed under the name *Guttulinopsis* sp. (Raper, 1960, pp. 387–388, Figs. 24–30) a slime mold isolated from forest soil that produced thin, upright columns of encysted myxamoebae 1.0–1.5 mm in height that showed no demarcation into a sorus, or spore-bearing area, and a stalk, or supporting structure. It is no longer believed that this can represent the genus *Guttulinopsis* as conceived and reported by E. W. Olive. Instead, it is here included as one of the smaller representatives of the unassigned isolates, next to be considered.

Unassigned Isolates. Included here are a number of isolates from animal dungs or from leaf litter that are characterized by fructifications that assume the form of simple or branched columns of varying dimensions and consist of uniformly encysted cells that show no evident separation into sporulating and supportive areas. In some isolates the terminal portion of the columns are often somewhat swollen and bulbous, but the composition of such enlargements is quite like the remainder of the structures. In others the columns tend to taper upward to more or less blunt points. Branching is not uncommon in most of the isolates and is quite characteristic of some, especially in more luxuriant cultures. The columns are not enveloped by membranes and, if newly formed, quickly disassociate when placed in water; older columns may retain their form for a limited period due to the presence of a dehydrated intercellular slime.

The isolates do seem to fall into two fairly distinct groups based upon the size of the myxamoebae and resting cells and upon the overall dimensions attained by their fructifications, the two characteristics generally showing a positive correlation. The myxamoebae are all of the limax-type when actively moving, and typically show one or more contractile vacuoles (not evident in some isolates under some conditions) and usually a single nucleus with a centrally positioned nucleolus.

Of the smaller forms, one was isolated from eagle dung while the others came from soil or decaying leaves. Of the larger forms, all were isolated from dungs, principally those of horse and cow.

Thus far it has not been possible to identify any of these with previously described genera of the cellular slime molds. We have considered that one or more of the larger forms may be assigned to *Copromyxa protea* (Fayod,

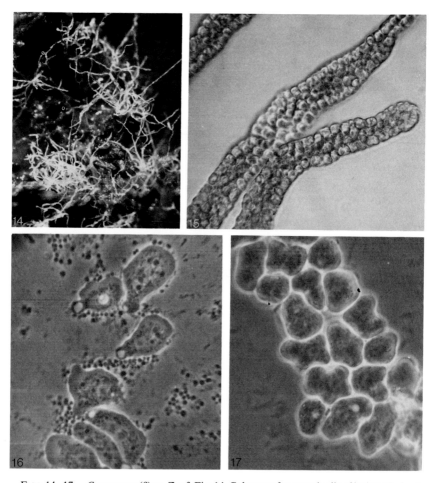

FIGS. 14–17. *Copromyxa* (?) sp. Zopf. Fig. 14. Columns of encysted cells of isolate H$_7$ developed upon sterilized horse dung partially submerged in 0.1% lactose-0.05% yeast extract agar, × 3.5. Fig. 15. Columns showing characteristic cellular structure and absence of an enveloping wall or membrane, × 294. Fig. 16. Vegetative amoebae showing typical limax-form, contractile vacuoles and nuclei with single nucleoli, × 350. Fig. 17. Portion of a thin column enlarged to show the irregular pattern of the encysted cells, × 862. Fig. 16: phase microscopy.

1883; Zopf, 1885) upon the basis of its reported fructifications, which took the form of "spindles, horns, clubs or little knobs" 1- to 3-mm high and showed no demarcation into spore-forming and stalk-forming areas. One of our larger isolates is illustrated in Figs. 14–17. However, we have not yet seen encystment stages similar to the reniform or the encrusted globose cysts illustrated by Fayod (1883) for *Guttulina protea*, under which name this organism was first described. The possible presence of two distinct limax-type amoebae in his cultures should not be overlooked.

Whereas the aforementioned cellular slime molds can be arbitrarily grouped according to cell size, striking differences are apparent among the fructifications of both the smaller and the larger forms, and it is questionable whether all representatives of either group can be accommodated within a single genus. We are now attempting to resolve this uncertainty.

There yet remains one unique undescribed cellular slime mold that cannot be fitted into any of the genera or subgroups previously considered, but would be included within the subclass Acrasidae upon the basis of its limax-like myxamoebae, its pattern of nonaligned cell aggregation, and the form of the sorocarps it produces. Sent to me several years ago by Professor L. S. Olive, it superficially resembles the form, "Pan-35," with globose white sori mentioned earlier as questionably representing a species of *Guttulina* but is in reality quite different from it. The myxamoebae first collect into low rounded mounds from which then develop narrow tapered columns of cells encased by thin hyaline walls or tubelike membranes. When a column has attained its ultimate length of 1.0–1.5 mm this covering ruptures or is dissolved at its apex and the cells formerly contained within the column, and in the bulblike reservoir beneath it, begin to ascend as if under pressure and collect into a terminal, globose unwalled sorus while the stalk beneath the enlarging spore mass may become virtually empty and appear only as a supporting hyaline tube. The structure withal suggests a miniature fountain and we propose to describe it under the generic name *Fonticula*.

IV. SUBCLASS DICTYOSTELIDAE

The trophic stage of this subclass consists of small, independently feeding and dividing uninucleate myxamoebae that subsequently aggregate to form multicellular pseudoplasmodia from which later develop sorocarps consisting of well-defined, globose to citriform sori borne on sorophores of varying patterns and dimensions. Such sorophores may be branched or unbranched and, except in the genus *Acytostelium*, are composed of strongly vacuolate cells compacted within a continuous enveloping cellulosic tube. Spores and stalk cells are formed by divergent differentiation of single myxamoebae. Pseudoplasmodia arise by the aggregation of large but variable numbers of myxamoebae which, in converging toward central points, collect into definite and generally conspicuous inflowing streams; once formed, the pseudoplasmodia may in a few species migrate freely across the agar surface before forming sorocarps, but more commonly such migration is lacking or in the larger forms is accompanied by continuous stalk formation. The hyaline myxamoebae possess filose pseudopodia, one or more contractile vacuoles, and a single nucleus with two or more peripheral nucleoli. Feeding is accomplished by the engulfment and subsequent digestion of bacterial cells with which these slime molds are regularly associated in nature and upon which they are routinely cultivated in the laboratory (Raper, 1951; Sussman, 1956; Gerisch, 1959; Hohl and Raper, 1963a,b,c).

Slime molds of this subclass, which includes the more commonly studied cellular slime molds, are of common occurrence in nature and have been isolated from a wide variety of substrata (Raper, 1951) and from collection sites throughout the world (Cavender and Raper, 1965a,b, 1968). First thought to be coprophilous in habit, they are now known to occur in almost any situation where vegetable matter is undergoing aerobic decomposition, having been isolated from decaying mushrooms, spoiled vegetables, decomposing grass, rotting wood, cultivated soils, and more especially from the surface litter of deciduous forests and the topmost layers of the underlying soil. Of recent years, several new species have been isolated, principally by Professor James C. Cavender, that appear to be indigenous to tropical and subtropical soils.

For a general review of the cellular slime molds, with special reference to developmental processes in the Dictyosteliales as then known, the reader is referred to Bonner (1967).

A.   Order Dictyosteliales

The order possesses the characteristics of the subclass.

KEY TO FAMILIES AND GENERA OF DICTYOSTELIALES

1. Sorocarps variable in size, ranging from <1 mm in some species to several centimeters in others; sorophores composed of strongly vacuolate cells contained within an enveloping cellulosic sheath or tube . . . . **(Dictyosteliaceae)**   2

1'. Sorocarps consistently small and delicate, not exceeding 1.5 mm; sorophores very thin (1–2 μm diameter) and consisting of acellular cellulosic tubes. Sorophores unbranched and bearing minute dropletlike sori . . . . **(Acytosteliaceae) Acytostelium**

  2(1) Sorophores unbranched or bearing branches without definite arrangement; sori unwalled, consisting of spores suspended in droplets of slime, terminal on main axes and on branches when present . . . . **Dictyostelium**

2′(1) Sorophores regularly branched with branches arranged in definite whorls; sori borne as in the preceding . . . . **Polysphondylium**

2″(1) Sorophores unbranched or sometimes bearing a whorl of branches, the main axes and branches expanded above to form cuplike structures that contain the sori

. . . . **Coenonia**

1. Dictyosteliaceae (Rostafinski, 1875)

Aggregated myxamoebae show markedly divergent differentiation during sorocarp formation; some cells become strongly vacuolate structural elements of the supportive stalk, the majority normally differentiating as spores.

Dictyostelium (Brefeld, 1869). The sorocarps are unbranched or irregularly branched, consisting of cellular supportive stalks bearing globose to citriform dropletlike sori. They are variable in dimensions and patterns; in some species they are consistently unbranched, in others irregularly branched, and in still others they show both conditions, the extent of branching (if present) often being strongly influenced by the cultural conditions. In all species the stalk or sorophore consists of a continuous cellulosic tube within which entrapped cells become strongly vacuolate to provide rigidity for the developing sorocarp (Raper and Fennell, 1952). The main axes and branches (if present) typically bear terminal unwalled sori of proportionate size; in two species they bear sessile lateral sori along the length of the sorophore. The sori consist of capsule-shaped or, in two species, globose spores suspended in droplets of slime, which are white or variously pigmented.

Dictyostelium, known since the description of *D. mucoroides* by Brefeld in 1869, is the oldest, largest, and most studied genus of the Acrasiomycetes. As reported by Brefeld, *D. mucoroides* was thought to form a true plasmodium prior to fructification, but this misinterpretation was corrected by van Tieghem (1880) who first recognized the persistent cellular nature of this and some additional cellular slime molds, including *Acrasis* from which the class takes its name.

The genus consists of twenty described species of which all but two or three are currently known in laboratory culture. Among the species comprising the genus, *Dictyostelium mucoroides* with milk-white sori (Fig. 25) and sorocarps of intermediate size, sometimes branched, is the most abundant in nature and is, apparently, worldwide in distribution. Strains capable of producing macrocysts, a multicellular encystment stage (Fig. 26), are occasionally encountered (Blaskovics and Raper, 1957). *Dictyostelium purpureum* (E. W. Olive, 1901, 1902) resembles *D. mucoroides* in many ways but differs in possessing dark purple sori (Fig. 27). *Dictyostelium discoideum* Raper (1935), characterized by freely migrating pseudoplasmodia and

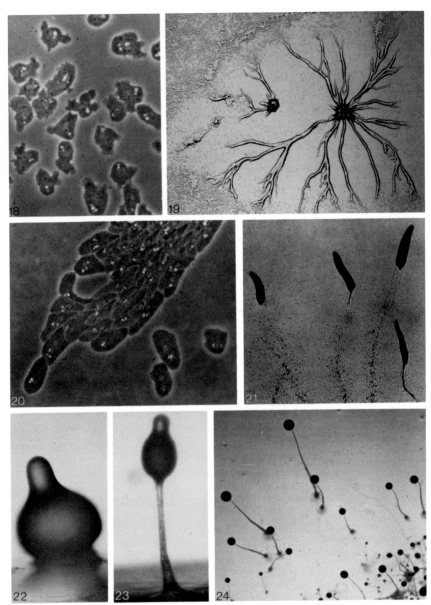

FIGS. 18–24. *Dictyostelium discoideum* Raper. Fig. 18. Postfeeding myxamoebae showing lack of cell orientation prior to aggregation, × 231. Fig. 19. A developing pseudoplasmodium showing characteristic streams of inflowing myxamoebae; the smaller organization at left resulted from the severance of a major stream of the larger aggregation, × 8. Fig. 20. Distal end of a developing stream showing characteristic orientation of aggregating myxamoebae, × 231. Fig. 21. Migrating pseudoplasmodia moving away from the colony in which they developed, × 5. Figs. 22 and 23. Successive stages in the development of a sorocarp—stalk formation occurs in the apical papilla, × 38. Fig. 24. Mature sorocarps of different dimensions showing relatively constant proportion of sori and sorophores, × 6. Figs. 18 and 20: phase microscopy.

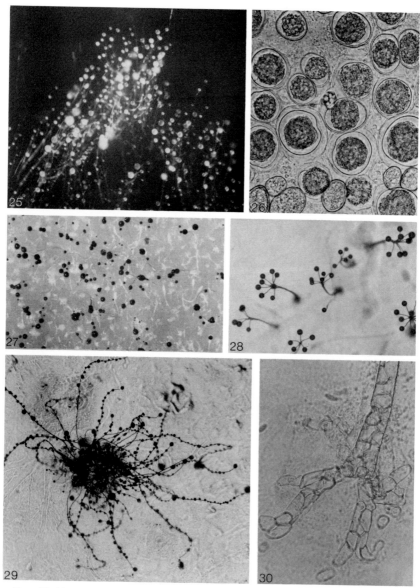

FIGS. 25–30. Additional species of *Dictyostelium* showing characteristic fructifications and structures. Fig. 25. *Dictyostelium mucoroides* Bref. showing sorocarps with milk-white sori developing toward a light source, surface illumination, × 3.5. Fig. 26. Mature macrocyst of the same species, × 186. Fig. 27. *Dictyostelium purpureum* Olive showing dark purple sori, surface illumination, × 3. Fig. 28. *Dictyostelium polycephalum* Raper showing typical clustered (coremiform) sorocarps, transmitted light, × 21. Fig. 29. *Dictyostelium rosarium* R. and C. showing multiple sessile sori borne throughout the length of the sorophore, transmitted light, × 6. Fig. 30. *Dictyostelium rhizopodium* R. and F. showing the lower portion of a sorophore with its characteristic crampon base, × 210.

unbranched sorocarps with basal disks and beautifully tapered upright sorophores (Figs. 18–24), has been studied most carefully and accounts for more than half of all published papers relating to the Acrasiomycetes. Of the other species, some that show strikingly different characteristics include *D. polycephalum* (Raper, 1956) with thin migrating pseudoplasmodia and small coremiform fructifications (Fig. 28), *D. rosarium* (Raper and Cavender, 1968) with many lateral sessile sori (Fig. 29), and *D. rhizopodium* (Raper and Fennell, 1967) with sorophores arising from cramponlike bases (Fig. 30) and with sori and sorophores lightly pigmented. While quite different in size and in the pattern of their sorocarps, *D. rosarium* and *D. lacteum* van Tieghem (1880) are unique in possessing globose spores rather than the capsule-shaped reproductive cells seen in all other species.

Most of the known species of *Dictyostelium* were first isolated from dung, soil, or other natural substrates of temperate origin. In contrast, *D. laterosorum* and *D. dimigraformum* (Cavender, 1970), *D. deminutivum* (Anderson et al., 1968), and the four species with cramponlike bases described by Raper and Fennell (1967) are known only from tropical or subtropical sources.

Polysphondylium (Brefeld, 1884).   This genus, closely related to *Dictyostelium*, differs from it in possessing sorocarps with multiple and typically regularly spaced whorls of side branches. These arise by the periodic abstriction of small masses of myxamoebae from the posteriormost part of the rising cell mass, or *sorogen*, followed by the cleavage of such masses into a number of segments, each of which then develops a small branch which in appearance resembles a diminutive fructification anchored to the primary sorophore (Raper, 1960). Each branch bears a sorus, as does the main stem, the terminal sorus being consistently larger than any of the branch sori except in one undescribed species or variety. The number and regularity of the whorls may vary markedly in different isolates and in response to changes in the cultural environment.

Of the three described species, the type, *Polysphondylium violaceum* (Brefeld, 1884), is the most robust and, as the name implies, shows a violet pigmentation in the sori and sometimes in the sorophores as well. It is especially common in decaying litter and surface soil of deciduous forests and appears to be worldwide in distribution. Sorocarps of the other two species, *P. pallidum* and *P. album*, both described by E. W. Olive (1901, 1902), bear white sori, and are of smaller dimensions and more delicate proportions. Olive recognized the latter species upon the basis of its larger sori, but he apparently had only one isolate of this taxon for comparison. In our studies, involving scores of isolates from throughout the world, we have been unable to confirm sorus size as a satisfactory criterion whereby the nonpigmented Polysphondylia can be separated into different taxons

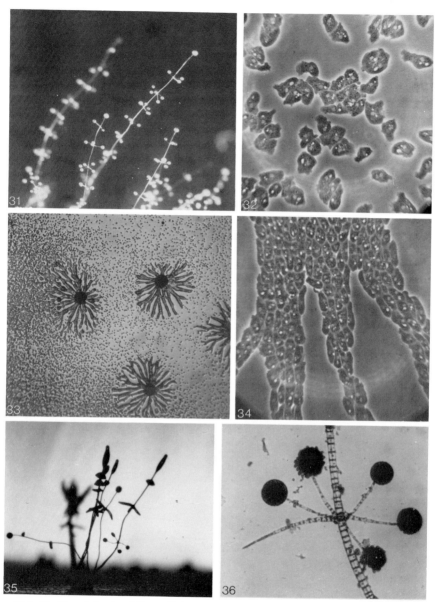

FIGS. 31–36.   *Polysphondylium pallidum* Olive. Fig. 31. Typical sorocarps with whorls of side branches, surface light, × 7. Fig. 32. Postfeeding nonoriented myxamoebae, × 231. Fig. 33. Developing pseudoplasmodia, × 7. Fig. 34. Enlarged view of a portion of the same showing uniform orientation of the converging cells, × 231. Fig. 35. A cluster of developing sorocarps; note progressive stages in the formation of whorls of side branches on structure at right, × 13. Fig. 36. Enlarged view of a single whorl showing attachment of branches to the central axis, × 60. Figs. 32 and 34: phase microscopy.

and have, perhaps as a matter of convenience, included all of them in *P. pallidum*. Developmental stages of a typical isolate are shown in Figs. 31–36.

Coenonia (van Tieghem, 1884). Known only from van Tieghem's original description of *C. denticulata* (1884), this genus must represent a most remarkable slime mold. As described, the sorocarps consisted of erect cellular stalks that arose from irregularly bifurcate holdfasts (crampons) and bore yellow, gelatinous globules of spores in terminal cuplike enlargements with finely dentate rims. The fructifications were unbranched in slide cultures, but larger structures developed on the surface of decaying beans showed a verticil of three equidistant branches, such branches being of the same structure as the stalk and each terminated by a small cupule bearing a globule of spores. Each peripheral cell of the stalk and the lower cells of the cupule, but not those of the crampon, bore a small upturned toothlike protuberance, while the marginal cells of the cupule projected their membranes between the spores as thin, more or less long and regular teeth. The spores with yellow membranes were spherical and 6 to 8 μm in diameter.

Pending its rediscovery, *Coenonia* is placed in the Dictyosteliaceae upon the bases of its cellular sorophore and crampon base, as was done by E. W. Olive (1901, 1902). Indeed, the crampon may approximate that of *D. rhizopodium*. Yet the sorocarp must differ markedly from that of any cellular slime mold now known in laboratory cultures, and it is difficult to visualize just how the terminal cupule with its enclosed sorus could be produced by a developmental process closely resembling morphogenesis in *Dictyostelium* and *Polysphondylium*. Additionally, no other species is known in which the cells of the stalk (not to mention those of the cupule rim) show any evidence of toothlike projections. Still, there can be no question but that Van Tieghem was describing a cellular slime mold, for when still amoeboid cells from the prespore globule were removed from a developing fructification and replaced in a nutrient drop the constituent cells soon produced a new holdfast, a new stalk, and finally a new complete but much smaller fructification!

When and if *Coenonia* is rediscovered and accorded careful study, it seems not unlikely that a new family, the Coenoniaceae, may be needed to include it.

2. Acytosteliaceae (Raper, 1958)

Aggregated myxamoebae differentiate *only* as spores, the stalk consisting of a thin acellular cellulosic tube.

Acytostelium (Raper, in Raper and Quinlan, 1958). In this genus, the sorocarps are uniformly small and delicate, consisting of very thin, unbranched acellular sorophores bearing diminutive, terminal globose sori. The sorophores are tubular with very narrow lumens, cellulosic in composition, 1–2 μm in diameter, and ranging from 200 μm to 1.0 mm in length in one species and often longer but not exceeding 1.5 mm in another. Sori very small, unwalled and, as in *Dictyostelium* and *Polysphondylium*, consisting of spores suspended in droplets of slime.

Two species are known: *A. leptosomum* Raper (Raper and Quinlan, 1958) which is the type upon which the genus and the family, the Acytosteliaceae, was based, and *A. ellipticum*, recently discovered and described by Cavender (1970). The former species typically shows comparatively large cell aggregations with prominent inflowing streams, and, when fruiting, such pseudoplasmodia normally give rise to large numbers of sorocarps that may reach up to 1.5 mm in length and bear small sori that contain spherical spores (Figs. 37–40). The latter species forms smaller aggregations, typically without stream formation, and, when fruiting, gives rise to fewer and generally smaller sorocarps (not exceeding 1.0 mm in length) and with elliptical rather than globose spores.

The presence of thin, acellular tubelike stalks in *Acytostelium* suggests to L. S. Olive (1970) that it is a primitive genus of cellular slime molds, probably derived from some *Protostelium*-like ancestral type. Perhaps this is true. On the other hand, it can be argued that a slime mold which wastes no cells in stalk formation but conserves all of its myxamoebae for spore production represents a more advanced type than either *Dictyostelium* or *Polysphondylium*.

REFERENCES

Anderson, J. S., D. I. Fennell, and K. B. Raper. (1968). *Dictyostelium deminutivum*, a new cellular slime mold. *Mycologia* **60**:49–64.

Blaskovics, J. C., and K. B. Raper. (1957). Encystment stages of *Dictyostelium*. *Biol. Bull.* **113**:58–88.

Bonner, J. T. (1967). "The Cellular Slime Molds," 2nd ed. Princeton Univ. Press, Princeton, New Jersey.

Brefeld, O. (1869). *Dictyostelium mucoroides*. Ein neuer Organismus aus der Verwandschaft der Myxomyceten. *Abh. Senckenb. Naturforsch. Ges.* **7**:85–107.

Brefeld, O. (1884). *Polysphondylium violaceum* und *Dictyostelium mucoroides* nebst Bemerküngen zur Systematik der Schleimpilze. *Unters. Gesammtgeb. Mykol.* **6**:1–34.

Cavender, J. C. (1970). *Dictyostelium dimigraformum, Dictyostelium laterosorum* and *Acytostelium ellipticum*. New Acrasieae from American tropics. *J. Gen. Microbiol.* **62**:113–123.

Cavender, J. C., and K. B. Raper. (1965a). The Acrasieae in nature. I. Isolation. *Amer. J. Bot.* **52**:294–296.

Cavender, J. C., and K. B. Raper. (1965b). The Acrasieae in nature. II. Forest soil as a primary habitat. *Amer. J. Bot.* **52**:297–302.

Cavender, J. C., and K. B. Raper. (1968). The occurrence and distribution of Acrasieae in forests of subtropical and tropical America. *Amer. J. Bot.* **55**:504–513.

Cienkowsky, L. (1873). *Guttulina rosea. Trans. Bot. Sect. 4th Meet. Russian Naturalists at Kazan.*

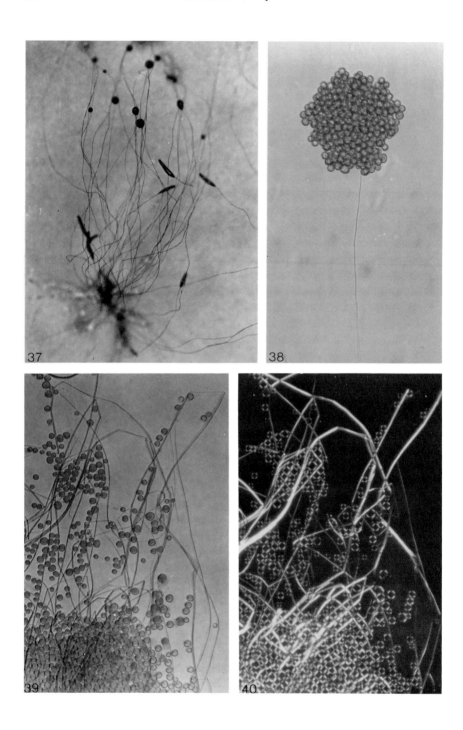

Fayod, V. (1883). Beitrag zur Keuntniss Niederer Myxomyceten. *Bot. Z.* **41**:169–177.

Gerisch, G. (1959). Ein Submerskulturverfahren für entwicklüngsphysiologische Untersuchungen an *Dictyostelium discoideum*. *Naturwissenschaften* **46**:654–659.

Harper, R. A. (1926). Morphogenesis in *Dictyostelium*. *Bull. Torrey Bot. Club* **53**:229–268.

Hohl, H-R., and K. B. Raper. (1963a). Nutrition of cellular slime molds. I. Growth on living and dead bacteria. *J. Bacteriol.* **85**:191–198.

Hohl, H-R., and K. B. Raper. (1963b). Nutrition of cellular slime molds. II. Growth of *Polysphondylium pallidum* in axenic culture. *J. Bacteriol.* **85**:199–206.

Hohl, H-R., and K. B. Raper. (1963c). Nutrition of cellular slime molds. III. Specific growth requirements of *Polysphondylium pallidum*. *J. Bacteriol.* **86**:1314–1320.

Olive, E. W. (1901). Preliminary enumeration of the Sorophoreae. *Proc. Amer. Acad. Arts Sci.* **37**:333–344.

Olive, E. W. (1902). Monograph of the Acrasieae. *Proc. Boston Soc. Natur. Hist.* **30**:451–523.

Olive, L. S. (1962). The genus *Protostelium*. *Amer. J. Bot.* **49**:297–303.

Olive, L. S. (1964). A new member of the Mycetozoa. *Mycologia* **56**:885–896.

Olive, L. S. (1965). A developmental study of *Guttulinopsis vulgaris* (Acrasiales). *Amer. J. Bot.* **52**:513–519.

Olive, L. S. (1967). The Protostelida—a new order of the Mycetozoa. *Mycologia* **59**:1–29.

Olive, L. S. (1970). The Mycetozoa: A revised classification. *Bot. Rev.* **36**:59–89.

Olive, L. S., and C. Stoianovitch. (1960). Two new members of the Acrasiales. *Bull. Torrey Bot. Club* **87**:1–20.

Olive, L. S., and C. Stoianovitch. (1966a). *Protosteliopsis*, a new genus of the Protostelida. *Mycologia* **58**:452–455.

Olive, L. S., and C. Stoianovitch. (1966b). *Schizoplasmodium*, a mycetozoan genus intermediate between *Cavostelium* and *Protostelium*; a new order of Mycetozoa. *J. Protozool.* **13**:164–171.

Olive, L. S., and C. Stoianovitch. (1971a). A new genus of protostelids showing affinities with *Ceratiomyxa*. *Amer. J. Bot.* **58**:32–40.

Olive, L. S. and C. Stoianovitch. (1971b). *Planoprotostelium*, a new genus of protostelids. *J. Elisha Mitchell Sci. Soc.* **87**:115–119.

Raper, K. B. (1935). *Dictyostelium discoidem*, a new species of slime mold from decaying forest leaves. *J. Agr. Res.* **50**:135–147.

Raper, K. B. (1951). Isolation, cultivation and conservation of simple slime molds. *Quart. Rev. Biol.* **26**:169–190.

Raper, K. B. (1956). *Dictyostelium polycephalum* n. sp.: A new species with coremiform fructifications. *J. Gen. Microbiol.* **14**:716–732.

Raper, K. B. (1960). Levels of cellular interactions in amoeboid populations. *Proc. Amer. Phil. Soc.* **104**:579–604.

Raper, K. B., and J. C. Cavender. (1968). *Dictyostelium rosarium*: A new cellular slime mold with beaded sorocarps. *J. Elisha Mitchell Sci. Soc.* **84**:31–47.

Raper, K. B., and D. I. Fennell. (1952). Stalk formation in *Dictyostelium*. *Bull. Torrey Bot. Club* **79**:25–51.

FIGS. 37–40. *Acytostelium leptosomum* Raper. Fig. 37. A cluster of sorocarps that have developed from a single cell aggregation, × 25. Fig. 38. Terminal portion of a single sorocarp showing the sorus (stained and flattened) and the very thin supporting stalk, × 270. Fig. 39. Stained preparation of spores and sorophores as seen with normal light, × 270. Fig. 40. The same as seen with polarized light—the strong birefringence results from the cellulosic nature of the spore walls and sorophores, × 270.

Raper, K. B., and D. I. Fennell. (1967). The crampon-based Dictyostelia. *Amer. J. Bot.* **54**: 315–328.

Raper, K. B., and M. S. Quinlan. (1958). *Acytostelium leptosomum*: A unique cellular slime mold with an acellular stalk. *J. Gen. Microbiol.* **18**:16–32.

Rostafinski, J. T. (1875). "Śluzowce (Mycetozoa). Monografia." Paris.

Schroeter, J. (1897). Myxogasteres. *In* "Engler and Prantl, Natürlichen Pflanzenfamilien," Vol. I, No. 1, pp. 1–41.

Sussman, M. (1956). The biology of the cellular slime molds. *Annu. Rev. Microbiol.* **10**:21–50.

van Tieghem, P. (1880). Sur quelques Myxomycètes á plasmode agrégé. *Bull. Soc. Bot. Fr.* **27**:317–322.

van Tieghem, P. (1884). *Coenonia*, genre nouveau de Myxomycètes á plasmode agrégé. *Bull. Soc. Bot. Fr.* **31**:303–306.

Zopf, W. (1885). "Die Pilzthiere oder Schleimpilze. Nach dem Neuesten Standpunkte bearbeitet." Breslau. (Separatabdruk aus der Encyklopaedie der Naturwissenschaften.)

Myxomycota
Myxomycetes

ERRATA

THE FUNGI: An Advanced Treatise

G. C. Ainsworth, Frederick K. Sparrow, and
Alfred S. Sussman, editors

Volume IV B

A Taxonomic Review with Keys: Basidiomycetes and Lower Fungi

Page	Location	Correction
5	line 3	p. 12 *should read* p. 10
5	line 6	p. 53 *should read* p. 51
5	line 8	p. 83 *should read* p. 77
5	line 12	p. 65 *should read* p. 71
5	line 23	p. 101 *should read* p. 95
5	line 26	p. 112 *should read* p. 106
5	line 28	p. 75 *should read* p. 71
6	line 2	p. 191 *should read* p. 185
6	line 4	p. 244 *should read* p. 238
6	line 31	p. 251 *should read* p. 245
6	line 35	p. 310 *should read* p. 304
7	line 1	p. 323 *should read* p. 317
7	line 2	p. 307 *should read* p. 301
247	title	title *should read* Uredinales

CHAPTER 3

Myxomycetes[1]

Constantine J. Alexopoulos

Department of Botany
University of Texas at Austin
Austin, Texas

I.   GENERAL CHARACTERISTICS

The assimilative stage of this class is a free-living, acellular, multinucleate, amoeboid mass of protoplasm, the plasmodium, enclosed by an amorphous gelatinous sheath, varying from microscopic to an extensive system of branching and anastomosing veins in which the protoplasm streams rhythmically and reversibly. Under favorable conditions the plasmodium becomes converted into one or many fruiting bodies (sporophores). Reproduction occurs by spores: in the exosporous forms they are borne singly on hairlike stalks, in the endosporous forms they are found inside stalked or sessile spore cases which are seated on a horny, membranous, spongy, or calcareous base, the hypothallus. Spores on germination produce one to four naked myxamoebae or flagellated swarm cells, the latter with one or two (rarely more) anterior whiplash flagella, invariably with at least two basal bodies. Asexual reproduction occurs by binary fission of myxamoebae or fragmentation of the plasmodium; sexual reproduction takes place by fusion of compatible myxamoebae or swarm cells forming zygotes which grow into plasmodia. Myxomycetes are homothallic or heterothallic. Mitosis is astral in myxamoebae, intranuclear in zygotes, and intranuclear and synchronous in plasmodia. Under conditions unfavorable for growth or metabolism, the myxamoeba encysts forming a microcyst, and the plasmodium either fragments into a number of disconnected cysts or becomes transformed into a horny mass, the sclerotium, consisting of usually multinucleate cell-like units, the spherules, capable of reforming the plasmodium under favorable conditions.

[1]Supported by National Science Foundation Grant GB-6812X. I would like to thank Dr. Raymond W. Scheetz for executing the line drawings that appear in this chapter.

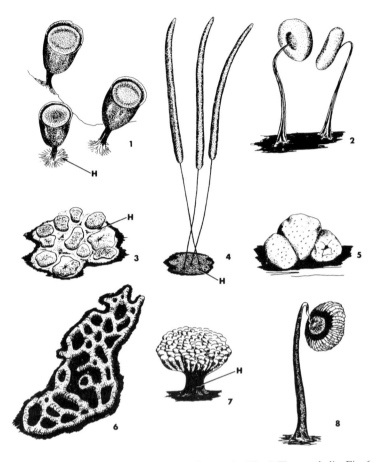

PLATE II. Sporophores. Figs. 1, 2, 3, 4, and 8. Sporangia. Fig. 5. Three aethalia. Fig. 6. Plasmodiocarp. Fig. 7. Pseudoaethalium. Hypothallus. Fig. 1. Disclike under individual sporangia. Fig. 3. Calcareous. Fig. 7. Massive, spongy.

(Plate II, Figs. 1–4, 8); curved, elongated, or dictyoid plasmodiocarps (Plate II, Fig. 6); massive aethalia (Plate II, Fig. 5); or compact pseudo-aethalia (Plate II, Fig. 7). Sporangia range in size from 4.5 to 25 mm in total height if stalked, and from 50 μm to 2.5 mm in diameter if sessile. A well-developed plasmodium may form hundreds of sporangia which may be separated from each other, crowded, or even heaped. When a plasmodium of a sporangiate species has reached the fruiting stage, the protoplasm becomes separated into a number of pulvinate masses, each of which assumes the shape of the mature fructification and eventually sporulates. Plasmodiocarps differ from sessile sporangia only in shape. When the plasmodium is ready to fruit, it may break up into several masses which retain the shapes of the portions of the plasmodial veins from which they originated and each becomes enveloped by a peridium before sporulating. Thus plasmodiocarps may have a variety of shapes even in the same fruiting, varying from sporangiumlike spheres to a network of tubes. An aethalium is usually a massive structure, sometimes reaching a diameter of 20 cm, but usually much smaller. An entire plasmodium is often converted into a single aethalium or pseudoaethalium. The latter is a mass of closely compacted sporangia which simulates an aethalium but in which individual sporangia are distinguishable at or close to maturity. Pseudoaethalia may reach a diameter of 20 cm or more, or, in the case of elongated structures, a length of more than 10 cm.

D. The Peridium

The sporophores of all endosporous Myxomycetes are covered by a thin, delicate, early-evanescent membrane or a more substantial, persistent, sometimes tough wall (Plate III, Figs. 2, 5–7). In either case the enveloping structure is called the peridium and may consist of only one or as many as three layers which may be closely appressed or widely separated and which are often of different consistency. The texture of the peridium, the presence or absence of lime on its surface and the mode of dehiscence are important taxonomic characters. In sporangia in which the peridium evanesces at an early stage in its development, a small peridial collar may persist at the sporangial base. In the Cribrariaceae, in which much of the peridium is typically in the form of a net, so-called dictydine granules are present on the nodes and their threadlike connections. The granules, the nature of which is unknown, are often dark and crowded; they vary from 0.5 to 3 μm in diameter.

E. The Stalk

The character of the stalk and sometimes its length are important taxonomic characters (Plate IV, Figs. 1–5). In a number of species, both stalked and sessile sporangia or short plasmodiocarps are found intermixed. The stalk, when represented by a definite structure, is usually cylindrical or subulate; in a few minute species it is hairlike. In many Myxomycetes, the stalk is indefinite, i.e., it is only an extension of the hypothallus. The stalk may be calcareous or not. In the Stemonitales, the stalk is cylindrical, hollow, or more or less filled with strandlike material (Plate IV, Figs. 2, 4). In other groups, the stalk is stuffed with granular material or sporelike cells (Plate IV, Figs. 1, 5). These two types of stalk (hollow or stuffed) are the result of two types of development (see diagnosis of subclasses Stemonitomycetidae and Myxogastromycetidae). Under certain microenvironmental conditions,

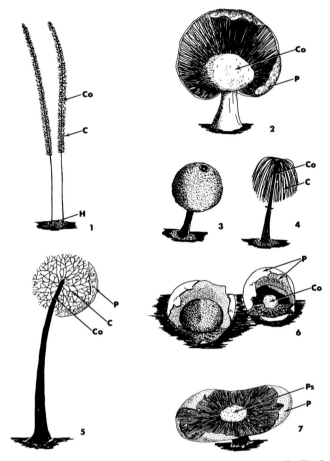

PLATE III.   Columella. Figs. 1, 4, 5. Noncalcareous extension of the stalk. Fig. 2. Calcareous extension of the stalk. Fig. 6. Calcareous columella in sessile sporangia. Figs. 3, 4. Columella with terminal apical disc from which capillitium is pendent (*Enerthenema*). Fig. 7. Pseudocolumella.

normally stalked sporangia may be sessile. Usually, however, some stalked sporophores are present together with the sessile.

F.   Columella and Pseudocolumella

The stalk may stop at the base of the peridium or may continue into the sporangium as a columella. The columella may resemble the stalk closely in structure, as in the Stemonitales (Plate III, Fig. 1), or may differ radically from it. In *Didymium iridis,* for example, the stalk is noncalcareous, whereas the columella, which appears to represent the tip of the stalk within the

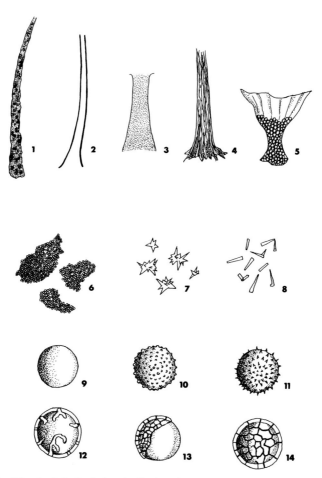

PLATE IV. Figs. 1–5. Stalks in longitudinal section. Fig. 1. Stuffed with granular material (*Echinostelium*). Fig. 2. Hollow (*Stemonitis*). Fig. 3. Calcareous throughout (several genera). Fig. 4. Partially filled with strands (*Comatricha*). Fig. 5. Filled with sporelike cells (*Arcyria*). Figs. 6–8. Lime. Fig. 6. Amorphous, granular. Fig. 7. Crystalline stellate. Fig. 8. Crystalline prismatic. Figs. 9–14. Spores. Fig. 9. Smooth. Fig. 10. Verrucose. Fig. 11. Spiny. Fig. 12. With broken reticulation. Fig. 13. With partial reticulation. Fig. 14. Completely reticulate.

sporangium, consists entirely of white lime. Sessile as well as stalked sporophores may possess a columella. In the former, the columella is usually represented by a slightly to greatly thickened sporangial base or even by a columnar structure attached to the base and extending upward into the sporangium (Plate III, Fig. 6). Color, size, shape, and calcareousness of the columella are characters often used to distinguish among species but these characters are not always constant, and care should be taken in ascribing too much importance to them in all cases. With this as well as with other characters, what is important taxonomically in some taxa (genera, families) may not be in others.

A pseudocolumella is represented by a mass of lime in the center of the sporophore away from the tip of the stalk or the base of the sporophore (Plate III, Fig. 7), but sometimes connected to the peridium at various points through calcareous extensions from the central mass.

G.   *Capillitium and Pseudocapillitium*

The capillitium is a system of separate free threads, the elaters; of separate threads attached at one or both ends to other structures (peridium, columella); or a complete network (Plate I, Figs. 4–9). It is deposited in preformed vacuoles within the sporophore at about the time of spore cleavage. Presence or absence and structure of capillitium are among the most important taxonomic characters in the classification of the Myxomycetes. In a number of species, particularly in the order Trichiales, the capillitium is elastic, carrying with it spores enmeshed in its network to a considerable height above the fructification.

Typically, the capillitial threads are more or less uniform although they often become attenuated at the tips. Pseudocapillitium may be in the form of threads, bristles, membranes, or perforated plates (Plate I, Figs. 10, 11). If it is threadlike, its threads vary in width at different points and are distinctly irregular.

H.   *Lime*

Calcium carbonate in the form of calcite is characteristically deposited on the stalk, columella, peridium, or capillitium of many species. Its presence or absence from any or all of these structures is fairly constant and has been used extensively in classification. Nevertheless, it appears that environmental conditions do affect the amount of lime that is deposited and limeless forms of typically calcareous fructifications are not uncommon. In most taxonomic keys, the lime of the Myxomycetes is described as crystalline or amorphous, the latter usually being found in the form of granules. There is some reason to believe that at least some of the so-called amorphous lime is, in reality, crystalline or in the stage of incipient crystallization. Various pigments are often incorporated in the lime so that the latter may be blue, purple, yellow, orange, or red instead of the more usual white. Capillitial lime, if present, is generally but not always concolorous with that on the peridium.

Lime on the peridium may be deposited in the form of granules, scales, or crystals (Plate IV, Figs. 6–8), or may be compacted into a smooth or rough shell, the former sometimes porcelainlike (Plate III, Fig. 6). In the capillitium, lime is most often seen as calcareous nodules at the junctions of a capillitial network (Plate I, Fig. 5). In *Badhamia*, however, the capillitium consists entirely of limy tubules (Plate I, Fig. 9). A limy stalk (Plate IV, Fig. 3) may be calcareous throughout or may be frosted with lime.

I. The Spores

Spore color, shape, size, and wall markings are among the most constant characteristics of Myxomycetes. Spores may be pallid, brightly colored, or dark by transmitted light; they are mostly globose, but are oval in some species. Though mostly separate, spores are sometimes aggregated in loose or tight clusters; they vary from 4 to 20 or more microns in diameter, and their walls may be smooth, pitted, spiny, warty, reticulate, or a combination of these characters (Plate IV, Figs. 9–14).

II. PHYLOGENY

Various authors, among them de Bary (1887), Lister (1925), Hagelstein (1944), and Olive (1970) refer to the plasmodial slime molds as Mycetozoa, thus stressing their protozoan affinities. Others, notably Schroeter (1897), Macbride (1922), Jahn (1928), and Martin (1949), prefer to use the name Myxomycetes to point to a possible link with the fungi. If the fungi, as some mycologists believe, are descendants of protozoan ancestors, the whole controversy about the origin of the Myxomycetes becomes purely academic.

The chief arguments supporting a relationship of the Myxomycetes to the Protozoa have been (1) their amoeboid, naked assimilative stage, and (2) their phagotrophic nutrition. Those supporting the fungal affinities of the group (Martin, 1960) remind us, however, that there are a number of undoubted fungi, such as *Coelomomyces*, whose assimilative stage is devoid of cell walls and whose protoplasts are multinucleate and coenocytic. They also point to the fact that *Physarum polycephalum* has been growing continuously in nonparticulate, liquid, axenic culture in Rusch's (1970) laboratory for many years thus proving that phagotrophic nutrition is not absolutely essential for slime mold metabolism. Additional arguments in favor of a fungal relationship of the Myxomycetes are (1) the possible homology of the plasmodial sheath to the hyphal cell wall of coenocytic fungi, and (2) the formation of spores, possibly with walls of chitin and cellulose, in definite fruiting bodies.

The most recent evidence from cytology has failed to solve the riddle of myxomycete ancestry. The myxomycete mitochondria, in common with those of many Protozoa, typically possess tubular cristae, whereas those of most fungi possess platelike cristae. Oomycetes, however, also have tubular cristae in their mitochondria. Although both astral and intranuclear divisions have been noted in both Protozoa and fungi, in neither group have the two forms of division been discovered to occur in the same life cycle as they do in the Myxomycetes. Biochemical studies which could throw significant light to this controversy are lacking.

It is obvious that the data at hand fail to clarify the relationship of the slime molds to any other group of living organisms except possibly the Protostelida, a group which, even at this point of our knowledge, appears to be a heterogenous one.

Concerning the evolutionary development within the Myxomycetes, two lines of evidence appear to be fruitful in phylogenetic theory: (1) type of sporophore development (Plate V) and (2) plasmodial type. It is probably no coincidence that plasmodial type correlates to a considerable extent with type of development (Alexopoulos, 1969). The subhypothallic type of development is common to the Echinosteliales, Trichiales, Physarales, and probably the Liceales. For the last group there are no data more recent than de Bary's available. All these orders are considered here to form a single evolutionary series, with the Echinosteliales, all of which possess a protoplasmodium, at the bottom, and the Physarales, all with a phanero-

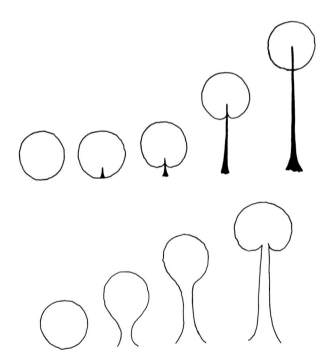

PLATE V.   Sporophore development. Epihypothallic (above); subhypothallic (below).

plasmodium, at the top. They are all placed in the subclass Myxogastromycetidae.

The epihypothallic type of development (Plate V, top figures) is found only in the Stemonitales. This group, the only one with an aphanoplasmodium, appears to have branched off the main developmental line at an early stage in the evolution of the Myxomycetes as Ross (1957) has postulated. These are separated in the subclass Stemonitomycetidae following Ross (1973).

The Ceratiomyxales are something of a puzzle. The stalked spores (which may very well be stalked monosporous sporangia) seem to develop in a subhypothallic manner (Famintzin and Woronin, 1873) with stalk and spore wall covered by the same sheath as in the protostelids, the Echinosteliales and, indeed, all the Myxomycetes except the Stemonitales.

The following diagram, modified from Alexopoulos (1969) will serve to illustrate a possible evolutionary scheme for the Myxomycetes.

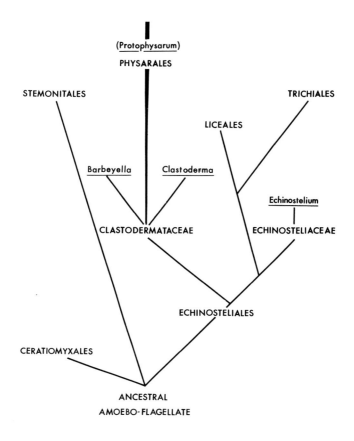

III. IMPORTANT LITERATURE

Alexopoulos, C. J. (1960). Gross morphology of the plasmodium and its possible significance in the relationship among the Myxomycetes. *Mycologia* **52**:1–20.

Alexopoulos, C. J. (1963). The Myxomycetes. II. *Bot. Rev.* **29**:1–78.

Alexopoulos, C. J. (1966). Morphogenesis in the Myxomycetes. *In* "The Fungi" (G. C. Ainsworth and A. L. Sussman, eds.), Vol. 2, Ch. 8, pр. 211–234. Academic Press, New York.

Alexopoulos, C. J. (1969). The experimental approach to the taxonomy of the Myxomycetes. *Mycologia* **61**:219–239.

Alexopoulos, C. J., and T. E. Brooks. (1971). Taxonomic studies in the Myxomycetes. III. Clastodermataceae: a new family of the Echinosteliales. *Mycologia* **63**:925–928.

de Bary, A. (1887). "Comparative Morphology and Biology of the Fungi, Mycetozoa and Bacteria." Clarendon Press, Oxford.

Famintzin, A. S., and M. S. Woronin. (1873). Über zwei neue Formen von Schleimpilzen: *Ceratium hydnoides* Alb. et Schw. und *Ceratium porioides* Alb. et Schw. *Mem. Acad. Imp. Sci. St. Petersbourg, VII Ser.* **20**:1–16.

Gray, W. D., and C. J. Alexopoulos. (1968). "The Biology of the Myxomycetes." Ronald Press, New York.

Hagelstein, R. (1944). "The Mycetozoa of North America." Published by the author. Mineola, New York.

Hattori, H. (1964). "Myxomycetes of Nasu District." Japanese Imperial Household, Tokyo. (Text in Japanese.)

Jahn, E. (1928). Myxomycetes. *In* "Die Natürlichen Pflanzenfamilien" (H. G. A. Engler und K. A. E. Prantl, eds.), Band **2**, pp. 304–339. Engelmann, Leipzig.

Lister, A. (1925). "Monograph of the Mycetozoa," (revised by G. Lister), Ed. 3. British Museum (Nat. Hist.), London.

Macbride, T. H. (1922). "North American Slime-Moulds," Ed. 2. Macmillan, New York.

Macbride, T. H., and G. W. Martin. (1934). "The Myxomycetes." Macmillan, New York.

Martin, G. W. (1940). The Myxomycetes. *Bot. Rev.* **6**:356–388.

Martin, G. W. (1949). Myxomycetes. *N. Amer. Fl.* **1**, No. 1:1–190.

Martin, G. W. (1960). The systematic position of the Myxomycetes. *Mycologia* **52**:119–129.

Martin, G. W., and C. J. Alexopoulos. (1969). "*The Myxomycetes.*" Univ. of Iowa Press, Iowa City, Iowa.

Nannenga-Bremekamp, N. E. (1967). Notes on Myxomycetes. XII. A revision of the Stemonitales. *Konikl. Nederl. Akad. Wetensch. C* **70**:201–216.

Olive, L. S. (1967). The Protostelida—a new order of the Mycetozoa. *Mycologia* **59**:1–29.

Olive, L. S. (1970). The Mycetozoa: A revised classification. *Bot. Rev.* **36**:59–89.

Ross, I. K. (1957). Capillitial formation in the Stemonitaceae. *Mycologia* **49**:809–819.

Ross, I. K. (1973). Stemonitomycetidae, a new subclass of the Myxomycetes. *Mycologia* **65**:483–485.

Rostafinski, J. (1874–76). *Śluzowce* (Mycetozoa) Monografia et App. Parıs.

Rusch, H. P. (1970). Some biochemical events in the life cycle of *Physarum polycephalum*. *In* "Advances in Cell Biology" (D. M. Prescott, ed.), Vol. I, pp. 297–328. Appleton-Century-Crofts, New York.

Schinz, H. (1920). Myxogasteres. *In* Rabenhorst's "Kryptogamen Flora," (L. Rabenhorst, ed.). Vol. I Abt. 10 Akad. Verlagsges, Leipzig.

Schroeter, J. (1897). Myxogasteres. *In* Engler und Prantl, "*Die Natürlichen Pflanzenfamilien*" (H. G. A. Engler and K. A. E. Prantl, eds.) Vol. I. Abt. 1, pp. 8–35. Engelman, Leipzig.

KEY TO SUBCLASSES OF MYXOMYCETES

1. Spores borne singly at the tips of hairlike stalks, on columnar, dendroid, or morchelloid sporophores . . . . **Ceratiomyxomycetidae**   p. 51

1'. Spores borne in masses, within various types of sporophores, peridium persistent or early evanescent 2

    2(1') Sporophore development subhypothallic, the plasmodial protoplast rising internally through the developing stalk in stipitate forms; peridium continuous with stalk and hypothallus; spores pallid, bright-colored, ferruginous, purple-brown, or black; assimilative stage of various types, but never a true aphanoplasmodium

    . . . .**Myxogastromycetidae**   p. 51

    2'(1') Sporophore development epihypothallic; stalk, when present, secreted internally, hollow, or partially filled with strands; spores violet-brown, lilac, ferruginous, or pallid by transmitted light; lime, if present, never on the capillitium; assimilative stage an aphanoplasmodium . . . .**Stemonitomycetidae**   p. 58

A. Subclass Ceratiomyxomycetidae[3]

In this subclass the sporophores are columnar, dendroid, or morchelloid consisting of a spongy matrix on the surface of which the protoplast cleaves into individual, uninucleate, amoeboid cells (protospores), each of which elevates a stalk bearing a single spore at the tip. The spores emerge upon germination, each releases a single quadrinucleate amoeboid protoplast which, after passing through a thread phase, cleaves into eight cells which become flagellate and disperse. Olive (1970) places these organisms in the protostelids.

Ceratiomyxomycetidae contain a single order, Ceratiomyxales, with a single family, Ceratiomyxaceae, and the single genus, Ceratiomyxa, with three species. *Ceratiomyxa fruticulosa* [Plate VI, Fig. 1 (see p. 60)] is cosmopolitan and the only one that has been studied developmentally. *Ceratiomyxa sphaerosperma* and *C. morchella* are tropical or subtropical.

B. Subclass Myxogastromycetidae[4]

Spores are borne internally in various types of sporophores. Upon germination, each produces one or more myxamoebae or swarm cells; sporophore development is subhypothallic in which the hypothallus is deposited on the surface of the plasmodium with the protoplasm rising below it and forming one to many stipitate or sessile sporophores of various types in which hypothallus, stalk, if present, and peridium are continuous (Plate V); the peridium is persistent, membranous, coriaceous, or calcareous, sometimes tough, or represented by a delicate membrane which evanesces by

[3]Martin, G. W. *In* Ainsworth, G. C. (1961). "Dictionary of the Fungi," 5th ed., p. 497.

[4]Martin, G. W. *In* Ainsworth, G. C. (1961). "Dictionary of the Fungi," 5th ed., p. 497. Emend. Ross, I. K. (1973). *Mycologia* **65**:483–485.

the time the spores are mature, leaving the spore mass exposed. The stalk, when present, is hollow or stuffed with granules, debris, or sporelike cells, and is calcareous or not; the capillitium, when present, is of tubular, rarely solid, threads in the form of elaters, or of branched and anastomosing filaments often forming a network. Lime, when present, is granular or crystalline; the crystals are stellate or prismatic. Spores are variously pigmented, sometimes colorless; the assimilative stage is of various types, but is never a true aphanoplasmodium.

KEY TO ORDERS OF MYXOGASTROMYCETIDAE

1. Spores in mass pallid, white or bright-colored, sometimes brown, rarely black; by transmitted light pallid to bright-colored, yellow-brown, smoky-brown or dingy, but never purple-brown 2

1′. Spores in mass black, violet-brown, dark purple-brown, ferruginous, deep red or purple; by transmitted light usually purple-brown or deeply tinted, rarely pallid
. . . . **Physarales** p. 52

    2(1) Peridium persistent in whole or in part; true capillitium lacking; pseudocapillitium when present of tubular, irregular filaments or of perforated plates which may fray out into threads . . . . **Liceales** p. 55

    2′(1) True capillitium typically present; if lacking then peridium early fugacious and sporocarps minute 3

3(2′) Stalked, minute, rarely exceeding 0.5 mm in total height (up to 1.5 mm in *Clastoderma*); peridium sometimes persistent in part or in whole; stalk filled with granular material; columella typically present, rarely lacking; capillitium, when present, scanty or forming an open globose net or branching and anastomosing; spores in mass white, yellowish, pinkish, gray, or brown . . . . **Echinosteliales** p. 56

3′(2′) Sessile or stalked, larger; columella absent; capillitium usually abundant, of individual threads, or a network, the threads sometimes delicate and solid, but mostly coarse, hollow, and conspicuously sculptured; spores in mass white, ochraceous, yellow, orange, or red (black in *Listerella*) . . . . **Trichiales** p. 57

1. Order Physarales

The peridium is typically calcareous in well-developed specimens except in *Protophysarum*; lime is present or absent in the hypothallus, stalk, columella, and capillitium when such structures are present. Spores in mass are dark purple-brown or black; by transmitted light purple-brown, brown, or violaceous, rarely pallid. The assimilative stage is a phaneroplasmodium which varies from minute, in *Protophysarum*, to very extensive.

This is a large order containing over half of the known species of Myxomycetes. The almost universal presence of lime on the peridium is the distinguishing feature of the order although this is a character which varies considerably with the environment so that species which are typically calcareous may be almost completely devoid of lime under certain conditions. The monotypic genus *Protophysarum*; in which no lime is secreted, is included here because of its violaceous spores, its subhypothallic development, and its phaneroplasmodium, minute though this is.

KEY TO FAMILIES AND GENERA OF PHYSARALES

1. Peridial lime granular, absent in *Protophysarum*; capillitium calcareous, rarely limeless, usually consisting either of calcareous tubules of nearly uniform diameter or of limeless, slender tubules connecting calcareous nodes . . . . **Physaraceae** 2

> The chief character of the family is the presence of lime inside the fructification (except in *Protophysarum*) in the form of calcareous capillitial tubules or nodes, calcareous plates, or limy trabeculae which extend inward from the peridium; not rarely, lime is concentrated in the center of the sporophore in the form of a sphere or elongated body, usually referred to as a pseudocolumella; the columella when present may also be limy as may the stalk.

1′. Peridial lime granular or crystalline, capillitium consisting of typically limeless threadlike, dark to pallid tubules, these occasionally exhibiting inclusions of crystalline lime, sometimes scanty, rarely lacking . . . . **Didymiaceae** 10

> The limeless capillitum is the chief characteristic of this family. The occasional crystalline inclusions in the capillitial threads of a few species will offer no taxonomic difficulty even to the uninitiated.

2(1) Lime absent from both peridium and capillitium; sporangia minute, not exceeding 0.75mm in height; peridium membranous, shiny
. . . . **Protophysarum** Alexop. & Blackwell (in ed).

> A single species: *P. phloiogenum* Alexop. & Blackwell (in ed.), known only from moist chamber cultures on bark from a few *Ulmus americana* trees in Boulder, Colorado.

2′(1) Lime present as described for the family 3

3(2′) Sporophores aethalioid, both pseudocapillitium and capillitium present . . . . **Fuligo**
(Plate VI, Fig. 12)

> A small genus with 5 species. *Fuligo septica* is one of the most common Myxomycetes; its aethalia are the largest of all myxomycete sporophores. *Fuligo cinerea* is another widely distributed species.

3′(2′) Sporophores sporangiate, or plasmodiocarpous, rarely aethalioid; pseudocapillitium absent 4

4(3′) Sporophores with calcareous, platelike, internal septa or spikelike trabeculae extending inward from the peridium 5

4′(3′) Calcareous septa or trabeculae typically absent, the latter sometimes present . . 6

5(4) Sporophores typically plasmodiocarpous sometimes sporangiate, divided into segments by vertical calcareous, sometimes branched plates; capillitium bearing spines, these often hooked . . . . **Cienkowskia**

> A single cosmopolitan but not too common species, *C. reticulata*. In its more luxuriant development the species forms extensive reticulate plasmodiocarps.

5′(4) Sporophores typically sporangiate, thimble-shaped, sometimes plasmodiocarpous; calcareous, spikelike trabeculae extending inward from the peridium . . . . **Physarella**

> A single distinctive species, *P. oblonga*; the sporangia vary from greenish yellow to white.

6(4′) Capillitium duplex, a limy network interlaced with and connected to a limeless one; sporangial walls smooth, shiny . . . . **Leocarpus**

> A single, easily recognized species, *L. fragilis*.

6′(4′) Capillitium homogenous 7

7(6′) Sporophores cylindrical, pendent . . . . **Erionema**
    A single species, *E. aureum* confined to S. E. Asia.

7′(6′) Sporophores sporangiate to plasmodiocarpous, rarely approaching aethalioid, very
rarely pendent   8

    8(7′) Capillitium a network of calcareous tubules, of nearly uniform diameter sometimes
    interrupted and the parts connected by very short limeless strands . . . . . **Badhamia**
        About 20 species a third of which produce their spores in clusters. *Badhamia utricularis*
        is cosmopolitan. Some other common and widely distributed species are: *B. obovata,*
        B. panicea, and B. versicolor.

    8′(7′) Capillitium a network of limeless tubules connecting calcareous nodes   9

9(8′) Dehiscence circumscissile, often by a preformed lid; lower portion remaining as a deep
goblet   . . . . **Craterium**
                                           (Plate VI, Fig. 15)
    Six species. *Craterium leucocephalum* is common and cosmopolitan. *Craterium para-*
    guayense with violet sporangia is tropical or subtropical.

9′(8′) Dehiscence irregular or lobate . . . . **Physarum**
                                         (Plate VI, Fig. 13)
    The largest and most varied genus of the Myxomycetes with over 80 described species
    many of which are cosmopolitan. *Physarum cinereum* forms large bluish patches on
    well-watered city lawns. *Physarum gyrosum* produces rosettes which when massed
    approach an aethalium. *Physarum polycephalum* forms irregular, gyrose sporangia;
    it grows easily in culture and is used extensively as an experimental organism. *Physa-*
    rum nicaraguenese, P. javanicum, and *P. rigidum* are tropical or subtropical.

    10(1′) Peridial lime granular   11

    10′(1′) Peridial lime crystalline 13

11(10) Peridial lime scanty; capillitium netted; columella typically absent . . . . . **Wilczekia**
    A doubtful genus, difficult to separate from *Diderma*. A single species, *W. evelinae*

11′(10) Peridial lime abundant; capillitium radiating from the columella or the columellalike
peridial base; limy columns sometimes present 12

    12(11′) Sporangia bearing limy peglike protuberances on their surfaces . . . . **Physarina**
    Two rare species, *P. echinocephala* and *P. echinospora,* neither known from the U.S.A.

    12′(11′) Sporangia without peglike protuberances, middle crystalline peridial layer some-
    times present . . . . **Diderma**
                                           (Plate VI, Fig. 16)
    A large genus of some 35 species some of which are predominantly alpine. Among the
    latter are *D. alpinum, D. lyallii, D. montanum,* and *D. niveum.* Among the more wide-
    spread species are *D. spumarioides, D. globosum, D. floriformis, D. testaceum, D.*
    effusum, and *D. hemisphaericum.*

13(10′) Sporophore an aethalium . . . . **Mucilago**
    A single cosmopolitan species, *M. crustacea,* which often fruits on living plants.

13′(10′) Sporophores sporangiate or plasmodiocarpous 14

    14(13′) Lime crystals scattered or united into a layer, but not forming scales
    . . . . **Didymium**
    A large genus with many common species such.as *D. nigripes, D. iridis, D. squamulosum,*
    D. minus, D. clavus, and *D. melanospermum. Didymium leoninum,* with very large lime
    crystals, is known only from the Far East and from Jamaica.

14'(13') Lime crystals united into scales . . . . **Lepidoderma**

A small genus of six species, most of which are predominantly montane. The genus has been most recently discussed by Kowalski (Mycologia **63**:490–516, 1971).

2. Order Liceales

The peridium is usually persistent in some forms and evanescent in some aethalia. The capillitium is lacking and the pseudocapillitium is present or absent. Spores in mass appear yellow, gray, or brown; by transmitted light colorless to pallid, smoky to yellow-brown.

KEY TO FAMILIES AND GENERA OF LICEALES

1. Sporophores sporangiate or plasmodiocarpous, generally small, often minute; pseudocapillitium and dictydine granules lacking . . . . **Liceaceae**

A single genus, *Licea*, with about 20 species, most of which produce minute sporangia from protoplasmodia on bark or other substrata in moist chamber culture. Many species are probably cosmopolitan, but go undetected because of their minute size. *Licea biforis, L. fimicola, L. kleistobolus, L. operculata,* and *L. pedicellata* are the easiest to recognize; all except *L. fimicola* are cosmopolitan.

1'. Sporophores sporangiate, pseudoaethalioid, or aethalioid; pseudocapillitium or dictydine granules may be present 2

2(1') Dictydine granules lacking . . . . **Reticulariaceae** 3

The sporophores are usually aethalia or pseudoaethalia, often of considerable size, but forms in which the sporangia are free, clustered, or connate are also encountered. The usually present pseudocapillitium is of various types: columnar, tubular, membranous, platelike, or bristlelike.

2'(1') Dictydine granules present; peridium usually netted, interstices fugacious except in *Lindbladia* . . . . : . . . . **Cribrariaceae** 6

A distinctive family typified by the presence of characteristic dictydine granules, of unknown chemical composition, on all parts of the sporophores including the spores, and by the absence of capillitium. The dictydine granules are inconspicuous in *Lindbladia* but easily discernible in the other two genera.

3(2) Sporophore sporangiate, the sporangia usually massed into a pseudoaethalium . . . 4

3'(2) Sporophore a true aethalium 5

4(3) Sporangial walls persistent . . . . **Tubifera**
(Plate VI, Fig. 2)

Tubifera bombarda, with a brushlike pseudocapillitium of stiff hairs is strictly tropical; *T. ferruginosa* and *T. microsperma* are more common and widely distributed.

4'(3) Sporangia closely compacted into a palisade layer, the walls disappearing at maturity except at the corners . . . . **Dictydiaethalium**

Two species. *Dictydiaethalium plumbeum* is cosmopolitan and the more common.

5(3') Pseudocapillitium of colorless, branching, tubular threads (Plate I, Fig. 10); spores in mass gray or ochraceous, almost colorless by transmitted light . . . . **Lycogala**
(Plate VI, Fig. 3)

Four species of which *L. epidendrum* is extremely common and cosmopolitan; *L. conicum*, with a conical sporophore, is cosmopolitan, but is rarely collected.

5'(3') Pseudocapillitium of perforated plates (Plate I, Fig. 11) which may be shredded into
   filaments; spores in mass yellow, brown, or olivaceous . . . . **Reticularia**
           Mostly massive aethalia; 7 species of which *R. lycoperdon* and *R. splendens* are the most
           common and easiest to recognize.

   6(2') Sporangia usually appressed into a pseudoaethalium or aethalium; peridial net, if
      present, and dictydine granules inconspicuous . . . . **Lindbladia**
           A single, extremely variable species, *L. tubulina.*

   6'(2') Sporangia free but sometimes crowded; peridial net and dictydine granules conspi-
      cuous 7

7(6') Peridial net relatively uniform with nodes at the junction of short threads
   . . . . **Cribraria**
                                                              (Plate VI, Fig. 4)
           A difficult genus with over 25 species many of which overlap in their characteristics.
           Cribraria intricata, C. tenella, and *C. violacea* are cosmopolitan; many other species
           are widely distributed as well.

7'(6') Vertical threads of peridial net relatively thick, connected by delicate horizontal filaments
   . . . . **Dictydium**
           Three species, of which *D. cancellatum* is cosmopolitan and very common (Plate II,
           Fig. 8).

3. Order Echinosteliales

In this order, the sporophores are sporangiate, minute, stalked, and do
not exceed 1.5 mm in total height; generally they are much smaller. The
stalk is translucent and stuffed with granular material; a columella is
typically present. Capillitium is present or absent. When present it usually
originates at the tip of the columella and varies from a single sparingly
branched thread to a system of branched and anastomosing threads. The
spores in mass are white, cream-colored, yellow, pink, gray, or brown;
their walls typically, but not always, bear areolae. They are smooth or
minutely spinulose. The assimilative stage is a protoplasmodium.

KEY TO FAMILIES AND GENERA OF ECHINOSTELIALES

1. Spores in mass white, cream-colored, yellow, pink, gray, or rarely pink-brown; peridium
   delicate, evanescent at an early stage . . . . **Echinosteliaceae**
           A single genus, *Echinostelium,* with seven species rarely seen because of their size.
           Echinostelium minutum is cosmopolitan.

1'. Spores in mass brown; peridium persistent, either as a whole or in fragments which cling
   to the tips of the capillitium . . . . **Clastodermataceae** 2

   2(1') Peridium tough, persistent, splitting stellately and remaining attached at base as a
      cup; spores brown by transmitted light . . . . **Barbeyella**
           A single, rare species *B. minutissima,* usually sporulating on bryophytes on the bark of
           trees.

   2'(1') Peridium delicate, fragmenting, the fragments clinging to the tips of the capillitial

threads . . . . **Clastoderma**
Clastoderma debaryanum is widely distributed and easy to identify; *C. pachypus*, the
only other known species, is known only from its type locality in Holland.

4. Order Trichiales

The sporophores of Trichiales are sporangiate to plasmodiocarpous
and the peridium persistent or early evanescent. There is no columella and
the capillitium consists of solid or tubular threads consisting of separate
elaters or of a loose or tight network. The spores are yellow to brown or
red in mass.

KEY TO FAMILIES AND GENERA OF TRICHIALES

1. Capillitial threads typically slender, rarely exceeding 2 μm in diameter, appearing solid,
   smooth or ornamented . . . . **Dianemaceae**   2
   The capillitial threads are usually attached to the top and bottom of the peridium. Some
   species are alpine.

1'. Capillitial threads typically coarser, rarely below 2 μm in diameter, tubular, generally
   conspicuously ornamented . . . . **Trichiaceae**   5

   2(1) Sporangia black; capillitium bearing annular, beadlike thickenings . . . **Listerella**
   A single species, *L. paradoxa*, whose taxonomic position is uncertain.

   2'(1) Sporangia rarely black; capillitium without annular thickenings   3

3(2') Capillitial threads essentially straight, attached to peridial walls . . . . **Dianema**
   Seven species of rarely collected Myxomycetes. *Dianema corticatum* is the best known
   and most widespread. *Dianema nivale*, *D. aggregatum*, and *D. subretisporum* are alpine
   species.

3'(2') Capillitial threads usually coiled, mostly free from the peridium   4

   4(3') Spores clustered . . . . **Minakatella**
   A single rare species, *M. longifila*.

   4'(3') Spores free . . . . **Calomyxa**
   (Plate VI, Fig. 5)
   A single widely distributed species, *C. metallica*, frequently appearing on bark in moist
   chamber culture.

5(1') Capillitial threads variously ornamented or nearly smooth, without well-defined spiral
   bands except in *Arcyria leiocarpa*   6

5'(1') Capillitial threads bearing conspicuous spirals   11

   6(5) Capillitium of short or long elaters, the latter rarely united to form a net   7

   6'(5) Capillitial threads typically united into a network   8

7(6) Sporangia scattered or crowded, but not heaped; peridium usually thick, capillitium
   without spirals . . . . **Perichaena**
   (Plate VI, Fig. 6)
   About 10 species of which *P. corticalis*, *P. depressa*, and *P. chrysosperma* are the most
   common, frequently encountered on bark in moist chamber culture.

7'(6) Sporangia typically heaped; peridium thin, translucent, shiny, or irridescent; capillitium
   with indistinct spirals . . . . **Oligonema**

Three species. *Oligonema schweinitzii* is the most common and most widely distributed.

8(6′) Peridium persistent, especially below; capillitium not elastic 9

8′(6′) Peridium evanescent above, leaving a well-defined, shallow or deep cup (calyculus); capillitium typically elastic, sometimes greatly so . . . . **Arcyria**
(Plate VI, Fig. 7)

Twenty or more species. *Arcyria cinerea*, *A. denudata*, *A. incarnata*, and *A. nutans* are cosmopolitan and among the most common of Myxomycetes. In *A. leiocarpa* the capillitial threads bear well-developed spirals. *Arcyria versicolor* is mainly montane.

9(8) Capillitium with inconspicuous spiral markings or nearly smooth . . . . . . **Calonema**
Two species. *Calonema aureum* is the type; the position of *C. luteolum* in this genus is debatable.

9′(8) Capillitium without spirals 10

10(9′) Capillitium bearing warts or spines . . . . **Arcyodes**
A single species, *A. incarnata*.

10′(9′) Capillitium bearing thick rings . . . . **Cornuvia**
A single species, *C. serpula*.

11(5′) Capillitial threads attached to sporangial walls by their penicillate tips . . **Prototrichia**
A single, mostly montane species, *P. metallica*.

11′(5′) Capillitium essentially free of sporangial walls 12

12(11′) Capillitium in the form of short or long, free elaters 13

12′(11′) Capillitium forming a tight network . . . . **Hemitrichia**
Eleven species. *Hemitrichia clavata*, *H. stipitata*, and *H. serpula* are the most commonly encountered; the first is a temperate zone species, the other two cosmopolitan.

13(12) Sporangia opening by a preformed lid; elaters conspicuously spiny . . . **Metatrichia**
Two species of which *M. vesparium*, the waspnest slime mold, is among the most common Myxomycetes.

13′(12) Sporangia usually dehiscing irregularly, or if by a preformed lid then elaters not prominently spiny (Plate I, Fig. 6) . . . . **Trichia**
(Plate VI, Fig. 8)

Fourteen species, many of which are common, i.e., *T. favoginea*, *T. scabra*, *T. varia*, *T. decipiens*, *T. botrytis*, and *T. floriformis*. *Trichia crateriformis* has a preformed lid and approaches *Metatrichia* but the peridium is different.

C.   Subclass Stemonitomycetidae

Spores are borne internally in various types of sporophores; upon germination, each produces one or two, rarely more, swarm cells. Sporophore development is epihypothallic in which the plasmodium deposits a hypothallus on the substratum and becomes concentrated above it in one or more masses which develop into usually stipitate, sometimes sessile sporophores. Sporophores are usually sporangiate, sometimes aethalioid or pseudoaethalioid, rarely subplasmodiocarpous; the peridium is sometimes persistent, but more often evanescent from a very early stage; the stalk, when present, is secreted internally, the protoplasm being elevated on it and forming the sporangium, hollow, or partially filled with strands, calcareous in *Diachea,* stuffed and waxy in *Elaeomyxa* which is tentatively included here pending developmental studies. True capillitium is typically present, of simple, branched, or anastomosing threads, usually free from the peridium, but occasionally attached to it, arising from the sporophore base or, more often, from the columella which is typically present. Lime, if present, is never on the capillitium, but confined to the hypothallus, stalk, columella, and occasionally to the base of the peridium. The spores are violet-brown, lilac, ferruginous, or pallid by transmitted light; the assimilative stage is typically an aphanoplasmodium.

There are a single order, Stemonitales, and a single family, Stemonitaceae, with the characters of the subclass.

5. Order Stemonitales[6]

KEY TO GENERA OF STEMONITACEAE

1. Sporophores aethalioid or pseudoaethalioid 2

1′. Sporophores sporangiate, sporangia rarely connate 4

    2(1) Sporophore an aethalium 3

    2′(1) Sporophore a pseudoaethalium . . . . **Schenella**
        Two rare species, *S. simplex* and *S. microspora*, known only from California.

3(2) Capillitial threads united into a net with many chambered vesicles at the nodes
   . . . . **Brefeldia**
    A single species, *B. maxima*.

3′(2) Capillitium dendroid, devoid of vesicles . . . . **Amaurochaete**
    Five species, of which *A. atra* and *A. tubulina* are the most common.

    4(1′) Lime present on some part of the fructification, sometimes scanty in the form of crystals at the base of the sporangium 5

    4′(1′) Lime totally absent 6

5(4) Columella and stalk, if present, calcareous . . . . **Diachea**
        (Plate VI, Fig. 14)
    A distinctive genus characterized by the thin, iridescent, long-persistent peridium and the distinctly calcareous columella. *Diachea leucopodia* is cosmopolitan; the other five species are less widely distributed. The genus is placed by some in the Physarales because of the presence of lime, but its general aspect is stemonitaceous. Developmental studies are lacking, however.

5′(4) Lime crystals usually present at the base of the sporangium and hypothallus; stalk and columella, when these are present, limeless . . . . **Leptoderma**
    A single species, *L. iridescens*, sometimes referred to the Physarales because of the presence of lime crystals.

    6(4′) Sporangial surface gelatinous when wet . . . . **Colloderma**
    Two species, *C. oculatum* and *C. robustum*, both usually associated with mosses and lichens on dead wood.

[6]For a somewhat different treatment of this order, see Nannenga-Bremekamp, N. E. (1967). *Konikl. Nederl. Akad. Wetensch. C.* **70**:201–216.

6'(4') Sporangial surface not gelatinous when wet 7

7(6') Wax present in the stalk and often in the other parts of the sporophore
. . . . **Elaeomyxa**
The stalk is stuffed with granular matter indicating a subhypothallic development; if so, the genus does not belong to the Stemonitales and should be transferred elsewhere. No developmental studies have been made. Both species, *E. cerifera* and *E. miyazakensis*, are rare.

7'(6') Wax absent 8

8(7') Sporangia sessile or with short thick stalks; columella absent . . . . . **Diacheopsis**
Four rare species. The genus is primarily alpine.

8'(7') Sporangia stipitate or, if sessile, columella then present 9

9(8') Stalk translucent, hollow, typically yellow at the base . . . . **Macbrideola**
All five species have minute sporangia and are known only from moist chamber culture on bark. *Macbrideola decapillata*, as defined by Alexopoulos (*Mycologia* **59**:113, 1967) is the most widely distributed.

9'(8') Stalk opaque, usually tough and shiny 10

10(9') Peridium tough, metallic, long persistent . . . . **Lamproderma**
(Plate VI, Fig. 9)
The genus has been most recently discussed by Kowalski (*Mycologia* **60**:756–867, 1968). The most common of the 18 species he recognizes are *L. arcyrioides*, *L. arcyrionema*, and *L. scintillans*. Most species are montane.

10'(9') Peridium early evanescent or, if persistent, thin, membranous, delicate 11

11(10') Capillitium pendent from an apical cup which is an enlargement of the tip of the columella . . . . **Enerthenema**
A distinctive genus with but three species, of which *E. papillatum* is common. (Plate III, Figs. 3 and 4).

11'(10') Columella without apical cupulate disc 12

12(11') Capillitium terminating in a surface net with no or few free ends . . . **Stemonitis**
(Plate VI, Fig. 11)
A difficult genus with 16 species. Most distinctive are *S. fusca*, *S. axifera*, and *S. splendens*; all three are cosmopolitan and common (Plate II, Fig. 4).

12'(11') Surface net absent or, if present, usually with many free ends (Plate III, Fig. 1)
. . . . **Comatricha**
(Plate VI, Fig. 10)
Closely resembling *Stemonitis* from which it is often difficult to separate without knowledge of capillitial development. Most common species is *C. typhoides*. Some others are *C. elegans*, *C. nigra*, *C. laxa*, *C. irregularis*, and *C. fimbriata*. *Comatricha suksdorfii* and *C. pacifica* are common in the western mountains of the United States.

PLATE VI. Various sporophores. Fig. 1. *Ceratiomyxa fruticulosa* simple columns. Fig. 2. *Tubifera ferruginosa*, young pseudoaethalium. Fig. 3. *Lycogala epidendrum* aethalia. Fig. 4. *Cribraria argillacea* sporangia. Fig. 5. *Calomyxa metallica* sporangia. Fig. 6. *Perichaena depressa* plasmodiocarps. Fig. 7. *Arcyria incarnata* sporangia. Fig. 8. *Trichia varia* sporangia. Fig. 9. *Lamproderma pyriforme* sporangia. Fig. 10. *Comatrichia typhoides* sporangium. Fig. 11. *Stemonitis axifera* sporangia. Fig. 12. *Fuligo septica* aethalia. Fig. 13. *Physarum rubiginosum* sporangia. Fig. 14. *Diachea leucopodia* sporangia. Fig. 15. *Craterium leucocephalum* sporangia. Fig. 16. *Diderma asteroides* sporangia.

CHAPTER 4

Mastigomycotina (Zoosporic Fungi)

F. K. SPARROW

Department of Botany
University of Michigan
Ann Arbor, Michigan

I. INTRODUCTION

At least six distinct groups of organisms, the Plasmodiophoromycetes, Saprolegniales, Lagenidiales, Peronosporales, Chytridiomycetes, and Hyphochytridiomycetes, are arbitrarily brought together in this subdivision on the sole basis that the nonsexual propagative spore by the production of one or two flagella is adapted for locomotion in a liquid, usually water. The diversity of origin of the various members can be detected most readily (but not solely) by the number, position, and type of flagella present. In eukaryotic organisms such as these, flagella are not mere extensions of the body of the spore but discrete organelles having their origin in the centriolar system and possessing the "9 + 2" arrangement of the component fibrils found in other eukaryotic (plant and animal) flagella. In view of the universal observations of centrioles originating by bipartitioning from a preexisting centriole (Picket-Heaps, 1969) it can be expected that two such structures will always be present in the spore body and such seems to be the case. In some fungus zoospores both centrioles function to produce flagella; in others just one. Two flagella are the maximum found in fungus spores. These may be one anterior, one posterior, two apically, or two "laterally" attached. It appears characteristic of biflagellate fungus spores that while the two flagella may be apically (Fig. 1, E–F) or laterally (Fig. 1, F–G) attached, they soon become oppositely directed when in motion (Couch, 1941). Anteriorly uniflagellate (Fig. 1, D–E) and posteriorly uniflagellate (Fig. 1, A–D) types are well known.

Recognition of the number and placement of the flagella on the spore as being of significance in detecting relationships among zoosporic fungi

F. K. Sparrow

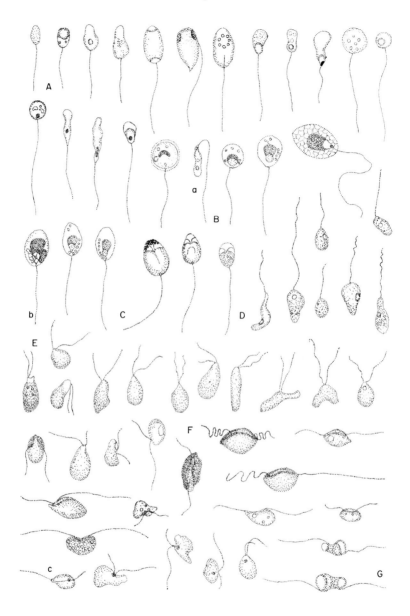

Fig. 1.   Zoospore types. A–D, Posteriorly uniflagellate. A–B, Chytridiales. B–C, Blastoc-ladiales. C–D, Monoblepharidales. D–E, Anteriorly uniflagellate: Hyphochytriales. E–F, Anteriorly biflagellate primary zoospores: Saprolegniales. F–G, Laterally biflagellate zoo-spores: Saprolegniales, Leptomitales, Lagenidiales, and Plasmodiophorales. (From Sparrow, 1960, freehand. Read from left to right; letters limit the various types.)

was emphasized by Scherffel (1925) and Sparrow (1935). Although flagella have been covered in some detail by Kohl in this series (Volume I, Chapter 4 1965) they are considered to be so essential to the identification of these fungi as to warrant some repetition here.

Couch (1938, 1941) and Vlk (1939) showed that as with certain other motile cells (Fischer, 1894) there were two types of flagella formed. One, always forwardly directed, was of the "tinsel" type ("Flimmergeissel") with lateral hairs borne singly down both sides of the main axis (Fig. 2, H) almost to the tip. The other, usually posteriorly directed lacked such hairs and was prolonged at its tip into a thin end-piece. This flagellum was designated as the "whiplash" type ("Peitchengeissel") (Fig. 2, A – G). Details of these types, first-named by Fischer, are discussed by Kohl (1965). More recent work on fine structure of both uni- and biflagellate zoospores as revealed by electron microscopy (EM) may be found in Madelin (1966), and in contributions to be cited in connection with specific types of zoospores formed by the taxa mentioned.

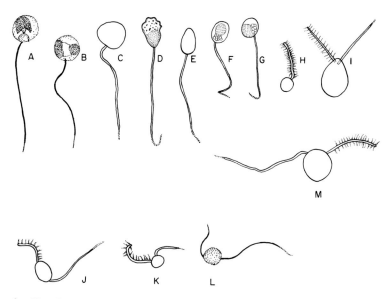

Fig. 2. Flagellar types. A–G, Zoospores of the Chytridiales, Monoblepharidales, and Blastocladiales; stained to show whiplash type of posterior flagellum. H, Zoospore of *Rhizidiomyces* showing tinsel type of anterior flagellum of Hyphochytriales. I, Primary zoospore of Saprolegniales, showing anterior tinsel and posterior whiplash flagella. J–K, Biflagellate secondary zoospore of *Lagenidium* and *Olpidiopsis* of the Lagenidiales, showing anterior tinsel and posterior whiplash flagella. L, Biflagellate zoospore of Plasmodiophorales showing flagella, both of the whiplash type. M, Biflagellate secondary zoospore of *Achlya*. (From Sparrow, 1960.)

II. PLASMODIOPHOROMYCETES[1]

The zoospore with two apically attached, unequal, oppositely directed flagella (shorter anterior) (Fig. 1, c) formed by this group is unique in its lack of any tinsel flagellum (Fig. 2, L). In three species, both the shorter anteriorly directed and posteriorly directed flagella are of the whiplash type, whereas in three others only the longer trailing one is of this type, the shorter being blunt ended (Karling, 1968). Furthermore, Keskin (1964) maintains that the flagella of *Polymyxa betae* are not associated with centrioles in the body of the spore but arise in proximity to the nucleus and a Golgi body. It is further stated by him that the short flagellum lacks a tailpiece. The EM photographs given in Keskin's paper and that of the flagellum in Karling (1968) are not too convincing evidences of these anomalies.

These features of the zoospore, together with the peculiar "promitotic" figures of intranuclear division found in the sporangial and cystogenous plasmodia and followed by typical mitotic (but not intranuclear) divisions in zoosporogenesis and cystogenesis, as well as formation of thin-walled phycomycetous zoosporangia, all tend to support Karling's (1968) position that this group is best included in a "separate division or class of fungi and (should be placed) at the bottom of the so-called biflagellate series of Phycomycetes."

III. SAPROLEGNIALES

In this biflagellate group one flagellum (anteriorly directed) is always of the tinsel type; the other, equal or nearly equal in length, trails and is of the whiplash type. A peculiarity of this group is the remarkable production in succession of two types of zoospores (or their equivalent), termed "dimorphism." Thus, in *Saprolegnia* the first-formed swarmer is "pip-," "pear-," or "sausage-shaped," with its two flagella anteriorly attached but in motion oppositely directed (Fig. 1, E–F). This spore is a poor swimmer and soon rounds up, absorbs its flagella into its body (Crump and Branton, 1966) and encysts. From this cyst there emerges after 1–4 hours a protoplast which remains at the collar of a short exit papilla while it develops its flagella. This zoospore (commonly termed the "secondary zoospore") possesses a grapeseedlike or "reniform" body and from a shallow dorsal groove running

---

[1]When in the course of a paper on the interrelationships of the Aquatic Phycomycetes (Sparrow, 1958), I proposed the class names Chytridiomycetes, Hyphochytridiomycetes, and Plasmodiophoromycetes, I was not aware that Cejp (1957) in a mycology text, in Czech, had similarly named and applied the first and third of these names. It is always good to have corroboration, even unknowingly, by a distinguished colleague!

nearly the length of the body two oppositely directed "laterally" attached flagella emerge (Fig. 3, A, C). Manton *et al.* (1951, 1952), Meier and Webster (1954), Gay and Greenwood (1966), and others have made EM studies of these spores. They find that the posterior flagellum of the secondary zoospore is finlike and this, together with its somewhat greater length, might account for the spore's stronger swimming capacities. Cysts of both primary and secondary zoospores have been found to have outgrowths on their outer wall surfaces. In primary zoospore cysts these tend to be simple hairs or tufts of hairs, whereas doubleheaded "boat hooks" are present on cysts of secondary zoospores.

This dimorphism (less accurately termed "diplanetism" in the older literature) is a remarkable phenomenon and is difficult to interpret. Traces of the two swarm stages, by encystment, etc., of primary spores are found in nearly all genera and families commonly grouped in the Saprolegniales. Indeed, their presence alone is strong evidence of saprolegniaceous relationship regardless of whether or not sex organs are present. There seems to be discernible within the group a marked trend toward suppression of the weak-swimming primary zoospore. Evidence of the latter can be traced, however, in the Leptomitaceae but not in the Rhipidiaceae of Leptomitales which seems better placed in the Peronosporales (Sparrow, 1971).

IV. LAGENIDIALES-PERONOSPORALES COMPLEX

These groups, which with the preceding have been in the past collectively termed "Oomycetes," all form a zoospore of the "laterally biflagellate" type similar to the secondary ones of Saprolegniales. There seems to be no traces of a primary zoospore stage. It has become usual in fact to equate the zoospore of these taxa with the secondary ones of the Saprolegniales and, because of similar but not identical sexual stages, to regard all these biflagellate groups as related. Perhaps this is correct. Since, however, over the years no real evidence for a primary swarm stage has been forthcoming for members of the Lagenidiales-Peronosporales Complex, the question may well be raised as to whether they ever possessed it. Possibly a similar spore type (the "laterally biflagellate" one) has arisen independently in the various groups in response to the common need for an efficient swimmer. This common type in both mono- and dimorphic species is shown in Fig. 3.

Although all members of this complex have laterally biflagellate zoospores, these do not look alike either in body structure, contents, relative flagellar lengths, or size (Fig. 1, F–G). The best known are the relatively large zoospores belonging to the Peronosporales, especially those of *Phytophthora*,

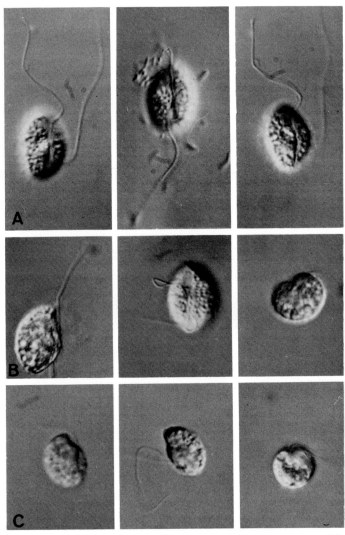

FIG. 3. Secondary zoospores. A (Left to right in row): Secondary zoospores of *Saprolegnia*; groove views; fixed material. B (Left to right in row): Same of *Pythium aphanidermatum*, lateral view; groove view; end view; fixed material. C (Left to right in row): Same of *Aphanomyces cochlioides*; lateral view; lateral view; end view; fixed material. All × 1400. Phase contrast. (From Ho *et. al.,* 1968b.)

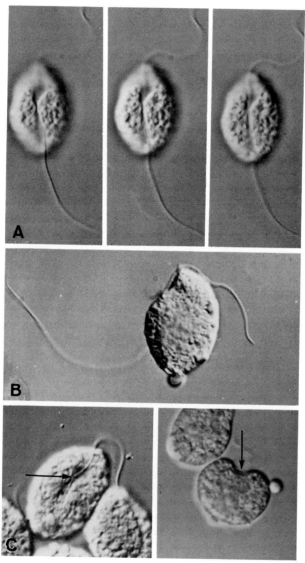

FIG. 4. Zoospores of *Phytophthora megasperma* var. *sojae*. A (Left to right): Groove view showing overarching of central portion of groove by zoospore body. Photos at successively lower points of focus down to base of groove. B. Lateral view with groove on lefthand side. Note conspicuous flattening of the groove side towards anterior end. C (Left to right): Arrow showing point of insertion of flagella. Righthand photo. End view with groove (arrow) in optical section, the typical reniform shape so often described. All × 1440. Phase contrast. (From Ho *et. al.*, 1968b) Fixed material.

since these and related genera attract the attention of the phytopathologist as well as mycologist. One of the best accounts of the *Phytophthora* zoospore, and certainly the best light photomicrographs, are in Ho and Hickman (1967), and Ho *et al.* (1968b). Ultrastructure of zoospore has been more widely studied by Ho *et al.* (1968a) and Vujičič *et al.* (1968). Colhoun (1966) has also reviewed the structure of zoospore and especially flagellar structure.

In *Phytophthora* (Fig. 4) the zoospore (Ho *et al.*, 1968b) body is ovoid, bluntly pointed at one or both ends; or, less often, reniform. Observations on moving and fixed zoospores in liquid mounts viewed with interference contrast microscopy showed that *Phytophthora megasperma* var. *sojae* zoospores are ovoid, bluntly pointed at the anterior end with a longitudinal groove shallow at each end and deeper in the center and where it seems overarched by outgrowths of the spore body. The two oppositely directed flagella are inserted at one point within the deep portion of the groove and slightly near the anterior apex. The latter is conspicuously flattened on the grooved surface and bluntly pointed rather than rounded as is the posterior end. Such a spore when seen in end view appears reniform because of the shallow groove. A large vacuole is just below the groove and towards the anterior end. According to Ho and Hickman (1967) the posterior flagellum is from four to five times longer than the anterior one. When the spore encysted "beads" appeared at the base of each flagellum. These moved out a varying distance on the flagella which then steadily shortened and finally disappeared into the beads which were then sluffed off the body. The ultrastructure of this zoospore was investigated by Ho *et al.* (1968a). They found among other things that the two flagella arose from a protuberance midway along a ridge projecting from the base of the groove and lying across it. The two flagella have their origin in a pair of centrioles. A concentration of tubules around the groove of the zoospore is believed to provide mechanical support in the area from which flagella arise. The pear-shaped central nucleus has its narrow end toward the groove region and the kinestosomes of the two flagella.

Zoospores of other Peronosporaceae and Lagenidiaceae are probably essentially like those of *Phytophthora* but a more extensive coverage of genera is needed (Fig. 3, B, *Pythium*). The groups forming numerous relatively small zoospores in their sporangia (as compared with the above), especially in the case of the Olpidiopsidaceae, and also in the Sirolpidiaceae, and Thraustochytriaceae, are in need of more precise flagellar studies. In some species of *Olpidiopsis* they are described with equal, others with unequal ("heterocont") flagella. Some species of *Rozellopsis* have been described with the anterior flagellum shorter than the posterior, others just the reverse; in *Petersenia* they have been described or figured as nearly equal or equal. In the marine Sirolpidiaceae (including *Haliphthorus*), *Sirolpidium zoöphthorum* Vishniac (1955) has markedly heterocont zoospores with the short anterior flagellum of the tinsel type, the posterior of the whiplash (by EM), whereas the less precise light microscope observations on *S. bryopsidis* by Sparrow (1934) reveal it to have zoospores with nearly equal flagella as are those of *Pontisma*. Zoospores with nearly equal flagella, having longer anterior tinsel and posterior whiplash, have been described for the marine genus *Thraustochytrium* (Gaertner, 1964). Amon and Perkins (1968; Perkins and Amon, 1969) have confirmed zoospore production in *Labyrinthula* and show the biflagellate spore to have a long tinsellated anterior, a short whiplash posterior flagellum, and a functional eye-spot. Details of fine structure are given. Their account of zoosporogenesis (which they believe probably to involve a meiotic process) by successive bipartitioning of the protoplast, as in the marine biflagellate fungus *Schizochytrium* and certain chlorosphaeriaceous algae and phytoflagellates, rather than by repeated nuclear divisions followed by multiple cytokineses, seems to indicate we have to do with yet another distinct group of biflagellate fungi, possibly allied to the Thraustochytriaceae[2] or algae.

V. CHYTRIDIOMYCETES

The chytridiomycete motile structures are the best known of all types due to the intensive work, with both light and electron microscopy, of Koch (1968). Turian and Kellenberg (1956, gametes), Cantino et al. (1963, 1968), Fuller (1966), Reichle and Fuller (1967), Fuller and Reichle (1968), Olson and Fuller (1968), Temmink and Campbell (1969), and others. This spore is typically, posteriorly uniflagellate although occasionally chytrids are found in which the body seems to have undergone a 90 degree turn with relation to flagellar attachment resulting in the latter place appearing anterior (Fig. 1, A, a). The flagellum is of the whiplash type and in the Chytridiales there is often present a prominent refractive globule. The latter gives a highly characteristic appearance to the spore rendering it, along with its frequent hopping movement alternating with periods of swimming, a readily recognizable object free in the water. No such globule is usually found in other Chytridiomycetes. In the Blastocladiales a centrally located "subtriangular" body of low refractivity, the so-called "nuclear cap," is frequently visible, even under the light microscope. Furthermore, in some spores a conspicuous "side body" is visible (Figs. 1b, and 5 A). These structures have been minutely investigated by the aforementioned investigators, using EM, who find the nuclear cap of some, but not all, to be a membranated organelle with a high concentration of

[2]This family should be in its own order, **Thraustochytriales**, ord.nov. Fungi nonmyceliales ("chytrids" simulantes); tunicae L-galactosorum instructus. Pseudorhizoidea sine organelliis protoplasmicis; rudimentiis reproductiis emergentes. Zoosporae (raro aplanosporae) biflagellatae laterales.

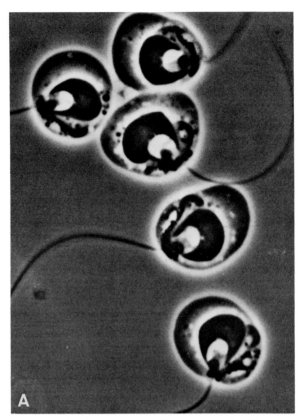

Fig. 5. Posteriorly uniflagellate zoospores. Posteriorly uniflagellate zoospores of *Blastocladiella emersonii* showing conspicuous nuclear caps; phase contrast. (From Fuller, 1966.)

ribosomes, whereas the side body is a large mitochondrion. There is, of course, a pair of centrioles associated with the flagellar apparatus, and Koch (1956), Fuller (1966), and Olson and Fuller (1968) have argued eloquently that the second centriole found in the chytrid spore is a vestigial flagellum base ("vestigial kinetosome"). Olson and Fuller feel that the presence of functional "props" in the nonfunctional vestigial kinetosome offers considerable support to Koch's notion of the biflagellate origin of these fungi. Which direction evolution is, in fact, proceeding in this structural complex is a fascinating question and the interested reader is referred to the aforementioned papers for further details. The Monoblepharidales (Fig. 1, C – D) (*Monoblepharella;* Fuller and Reichle, 1968) lack a membranated nuclear cap and massive mitochondrion.

VI. HYPHOCHYTRIDIOMYCETES

Hyphochytridiomycete zoospores have a single anterior flagellum (Fig. 2, H). Sometimes a prominent globule is present, sometimes not (Fig. 1, D – E). The uniqueness of this spore type was early pointed out by Couch (1938) as indicating the existence of a distinct group of fungi which in their body plan may resemble true Chytridiomycetes. EM work by Fuller and Reichle (1965) and Fuller (1966) reveals no membranated nuclear cap but a concentration of ribosomes around the nucleus. The mitochondria are found to surround the ribosome-rich area but are not associated with the kinetosome and its parts. An interconnected pair of centrioles of unequal length is associated with the nucleus, the longer terminating in a plate beyond which the strands of the flagellum are found.

In all these zoospores, of whatever affinities the same basic flagellar ultrastructure ("9 + 3" fibrils) is found as in other flagellated eucaryotic organisms and as Fuller (1966) states, "... we no longer need to assume that motile spore structure in fungi is the same as that of other flagellate organisms. The basic structure of the flagellum and its associated organelles *is* like that of other flagellates."

VII. KEY

KEY TO CLASSES AND CERTAIN ORDERS OF MASTIGOMYCOTINA[3]

1. Zoospores unequally biflagellate, the flagella anteriorly attached to spore body; one or both flagella of the whiplash type, never of the tinsel type; thallus always plasmodial, with promitotic ("cruciform") division figures as well as mitotic ones; thin-walled zoosporangia and groups of thick-walled resting spores formed . . . . **Plasmodiophoromycetes**

1'. Zoospores posteriorly uniflagellate, or equally or unequally biflagellate and variously positioned, the anterior always of the tinsel type, the posterior of the whiplash type; thallus unwalled or walled; no promitotic figures 2

    2(1') Zoospores posteriorly or anteriorly uniflagellate 3

    2'(1') Zoospores biflagellate . . . . **Oomycetes** 4

3(2) Zoospores posteriorly uniflagellate . . . . **Chytridiomycetes** p. 95

3'(2) Zoospores anteriorly uniflagellate . . . . **Hyphochytridiomycetes** p. 107

    4(2') Zoospores dimorphic or with evidences of it . . . . . . **Saprolegniales** p. 126
    . . . . **Leptomitales** p. 154

    4'(2') Zoospores with no evidences of dimorphism; the flagella "laterally" attached
    . . . . **Lagenidiales** p. 162
    . . . . **Peronosporales** p. 165

[3]Zoospore structure is the primary consideration here; there are of course other features characterizing these groups.

REFERENCES

Amon, J. P., and F. O. Perkins. (1968). Structure of *Labyrinthula* sp. zoospores. *J. Protozool.* **15**:543–546.

Cantino, E. C., J. S. Lovett, L. V. Leak, and J. Lythgoe. (1963). The single mitochondrion, fine structure, and germination of the spore of *Blastocladiella emersonii*. *J. Gen. Microbiol.* **31**:393–404.

Cantino, E. C., L. C. Truesdell, and D. S. Shaw. (1968). Life history of the motile spore of *Blastocladiella emersonii*: A study in cell differentiation. *J. Elisha Mitchell Sci. Soc.* **84**:125–146.

Cejp, D. (1957). "Houby". I. Česk. Akad., Prague.

Colhoun, J. (1966). The biflagellate zoospore of aquatic Phycomycetes with particular reference to *Phytophthora spp. In* "The Fungus Spore" (M. F. Madelin, ed.), Colston Pap. No. 18, p. 85. Butterworth, London.

Couch, J. N. (1938). Observations on cilia of aquatic phycomycetes. *Science* **88**:476.

Couch, J. N. (1941). The structure and action of the cilia in some aquatic Phycomycetes. *Amer. J. Bot.* **28**:704–713.

Crump, E., and D. Branton. (1966). Behavior of primary and secondary zoospores of *Saprolegnia* sp. *Can. J. Bot.* **44**:1393–1400.

Fischer, A. (1894). Ueber die Geisseln einiger Flagellaten. *Jahrb. Wiss. Bot.* **26**:187–235.

Fuller, M. S. (1966). Structure of the uniflagellate zoospores of aquatic Phycomycetes. *In* "The Fungus Spore" (M. F. Madelin, ed.), Colston Pap. No. 18, p. 67. Butterworth, London.

Fuller, M. S., and R. E. Reichle, (1965). The zoospore and early development of *Rhizidiomyces apophysatus*. *Mycologia* **57**:946–961.

Fuller, M. S., and R. E. Reichle. (1968). The fine structure of *Monoblepharella* sp. zoospores. *Can. J. Bot.* **46**:279–283.

Gaertner, A. (1964). Electronmikroskopische Untersuchungen zur Struktur der Geisseln von *Thraustochytrium spec. Veröff. Inst. Meeresforsch. Bremerhaven* **9**:25–30.

Gay, J. L., and A. D. Greenwood. (1966). Structural aspects of zoospore production in *Saprolegnia ferax*, with particular references to the cell and vacuolar membranes. *In* "The Fungus Spore", (M. F. Madelin, ed.), Colston Pap. No. 18, p. 95 Butterworth London.

Ho, H. H., and C. J. Hickman. (1967). Asexual reproduction and behavior of zoospores of *Phytophthora megasperma* var. *sojae. Can. J. Bot.* **45**:1963–1981.

Ho, H. H., K. Zachariah, and C. J. Hickman. (1968a). The ultrastructure of zoospores of *Phytophthora megasperma* var. *sojae. Can. J. Bot.* **46**:37–41.

Ho, H. H., C. J. Hickmañ, and R. W. Telford. (1968b). The morphology of zoospores of *Phytophthora megasperma* var. *sojae* and other Phycomycetes. *Can. J. Bot.* **46**:88–89.

Karling, J. S. (1968). "The Plasmodiophorales," 2nd ed. Hafner, New York.

Keskin, B. (1964). *Polymyxa betae* n. sp., ein Parasite in den Wurzeln von *Beta vulgaris* Tournefort, besonders während der Jugendentwickelung der Zuckerrübe. *Arch Mikrobiol.* **49**: 348–374.

Koch, W. J. (1956). Studies of the motile cells of chytrids. I. Electron microscope observations of the flagella, blepharoplast and rhizoplast. *Amer. J. Bot.* **43**:811–819.

Koch, W. J. (1968). Studies of the motile cells of chytrids. V. Flagellar retraction in posteriorly uniflagellate fungi. *Amer. J. Bot.* **55**:841–859. (References to Koch's other papers on motile cells may be found here.)

Kohl, A. P. (1965). Flagella. *In* "The Fungi" (G. C. Ainsworth and A. S. Sussman, eds.), Vol. I, p. 77 Academic Press, New York.

Madelin, M. F., ed. (1966). "The Fungus Spore," Colston Pap. No. 18. Butterworth, London.

Manton, I., B. Clarke, and A. D. Greenwood. (1951). Observations with the electron microscope on a species of *Saprolegnia*. *J. Exp. Bot.* **2**:321–331.

Manton, I., B. Clarke, A. D. Greenwood, and E. A. Flint. (1952). Further observations on the structure of plant cilia by a combination of visual and electron microscopy. *J. Exp. Bot.* **3**:204–215.

Meier, H., and J. Webster. (1954). An electron microscope study of cysts in the Saprolegniaceae. *J. Exp. Bot.* **5**:401–409.

Olson, L. W., and M. S. Fuller. (1968). Ultrastructural evidence for the biflagellate origin of the uniflagellate fungal zoospore. *Arch. Mikrobiol.* **62**:237–250.

Perkins, F. O., and J. P. Amon. (1969). Zoosporulation in *Labyrinthula sp.*, an electron microscope study. *J. Protozool.* **16**:235–257.

Picket-Heaps, J. D. (1969). The evolution of the mitotic apparatus: An attempt at comparative ultrastructural cytology in dividing plant cells. *Cytobios* **3**:257–280.

Reichle, R. E., and M. S. Fuller. (1967). The fine structure of *Blastocladiella emersonii* zoospores. *Amer. J. Bot.* **54**:81–92

Scherffel, A. (1925). Endophytische Phycomyceten—Parasiten der Bacillariaceen und einige neue Monadinen. Ein Beitrag zur Phylogenie der Oomyceten (Schröter). *Arch. Protistenk.* **52**:1–141.

Sparrow, F. K. (1934). Observations on marine Phycomycetes collected in Denmark. *Dan. Bot. Ark.* **8**(6):1–24.

Sparrow, F. K. (1935). The interrelationships of the Chyridiales. *Proc. Zesde. Int. Bot. Congr.* Amsterdam 1935 Vol. 2, p. 181.

Sparrow, F. K. (1958). Interrelationships and phylogeny of the aquatic Phycomycetes. *Mycologia* **50**:797–813.

Sparrow, F. K. (1960). "Aquatic Phycomycetes". 2nd rev. ed. Univ. of Michigan Press. Ann Arbor.

Sparrow, F. K. (1971). A general review of Biflagellate Fungi. *Abstr., I. Int. Mycolo. Congr.*, Exeter. 1971 p. 90.

Temmink, J. H. M., and R. N. Campbell. (1969). The ultrastructure of *Olpidium brassicae*. II. Zoospores. *Can. J. Bot.* **47**:227–231.

Turian, G., and E. Kellenberg. (1956). Ultrastructure du corps paranucléaire, des mitochondries et de la membrane nucléaire des gametes d'*Allomyces*. *Exp. Cell. Res.* **11**:417–422.

Vishniac, H. S. (1955). The morphology and nutrition of a new species of *Sirolpidium*. *Mycologia* **47**:633–645.

Vlk, W. (1939). Ueber die Geisselstruktur der Saprolegniaceenschwarmer. *Arch. Protistenk.* **92**:157–160.

Vujičič, R., J. Colhoun, and J. A. Chapman. (1968). Some observations on the zoospores of *Phytophthora erythroseptica*. *Trans. Brit. Mycol. Soc.* **51**:125–127.

Mastigomycotina
Plasmodiophoromycetes

CHAPTER 5

Plasmodiophoromycetes

GRACE M. WATERHOUSE

Formerly of the
Commonwealth Mycological Institute
Kew, Surrey, England

I. CLASS PLASMODIOPHOROMYCETES

In this class, the thallus is a naked holocarpic plasmodium with plasmodial movement and feeding. Except for reinfection phases (in water), the life cycle occurs entirely within a plant host. The sporangial plasmodium forms sporangia aggregated in loose masses or in sporangiosori composed of numerous or few, small or large, walled, uni-, bi- or multi-nucleate sporangia, each producing single, few, or many secondary zoospores. The zoospores are anteriorly biflagellate, heterokont, with one flagellum long and whiplash and directed backwards, the other being very short and smooth (probably whiplash also). They sometimes have a vacuole, are capable of amoeboid movement, and are 1.5–7 μm in diameter, but mostly 3–5 μm; these possibly fuse (but remain binucleate) before infecting the host to form a cystogenous plasmodium. Cyst formation is probably preceded by karyogamy followed by meiosis. The cysts are free or united into cystosori, often of a characteristic shape; each cyst is spherical with a smooth or ornamented wall, producing on germination one (occasionally more) anteriorly biflagellate, heterokont, uninucleate primary zoospore or myxamoeba. Vegetative nuclear divisions in the plasmodium show very characteristic, unique, intranuclear figures ("promitosis" or "cruciform" divisions). Heavy infection often may cause considerable host hypertrophy, but in some groups hypertrophy does not occur though there may be some cell enlargement.

There is one order, Plasmodiophorales, and one family, Plasmodiophoraceae (Karling, 1968).

A. General Characters

These are microscopic fungi occurring as obligate parasites of freshwater algae and fungi and of angiosperms (usually in the roots). Infection of the host is achieved by an amoeboid primary zoospore (Williams *et al.*, 1971). The zoospores penetrate the host anterior end first either by insertion of an infection tube, stylet, or peg (*Polymyxa betae*) through which the proto-plasm passes, or by passage of the myxamoeba through the wall. The myxamoebae form plasmodia within the host, enlarge, fuse, and multiply, usually inciting the invaded host cells to grow and in angiosperms often to repeated cell divisions, simultaneously with those of the parasite, thus causing hypertrophy. Amoebae and small plasmodia appear to migrate from cell to cell. It is not known with certainty in all genera whether the large multinucleate plasmodia eventually found are formed by the fusion of amoebae or small plasmodia from multiple infections or from plas-modial divisions and subsequent enlargement, or whether large plasmodia multiply by dividing into uni- or multinucleate units which enlarge to form new plasmodia before sporogenesis. Fusions of plasmodia have been demonstrated in *Plasmodiophora brassicae* (Tommerup and Ingram, 1971). Undoubtedly nuclear divisions take place but all the evidence is from fixed and stained material. The observation of schizogony in living material has been reported only in *Polymyxa* and *Woronina*. Cystogenesis is preceded by mitoses distinct from the promitoses in the vegetative plasmodia.

At length the hypertrophy, e.g., in *P. brassicae*, may be extensive and the very numerous host cells are filled with sporogenous plasmodia. These are transformed into masses of sporangia by the rounding off of protoplasmic units and the development around them of thin walls. They are released by the disintegration of the tissues or germinate in situ. In *Polymyxa* and *Woronina* a discharge tube penetrates to the exterior.

The penetration of the host by the zoospores from these sporangia and the development of the plasmodium within the host is as before. It may become another zoosporogenous plasmodium (Macfarlane, 1970) or a cystogenous one. The position of fusion and meiosis in the life cycle has been and still is a much debated point. Different workers have suggested the occurrence of fusion at various stages, namely (1) between primary zoospores or amoebae, (2) between secondary zoospores or amoebae, (3) as karyogamy within the cystogenous plasmodium, with meiosis following soon after fusion but most often just before cyst formation. However, most workers (but, see Tomerup and Ingram, 1971) in the last 20 years have been unable to find any fusions that they would regard as sexual, and three cytologists regarded their nuclear figures as inconclusive. Never-theless, there is no doubt that the nuclear figures in the sporangioplasmodium appear to be different from those in the cystogenous plasmodium. The former are mitotic.

About the time of the last nuclear divisions, which are synchronous, signs of cleavage begin to appear dividing the plasmodium into (usually uninucleate) segments. These, after rounding off, become enclosed in chitinous walls that are thicker than those of the sporangia and in some species are ornamented with warts or spines. What causes the cysts to remain stuck together in cystosori in most genera while they are free in *Plasmodiophora* is not known. They are set free by the death and disintegration of the tissues, are very resistant, and may survive in the soil for years (at least seven in the case of club root). Germination of the cysts (Macfarlane, 1970) is promoted by the presence of the host or its metabolic products.

B. Genera

Plasmodiophoraceae Zopf 1884[1]

Woronina Cornu 1872

　　Type species *W. polycystis* Cornu

Plasmodiophora Woron. 1877

　　Type species *P. brassicae* Woron.

Tetramyxa Goebel 1884

　　Type species *T. parasitica* Goeb.

Sorosphaera Schroet. 1886

　　Type species *S. veronicae* (Schroet.) Schroet.

Spongospora Brunch. 1887

　　Type species *S. subterranea* (Wallr.) Lagerh. syn. *S. solani* Brunch.

Ligniera Maire & Tyson 1911

　　Type species *L. verrucosa* Maire & Tison

Sorodiscus Lagerh. & Winge 1912

　　Type species *S. callitrichis* Lagerh. & Winge

Polymyxa Ledingh. 1939 (Feb.)

　　Type species *P. graminis* Ledingh

Octomyxa Couch, Leitn. & Whiff. 1939 (Dec.)

　　Type species *O. achlyae* Couch et al.

Doubtful genus

Membranosorus Ostenf. & Peters.

　　Type species *M. heterantherae* Ostenf. & Peters.

C. Key

KEY TO THE PLASMODIOPHORALES

1. Cysts (resting sporangia) not united into cystosori but lying free in the host cell, smooth, spiny, or bristly . . . . **Plasmodiophora**
　　P. brassicae (club root) widespread on crucifers (see Colhoun, 1958)

1′. Cysts united into clusters or cystosori 2

1″. Cysts absent 6

　　2(1′) Cysts united in units of 2–8 (occasionally single) 3

[1]See Karling (1968).

2′(1′) Cysts mostly united in loose or compact cystosori 4

3(2) Cysts in pairs or fours; parasitic in aquatic angiosperms . . . . **Tetramyxa**

3′(2) Cysts in groups of 8 (occasionally 6); parasitic in Saprolegniales . . . . . **Octomyxa**

    4(2′) Cystosori predominantly spherical to subspherical or ellipsoidal, hollow

    . . . . **Sorosphaera**

    4′(2′) Cystosori predominantly disk-shaped, two-layered . . . . **Sorodiscus**

                        (? including **Membranosorus**)

    4″(2′) Cystosori spongelike, ovoid, subspherical or irregular, traversed by canals and

    fissures. . . . .**Spongospora**

        Spongospora subterranea (powdery scab) widespread on potatoes; *S. s.* f. sp. *nasturtii*

        (crook root of watercress)

    4‴(2′) Cystosori very variable in form or of indefinite shape 5

5(4‴) Little or no hypertrophy of the host (angiosperms); sporangia 4–10 (–? 15) μm, nume-

rous . . . . **Ligniera**

5′(4‴) In fungi or algae; sporangia 8–35 μm, single or numerous . . . . **Woronina**

5″(4‴) Little or no hypertrophy (angiosperms); sporangia larger, elongated lobed, constricted

or segmented, exit tubes long . . . . **Polymyxa**

    P. graminis in cereals in Canada (Ledingham, 1939); *P. betae* in beet in Italy.

6(1″) Plasmodia only; in *Triglochin* with extensive hypertrophy . . . . . . **Tetramyxa**

6′(1″) Sporangia of two sizes, 4–6 μm and 15 μm; no hypertrophy, only swelling, hosts

numerous . . . . **Ligniera**

II. DISCUSSION

This is a small unique order of closely related genera with no clear affinities with any other group. A unique and characteristic feature found throughout the group is the type of nuclear division, the so-called cruciform division or promitosis, occurring in the vegetative plasmodium. In stained material the elongated nucleolus and an equatorial band of chromatin has a crucifix shape. Nowhere else in the plant kingdom does this appearance occur in mitosis, but divisions of a similar nature are to be found in *Amoeba*.

The fact that all members are obligate parasites probably obscures some features that might have indicated relationships; for example, the mode of feeding in a host cannot be compared with that of a free-living plasmodial organism, such as a myxomycete. Electron microscope studies of the plasmodium may reveal structures indicating relationships with other groups. The detection of Golgi in the zoospore (Kole and Gielink, 1962, 1963; Keskin, 1964) and plasmodium (Williams and Yukawa, 1967) would link the Plasmodiophorales with those biflagellate Mastigomycotina which also have Golgi, rather than with groups which do not. The presence of a thin-walled sporangium (or spore), producing one or more zoospores, and of a resting sporangium (or spore), the cyst, likewise producing one or more zoospores also relates this group to the biflagellate phycomycetes, especially those with small zoospores, rather than to the myxomycetes (Sparrow, 1958). The zoospores are 1.5–7 μm, mostly 3–5 μm. The walls of the cysts are two-layered and have a basis of chitin, thus differing from those of myxomycete spores which are two- or three-layered, the inner wall being of cellulose (Schuster, 1964).

Until more is known of the life cycles and the position of fusion and meiosis, if present, any discussion of relationships, full data on which are given by Karling (1968), is unprofitable.

Classification within the group has been based primarily on cyst morphology. This is the phase nearly always present and clearly seen whereas sporangia are often lacking or have been missed. In *Plasmodiophora*, towards the end of plasmodial cleavage the developing cysts are stuck together by some product of the disintegrating protoplasm. Soon, however, this adhesive dissolves away and the cysts lie free in the host cells (Fig. 1). In all other genera the cysts remain attached in smaller or larger groups, sometimes of quite characteristic shapes. In *Tetramyxa* (Fig. 2) and *Octomyxa* (Fig. 3) the groups are small (one to eight cysts) and the number in a group variable. While in *Tetramyxa* tetrads are the largest found, in *Octomyxa* octads predominate. This might be considered rather slender grounds for generic distinction but in addition the former genus has been found only in aquatic angiosperms and the latter only in saprolegniaceous fungi. Sporangia have not been found in *Tetramyxa*; until they are, these genera would best be left at present.

In the remaining genera the cystosori are larger and in *Sorosphaera*, *Sorodiscus*, and *Spongospora* of characteristic shapes (Figs. 4–6) as their names imply. In *Ligniera*, *Woronina*, and *Polymyxa* the cystosori are variable in shape and size and the generic differentiation criteria are sporangial form and size and host range (Figs. 7–9).

Although differences in cyst form and aggregation provide a convenient means for the identification of these organisms, whether these characters constitute good generic criteria has been questioned. Palm and Burk (1933) summarized earlier views on their value and as a result of finding a variety of soral forms in one host, *Veronica americana*, which were of the *Sorosphaera*, *Sorodiscus*, *Ligniera*, and *Spongospora* types, these workers suggested merging all four genera into *Sorosphaera*, together with *Ostenfeldiella* and *Clathrosorus* [both rejected by Cook (1933) and Karling (1968)] and *Membranosorus*, which is difficult to place owing to its very variable cystosori. *Ligniera* appears to be different in that it causes little or no hypertrophy, i.e., no cell multiplication though there is a swelling of the cells, in its hosts (which are numerous), but there seems to be some reason for merging the remaining genera. However, while life cycles are incompletely known and details often missing the present classification, which finds

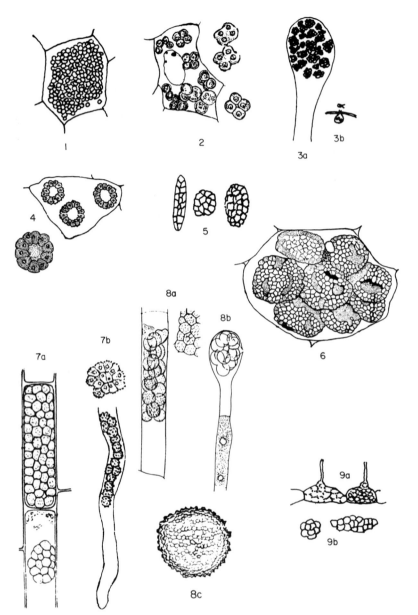

FIG. 1. Plasmodiophorales. *Plasmodiophora brassicae* cysts, × 350. Fig. 2. *Tetramyxa parasitica* cysts, × 648 (from Maire and Tison, 1911). Fig. 3. *Octomyxa achlyae* (a) cysts in hyphal tip, × 350, (b) sporangium dehiscing through hyphal wall. Fig. 4. *Sorosphaera veronicae* cystosori, × 280 (from Maire and Tison, 1909). Fig. 5. *Sorodiscus callitrichis* cystosori, × 420. Fig. 6. *Spongospora subterranea* groups of cystosori, × 420 (from Osborn, 1911). Fig. 7. (a) *Ligniera junci* cystosori, × 648 (from Maire and Tison, 1911), (b) *L. verrucosa* cystosori, × 648 (from Maire and Tison, 1911). Fig. 8. *Woronina polycystis* sporangia (a) before dehiscence, and (b) after, (c) cystosorus, all × 238 (from Cornu, 1872). Fig. 9. *Polymyxa graminis* (a) sporangiosori, (b) cystosori, both × 420.

current acceptance, even if artificial, serves the useful purpose of enabling identifications to be made.

Within genera, distinction between species is made on cyst ornamentation and size, host range, and zoosporangial form and size.

Klein and Cronquist (1967) suggest that *Woronina* and *Polymyxa* should be separated from the Plasmodiophorales and with the old Woroninaceae placed in the Lagenidiales because of the cellulose walls and zoospore structure, claiming that if this is done the Plasmodiophorales are more nearly allied to the Myxomycetes. Goldie-Smith (1956), however, made specific tests on *Woronina pythii* and could detect no cellulose. As far as zoospore structure is concerned all species investigated recently reveal no differences between these two genera and other Plasmodiophorales. Sparrow (1960) retains these two genera in the Plasmodiophorales on the grounds of clearly demonstrated promitosis, the formation of zoosporangia, and zoospore flagellation.

REFERENCES

Colhoun, J. (1958). Club root of Crucifers caused by *Plasmodiophora brassicae* Woron. *Phytopathol. Pap.* 3:1–108.

Cook, W. R. I. (1933). A monograph of the Plasmodiophorales. *Arch. Protistenk.* 80:179–254.

Cornu, M. (1872). Monograph des Saprolegniées; étude physiologique et systématique. *Ann. Sci. natur.: Bot.* [5] 15:1–198.

Goldie-Smith, E. K. (1956). A new species of *Woronina* and *Sorodiscus cokeri* emended. *J. Elisha Mitchell Sci. Soc.* 72:348–356.

Karling, J. S. (1968). "The Plasmodiophorales," 2nd rev. ed. Hafner, New York.

Keskin, B. (1964). *Polymyxa betae* n. sp., Parasit von *Beta vulgaris* Tournefort. *Phytopathol. Z.* 49:348–374.

Klein, R. M. and A. Cronquist. (1967). A consideration of the evolutionary and taxonomic significance of some biochemical, micromorphological, and physiological characters in the Thallophytes. *Quart. Rev. Biol.* 42:105–296.

Kole, A. P., and A. J. Gielink. (1962). Electron microscope observations on the resting spore germination of *Plasmodiophora brassicae. Proc. Kon. Ned. Akad. Wetensch., Ser. C* 65: 117–121.

Kole, A. P., and A. J. Gielink. (1963). The significance of the zoosporangial stage in the life cycle of the Plasmodiophorales. *Neth. J. Plant Pathol.* 69:258–262.

Ledingham, G. A. (1939). Studies on *Polymyxa graminis*, n. gen. n. sp., a plasmodiophoraceous root parasite of wheat. *Can. J. Res., Sect. C* 17:38–51

Macfarlane, I. (1970). Germination of resting spores of *Plasmodiophora brassicae. Trans. Brit. Mycol. Soc.* 55:97–112.

Maire, R., and A. Tison. (1909). La cytologie des Plasmodiophoracées et la classe des phytomyxinae. *Ann. Mycol.* 7:226–253.

Maire, R., and A. Tison. (1911). Nouvelles recherches sur les Plasmodiophoracées. *Ann. Mycol.* 9:226–246.

Osborn, T. G. (1911). *Spongospora subterranea* (Wallroth) Johnson. *Ann. Bot. (London)* 25: 327–341.

Grace M. Waterhouse

Palm, B. T., and M. Burk. (1933). The taxonomy of the Plasmodiophoraceae. *Arch. Protistenk.* **79**:263–276.

Schuster, F. (1964). Electron microscope observations on spore formation in the true slime mold *Didymium nigripes*. *J. Protozool.* **11**:207–216.

Sparrow, F. K. (1958). Interrelationships and phylogeny of the aquatic phycomycetes. *Mycologia* **50**:797–813.

Sparrow, F. K. (1960). Plasmodiophorales. *In* "Aquatic Phycomycetes," 2nd rev. ed., pp. 768–791. Univ. Michigan Press, Ann Arbor, Michigan.

Tommerup, I. C. and D. S. Ingram (1971). The life-cycle of *Plasmodiophora brassicae* Woron. in *Brassica* tissue cultures and in intact roots. *New Phytol.* **70**:327–332.

Williams, P. H., and Y. B. Yukawa. (1967). Ultrastructural studies on the host-parasite relations of *Plasmodiophora brassicae*. *Phytopathology* **57**:682–687.

Williams, P. H., S. J. Aist, and J. R. Aist. (1971). Response of cabbage root hairs to infection by *Plasmodiophora brassicae*. *Can. J. Bot.* **49**:41–47.

Mastigomycotina
Chytridiomycetes
and
Hyphochytridiomycetes

CHAPTER 6

Chytridiomycetes
Hyphochytridiomycetes

F. K. SPARROW

Department of Botany, University of Michigan, Ann Arbor, Michigan

I. CLASS CHYTRIDIOMYCETES[1]

The thallus in this class is coenocytic, holocarpic (lacking sterile parts), or eucarpic (with sterile parts and reproductive rudiment), monocentric (one reproductive rudiment), polycentric (more than one), rhizoidal or mycelial, and its wall are frequently chitinous. The zoospores have a single posteriorly directed flagellum of the whiplash type with a conspicuous nuclear cap and sometimes a single basal mitochondrion. Sexual reproduction is varied; the zygote forms an encysted structure or a diploid plant.

A. General Characteristics

These are primarily minute microscopic fungi occurring as parasites (rarely epiphytes) of various freshwater plants and animals (algae, other aquatic fungi, spores, mosquito larvae, rotifers and their eggs, etc.), terrestrial vascular plants, and as saprobes on a wide variety of debris of plant and animal origin. A few are found in marine waters.

Great numbers of kinds of chytrid saprobes have been recovered from soil and water throughout the world by use of simple baiting techniques. Soil and water samples are merely put in petri dishes, covered with sterile water and bits of "bait," such as snake skin casts, decalcified shrimp skeleton, defatted hair, sterile pine pollen, boiled cellophane, split hempseed, etc., added.

The thallus may consist of a simple sphere, without specialized vegetative parts (often naked in endoparasites), which becomes converted as a whole into a reproductive organ ("holocarpic") (Plate I, A–C, I, J). It may, in

[1]Sparrow, F. K., *Mycologia* **50**:811, 1958.

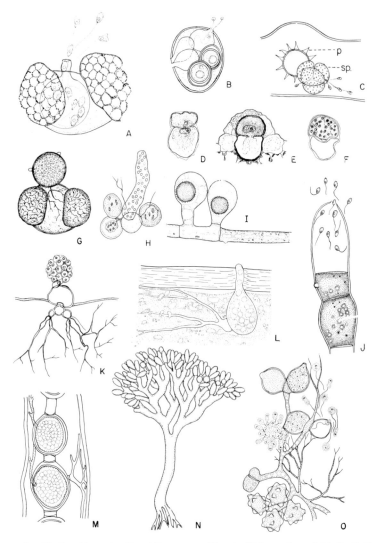

PLATE I. Thallus Types in Chytridiomycetes. Fig. A. Holocarpic endobiotic thallus of *Olpidium maritimum* in pine pollen; the thallus has been converted into a sporangium which is discharging its zoospores. Fig. B. Holocarpic thalli of *Olpidium rotiferum* in rotifer egg; some have formed sporangia, others resting spores. Fig. C. Holocarpic thallus of *Micromyces zygogonii* in cell of alga *Mougeotia*; thallus has formed a spiny prosorus which in turn has produced a sorus of zoosporangia. Fig. D. Holocarpic thallus of *Synchytrium ranunculi* has formed a smooth 1-nucleated prosorus from which a vesicle, the future sorus, is being formed above. Fig. E. The large prosorus nucleus is now in the vesicle which in Fig. F is being divided into a sorus of sporangia. Fig. G. Eucarpic thallus of *Rhizophydium sphaerotheca* on a pine pollen grain; endobiotic rhizoids have been produced from the epibiotic part which is now a mature sporangium. Fig. H. Eucarpic, interbiotic thallus of *Polyphagus euglenae* parasitic on encysted *Euglena*; the spherical thallus has functioned as a prosporangium and formed as an outgrowth a long saccate zoosporangium. Fig. I. Holocarpic endobiotic thalli of *Rozella septigena* parasitic in hypha of *Saprolegnia*; horizontal segment at right has thallus which fills host cell initiating zoosporangial development; each swollen lateral branch has a resting spore of parasite in it. Fig. J. Holocarpic endobiotic thalli of *Rozella allomycis* parasitic in *Allomyces*; the

ascertaining whether discharge is operculate or inoperculate. Four types of dehiscence mechanisms are recognized.

Type I: *Exooperculation* (Plate II). The distal wall of the discharge papilla develops a circumlinear weakening which delimits a convex cap or operculum. Immediately beneath the latter is an invaginated concave mass of nonsporogenous, gelatinous substance. Upon dehiscence of the operculum this material exudes first, followed by the emerging mass of passive zoospores. Outside, the spores form a quiescent mass with the operculum hinged to the rim of the discharge pore or somewhere on the periphery of the mass. After a brief period of rest, the zoospores become centripetally motile (i.e., peripheral ones first exhibiting movement) and quickly dart away. The operculum is a constant distinct and persistent structure derived from the distal wall of the papilla (and hence, wall material of the sporangium). This has been called "true operculation."

Type II: *Endooperculation*. There are two types recognized. In Type II–A (Plate III), the apex of the papilla inflates slightly and its distal wall dissolves in the medium. The pore is then plugged up with a clear mass of gelatinous substance (presumably from cytoplasm). Beneath the plug the granular protoplasm retracts slightly and there is formed on its surface a dense, convex zone of condensed cytoplasm which solidifies and becomes an endooperculum whose walls are continuous with the inner layer of the side wall of papilla. The gelatinous plug rests on the outer surface of the endooperculum. It disappears well before zoospore formation. Beneath the endooperculum there develops, as in Type I, a concavity of nonsporogenous, genlatinous substance. Upon spore discharge the zoospores are released en masse and the dehisced cap carried away as in Type I. Dogma points out that this type of endooperculation can easily be mistaken for exooperculation if early stages in papilla formation are not observed, since the lid is often only very slightly sunken in the discharge papilla. The persistent collarlike ring of papilla side wall may give a clue as to discharge type when the sporangium is at rest after dissolution of the distal underplug. Even here, however, this collar may disappear as in *Karlingia granulata* and an appearance of exooperculation results. It is still regarded, however, as endooperculation even though it approaches exooperculation.

In Type II–B (Plate IV), early stages are like Type II–A. When discharge tubes are formed the endooperculum is convex and generally sunken in the lower half of the tube. When discharge papillae are sessile on the wall of the sporangium the endooperculum is *concave* and invaginated into the sporogenous contents. In both cases the exit pore is plugged with gelatinous material which covers the upper surface of the lid and dissolves before zoospore delimitation. In contrast to Type II–A, *no* concavity of clear nonsporogenous material develops beneath the lid. Here, the motile zoo-

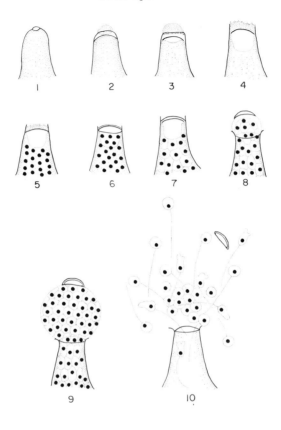

PLATE III.   Type II (Endooperculation-A) of Dehiscence Mechanism in Rhizophlyctoid
Fungi (*Karlingia*). Fig. 1. Papilla with blisterlike, gelatinizing apex and homogeneously granular
contents. Fig. 2. Later stage in gelatinization of apex of papilla. Sporogenous, granular content
has retracted downward and with a convex zone of condensed cytoplasmic material on top.
Fig. 3. Papilla with gelatinous plug in exit orifice. Aforementioned convex zone has become
solid and continuous with papillar side wall. A concavity of gelatinous substance beginning
to invaginate from underneath the endooperculum. Fig. 4. Papilla with fully invaginated
mass of gelatinous substance. Gelatinous plug is beginning to dissolve. Zoospore globules are
condensing. Fig. 5. Fully developed papilla. Gelatinous plug has dissolved almost completely.
Zoospore globules are fully formed. Fig. 6. Same as Fig. 5 but in surface view to show line of
dehiscence of endooperculum. Gelatinous plug is already gone. Note collarlike remnants
of papillar wall above point of attachment of endooperculum. Fig. 7. same as Fig. 6 but in
optical section and with fully delimited zoospore units. Fig. 8. Early stage in dehiscence and
discharge. Zoospores released *en masse* and passively. Exuded gelatinous substance envelops the
zoospore mass. Fig. 9. Later stage in zoospre liberation. Endooperculum caps the globular
mass of immobile zoospores. Fig. 10. Dispersal of zoospores. Detached endooperculum
nearby. (From Dogma, 1970.)

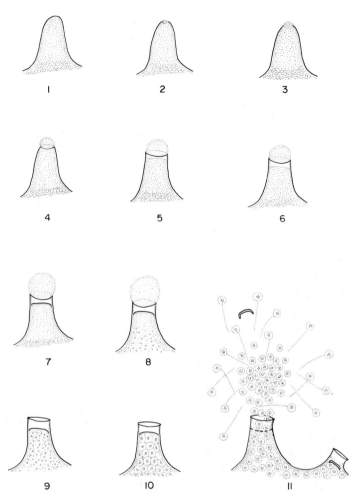

PLATE IV. Type II (Endooperculation-B) of Dehiscence Mechanism in Rhizophlyctoid Fungi (*Karlingia*). Figs. 1–4. Stages in gelatinization of apex of discharge papillae. Fig. 5. Fully formed gelatinous plug in exit orifice. The sporogenous content has retracted downward and with a dense zone of condensed cytoplasmic material on top. Fig. 6. Aforementioned zone thickens, solidifies and becomes continuous with papillar side wall. Gelatinous plug extending down to upper surface of endooperculum is beginning to dissolve. Figs. 7–8. Dissolution and expansion of gelatinous plug. In Fig. 8, zoospore globules are beginning to condense. Fig. 9. Fully developed discharge papilla with convex endooperculum sunken in exit tube. Gelatinous plug is gone. Zoospore globules are fully formed. Note absence of a distinct concavity of gelatinous substance beneath endooperculum. Fig. 10. Same as Fig. 9 but with zoospore units fully delimited moments before dehiscence. Fig. 11. Explosive discharge of motile zoospores. Endo operculum of papilla on left is expelled violently, that of one on right is displaced and sucked into sporangial cavity. Note region of attachment of endooperculum in both papillae. (From Dogma, 1970.)

spores and cap are violently expelled to the outside and if there is more than one orifice the sudden release of pressure is likely to suck inward the other endoopercula which are then whirled around with the residual zoospores in the sporangium. Initial stages in the discharge should be followed since the presence of the lids, which are often very thin, is hard to detect once discharge is under way. Dogma notes that the difference in behavior of emerging zoospores (forming a motionless group outside or at once swimming away) is correlated with the presence or absence of the concave mass of nonsporogenous gelatinous substance beneath the endooperculum.

Type III. *Inoperculation* (Plate V). Here, the distal wall of the discharge papilla region becomes thin and there early develops beneath it a conspicuous clear concave mass of nonsporogenous material. At discharge this emerges and expands as the first zoospores are passively liberated into it and after a short period of quiescence the zoospores in the mass become motile centripetally and dart away. Fragments of the distal papillar wall may sometimes persist.

Since parallelism in body types is frequent among chytrids the differences in dehiscence mechanisms and associated characters become important in identification. Dogma has found these discharge types to be constant in repeated zoosporangial generations of single-spore or unisporangial isolates in axenic or unichytrid cultures. Furthermore, the germ sporangia of resting spores are identical with those of ordinary sporangia in this respect. Endooperculate genera are keyed out here as "operculate" for convenience in identification.

Zoospore liberation in some of the Blastocladiales may involve endooperculation, but in most members of this order and in all of the Monoblepharidales it is inoperculate.

Sexual reproduction may be by copulation of iso- (Plate VI, A, J–L) or anisoplanogametes (Plate VI, M, N), fusion of egg and sperm (Plate VI, P, Q), gametangial copulation (Plate VI, B–F, O), or fusion of two vegetative plants ("somatogamy") which may often form a third structure, a resting spore (Plate VI, G–I). In some, planogametic fusion results in a motile zygote which may germinate to form a diploid plant; in others, the zygote forms a thick-walled resting body. Meiosis occurs upon germination of diploid resting spores where a diploid plant is formed and, presumably, in the encysted zygote at germination in other cases. Resting spores may also be asexually formed. Furthermore, ordinary sporangia may become quiescent and capable of withstanding considerable desiccation particularly in soil inhabitants such as *Karlingia rosea*. Several kinds of life cycles are known, some involving isomorphic (*Allomyces arbuscula*), and others heteromorphic (*Urophlyctis*) alternation of generations.

Important diseases of crop plants are caused by certain obligately parasitic members of the chytrids, notably Potato Wart (*Synchytrium endobioticum*),

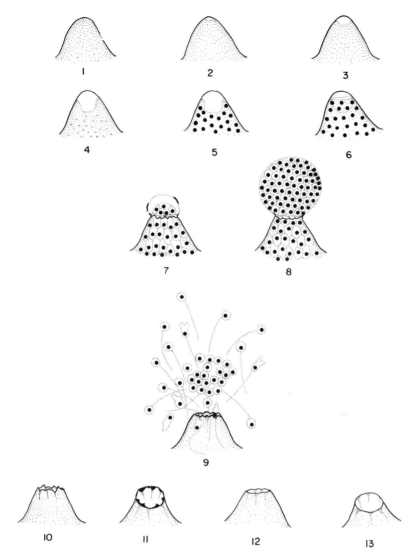

PLATE V.   Type III (Inoperculation) of Dehiscence Mechanism in Rhizophlyctoid Fungi (*Rhizophlyctis*). Figs. 1–3. Stages in a concave invagination of nonsporogenous, gelatinous substance into the homogeneously granular, sporangial contents. Fig. 4. Fully developed discharge papilla with ∪-shaped mass of gelatinous substance beneath the apical papillar wall. Zoospore globules are beginning to condense. Fig. 5. Papilla at sporangial maturity. Zoospore globules are fully formed. Fig. 6. Same as Fig. 5 but in surface view to show invaginated nature of mass of gelatinous substance. Fig. 7. Initial stage in inoperculate dehiscence. Zoospores are released *en masse* and passively. Remnants of ruptured apical papillar wall are adhering as "scales" around the exuded gelatinous substance enveloping the zoospore mass. Fig. 8. Later stage in zoospore liberation. Enveloping gelatinous substance thins out as more zoospores are discharged. Fig. 9. Dispersal of zoospores. Figs. 10–13. Side and oblique view of discharge papillae after zoospore dispersal. Figs. 10–11. Exit orifice with irregular, serrate rim. Figs. 12–13. Exit orifice with more or less smooth rim. (From Dogma, 1970.)

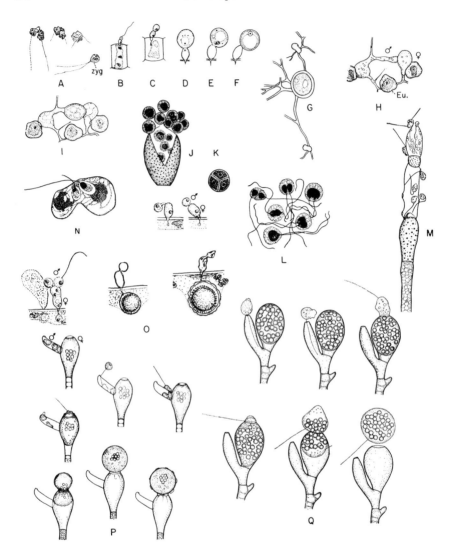

PLATE VI.   Sexuality in Chytridiomycetes. Fig. A. Isogamous planogametes of *Synchytrium*. Figs. B–F. Stages in formation of zygotic resting spore in *Rhizophydium ovatum* on *Stigeoclonium*. Fig. G. Fusions of rhizoids of 2 plants of *Siphonaria variabilis* to produce a central resting-spore plant. Figs. H, I. Conjugation of two thalli of *Polyphagus laevis*, parasitic on *Euglena*. In Fig. I, the zygotic resting spore has formed in tip of conjugation tube of "male" plant. Figs. J–L. Isogamy in *Allomyces neomoniliformis*. In Fig. J, the stained cysts emerge from resting spore of sporophyte plant. Fig. K. One cyst (reduced gametophyte) whose contents have formed a tetrad of gametes. Fig. L. Stained preparations showing fusions of isoplanogametes. Fig. M. Anisogamous planogametes of *Allomyces macrogynus* emerging and fusing. Fig. N. Stained preparation of anisogamous planogametes of *Allomyces arbuscula* fusing. Fig. O.

Brown Spot of maize (*Physoderma maydis*), and Crown Wart of alfalfa (*Urophlyctis alfalfae*). Aquatic chytrids are known to induce epidemics among such primary producers as algae. Among the Blastocladiales, the Coelomomycetaceae is composed of endoparasites of mosquito larvae; the genus *Catenaria* contains species parasitic on fungi and eggs of animals, as well as on algae. In general, the Blastocladiales and all of the Monoblepharidales appear to function as detritus decomposers in the aquatic and soil ecosystems.

KEY TO ORDERS OF CHYTRIDIOMYCETES

1. Thallus variously constructed, always penetrating substratum by means of rhizoids or hyphae, or entirely within it 2

1′. Thallus uniaxial, eucarpic with proximal basal or subbasal disclike holdfast on surface only of substratum, i.e., epiphytic or epizooic; distal part composed of upper sporogenous and lower nucleated vegetative region which persists after sporulation and is capable of sporulation . . . . **Harpochytriales** p. 104

   2(1) Thallus holocarpic, eucarpic, monocentric, or polycentric, variously developed; zoospores often with a conspicuous globule, germination monopolar

   . . . . **Chytridiales** p. 95

   2′(1) Thallus nearly always differentiated into a well-developed hyphal-like vegetative system bearing numerous reproductive bodies, occasionally monocentric; zoospores without a conspicuous globule, germination bipolar 3

3(2′) Thallus usually bearing a thick-walled, usually punctate or ornamented, asexually formed resting spore at some stage of its life history; sexuality varied by iso- or anisoplanogametes; not oogamous . . . . **Blastocladiales** p. 104

3′(2′) Thallus without above type of resting spore; oogamous, with motile sperms and encysted oospores . . . . **Monoblepharidales** p. 105

B. Order Chytridiales

The Chytridiales are microscopic parasitic or saprobic primarily aquatic fungi of simple body plan. They are monocentric or polycentric and their vegetative parts, where present, are rhizoidal and almost never mycelial. The zoosporangium is inoperculate, exo- or endooperculate. and the zoospore frequently has a conspicuous droplet. Germination is unipolar and the resting spore is thick walled and generally fills its container

Cluster of 4 figures of sexual reproduction and endobiotic formation of zygotic resting spore in *Chytridium sexuale*. Fig. P. Cluster of sequential figures showing oogamous sexual reproduction in *Monoblepharis polymorpha*. In the second row the zygote is shown emerging and forming an oospore attached to orifice of oogonium. Fig. Q. Cluster of 6 figures showing sequence of oogamous reproduction in *Monoblepharella taylori*. It can be seen that the flagellum of the sperm remains functional and propels the zygote away from the oogonium. The zygote finally encysts. (Fig. O, from Koch, 1951. All others may be found in Sparrow, 1960 where original sources are given.)

which consists of an encysted sporangium or a special asexually or sexually formed body. At germination this functions as a sporangium or pro-sporangium.

In identifying "Chytrids" the numerous papers of Karling (1966, 1967, a,b, 1968), Canter (1960, 1966, 1968), Willoughby, 1962, 1965), and others, which have appeared since Sparrow (1960) should be consulted.

KEY TO FAMILIES AND IMPORTANT GENERA OF CHYTRIDIALES[2]

1. Sporangium opening by the deliquescence or rupturing of one or more papillae 2

1'. Sporangium opening by one or more opercula or endoopercula 50

    2(1) Thallus holocarpic, endobiotic 3

    2'(1) Thallus eucarpic, epi-, endobiotic or both 12

3(2) Thallus forming a single sporangium . . . . **Olpidiaceae** 4

3'(2) Thallus forming more than 1 sporangium 8

    4(3) Sporangium not filling the host cell 5

    4'(3) Sporangium almost or completely filling host cell and assuming its shape . . . 7

5(4) Sporangium predominantly forming 1 discharge tube 6

5'(4) Sporangium predominantly forming more than 1 discharge tube
. . . . **Pleotrachelus** p. 158
    Pleotrachelus fulgens in *Pilobolus*; *P. petersenii* in moss rhizoids.

    6(5) Sporangia generally scattered; resting spore usually filling its container; in a wide variety of living and dead materials . . . . **Olpidium** p. 128
    Olpidium endogenum in Conjugatae; *O. gregarium* in rotifer eggs; *O. maritimum* in seawater.

    6'(5) Sporangia formed in dense clusters; resting spores lying loosely in containers
. . . . **Dictyomorpha** p. 157
    Dictyomorpha dioica Mullins. (1961) parasitic in watermolds as *Pringsheimiella*, a synonym, in Sparrow (1960).

7(4') Sporangium almost filling host cell; walls never completely fusing with those of host
. . . . **Plasmophagus** p. 164
    Plasmophagus oedogoniorum in somewhat swollen cells of *Oedogonium sp.*

7'(4') Sporangium completely filling reproductive organ or hypertrophied part of host; walls of parasite and host fused, at least laterally . . . . **Rozella** p. 165
    Rozella allomycis in hyphae and reproductive organs of *Allomyces* [Plate I, J]; *R. blastocladiae* in *Blastocladia*; *R. coleochaetis* in *Coleochaete*; *R. septigena* in *Achlya, Saprolegnia* [Plate I, I].

    8(3') Thallus converted into a linear series of sporangia . . . . **Achlyogetonaceae** 9

    8'(3') Thallus converted into a prosorus, or sorus surrounded by a common membrane, or resting spore . . . . **Synchytriaceae** 10

---

[2]The page numbers following generic names in the keys in this chapter refer to Sparrow (1960) where descriptions of genera and species up to 1957 can be found.

9(8) Zoospores encysting at orifice of sporangial discharge tube . . . . **Achlyogeton** p. 183
Achlyogeton entophytum in *Cladophora.*

9'(8) Zoospores clustering at orifice of discharge tube but eventually swimming away without encystment . . . . **Septolpidium** p. 187
Septolpidium lineare, parasitic in diatoms.

10(8') Thallus large, stimulating multicellular galls on vascular plants; never amoeboid; sorus sessile or a prosorus; sporangial zoospores freed outside host cells
. . . . **Synchytrium**
Karling (1964) recognizes 6 subgen: and 200 spp. of which those fully known are keyed out. *Synchytrium endobioticum* (Potato Wart disease); *S. decipiens* on *Amphicarpa,* most widely distributed in North America. *Synchytrium* is considered by Karling (1964) to embrace *Micromyces.*

10'(8') Thallus in microscopic algae, sometimes causing a slight swelling; sorus sessile on prosorus or at tip of a tube; sporangia immobile or amoeboid; zoospores freed inside or outside host cell 11

11(10') Sorus sessile on prosorus, wall not divided into segments; sporangia small, spherical, amoeboid, or occasionally uniflagellate . . . . **Endodesmidium** p. 191
Endodesmidium formosum, parasitic in desmids.

11'(10') Sorus sessile on prosorus or at extramatrical tip of its discharge tube, always divided into a varying number of segments, not surrounded by a common soral membrane, "simple" (discharging flagellated zoospores) or "compound" (discharging flagellated sporangia which soon encyst and give rise to zoospores) . . . . **Micromyces** p. 192
Micromyces cristata var. *cristata* parasitic in desmids; *M. zygogonii, M. fischeri* in filamentous Conjugatae.

12(2') Thallus monocentric 13

12'(2') Thallus at least in some phases of life history polycentric 44

13(12) Thallus epi- and endobiotic or entirely endobiotic . . . . **Phlyctidiaceae** 14

13'(12) Thallus interbiotic or with no well-defined position with respect to substratum; sporangium, prosporangium, or resting spore formed wholly or in part from enlarged body of encysted zoospore . . . . **Rhizidiaceae** 33

14(13) Sporangium epibiotic or extracellular 15

14'(13) Sporangium and resting spore endobiotic; vegetative system rhizoidal; zoospore cyst usually evanescent 28

15(14) Body of encysted zoospore wholly or in part enlarging to form a sporangium, the latter usually sessile, with or without a sterile base 16

15'(14) Body of encysted zoospore either sessile and enlarging to form a prosporangium, or lying free in water or on gelatinous host sheath, not enlarging, and producing at tip of germ tube a sporangium or an appressorium which expands to form a sporangium; sterile base never formed 26

16(15) Epibiotic or extracellular part completely fertile; endobiotic part varied . . . 17

16'(15) Epibiotic part typically with a sterile septate base or a small knoblike structure on which sporangium rests; endobiotic part knoblike or rhizoidal 25

17(16) Endobiotic part a distinctly double-contoured tube, irregular sac, minute sphere, or papilla or digitation, never branched or catenulate or tapering 18

17'(16) Endobiotic part a single tapering rhizoid or a branching system of rhizoids, or catenulations arising from the tip of germ tube or a prolongation of it, or from an apophysis . . 21

18(17) Endobiotic part not digitate 19

18'(17) Endobiotic part a complex of blunt digitations arising from a central knob
. . . . **Loborhiza** p. 227
L. metzneri on *Volvox* sp.

19(18) Specialized parasite of *Gloeosporium* conidia with sessile gametangial plants with minute peglike endobiotic part and aerial very long-stalked meiosporangial plants producing zoospores . . . . **Caulochytrium**
Caulochytrium gloeosporii on conidia of *Gloeosporium*; Voos and Olive (1968), Voos (1969).

19'(18) Not as above 20

20(19') Resting spore completely filling its container . . . . **Phlyctidium** p. 210
Phlyctidium laterale, P. eudorinae on Green Algae; *P. keratinophilum* on keratin; *P. mycetophagum* on phycomycetes.

20'(19') Resting spore formed in distal part of its container . . . . **Septosperma** p. 225
Septosperma anomala, parasitic on chytrid sporangia.

21(17') Endobiotic part a tapering rhizoid or a branching system of rhizoids arising from tip of germ tube 22

21'(17') Endobiotic part apophysate generally with rhizoids or catenulations arising from it
. . . . 23

22(21) Sporangium and resting spore epibiotic; rhizoids developed to a varying degree
. . . . **Rhizophydium** p. 230
Rhizophydium sphaerotheca on pollen; *R. keratinophilum* on keratin; *R. planktonicum* on the diatom *Asterionella*; a large genus with poorly defined species, on a variety of living and dead plant and animal substrata.

22'(21) Sporangium and resting spore extracellular imbedded in gelatinous sheath of host colony; rhizoids forming a bushy tuft; parasitic on planktonic algae
. . . . **Dangeardia** p. 319
Dangeardia mammillata on *Pandorina, Eudorina*, etc.

23(21') Sporangium always with an apiculus formed by an unexpanded portion of zoospore cyst; discharge pore varied in position; resting spore endobiotic . . . **Blyttiomyces** p. 354
Blyttiomyces spinulosus on zygospores of *Spirogyra*; *B. helicus* on pine pollen.

23'(21') Sporangium without a single apiculus formed as above variously ornamented or smooth, discharge pore usually apical; resting spore epibiotic 24

24(23') Rhizoids, if present, of tapering elements only . . . . . **Phlyctochytrium** p. 323
Phlyctochytrium quadricorne on Green Algae; *P. nematodae* on nematodes.

24'(23'') Rhizoids from expanded catenulate apophyses . . . . **Polyphlyctis**
Polyphlyctis unispinum on Green Algae and dead grass leaves. Karling (1967a) (*Phlyctochytrium unispinum*, Paterson, 1956).

25(16') Sterile part an inconspicuous knob on which sporangium rests; endobiotic part knoblike
. . . . **Physorhizophidium** p. 358
Physorhizophidium pachydermum on *Amphora* and *Navicula*.

25'(16') Sterile part conspicuous and an intregal component or continuation of base of fertile portion, rarely lacking; endobiotic part rhizoidal or straplike . . . **Podochytrium** p. 359
Podochytrium clavatum on various freshwater diatoms. Includes *Rhizidiopsis*.

26(15′) Body of encysted zoospore sessile, enlarging to form a prosporangium; endobiotic part consisting of a series of intercommunicating broad lobes . . **Saccomyces** p. 364
Saccomyces endogenus on encysted *Euglena*.

26′(15′) Body of encysted zoospore not sessile 27

27(26′) Body of encysted zoospore producing at tip of a ± extended germ tube stubby endobiotic vegetative elements and a sporangium; zoospores lacking flagella, amoeboid
. . . . **Scherffeliomycopsis**
Scherffeliomycopsis coleochaetis in outer wall of *Coleochaete*, Geitler (1962).

27′(26′) Body of encysted zoospore producing at tip of germ tube an appressorium which expands to form sporangium; endobiotic part rhizoidal or apophysate and bearing a distal complex of stubby digitations; zoospores with flagella . . . . . **Scherffeliomyces** p. 365
Scherffeliomyces parasitans on *Euglena*, Green Algae. Includes *Coralliochytrium* Domjàn.

28(14′) Vegetative part rhizoidal; usually confined to one cell 29

28′(14′) Vegetative part an isodiametric coenocytic or septate tube, usually extending through several host cells 32

29(28) Sporangium spherical, pyriform or irregular, never strongly tubular; typically forming a single discharge tube or pore 30

29′(28) Sporangium strongly tubular; forming one or more discharge tubes
. . . . **Mitochytridium** p. 395

30(29) Sporangium formed from a localized primary swelling at tip of germ tube or from a secondary expansion of the more proximal part of the rhizoidal rudiments 31

30′(29) Sporangium formed by vesiculation of the dorsal side of rhizoids, strongly dorsiventrally differentiated . . . . **Phlyctorhiza** p. 391
Phlyctorhiza endogena in insect integuments.

31(30) Rhizoids or rhizoidal axes arising directly from the sporangium
. . . . **Entophlyctis** p. 369
Entophlyctis apiculata in *Chlamydomonas*; *E. helioformis* in dead Characeans; *E. aurea* on cellulosic materials.

31′(30) Rhizoids arising from an apophysis. . . . . **Diplophlyctis** p. 383
D. intestina in dead Characeans; *D. amazonense*, *D. sexualis* on cellulosic materials.

32(28′) Vegetative part broadly tubular, nonseptate . . . . **Rhizosiphon** p. 397
Rhizosiphon crassum, *R. anabaenae* in trichomes and heterocysts of Blue Green Algae.

32′(28′) Vegetative part a narrow septate tube . . . . **Aphanistis** p. 401
Aphanistis oedogoniorum in filaments and oogonia of *Oedogonium*.

33(13′) Body of encysted zoospore or aplanospore forming sporangium 34

33′(13′) Body of encysted zoospore forming rudiment of prosporangium or a germ tube which expands in part to form prosporangium 41

34(33) Vegetative system an unbranched double-contoured tube; aplanospores formed
. . . . **Sporophlyctidium** p. 406
Sporophlyctidium africanum on *Protoderma*.

34′(33) Vegetative system rhizoidal 35

35(34′) Rhizoidal system arising predominantly from a single rhizoidal axis on the sporangium
. . . . 36
35′(34′) Rhizoidal system arising from a thick-walled, cuplike basal portion of sporangium or

from an apophysis or as axes from several places on sporangium 38

   36(35) Sporangium wall persisting after zoospore discharge; predominantly smooth
. . . . 37

   36'(35) Sporangium wall deliquescing save for base at spore discharge; spiny-walled
. . . . **Solutoparies** p. 426
     Solutoparies pythii, parasitic on *Pythium*.

37(36) Main rhizoidal axis predominantly distinct and prolonged; resting spores asexually
formed or sexually by rhizoidal anastomosis . . . . **Rhizidium** p. 407
     Rhizidium mycophilum in gelatinous sheath of *Chaetophora*; *R. chitinophilum* on chitin;
     R. windermerense on planktonic *Gemellicystis*.

37'(36) Main rhizoidal axis predominantly short, indistinct, occasionally somewhat expanded;
resting spores always sexually formed usually as a third thallus at juncture of conjugants
. . . . **Siphonaria** p. 423
     Siphonaria variabilis on empty exuviae of aquatic insects. For sexuality, see Dogma,
     (1970).

   38(35') Rhizoids arising from a thick-walled cuplike basal portion of sporangium or from
an apophysis 39

   38'(35') Rhizoidal axes arising from several places on surface of sporangium
. . . . **Rhizophlyctis** p. 435
     Rhizophlyctis mastigotrichis on Blue Green Algae; *R. petersenii* on aquatic debris.

39(38) Rhizoids arising from a thick-walled, cuplike basal portion of sporangium
. . . . **Obelidium** p. 427
     Obelidium mucronatum in empty exuviae of aquatic insects; chitin bait, in soil.

39'(38) Rhizoids arising from an apophysis 40

   40(39') Sporangium spherical . . . . **Rhizoclosmatium** p. 431
     Rhizoclosmatium globosum in empty insect exuviae.

   40'(39') Sporangium irregularly lobed often distinctly stellate
. . . . **Asterophlyctis** p. 433
     Asterophlyctis sarcoptoides in empty exuviae of aquatic insects and on chitin sub-
     strata. Placed under *Diplophlyctis* by Dogma (1970).

41(33') Body of encysted zoospore enlarging into rudiment of a prosporangium 42

41'(33') Body of encysted zoospore producing a germ tube, part of which expands to form a
prosporangium . . . . **Endocoenobium** p. 459
     Endocoenobium eudorinae in *Eudorina* colonies.

   42(41) Early thallus development strongly radial with 5 radiating lateral rhizoids and a
sixth basal one, all provided with numerous short lateral protistan-capturing branches;
nonsexual and sexual reproduction as in *Polyphagus* . . . . **Arnaudovia**
     Arnaudovia hyponeustonica, a neustonic organism capturing planktonic *Phacotus*,
     Trachelomonas and *Strombomonas*, Valkanov (1963).

   42'(41) Early thallus development not strictly radially symmetrical and without capturing
organelles 43

43(42') Zoospores escaping from sporangium through an apical orifice as free-swimming
bodies; resting spore sexually formed in conjugation tube produced by 1 of 2 conjugating
thalli . . . . **Polyphagus** p. 449
     Polyphagus euglenae on encysted *Euglena* and *Chlamydomonas*; *P. parasiticus* on
     Tribonema.

43′(42′) Zoospores escaping from sporangium through a subapical or lateral orifice as non-motile or motile bodies or germinating in the sporangium; resting spore sexually formed in receptive thallus, conjugating thalli adnate . . . . **Sporophlyctis** p. 456
    Sporophlyctis rostrata on *Draparnaldia.*

  44(12′) Vegetative system predominantly rhizoidal and nonseptate except for special turbinate cells 45

  44′(12′) Vegetative system predominantly tubular, septate or nonseptate 49

45(44) Life history involving two independent plants, one a monocentric, epibiotic zoosporangium, the other a polycentric, endobiotic system bearing septate turbinate cells, on which many dark-colored thick-walled resting spores are formed; the latter germinating to form zoospores; parasites of vascular plants . . . .**Physodermataceae** 46

45′(44) Life history involving only a single plant on which turbinate cells, thin-walled zoosporangia and sometimes thick-walled resting spores may be formed
. . . . **Cladochytriaceae** 47

  46(45) Thallus bearing resting spores of limited extent, somewhat ribbonlike, confined within a lysigenous cavity in host on which it causes strong hypertrophy; rhizoids of epibiotic stage dense and bushy . . . . **Urophlyctis**
    Urophlyctis pulposa on Chenapodiaceae; *U. alfalfae* on *Medicago.* Sparrow (1962) has given the reasons for maintaining *Urophlyctis* distinct from *Physoderma.* Y. Lingappa (1959) has witnessed planogamete fusion in *U. pulposa.*

  46′(45) Thallus not bearing resting spores in a lysigenous cavity, the delicate rhizoids wandering from cell to cell in the often scarcely if at all hypertrophied host; rhizoids of epibiotic stage scant, stubby . . . . **Physoderma** p. 483
    Physoderma maculare on *Alisma* Sparrow, (1964); *P. maydis* on maize. Sparrow and Griffin (1964) have given details of structure and host range trials.

47(45′) Zoospores nonflagellate, strongly amoeboid, sporangia not proliferating
. . . . **Amoebochytrium** p. 476
    Amoebochytrium rhizidioides in gelatinous matrix of *Chaetophora.*

47′(45′) Zoospores flagellate, sporangia internally proliferous 48

  48(47′) Sporangia and rhizoids predominantly endobiotic, the sporangia terminal or intercalary, often apophysate, with a discharge tube; zoospores at discharge forming a temporary motionless cluster . . . . **Cladochytrium** p. 461
    Cladochytrium tenue in decaying plant tissues in water.

  48′(47′) Sporangia and rhizoids predominantly extramatrical, the sporangia without discharge tubes, borne at tips of rhizoids; zoospores at discharge swarming for a time in a vesicle at orifice . . . . **Physocladia** p. 475
    Physocladia obscura, saprophytic in staminate cones of pine in bog water.

49(44′) Vegetative system intra–and extramatrical, mycelioid, with 1 or more axes with rhizoidal holdfasts, occasionally septate, sporangia clustered, smooth-walled and tuberculate
. . . . **Polychytrium** p. 477
    Polychytrium aggregatum on chitinous substrata, often in bogs.

49′(44′) Vegetative system extramatrical or endobiotic, without secondary axes, tubular and septate throughout; sporangia smooth-walled, not clustered . . . . . **Coenomyces** p. 479
    Coenomyces consuens on marine Blue Green Algae.

  50(1′) Thallus monocentric, epi- and endobiotic, endobiotic or interbiotic, variously developed . . . .**Chytridiaceae** 51

50'(1') Thallus polycentric, completely endobiotic or endobiotic and extramatrical; sporangium exo- or endooperculate . . . . **Megachytriaceae** 65

51(50) Sporangium exooperculate 52

51'(50) Sporangium endooperculate 61

52(51) Sporangium epi- or interbiotic, formed from body of encysted zoospore or outgrowth from it or the main thallus axis, external to substratum
. . . . subfamily, **Chytridioideae** 53

52'(51) Sporangium usually within substratum, formed at tip of germ tube; resting spore always endobiotic . . . . subfamily, **Endochytrioideae** 62

53(52) Sporangium from all or part of expanded body of zoospore cyst, external to substratum
. . . . 54

53'(52) Sporangium formed as a lateral, walled-off outgrowth of distal external part of the cylindrical main axis of thallus; rhizoids broad and tubular the whole plant large (up to 800 μm × 650 μm) . . . . **Macrochytrium** p.565
Macrochytrium botrydioides, primarily on submerged fruits.

54(53) Rhizoids arising from single place on sporangium 56

54'(53) Rhizoids arising from more than 1 place on sporangium, with 1 or more discharge papillae 55

55(54') Sporangium with 1 discharge papilla, developing from expansion of epibiotic part of germ tube; zoospore cyst confluent with sporangium at maturity; rhizoids from endobiotic tip of germ tube and from sporangium . . . . **Allochytridium**
Allochytridium expandens, on vegetable debris, roadside puddle; Salkin, (1970).

55'(54') Sporangium with more than 1 papilla, developing from enlarged body of encysted zoospore; rhizoids from several main axes on sporangium . . . . **Karlingiomyces** p. 559
Karlingiomyces asterocystis on chitinous substrata in soil and water.

56(54) Sporangium with or without a simple apophysis, developing from all or part of zoospore cyst 57

56'(54) Sporangium with a compound apophysis . . . . . . **Catenochytridium** p. 555
Catenochytridium carolinianum on vegetable debris in water.

57(56) Sporangium completely fertile 58

57'(56) Sporangium frequently with a sterile base . . . . **Rhopalophlyctis** p. 553
Rhopalophlyctis sarcoptoides, in insect exuviae in water.

58(57) Resting spore 1-celled 59

58'(57) Resting spore 2-celled, the distal part fertile . . . . **Sparrowia**
Sparrowia parasitica, parasitic on oogonia of a water mold; Willoughby (1963).

59(58) Resting spore within substratum . . . . **Chytridium** p. 488
Chytridium olla, parasitic on oogonia and oospores of *Oedogonium*; *C. lecythii* on rhizopod *Lecythium*; *C. cocconeidis* on diatom *Cocconeis*.

59'(58) Resting spore external to substratum 60

60(59') Resting spore asexually formed, or if sexually, without a male plant and conjugation tube . . . . **Amphicypellus** p. 546 and **Chytriomyces** p. 537
Amphicypellus elegans on dead cells of *Ceratium*; *C. hyalinus* in insect exuviae and on chitin; *C. mammilifer* on pollen (Persiel, 1960).

60'(59') Resting spore sexually formed in receptive plant after conveyance of male gamete from a ± developed thallus by means of a conjugation tube
. . . . **Zygorhizidium** p. 547

Zygorhizidium willei on Conjugatae; *Z. planktonicum* on *Synedra, Asterionella.*

61(51′) Sporangium and resting spore epibiotic, the latter sexually formed probably as in *Zygorhizidium*, rhizoidal system from its base . . . . **Pseudopileum**
  Pseudopileum unum, parasitic on cysts of *Mallomonas* (Canter, 1963).

61′(51′) Sporangium and resting spore interbiotic, the latter sometimes sexually formed; rhizoidal axes several over surface of usually multipored sporangium . . . . . . **Karlingia**
  Karlingia rosea, very common on cellulosic substrata in soil-water culture; Johanson, (1944). Dogma (1970) from a study of nearly 100 (93) single-spored isolates of the ubiquitous, strongly cellulosic "*Karlingia rosea*" (*Rhizophlyctis rosea*), from throughout the world, finds that there in fact exists a series of fungi with similar appearing sporangial stages but with *different* types of resting spores. Some of the latter are sexually, others asexually formed. Since no resting spores were found by de Bary and Woronin in their *Rhizophlyctis rosea* it is suggested by Dogma that their species name be applied *only* to plants lacking these structures. Species with true resting spores should be placed in new taxa. Thirty of Dogma's single-spore isolates, although maintained over many cultural generations, failed to form true resting spores. The ready capacity of the ordinary sporangium in nearly all isolates to become converted into dormant, highly drought-resistant structures has undoubtedly contributed to the ubiquity of these fungi. Endooperculation was found to be a constant feature in all isolates. Perhaps the last word has yet to be said on this matter since Konno (1971) has stated positively that no endo- or exoopercula were found in *R. rosea* from Greenland!

62(52′) Rhizoidal system not catenulate 63

62′(52′) Rhizoidal system catenulate 64

63(62) Sporangium attached directly to the rhizoidal system . . . . . **Endochytrium** p. 568
  Endochytrium ramosum, in algae and vegetable debris.

63′(62) Sporangium arising as an outgrowth of an apophysis to which is attached the rhizoidal system . . . . **Nephrochytrium** p. 573
  Nephrochytrium stellatum, N. appendiculatum in dead Nitella.

64(62′) Rhizoids arising from a single (basal) place on a cylindrical and stalked sporangium . . . . **Cylindrochytridium** p. 577
  Cylindrochytridium johnstonii in decaying vegetable debris.

64′(62′) Rhizoids arising from several places on the spherical, oval or irregular unstalked sporangium . . . . **Truittella** p. 578
  Truittella setifera in cellulosic substrata.

65(50′) Thallus forming tenuous, strongly tapering rhizoids 66

65′(50′) Thallus forming a broadly tubular vegetative system, not tapering strongly distally . . . . **Megachytrium** p. 596
  Megachytrium westonii, parasitic on *Elodea.*

66(65) Rhizoids septate only where delimiting reproductive organs, profusely and extensively developed with irregularities, endooperculations sometimes present.
  . . . . **Nowakowskiella** p. 580
  Nowakowskiella elegans in vegetable debris; *N. hemisphaerospora*, cellulosic debris in soil and water.

66′(65) Rhizoids septate and constricted at intervals as well as delimiting reproductive organs . . . . **Septochytrium** p. 592
  Septochytrium variabile in vegetable debris.

C.   Order Harpochytriales

KEY TO GENERA OF HARPOCHYTRIALES

1. Mature thallus short, somewhat fusiform or cylindrical, consisting of a basal vegetative and an upper sporogenous part; zoospores discharged upon circumscissle dehiscence of pointed apex; sporangium renewed by internal proliferation . . . . **Harpochytrium**
    Harpochytrium hedinii, on algae, vegetable debris; Lagerheim, 1890; Emerson and Whisler, (1968).

1' Mature thallus an unbranched filament which becomes divided in basipetalous succession into H-shaped segments delimiting zoosporangia; zoospores discharged upon disarticulation of the H-shaped segments.   . . . . **Oedogoniomyces** p. 694
    Oedogoniomyces lymnaeae Kobayashi and Ookubo, 1954; Emerson and Whisler, 1968; on snail shells, fruits and seeds in water and recovered from dry soil.

D.   Order Blastocladiales

This order is freshwater or terricolous. The thallus is usually walled and predominantly polycentric or mycelial, although it is sometimes pseudo-septate, often with a ± differentiated basal hyphal segment bearing one or more reproductive structures or hyphae and proximal anchoring rhizoids. The life history sometimes shows similar (rarely dissimilar) asexual sporangial and resting spore-bearing, and sexual gametangia-bearing plants; thus, there is an alternation of generations. The planogametes as well as zoospores lack a single conspicuous globule, but usually have several small ones, often with a conspicuous nuclear cap. Germination is predominantly bipolar.

Blastocladiales occurs primarily as saprobes on vegetable debris, etc. One family is parasitic in mosquito larvae.

KEY TO FAMILIES AND IMPORTANT GENERA OF BLASTOCLADIALES

1. Thallus unwalled, branched or lobed, specialized endoparasites of mosquito larvae, converted into a mass of thick-walled variously ornamented resting spores which crack open and function as zoosporangia upon germination; thin-walled sporangia rare
    . . . . **Coelomomycetaceae** p. 635
        One genus, *Coelomomyces*; *C. stegomyiae* in *Aedes albopictus*, *A. aegypti*; *C. anophelescia* in *Anopheles spp.*

1'. Thallus walled, mono- or polycentric, bearing rhizoids; not specialized as above 2

2(1') Thallus polycentric, occasionally monocentric, with catenulate fertile swellings separated by sterile isthmuses   . . . . **Catenariaceae** 3

2'(1') Thallus of varying degrees of complexity, typically with a ± prominent basal part on which are 1 or more reproductive structures or hyphae and proximal rhizoidal, holdfast system   . . . . **Blastocladiaceae** 5

3(2) Thallus linearly organized with a strong central axis, resting spores loosely held within container; sporangia lacking endoopercula . . . . **Catenaria** p. 652

Catenaria anguillulae, in Angullulae, liver-fluke eggs and those of other microscopic animals, Characeae; *C. allomycis* in hyphae of *Allomyces spp.* (Pl. I, M)

3′(2) Thallus radially or diffusely organized 4

4(3′) Thallus monocentric or polycentric, when polycentric usually radially organized; zoospore germination unipolar; sporangia without strongly developed discharge tubes and lacking endoopercula; resting spore filling container . . . . . . **Catenophlyctis**
One species, *C. variabilis*, saprophytic in cellulosic and keratinic substrata; apparently monocentric forms resembling the chytrid *Phlyctorhiza* common; note that in contrast to other members of order, zoospore germination is chytridlike and unipolar. Karling (1965).

4′(3′) Thallus polycentric, diffusely organized; zoospore germination bipolar; sporangia with strongly developed discharge tubes with endoopercula; resting spores
. . . . **Catenomyces** p. 658
Catenomyces persicinus, saprobic on cellulosic substrata.

5(2′) Plant bearing a single reproductive rudiment . . . . **Blastocladiella** p. 660
Blastocladiella simplex; *B. stubenii*, on vegetable and animal debris in soil.

5′(2′) Plant with indeterminate number of reproductive rudiments 6

6(5′) Thallus consisting of a basal part and a few depauperate distal dichotomously branched hyphae without pseudoseptae, 1 of dichotomies usually somewhat aborted; resting spore loose in container, seemingly unpitted . . . . **Blastocladiopsis** p. 668
One species, *B. parva*, saprobic on debris in soil.

6′(5′) Thallus consisting either of a distinct basal part from which arise strongly developed dichotomously branched, pseudoseptate hyphae bearing reproductive rudiments, or a specialized basal part alone on which are reproductive rudiments; resting spore filling container, usually distinctly pitted 7

7(6′) Thallus with distinct basal part bearing proximal holdfasts, the distal part with pseudoseptate dichotomously branched hyphae bearing reproductive organs; sporangia with 1 or more discharge pores; alternation of generations often present . . . . **Allomyces** p. 669
Allomyces arbuscula, *A. macrogynus* (Plate VI, M), *A. javanicus; A. neomoniliformis* (Plate VI, J–L); common on organic debris in soils, water.

7′(6′) Thallus with well-developed basal part which may be simple, lobed or sparingly short-branched with proximal rhizoids, without pseudoseptae, sterile setae sometimes present; reproductive structures sessile on basal part, sporangia with 1 discharge pore; apparently lacking gametophyte plants . . . . **Blastocladia** p. 678
Blastocladia pringsheimii, B. ramosa (Plate I. N), on apples, twigs, etc., in water.

E. Order Monoblepharidales

This order consists of microscopic eucarpic, mycelial fungi with the hyphae being nonseptate or pseudoseptate. Its reproductive organs are cut off by true cross walls and the zoosporangia and sex organs occur on the same thallus. The zoospores often have an anterior group of refractive granules. Sexual reproduction is oogamous and the antherozoids are posteriorly uniflagellate and borne in antheridia. The oospheres are nonflagellate with one or more being borne in an oogonium; the fertilized egg becomes a thick-walled oospore.

Monoblepharidales occur as saprobes on twigs, fruits, etc., in water (*Monoblepharis*, *Gonapodya*) or on organic debris in primarily tropical soils (*Monoblepharella*).

KEY TO FAMILIES AND IMPORTANT GENERA OF MONOBLEPHARIDALES

1. Zygote undergoing a period of motility before encystment propelled by flagellum of male gamete; mycelium with or without pseudoseptae; oogonium with one or more oospheres
. . . . **Gonapodyaceae** 2

1′. Zygote (usually 1) remaining in oogonium or merely oozing to gametangial orifice where it encysts and remains attached; male gamete completely engulfed at fertilization; mycelium never pseudoseptate . . . . **Monoblepharidaceae**
    One genus *Monoblepharis*, p. 726. *Monoblepharis polymorpha;* (Plate VI, P) *M. insignis*, on twigs in water.

    2. Mycelium with pseudoseptae accompanied by constrictions which often produces a catenulate appearance; sporangia ovoid or podlike; gametangia internally proliferous, more than 1 often amoeboid female gamete in a gametangium
    . . . . **Gonapodya** p. 715
        Gonapodya prolifera, G. polymorpha on twigs, fruits, in water.

    2′ Mycelium lacking pseudosepta and constrictions; sporangia narrowly ovate; gametangia not proliferous, the female typically bearing one oosphere
    . . . . **Monoblepharella** p. 720
        Monoblepharella taylori (Plate VI, Q), *M. mexicana*, both in tropical soils recovered with hempseed bait.

II. CLASS HYPHOCHYTRIDIOMYCETES[3]

A. General Characteristics

Hyphochytridiomycetes are made up of microscopic aquatic fungi with body plans resembling those of Chytridiomycetes. They are holocarpic or eucarpic, monocentric or polycentric and their vegetative system is rhizoidal or hyphal-like having intercalary swellings and walls with cellulose and chitin or chitin only. The sporangia are inoperculate and the zoospores are anteriorly uniflagellate, maturing, partially or wholly inside or outside the sporangium. The resting spores are asexually or sexually formed by fusion of isoaplanogametes and upon germination they function as a sporangium. There is one order, the Hyphochytriales.

Members of this group, which strikingly resemble in body plan the Chytridiomycetes but differ from them in the structure and anterior flagellation of the zoospore and the occasional presence of cellulose in the wall, are found as parasites in freshwater and marine algae, aquatic phycomycetes, ascocarps of discomycetes and saprobs on a variety of dead plant and insect remains. One is on the marine zooplankter *Eurytemora*. Johnson (1957) has reported fusion of endophytic isoaplanagometes to form an endobiotic resting spore.

[3]Sparrow, (1958). Mycologia **50**:811.

B. Important Literature

T. W. Johnson, *Amer. J. Bot.* **44**:875–878 (1957); J. S. Karling, *Amer. J. Bot.* **30**:637–648 (1943); **31**:391–397 (1944); *Sydowia* **20**:137–143 (1967); *J. Elisha Mitchell Sci. Soc.* **84**:166–178 (1968); F. K. Sparrow, "Aquatic Phycomycetes," University of Michigan Press, Ann Arbor (1960).

KEY TO FAMILIES AND IMPORTANT GENERA OF HYPHOCHYTRIALES

1. Thallus holocarpic, endobiotic; zoospores formed outside or inside sporangium
. . . . **Anisolpidiaceae** 2

1′. Thallus eucarpic 3

2(1) Zoospores formed within the sporangium . . . . **Anisolpidium** p. 747
Anisolpidium sphacellarum in *Sphacelaria*; *A. ectocarpii* (Plate VII, M) in *Ectocarpus*; both marine algae; *A. saprobium* in pine pollen, Karling, (1968).

2′(1) Zoospores formed outside the sporangium . . . . **Canteriomyces** p. 750
Canteriomyces stigeoclonii in *Stigeoclonium, Draparnaldia.*

3(1′) Thallus monocentric, epibiotic, zoospores formed inside or outside sporangium
. . . . **Rhizidiomycetaceae**[4] 4

3′(1′) Thallus polycentric, zoospores partly or fully formed in sporangium or at mouth of a discharge tube . . . . **Hyphochytriaceae**
One genus, *Hyphochytrium* (p. 760) on higher fungi in algae, vegetable debris. *Hyphochytrium infestans* on a discomycete; *H. catenoides* (Plate VII, A, B) on vegetable debris; *H. hydrodictyii* on *Hydrodictyon* (Plate VII, C–E); *H. peniliae* in marine copepod *Eurytemora.*

4(3) Zoospores undergoing cleavage at orifice of a narrow discharge tube; thallus epibiotic and apophysate . . . . **Rhizidiomyces** p. 751
Rhizidiomyces apophysatus, (Plate VII, G–I) parasitic on eggs and oospores of water molds; pollen; flies as bait in soil.

4′(3) Zoospores formed within sporangium 5

5(4′) Sporangium apophysate with an apical discharge pore, epibiotic on oogonia and oospores of water molds . . . . **Rhizidiomycopsis** p. 757
Rhizidiomycopsis japonicus, (Plate VII, F) on oogonia of *Aplanes (Achlya).*

5′(4′) Sporangium resting directly on the egg of *Vaucheria*; nonapophysate, with a broad lateral pore . . . . **Latrostium** p. 758
Latrostium comprimens, (Plate VII, J–L) parasitic on eggs of *Vaucheria* lying in open oogonium

REFERENCES

Canter, H. M. (1960). Fungal parasites of the phytoplankton. V. *Chytridium isthmiophilum* sp. nov. *Trans. Brit. Mycol. Soc.* **43**:660–664.

Canter, H. M. (1963). Studies on British chytrids. XXIII. New species on chrysophycean algae. *Trans. Brit. Mycol. Soc.* **46**:305–320.

Canter, H. M. (1966). Studies on British chytrids. XXV. *Chytriomyces heliozoicola* sp. nov., a parasite of Heliozoa in the plankton. *Trans. Brit. Mycol. Soc.* **49**:633–638.

[4]*See also Elina* Artemtchuk in *Veröff. Meeresforsch. Bremerh.* **13**:231, 1972.

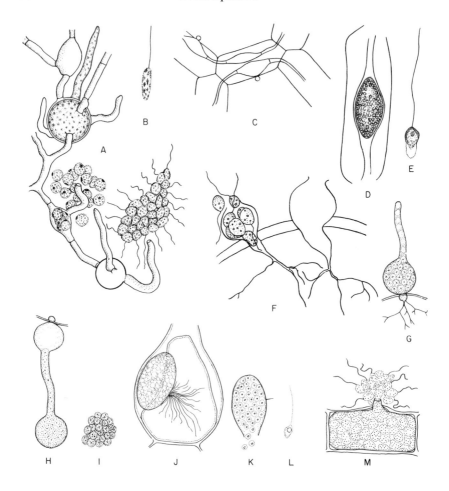

PLATE VII   Hyphochytridiomycetes. Fig. A. *Hyphochytrium catenoides*, polycentric tubular thallus on decaying maize. Portion of thallus showing (from lower right) formation of zoospores at orifice of discharge tube, quiescent spores, sporangium with discharge tubes and, at top, empty sporangium. Fig. B. Anteriorly uniflagellate zoospore of same. Figs. C–E. *Hyphochytrium hydrodictyii* in *Hydrodictyon*. Fig. C. Portion of endobiotic thallus showing epibiotic cysts of zoospore and primary and secondary swellings of thallus. Fig. D. Resting spore. Fig. E. Anteriorly uniflagellate zoospore. Fig. F. *Rhizidiomycopsis japonicus* on oogonium of *Aplanes*. One sporangium is discharging fully formed zoospores. Figs. G–I. *Rhizidiomyces apophysatus* on oogonia of *Achlya*. Fig. G. Mature epibiotic sporangium with discharge tube. Fig. H. Sporangium with contents emerging into vesicle formed a tip of discharge tube. Fig. I.

Canter, H. M. (1968). On an unusual fungoid organism, *Sphaerita dinobryoni* n. sp., living in species of *Dinobryon*. *J. Elisha Mitchell Sci. Soc.* **84**:56–61

Dogma, I. J., Jr. (1970). A developmental, morphologic, and taxonomic study of some monocentric Chytridiomycetes. Ph.D. Dissertation, University of Michigan, Ann Arbor.

Emerson, R., and H. Whisler. (1968). Cultural studies of *Oedogoniomyces* and *Harpochytrium*, and a proposal to place them in a new order of aquatic Phycomycetes. *Arch. Mikrobiol.* **61**:195–211.

Geitler, L. (1962). Entwicklung und Beziehung zum Wirt der Chytridiale *Scherffeliomycopsis coleochaetis* n. gen., n. spec. *Oesterr. Bot. Z.* **109**:250–275.

Johanson, A. E. (1944). An endo-operculate chytridiaceous fungus: *Karlingia rosea* gen. nov. *Amer. J. Bot.* **31**:397–404.

Karling, J. S. (1955). *Synchytrium ranunculi* Cook. *Mycologia* **47**:130–139.

Karling, J. S. (1964). "Synchytrium." Academic Press, New York.

Karling, J. S. (1965). *Catenophlyctis*, a new genus of the Catenariaceae. *Amer. J. Bot.* **52**: 133–138.

Karling, J. S. (1966). The chytrids of India with a supplement of other zoosporic fungi. *Sydowia* 6, Suppl. 3. 125 pp.

Karling, J. S. (1967a). Some zoosporic fungi of New Zealand. IV. *Polyphlyctis* gen. nov., *Phlyctochytrium* and *Rhizidium. Sydowia* **20**: 86–95.

Karling, J. S. (1967b). Some zoosporic fungi of New Zealand. VI. *Entophlyctis*, *Diplophlyctis*, *Nephrochytrium* and *Endochytrium. Sydowia* **20**:109–118.

Karling, J. S. (1968). Zoosporic fungi of Oceania. I. *J. Elisha Mitchell Sci. Soc.* **84**:166–178.

Kobayashi, Y., and M. Ookubo. (1954). On a new genus Oedogoniomyces of the Blastocladiaceae. *Bull. Nat. Sci. Mus.*, *Tokyo* [N. S.] **1**:59–66.

Kobayasi, Y., N. Hiratsuka, Y. Otani, K. Tubaki, S. Udagawa, and J. Sugiyama. (1971). Mycological studies of the Angmagssalik region of Greenland. *Bull. Nat. Sci. Mus.*, *Tokyo* [N. S.] **14**:1–96.

Koch, W. J. (1951). Studies in the genus *Chytridium*, with observations on a sexually producing species. *J. Elisha Mitchell Sci. Soc.* **67**:267–278.

Konno, K. (1971). Cited in Kobayasi *et al.* (1971).

Lagerheim, G. (1890). *Harpochytrium* und *Achlyella*, zwei neue Chytridiaceen Gattungen. *Hedwigia* **29**:142.

Lingappa, Y. (1959). Sexuality in *Physoderma pulposum* Wallroth. *Mycologia* **51**:151–158.

Mullins, J. T. (1961) The life cycle and development of *Dictyomorpha* gen. nov. (formerly *Pringsheimiella*), a genus of the aquatic fungi. *Amer. J. Bot.* **48**:377–387.

Paterson, R. A. (1956). Additions to the phycomycete flora of the Douglas Lake region. II. New chytridiaceous fungi. *Mycologia* **48**:270–277.

Persiel, I. (1960). Beschreibung neuer Arten der Gattung *Chytriomyces* und einiger seltener niederer Phycomyceten. *Arch. Mikrobiol.* **36**:283–305.

Salkin, I. F. (1970). *Allochytridium expandens*, gen. et sp. n.: Growth and morphology in continuous culture. *Amer. J. Bot.* **57**:649–658.

Sparrow, F. K. (1960). "Aquatic Phycomycetes," 2nd rev. ed. Univ. of Michigan Press, Ann Arbor.

Mature zoospores formed outside sporangium. Figs. J–L. *Latrostium comprimens* in oogonia of *Vaucheria*. Fig. J. Thallus within oogonium; reproductive rudiment on surface of host egg, rhizoids within. Fig. K. Discharging sporangium. Fig. L. Anteriorly uniflagellate zoospore. Fig. M. Endobiotic zoosporangium of *Anisolpidium ectocarpii* in marine *Ectocarpus*, the fully formed anteriorly uniflagellate zoospores emerging. (All may be found in Sparrow, 1960, except Figs. D–F, from Karling, 1955.)

Sparrow, F. K. (1962). *Urophlyctis* and *Physoderma*. *Trans. Mycol. Soc. Jap.* **3**:16–18.

Sparrow, F. K. (1964). Observations on chytridiaceous parasites of phanerogams. XIII. *Physoderma maculare* Wallroth. *Arch. Mikrobiol.* **48**:136–149.

Sparrow, F. K. (1969). Zoosporic marine fungi from the Pacific Northwest (U.S.A.). *Arch. Mikrobiol.* **66**:129–146.

Sparrow, F. K., and J. E. Griffin. (1964). Observations on chytridiaceous parasites of phanerogams. XV. Host range and species concept studies in *Physoderma*. *Arch. Mikrobiol.* **49**: 103–111.

Valkanov, A. (1963). *Arnaudovia hyponeustonica* n. gen, n. sp., einhochspezialisiter tierfangender Wasserpilz. *Arch. Protistenk.* **106**:553–564.

Voos, J. R. (1969). Morphology and life cycle of a new chytrid with aerial sporangia. *Amer. J. Bot.* **56**:898–909.

Voos, J. R., and L. S. Olive. (1968). A new chytrid with aerial sporangia. *Mycologia* **60**:730–733.

Willoughby, L. G. (1962). New species of *Nephrochytrium* from the English Lake District. *Nova Hedwigia* **3**:439–444.

Willoughby, L. G. (1963). A new genus of the Chytridiales from soil and a new species from freshwater. *Nova Hedwigia* **5**:335–340.

Willoughby, L. G. (1965). A study of Chytridiales from Victorian and other Australian soils. *Arch Mikrobiol.* **52**:101–131.

Mastigomycotina
Oomycetes

CHAPTER 7

Saprolegniales

M. W. DICK

Department of Botany
University of Reading
Reading, England

I. INTRODUCTION

The Saprolegniales consists of one large family and a number of smaller ones which will require careful taxonomic evaluation. The Saprolegniaceae is the best known and by far the largest family of the order with approximately 150 species. The other families are assigned to the Saprolegniales within the Oomycetes because of two attributes. They have been shown to possess biflagellate zoospores heterokont in morphology and possibly also in flagellar length, and these zoospores are always delimited *within* the zoosporangia. These two characters are not sufficient to define the order as at present constituted, and exceptions from one or another of the families have to be made when other characters are considered.

The Saprolegniales contain both holocarpic and eucarpic families. The vegetative thallus may or may not be septate. Zoosporangial discharge never involves vesicle formation but vesicle formation accompanying zoosporangial discharge may be absent in taxa of the other orders of Oomycetes which normally possess this character. The first-formed zoospore does not always have apical flagellar insertion: in *Haliphthoros* and the Thraustochytriaceae they are laterally inserted.

There may be significant differences in oospore structure between the orders of Oomycetes, but the important details are lacking or unknown in critical groups of taxa, e.g., the Ectrogellaceae and Haliphthoraceae within the Saprolegniales, and the Lagenidiales. For this reason the ordinal position of the Leptolegniellaceae can only be regarded as provisional because it may depend on oospore structure.

A recent report (Darley and Fuller, 1970) has suggested that the Thraustochytriaceae (including the invalid Schizochytriaceae Karling) may belong to the Labyrinthulales[1] and not the Oomycetes. It would be unwise to validate the Schizochytriaceae until we can be more certain that the species included therein are Oomycetes. A transfer of such magnitude influences opinions of the features to be considered of diagnostic value.

The families Leptolegniellaceae, Haliphthoraceae, Ectrogellaceae, and Thraustochytriaceae will be considered here as a matter of convenience. No ordinal diagnosis can be given to embrace these families and the Saprolegniaceae which would differ in any significant detail from that of the class. Nevertheless the Saprolegniaceae is a sufficiently distinct family to warrant ordinal status alone. The ordinal diagnosis given here recognizes this fact and does not attempt to resolve the obvious inconsistencies.

II. GENERAL MORPHOLOGICAL ACCOUNT

The vegetative thallus of the Saprolegniales may be either holocarpic or eucarpic. Holocarpy as defined by Sparrow (1960) is the state of vegetative growth in which the entire protoplasm is utilized in the production at one time of a number of sexual or asexual spores. After sporulation only the secreted cell wall of the fungus remains. In monocentric chytridlike forms this is fairly easily determined, but where the vegetative thallus is more or less filamentous and extensive it becomes increasingly difficult to determine whether a fungus is holocarpic or eucarpic. This becomes especially difficult when the fungus cannot or has not been cultured *in vitro*. Thus, genera such as *Brevilegniella* and *Leptolegniella* may be holocarpic, but the full extent of the thallus has never been traced.

Septation of the thallus certainly occurs in the Leptolegniellaceae. Such septation at the time of sporulation is not necessarily an argument against holocarpy since the time element may not be contravened. However, it is more usual, given a thallus which is more or less filamentous and septate, for certain segments to continue in a vegetative state at least for a time. Where basipetal development of zoosporangia occurs, the entire protoplasm may be utilized in spore production over a period of time as in certain *Aphanomyces* species (Scott, 1961). Thus, in the Saprolegniales the entire range of variants from true holocarpy (*Ectrogella*) to true eucarpy (*Saprolegnia*) may be found. As a taxonomic criterion it is most useful at the extreme ends of this range, and caution should be exercised in trying to categorize the thallus of species of the Leptolegniellaceae and Haliphthoraceae. To what extent these differences in thallus form reflect adaptations

---

[1]See also Gaertner (1972), Kazama (1972), and Perkins (1972) for references to features suggesting affinity to Labyrinthulales.

to a parasitic mode of nutrition, or to life in a marine environment is open to question.

The holocarpic or eucarpic nature of the thallus is therefore considered to be subordinate to the type of thallus construction. This may be filamentous, mycelial and nonseptate in the Saprolegniaceae, irregularly filamentous and occasionally septate in the Leptolegniellaceae, irregularly filamentous or lobate in the Haliphthoraceae, monocentric and holocarpic in the Ectrogellaceae, and monocentric and eucarpic in the Thraustochytriaceae. In the majority of filamentous forms tip growth is normal. Multidimensional increase is found in the Thraustochytriaceae and Ectrogellaceae, and may occur in the thalli of *Atkinsiella*. The Saprolegniaceae have the stoutest hyphae of any fungi, vegetative hyphae commonly exceeding 40 μm. However, the diameter in the region of the hyphal tip rarely exceeds 20 μm, while the basal hyphae may exceed 100 μm in diameter. Thus, in the Saprolegniaceae at least, there must be intersusception of new wall material accompanied by an increase in hyphal diameter proceeding in old hyphae.

On particulate substrata highly branched hyphae of fine diameter may be formed in the substratum. These have been termed rhizoids, but it is by no means certain that these rhizoids are homologous structures in the Thraustochytriaceae, Leptolegniellaceae, Haliphthoraceae, and Saprolegniaceae (Darley and Fuller, 1970; Dick, 1971; Sparrow and Gotelli, 1969). Rhizoidal fusions (interthallic rhizoids) are reported in the Thraustochytriaceae (Booth and Miller, 1968).

Zoospore development is either initiated in undifferentiated thalli or segments of thalli, or it may take place in elongate to subspherical terminal segments of a vegetative mycelium. The development sequence is presumed to be the same for all members of the Saprolegniales and is described by Sparrow (1960) and Gay and Greenwood (1966). In the majority of eucarpic species the development is initiated by the aggregation of dense cytoplasm and arrestment of linear extension at the hyphal tip. This is normally followed by the formation of a basal septum, but there is inconclusive evidence that in the Leptolegniellaceae zoospore delimitation is progressive and basipetal without the secretion of a basal septum (Dick, 1971). Septum formation is extremely rapid and the final sealing of the septum may be subject to pressure from the vegetative protoplasm, resulting in a peg which projects into the zoosporangium. Similar septation patterns occur in chlamydospore and oogonium development, and it is probable that some secondary wall deposits may occur (as in these latter structures) over the basal septum and possibly over the whole zoosporangium.

In all Saprolegniales, cleavage of the zoosporangial protoplasm always takes place by the coalescence of vesicles. These vesicles unite with the tonoplast of the central vacuole if this is present, and then with the plasmalemma. Thus segmentation appears to proceed from the central vacuole to the periphery. There is no evidence that the cleft vesicles are derived from the Golgi apparatus. When the plasmalemma is incorporated into the zoospore membrane, there is an immediate loss of turgor accompanied by a reduction of volume of about 10%. There is also a change in the curvature of the basal septum from concave to convex in relation to the subtending hypha.

It is not clear whether *Atkinsiella* follows this pattern at all, since Fuller *et al.* (1964), describe the zoosporangial protoplasm as condensing to the center *followed* by cleavage. The resulting spores are amoeboid and move along fine cytoplasmic strands within the zoosporangium before becoming flagellate. Similar fine cytoplasmic strands associated with zoospore initiation are reported for *Eurychasma*. Zoosporogenesis in *Thraustochytrium aureum* and *T. roseum* is described by Booth and Miller (1968) as centripetal, segmentation originating first adjacent to the zoosporangium wall. In *Schizochytrium,* repeated bisection of the protoplast results in the formation of *zoosporangia* from which zoospores are released. Divergent patterns of morphogenesis such as these make the inclusion of these organisms in the Saprolegniales highly suspect.

The dehiscence mechanisms of zoosporangia, flagellation of the first formed zoospores, and the spatial relationship of the first-formed cysts to the zoosporangia are variable features in all families of the Saprolegniales.

In most genera the zoosporangium develops one, or more rarely two or three papillae, which probably dissolve to release the zoospores. The precise manner in which the zoospores leave the zoosporangium is still in doubt, no hypothesis adequately explaining all observations. Zoospore motility can be ruled out since the flagella are not particularly active on discharge and zoospores frequently emerge backward with trailing flagella. The latter may become caught on the papilla lip and this could account for the formation of hollow spore balls at the orifices of *Achlya* zoosporangia. In some genera there may be tenuous protoplasmic threads joining the zoospores and in other genera the zoosporangia produce aplanospores. Plasmamembrane activity can also be eliminated since the plasmalemma is disrupted by zoospore formation. Surface energy (Webster and Dennis, 1967) cannot account for discharge when the diameter of the papilla exceeds the diameter of the zoospore as in *Scoliolegnia blelhamensis*. It is possible that there may be a mucilaginous or hemicellulosic inner-wall layer to the zoosporangium which has an infinite hydrophilic capacity. The pressure from hydration and differential viscosity could then account for the discharge of zoospores.

Zoospore encystment within the zoosporangium occurs in a number of taxa in different families. Thereafter, zoospore release may take place in many different ways. In *Eurychasma* and *Atkinsiella* the zoospores undergo a period of motility before encysting within the zoosporangia. In *Thraustotheca* the zoosporangial wall disintegrates to release the encysted spores. In *Calyptralegnia* the zoosporangium undergoes circumcissile dehiscence. This is followed by swelling of the encysted zoospores exposed to the external environment within the rigid cylindrical zoosporangial wall. The subsequent mutual displacement of the enlarged cysts enables some to drop out. In *Dictyuchus* the zoospore cysts produce individual exit papillae through the original zoosporangium wall. The zoosporangium may be deciduous in this genus.

In certain species of *Thraustochytrium* the zoosporangial wall gelatinizes and the naked spores separate before becoming flagellate.

In many species the type of dehiscence is affected by the environmental conditions.

Zoosporangial renewal in eucarpic taxa may be by internal proliferation through the basal septum, by lateral branching below the terminal zoosporangium, or by basipetal succession. Complex combinations of all the above occur. In *Thraustochytrium* development of the secondary zoosporangium may have started before the first zoosporangium has reached maturity.

The morphology of the oomycete zoospore and the phylogenetic implications of this morphology are involved topics confused by etymologically incorrect terminology and ambiguous epithets. The discussion centers around the repeated emergence and encystment of individual zoospores and the fact that the first-formed zoospore in certain genera differs morphologically from subsequently formed zoospores in an encystment-emergence sequence. Also this first-formed zoospore forms a distinctive cyst. The topic is discussed by Bessey (1950) and Sparrow (1960). The greatest complexity is reached in the Saprolegniaceae.

The terminology presented here retains the terms *primary zoospore* and *secondary zoospore* only in the morphological sense. There is no implied developmental or phylogenetic significance. The term *dimorphic* is used when both types occur in the life cycle, but monomorphic is not retained because it could be ambiguous if *Pythiopsis* and *Rhipidium* were being discussed. The terms *diplanetic* and *polyplanetic* should be restricted to repeated emergences of the secondary, or principal, zoospore form only, thus using these terms in their etymologically correct sense (contra Sparrow, 1960). It should be noted that in spite of contrary reports, there is no unequivocal evidence for repeated emergence of the primary, or auxiliary, zoospore form.

The primary zoospore is not well documented. There are no EM sections of this zoospore nor are there any interference microscopy photographs to show the precise manner of insertion of the flagella. These are reputed to be inserted apically without any accompanying groove in the zoospore body. The two flagella are almost equal in length. In *Leptolegnia* and possibly other genera this apical insertion may be the result of an amoeboid folding of an elongate protoplast with lateral or bipolar flagellar insertion. Compared with the secondary zoospore, the primary zoospore is a short-lived and feeble swimmer and may only serve to separate the zoospore mass. On encystment the flagella are retracted. Tufts of spines may be found on the outside of the developing cysts. It was originally suggested that these represented mastigonemes derived from the tinsel flagellum. J. L. Gay (personal communication) has found that they are derived from "bar" bodies developed in the zoospore initial. Except in one species of *Aphanomyces* (Scott, 1961) germination or reemergence is a result of a softening of a small area of the cyst wall. This softened patch may expand to form a papilla before rupturing to release the secondary zoospore.

The secondary zoospore is beaked, with a deep lateral groove in which the flagella are inserted. The anteriorly directed flagellum is short and of the tinsel type; the longer, posteriorly directed flagellum is of the whiplash type. The observed shape of this zoospore depends on the plane of view and may be more or less oval, bean shaped or reniform. For this reason the zoospore is not designated by its shape. It is a vigorous, smoothly moving zoospore which can remain motile for several hours. On encystment both flagella are normally cast off and the resulting cyst wall may bear characteristic hooked or spiny appendages. Reemergence is via an enzymatically softened pore or papilla in the wall, while germination is usually by a slender germ tube which widens gradually to form a hypha.

In the Haliphthoraceae a different type of germination has been reported in which an extremely slender filament of less than 0.5 μm diameter is developed. This filament abruptly expands to form the thallus rudiment. Zoospore germination in the other families is not fully described but may resemble the Haliphthoraceae in the Leptolegniellaceae and Ectrogellaceae. In the Thraustochytriaceae a polar development such as is found in the Rhipidiaceae occurs, the rhizoidal development being initiated first, followed by an expansion of the zoospore cyst to form the thallus.

Whatever the normal situation in a particular species, any encysted zoospore is capable of germination.

Dimorphism is reported in the Saprolegniaceae, Leptolegniellaceae, Halipthoraceae, and Ectrogellaceae, but is unknown in the Thraustochytriaceae. Only in the first two families is there undoubted evidence of both types of flagellated zoospore present in the life cycle. In the Ectrogellaceae specific diagnoses suggest that the first-formed zoospore may be of the secondary or principal form. In *Haliphthoros* and the Thraustochytriaceae it is of the secondary type with laterally inserted flagella.

Within the order as at present constituted there are three modal sizes for encysted zoospores. The volume of the zoospore cyst is usually less than that of the zoospore from which it was formed or the zoospore which may develop from it. These modal diameters are approximately $3-4$ μm (Ectrogellaceae and Thraustochytriaceae) $6-8$ μm (Leptolegniellaceae and Haliphthoraceae) and $10-12$ μm (Saprolegniaceae) but there is naturally some overlap and this does not constitute a diagnostic character on its own.

Chlamydospores (gemmae) may be found in the Saprolegniaceae and Haliphthoraceae. In the former they occur either singly or in catenulate chains, and they may approximate to zoosporangial or oogonial shapes. They are plerotic and have thick walls. They may germinate to give hyphae, or a short hypha bearing a zoosporangium; they may become transformed into zoosporangia, or oogonia if the species is parthenogenetic. In the Haliphthoraceae the chlamydospores are aplerotic and germinate to give new filaments.

The morphology of sexual reproduction in the Saprolegniales has been reviewed recently by Dick (1969a). Cytological and genetical evidence (Sansome, 1965; Bryant and Howard, 1969; Barksdale, 1966, 1968)[2] indicate that the thallus of the Saprolegniaceae is diploid. There is no such evidence for the Leptolegniellaceae, Ectrogellaceae, Haliphthoraceae, or Thraustochytriaceae. However, since similar evidence in support of gametangial meiosis is accumulating in other orders of Oomycetes (Leptomitales and Peronosporales) it is likely that the thallus is diploid in all Oomycetes. If any family is shown to have a haploid vegetative stage its taxonomic position will have to be reassessed.

The male and female gametangia are morphologically distinct and delimited by septa in the Saprolegniaceae. Delimitation of the female gametangium, or oogonium, occurs slightly in advance of delimitation of the male gametangium, or antheridium. Parthenogenetic development is frequent. Heterothallism is known only in a few species of *Achlya* and *Dictyuchus* and may be a secondary development from homothallic ancestors (Barksdale, 1960). There is an extensive literature on the hormones involved (see Machlis, 1966), and it is probable that similar hormones may act in homothallic species.

In the Leptolegniellaceae discrete gametangia may not be formed. The oospores develop in more or less undifferentiated segments of the thallus which are delimited by septa. In the Ectrogellaceae and Thraustochytriaceae oospore production is parthenogenetic or absent, development being from

[2]See detailed review by Dick and Win-Tin (1973).

otherwise undifferentiated vegetative thalli. Sexual reproduction is not known in the Haliphthoraceae.

Oosporogenesis is centrifugal in the Saprolegniaceae: the initiation of the oospheres follows a similar sequence to that of zoospores. Pluriovulate oogonia are therefore to be expected. It also follows that periplasm (residual oogonial cytoplasm not incorporated into oospheres) is absent. Oosporogenesis has not been described for the Leptolegniellaceae, but the pluriovulate oogonia and absence of periplasm indicate a similar morphogenesis.

In the Ectrogellaceae, *Pythiella* is described as having an oogonium in which the single oosphere is accompanied by periplasm. If this is so, such centripetal oosporogenesis is unique in the Saprolegniales and casts doubt on the taxonomic position of this genus. Kanouse (1927) stated: "In the Saprolegniales many oospores without periplasm are commonly formed in each oogonium. This fact at once precludes any possibility of including in that order any fungi with single oospores with periplasm . . . ." Otherwise the details of oosporogenesis are not known in the Ectrogellaceae or Thraustochytriaceae.

Where antheridial development occurs, the origin of the antheridial branch is characteristic of the species. Definitions are given in Dick (1969a). Discrete male gametes are never formed; thus, gametangial contact is essential for sexual reproduction. It is probable that hormonal activity not only directs growth responses, but also stimulates local cellulase production resulting in a limited softening of the gametangial walls (Thomas and Mullins, 1969). This may result in the development of a receptive papilla (*Leptolegnia caudata*) or, more usually, merely the growth and penetration of a fertilization tube from the antheridium into the oogonial cavity. The fertilization tube may branch and its tips penetrate the oospheres before rupturing to release male gametic nuclei.

The morphology of the developing oospore is of critical importance in the Saprolegniales. The oospore is defined (Dick, 1969a) as "a resting spore formed either as a result of a real or apparent sexual fusion, or parthenogenetically in a modified part of the vegetative system. This spore always shows a distinct redistribution of protoplasmic contents, especially those presumed to be food reserves, and this character may be used to distinguish an oospore from a chlamydospore in the absence of morphologically distinct sexual organs."

The oospore wall is of two layers; the outer epispore is the initial zygote membrane together with the oosphere membrane if such exists and any additional deposits which are physically, chemically, and physiologically indistinguishable from it. In the Saprolegniales this layer is always smooth and contains cellulose. The inner, endospore layer is usually thicker and stratified. It increases in thickness when the oospore is ruptured and probably becomes enzymatically degraded when the oospore germinates. Nothing is known of this layer in the Ectrogellaceae or Thraustochytriaceae. In the Saprolegniaceae the endospore is usually relatively thin, but in *Leptolegnia* it is very thick and the layers may be convoluted in *L. eccentrica.* In the *Leptolegniellaceae* the endospore is invariably complex. In *Leptolegniella* and *Brevilegniella* the endospore is separable and cellulosic. It does not apparently swell when the epispore is ruptured. The endospore (?) may be punctate (?) in certain species of *Aphanomycopsis.*

The cytoplasm of the mature oospore contains at least two types of reserve globules or vesicles. In the Saprolegniaceae there is a central ooplast (Howard, 1971) usually of granular appearance and usually with the granules in Brownian motion. The granules may be almost entirely absent in *Aphanomyces*, resulting in an optically inactive vesicle. Until detailed observations have been made of germinating oospores it is not possible to determine whether there is homology between the ooplast and the central globule of the Leptomitales or Peronosporales, but what little evidence there is suggests that they are not homologous. If this is so, then only the Saprolegniales possess an ooplast. On the other hand the Saprolegniaceae do not possess a central globule such as is found in the Peronosporales or the Leptolegniellaceae. The occurrence of these moieties is unrecorded in the Ectrogellaceae or Thraustochytriaceae.

In addition to the ooplast, the cytoplasm of the Saprolegniaceae contains a large quantity of lipidlike material which may be accumulated in one eccentric globule or distributed in many globules dispersed more or less evenly around the ooplast. There are many intergrades between these extremes; some have been given names such as centric, subcentric, subeccentric, and eccentric. The type of distribution appears to be more or less constant for a given species. Nothing is known of this moiety in the other families of the Saprolegniales but a parallel development may occur in the Leptomitaceae (Leptomitales).

Oospore germination is known only in the Saprolegniaceae and Leptolegniellaceae. In the latter family all that is known is that the oospores of *Leptolegniella* germinate to produce thalli which form oospores only (Scott *et al.,* 1963). In the Saprolegniaceae Ziegler (1948) has described four patterns of germination ranging from the production of a short hypha bearing a zoosporangium to branched mycelia with or without zoosporangia. The oospores of all taxa investigated require a resting period of from several days up to six months, depending on the species. It is also characteristic that only a very small percentage of apparently normal oospores will germinate. Dick (1972) and Dick and Win-Tin (1973) have suggested a possible explanation for this based on aneuploidy.

III.   DEVELOPMENT OF TAXONOMIC THEORY

The taxonomy of the Saprolegniales dates from 1824 when Nees von Esenbeck gave diagnoses based on zoosporangial discharge for *Saprolegnia* and *Achlya*. The family name was first used by Kützing in 1843 and a concise diagnosis for the Saprolegniaceae was first published by Pringsheim (1858).

The consolidation of the taxonomic concepts of genera in the Saprolegniaceae is due almost entirely to the researches of de Bary (1852, 1881, 1888). De Bary established that the characters used by Nees von Esenbeck were of generic significance and that the morphology of sexual reproduction could be used at the specific level. He established three further genera based on variations in zoosporangial behavior.

During this period there arose a controversy between de Bary (1881) and Pringsheim (1882) over the function of the fertilization tube in sexual reproduction. This controversy has been considered resolved, following the studies of Trow (1905) in which he asserted that true nuclear fusions occurred in the oospore. However, de Bary's general conclusion that functional sexuality has been lost, leaving only the morphology of sexuality, cannot be dismissed in view of the many parthenogenetic species in this order. Barksdale (1966) has provided some genetic analyses supporting true sexuality in the heterothallic species, but in some homothallic species genetic analysis may be the only way to determine whether de Bary was correct. The recent controversy over the position of meiosis highlights the significance and relevance of this debate in relation to speciation and the taxonomic reliability of features associated with sexual reproduction.

The taxonomy of the family was reviewed by Fischer (1892), Humphrey (1893), Schroeter (1893), Migula (1903), and von Minden (1912), without making any major changes from de Bary's 1888 account. By 1937 (Coker and Mathews, 1937), Coker had further developed de Bary's concept of the genus in this family, describing four new genera based on variations in zoosporangium formation and behavior. Differences in sexual reproduction were regarded as of subgeneric value. This view was accepted by Johnson (1956), but Seymour (1970) has refrained from using or erecting subgenera. Criticism of generic diagnoses based on characters of asexual reproduction date from Lechmere (1911). Emerson (1950) has reviewed one aspect of this problem suggesting that hybridization may have been involved. Nevertheless, as Coker (1914) originally retorted, variation due to extreme environmental conditions is not sufficient reason for rejecting these generic concepts.

In spite of this, generic concepts in the Saprolegniaceae are changing. Seymour (1970) allows diversity in zoosporangial renewal within the genus

Saprolegnia by uniting *Isoachlya* with this genus. Dick (1969b) erected the genus *Scoliolegnia* based on a combination of vegetative, asexual, and sexual characters. Within the genus *Achlya* species may be grouped using zoosporangial dimensions as well as and together with oospore characters. It is probable that some of the genera with eccentric oospores (*Thraustotheca, Geolegnia*, and *Brevilegnia*) are closely related to certain sections of the genus *Achlya*, ánd may show greater kinship therein, than do the sections of the genus *Achlya* themselves. It is therefore possible that the generic boundaries may have to be revised when more information is available.

While recent physiological reviews (Cantino, 1966) have pointed to an early separation of the Saprolegniaceae from the oomycete stock, there is little or no comparable information for the other families placed in the order. Within the Saprolegniaceae, Gleason *et al.* (1970) have shown that *Aphanomyces* and *Dictyuchus* have lost the ability to catabolize maltose, thus showing a parallel physiological evolution to the Leptomitaceae. However, the only other genera studied were *Achlya, Saprolegnia*, and *Leptolegnia*.

It is clear that our knowledge of the taxonomy of the Saprolegniaceae will only advance with a multivariate analytical approach. This in turn requires a search for additional taxonomic characters and a critical analysis of the methods of measurement, as has been done for the oospore (Dick, 1969a).

The first member of the order to be described, apart from Saprolegniaceae, was *Ectrogella* (Zopf, 1884). This genus was originally placed in the Olpidiaceae because of its thallus form. It was later transferred to the Saprolegniaceae, together with *Eurychasma*, when shown to produce biflagellate zoospores.

Sparrow, following Scherffel, regarded the biflagellate condition of the zoospore as of fundamental importance, and therefore placed *Thraustochytrium* (Sparrow, 1936) in the Saprolegniales. In the last four decades the remaining genera of the other families have been described and placed in the Saprolegniales on the basis of their biflagellate zoospores, thallus form, and position of zoosporogenesis. Problems in interpreting from the literature the morphological and morphogenetic development of these taxa preclude any useful discussion of the validity of taxonomic criteria. For example, it seems that very little, except zoospore size and host organism or substrate, separates *Eurychasma* (Ectrogellaceae) from *Atkinsiella* (Haliphthoraceae), yet zoospore size has never been given any prominence as a taxonomic criterion.

Sparrow (1968) would use wall behavior at discharge, zoospore behavior at time of discharge, and the presence or absence of internal proliferation as major taxonomic criteria within the Thraustochytriaceae. On the other hand, Booth and Miller (1968) have concluded "that sporangial size, sporangial wall thickness, size of zoospores, method of sporangial discharge, time of proliferating fundament appearance, nature of the rhizoidal system and colour of the thalli are either so variable or clinal as to be of little value in assigning .... isolates .... to an established taxon." It would appear that taxa within this family should be treated with considerable caution.

IV. SUMMARY

For the purposes of classification in this order, holocarpy or eucarpy is considered to be subordinate to the form of the thallus. The forms of the thallus and, in the Thraustochytriaceae, the polar development of this thallus from zoospores are the principal criteria on which the families are determined.

Zoosporogenesis, and the position and subsequent development of delimited zoospores in relation to the zoosporangium from which they came, have been confused in the past. Critical features may be lacking from published descriptions. In the families of the Saprolegniales, zoosporogenesis would be expected to be constant, while the latter could show limited variation. Whether this variation should extend to the formation of secondary zoospores within the zoosporangium without any prior encystment is doubtful.

The use of the zoospore in the taxonomy of the Oomycetes depends on a degree of precision in morphological description which is lacking for most groups of species. In particular, relative flagellum length, type of mastigoneme, the exact position of flagellar insertion, and the presence or absence of a groove in which the flagella are inserted are all uncertain. Similarly, the phylogenetic significance of the primary zoospore and of the formation of secondary zoospores as first-formed zoospores must be assessed before zoospore morphology can be used as a taxonomic criterion. If the secondary or principal zoospore form is uniform for the order, then zoospore size may prove to be a valuable character when used in conjunction with other characters at the family level.

Little is known of sexual reproduction in the Ectrogellaceae, Haliphthoraceae, or Thraustochytriaceae. At present these families are recognized on the basis of thallus construction and zoospore size only, but since these characters may reflect environmental influences, it is possible that some genera will need to be placed in other families or orders in the future.

When the morphology of sexual reproduction is considered in the context of a probable haplobiontic diploid life cycle with strong parthenogenetic tendencies, it is probable that the structure of the oospore will have more taxonomic significance than the morphology of the gametangia. The ecological pressures on the oospore are more likely to affect its presence or absence rather than its structure. The Saprolegniaceae is thus characterized by the presence of an ooplast in the oospore and a simple endospore. The Leptolegniellaceae has been erected for certain genera formerly included in the Saprolegniaceae in which the oospore apparently lacks an ooplast and has a complex endospore.

V. DIAGNOSTIC CHARACTERS IN THE SAPROLEGNIALES[3]

The families are separated as follows: filamentous or saccate, septate (Leptolegniellaceae only) or nonseptate, holocarpic or encarpic thallus; a thallus which is monocentric and of polar development only in Thraustochytriaceae; the degree of differentiation of the zoosporangia from the vegetative thallus; the modal zoospore cyst diameter, which is 3–4 μm (Ectrogellaceae and Thraustochytriaceae), 6–8 μm (Leptolegniellaceae and Haliphthoraceae), and 10–12 μm (Saprolegniaceae). Oospore structure is of value only in separating the Saprolegniaceae (with ooplast) and Leptolegniellaceae (with complex endospore).

Throughout the order, generic distinctions have been based on differences in the behavior of the delimited zoospore within and after release from the zoosporangium. In the Saprolegniaceae, the modal number of oospores and protoplasmic structure of the oospore are used. In the Leptolegniellaceae the degree of differentiation of gametangia can be used.

At the specific level, a degree of host specificity has been assumed for parasitic species. Otherwise, oospore diameter is one of the most reliable measurable taxonomic parameters in the Oomycetes, in spite of the wide dimensional ranges recorded. It is suggested that speciation has been linked with an increase in mean oospore diameter. Gametangial origins are also used.

A. *Thraustochytriaceae.*[4]

This family has a vegetative nonreproductive thallus which is monocentric, polar, and eucarpic. Its encysted zoospore diameter falls in the range of 3–5 μm and its first-formed zoospores are of the secondary or principal form.

SAPROLEGNIALES

The order has a vegetative, nonreproductive thallus which is never segmented and only rarely septate. Zoospore cleavage is centrifugal and takes place within the zoosporangium. Zoosporangial discharge never

---

[3]Note: The genera *Jaraia, Hamidia, Synchaetophagus* are not sufficiently known to be included here.

[4]See also Gaertner (1972), Kazama (1972), and Perkins (1972) for references to features suggesting affinity to Labyrinthulales.

involves vesicle formation and the secondary or principal zoospore f‹
is always developed outside the zoosporangium or from a previou
formed and encysted zoospore lodged within the zoosporangium.

Oosporogenesis is centrifugal and never periplasmic; the oogonia may
be plurovulate and the oospore possesses an ooplast.

B.   Saprolegniaceae

The vegetative, nonreproductive thallus is filamentous, normally eucarpic,
and never septate. Zoosporangia are delimited by a basal septum. The female
gametangia are always morphologically distinct consisting of oospores
with an ooplast, aggregated lipidlike globules, and endospore food reserve
material.

Encysted zoospore dimensions range from 8–15 μm.

C.   Leptolegniellaceae

Here, the vegetative nonreproductive thallus is more or less filamentous,
holocarpic, holocarpic and septate or eucarpic. The zoosporangium is
not always delimited by a basal septum. The encysted zoospore diameter
falls in the range from 5–10 μm. Gametangia are frequently not morphologically differentiated. The oospores have a distinct separable endospore
membrane but no (?) endospore food reserve material and the oospore
protoplast has a single central (eccentric) globule.

D.   Ectrogellaceae

This family has a vegetative nonreproductive thallus which is saccate,
nonseptate, and holocarpic. Encysted zoospore diameter falls in the range
of 3–6 μm.

E.   Haliphthoraceae

The Haliphthoraceae has a vegetative nonreproductive thallus which is
filamentous and holocarpic, or lobate to filamentous and eucarpic with
rhizoids delimited by septa. The encysted zoospore diameter ranges from
6–8 μm.

VI.   KEYS

KEYS TO THE SAPROLEGNIALES
KEY 1

1. Oospores possessing an ooplast with granular contents (rarely with very few granules).
   Thallus extensive, myceliar, vegetative hyphae rarely less than 5 μm diameter, nonseptate,

normally eucarpic. Morphologically distinct gametangia always differentiated. First-formed zoospore initials never maturing directly to the principal zoospore form
· **Saprolegniaceae Key 2** p. 131

1′. Oospores lacking an ooplast with granular contents. Thallus holocarpic, holocarpic and septate, eucarpic with very slender hyphae or saccate and eucarpic with rhizoids. Morphologically distinct gametangia rarely differentiated. First-formed zoospore initials sometimes maturing directly to the principal zoospore form · · · · · · · · · · · · · · 2

2(1′) Bipolar thallus eucarpic, monocentric, unlobed, with rhizoids. (Marine or halophytic fungi of doubtful affinity) · · · · · · · · · · · · · · · **Thraustochytriaceae** 12

2′(1′) Nonpolar thallus holocarpic, holocarpic and septate, filamentous or lobed, or eucarpic with very slender hyphae · 3

3(2′) Thallus holocarpic and saccate. Zoospore cyst diameter 3–5 mm. Oospores lacking or endospore membrane of oospores not separable from the epispore. (Probably an artificial grouping of little-known parasites.) · · · · · · · · · · · · · · · · · **Ectrogellaceae** 9

3′(2′) Thallus more or less filamentous or lobed. Zoospore cyst diameter 6–10 mm. Oospores lacking or endospore membrane of oospore separable from the epispore · · · · · · 4

4(3′) Oospores not known. Saprophytes, or parasites or invertebrates. (Marine fungi of doubtful affinity) · **Haliphthoraceae** 8

4′(3′) Oospores present or absent. Saprophytes, frequently keratinophilic, rarely parasites of algae. (Most species freshwater or terrestrial. May be more closely allied to the Lagenidiales) · **Leptolegniellaceae** 5

5(4′) Vegetative growth eucarpic, hyphae 3–5 μm diameter. Both zoosporangia and gametangia distinctly differentiated. First-formed zoospores encysting within the zoosporangium. Oogonia pluriovulate. Saprophytic and keratinophilic · · · · · · · · · **Aphanodictyon**

5′(4′) Vegetative growth holocarpic, more or less filamentous, rarely eucarpic, hyphae 5–20 μm diameter. Zoosporangia not, or only slightly differentiated from vegetative mycelia. Morphologically distinct gametangia usually absent, or if differentiated, then fungus parasitic or oogonia uniovulate. First-formed zoospores emergent, or if encysting within the zoosporangium, then gametangia not morphologically differentiated. Saprophytic or parasitic
· 6

6(5′) Zoospores sometimes encysting within the zoosporangium. No motile stage known.
· **Brevilegniella**

6′(5′) Zoospores emergent, biflagellate · · · · · · · · · · · · · · · · · · · 7

7(6′) Zoospores encysting at the mouth of the zoosporangium · · · · · · · **Aphanomycopsis**

7′(6′) Zoospores free-swimming on discharge · · · · · · · · · · · · · **Leptolegniella**

8(4) First-formed zoospores with laterally inserted flagella: emergent. Thallus irregularly filamentous, nonseptate · **Haliphthoros**

8′(4) First-formed zoospores with apically inserted flagella: commonly encysting within the zoosporangium. Thallus filamentous or lobed, sometimes with rhizoids cut off from the thallus by septa · **Atkinsiella**

9(3) Zoospore cysts spherical. (Parasitic in Bacillariophyta or fungi) · · · · · · · · · 10

9′(3) Zoospore cysts angular. (Parasitic principally in Phaeophyta or Rhodophyta) · · · 11

10(9) Parasites of diatoms · **Ectrogella**

10′(9) Parasites of phycomycetes · **Pythiella**

11(9′) Zoosporangial discharge by one or two broad discharge tubes · · · · · **Eurychasma**

11′(9′) Zoosporangial discharge by numerous narrow discharge tubes · · · · **Eurychasmidium**

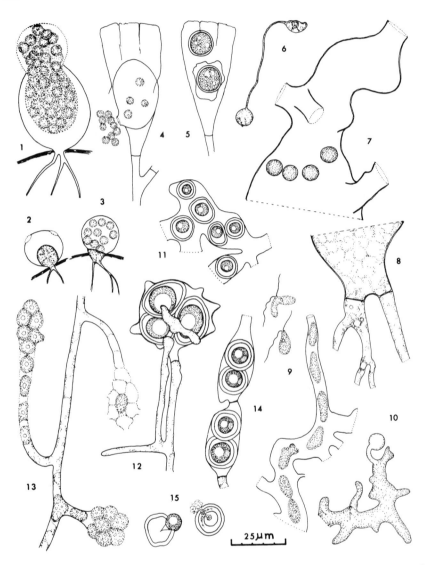

PLATE I.   Fig. 1. *Thraustochytrium roseum*. Thallus showing zoospore mass in process of discharge.   Fig. 2.   *Thraustochytrium aureum*. Thallus with discharged zoosporangium (with pores) new zoosporangial rudiment developing from the intramatrical rhizoids.   Fig. 3.   *Thraustochytrium* and *Atkinsiella* (Fig. 7).   Fig. 5.   *Ectrogella licmophorae*. Resting spores, the lower rudiment of the secondary zoosporangium is visible at the base.   Fig. 4.   *Ectrogella licmophorae*. Discharged thallus with encysted zoospores. Compare zoospore diameter with *Thraustochytrium* and *Atkinsiella* (Fig. 7).   Fig. 5.   *Ectrogella licmophorae*. Resting spores, the lower within an investing membrane probably derived from the thallus. Oospore structure in-

12(2) Zoosporangia formed indirectly from the monocentric thallus by cleavage of the vegetative rudiment. Zoosporangial membrane not formed by expansion of the zoospore cyst wall . . . . **Schizochytrium**

12'(2) Zoosporangium formed directly from the monocentric thallus. Zoosporangial membrane formed by expansion of the zoospore cyst wall 13

13(12') Zoosporangial wall thin (< 1.5 μm) or persistent (not disintegrating prior to zoospore discharge). Zoospores released through a pore or fissure. Zoospores flagellate and motile at the time of discharge (except in *Thraustochytrium proliferum*). Thallus proliferous or nonproliferous 14

13'(12') Zoosporangial wall thick (< 2.0 μm) and deliquescent. Zoospores released by dis-integration of the zoosporangial wall. Zoospores not motile by flagellar activity until after discharge. Thallus nonproliferous 16

14(13) Thallus apophysate. Zoospores discharged through a pore . . . **Japonochytrium**

14'(13) Thallus not apophysate. Zoospores discharged through a fissure 15

15(14') Thallus proliferous . . . . **Thraustochytrium A**
(5 spp.)

15(14'(Thallus nonproliferous . . . . **Thraustochytrium B**
(2 spp.)

16(13') Rhizoidal component of thallus present . . . . **Thraustochytrium C**
(4 spp.)

16'(13') Rhizoidal component of thallus absent . . . . **Althornia**

completely known.   Fig. 6.   *Haliphthoros milfordensis*. Germinating zoospore showing fine filament inflating to initiate the mature thallus.   Fig. 7.   *Atkinsiella dubia*. Discharged thallus with many exit papillae and a few encysted zoospores.   Fig. 8.   *Atkinsiella dubia*. Base of thallus showing rhizoidal development, separated from the sporogenous thallus by septa.   Fig. 9.   *Leptolegniella keratinophila*. Branched thallus discharging zoospores.   Fig. 10.   *Lepto-legniella keratinophila*. Young thallus with the outline of the zoospore from which it developed clearly visible.   Fig. 11.   *Leptolegniella keratinophila*. Branched thallus with oospores. Com-pare oospore structure and dimensions with *Aphanodictyon* (Fig. 12) and *Brevilegniella* (Fig. 14). Fig. 12.   *Aphanodictyon papillatum*. Hyphae with gametangia.   Fig. 13.   *Aphanodictyon papillatum*. Hyphae with zoosporangia. Zoospores encysting in situ, later germinating to give planonts. Note the apparent absence of a basal septum to the zoosporangia.   Fig. 14.   *Brevi-legniella keratinophila*. Intramatrical, intercalary, parthenogenetic gametangia.   Fig. 15. *Brevilegniella keratinophila*. Two oospores showing the separable nature of the endospore wall from the epispore wall and the apparently fluid filled cavity between. (Fig. 1 after Goldstein, 1963b; 2, 3 after Goldstein, 1963a; 4, 5 after Johnson, 1966; 6 after Fuller *et al.*, 1964; 7 after Atkins, 1954; 8 after Sparrow and Gottelli, 1969; 9, 10 after Huneycutt, 1952; 12 after Huneycutt, 1948; 13 after Sparrow, 1950; 14, 15 after Dick, 1961.)

A.   List of Taxa in the Ectrogellaceae, Thraustochytriaceae, and Haliphthoraceae[5]

ECTROGELLACEAE Sparrow ex Cejp 1959
  Ectrogella bacillariacearum Zopf 1884
  Ectrogella monostoma Scherffel 1925
  Ectrogella gomphonematis Scherffel 1925
  Ectrogella licmophorae Scherffel 1925
  Ectrogella perforans Petersen 1905
  Ectrogella eunotiae Friedmann 1952
  Ectrogella eurychasmoides J. and G. Feldmann 1955
  Ectrogella marina (Dangeard) J. and G. Feldmann 1955
  Eurychasma dicksonii (Wright) Magnus 1905
  Eurychasma sacculus Petersen 1905
  Pythiella vernalis Couch 1935
  Pythiella besseyi (Sparrow & Ellison) Sparrow 1960
  Eurychasmidium tumefaciens (Magnus) Sparrow 1936
THRAUSTOCHYTRIACEAE Sparrow ex Cejp 1959
  Thraustochytrium proliferum Sparrow 1936   (A)
  Thraustochytrium globosum Kobayasi and Ookubo 1953 (B)
  Thraustochytrium pachydermum Scholz 1958   (C)
  Thraustochytrium aureum Goldstein 1963   (A)
  Thraustochytrium motivum Goldstein 1963   (A)
  Thraustochytrium multirudimentale Goldstein 1963   (A)
  Thraustochytrium roseum Goldstein 1963   (C)
  Traustochytrium visurgense Ulken 1965   (C)
  Thraustochytrium aggregatum Ulken 1965   (B)
  Thraustochytrium striatum Schneider 1967   (C)
  Thraustochytrium kinnei Gaertner 1967   (A)
  Japonochytrium marinum Kobayasi & Ookubo 1953
  Schizochytrium aggregatum Goldstein & Belski 1964
  Althornia crouchii Jones & Alderman 1970
HALIPHTHORACEAE Vishniac 1958
  Haliphthoros milfordensis Vishniac 1958
  Atkinsiella dubia (Atkins) Vishniac 1958
  (? *Synchaetophagus balticus* Apstein 1911)
LEPTOLEGNIELLACEAE Dick 1971
  Aphanomycopsis bacillariacearum Scherffel 1925
  Aphanomycopsis desmidiella Canter 1949
  Aphanomycopsis saprophyticus Karling 1968
  Aphanomycopsis punctatus Karling 1968
  Aphanodictyon papillatum Huneycutt ex Dick 1948
  Leptolegniella keratinophila Huneycutt 1952
  Leptolegniella piligena Ookubo & Kobayasi 1955
  Leptolegniella exospora Kane 1966
  Leptolegniella marina (Atkins) Dick 1971
  Brevilegniella keratinophila Dick 1961

[5]Note: The relationship of fungi assigned to *Dermocystidium* and *Ostracoblabe* is not discussed in this chapter.

KEY 2: SAPROLEGNIACEAE

NOTES: This key should enable identification to groups of a very few species whether only oogonia or zoosporangia are available. A number of taxa, yet to be reexamined, are very poorly described and may have been wrongly assigned to particular genera. This key should therefore be used in conjunction with the checklist of species, which follows an unorthodox arrangement. Numbers following generic names (in bold face) refer to sections in the check list together with the number of species (in parentheses) in each section.

The subgenera of *Aphanomyces* (Scott, 1961) are illegitimate since the subgenus *Aphanomyces* does not contain the type species *A. stellatus* de Bary.

1. Zoosporangia absent 2

1'. Zoosporangia present 3

   2(1) Oogonia absent . . . . **Unidentifiable**

   2'(1) Oogonia present 25

3(1') Zoosporangia with spores in single rank distally, emergent and encysting at the mouth of the zoosporangium. Oogonia with one oospore, oogonia without pits 39

3'(1') Not the above combination of characters 4

   4(3') Most zoospores never encysting within or at the mouth of the zoosporangium, except occasionally in old cultures 5

   4'(3') Zoospores encysting within or at the mouth of the zoosporangium 11

5(4) Oospore absent. OR Modal number of oospores per oogonium 2, frequently 1 . . . 6

5'(4) Modal number of oospores per oogonium 3 or more 13

   6(5) Oospores absent. OR Oospore structure eccentric or subeccentric; or if centric or subcentric then with densely papillate oogonia. Oogonia without pits 7

   6'(5) Oospores structure centric or subcentric, oogonia rarely with occasional papillae. OR Oogonia with pits, except for *S. richteri* and some strains of *S. eccentrica* which have oogonia without pits and eccentric oospores. . . . . **"Cladolegnia"** **1c(7)**

7(6) Zoosporangia inflated, proliferation cymose at a wide angle, never internal; empty cysts very rare (zoospores frequently encysting on hyphae) . . . . **Pythiopsis** **6(2)**

7'(6) Zoosporangia filamentous or cylindrical, or if inflated then either with proliferation internal, or proliferation cymose at a narrow angle and empty cysts frequent 8

   8(7') Zoospores vigorously motile from the moment of discharge, never sluggish. Hyphae moderately stout and straight . . . . **Saprolegnia** sensu lato **1(22)** 14

   8'(7') Zoospore movement sluggish, taking a short time to establish motility (with amoeboid movement) after discharge, or even encysting without appreciable flagellar activity. Hyphae often slender or flexuous 9

9(8') At least some zoosporangia with a cluster of encysted zoospores at the mouth of the zoosporangium, mycelium slender and straight . . . . **Achlya** (–proto) 24

9'(8') Not as above, mycelium flexuous 10

   10(9') Undischarged zoosporangia with zoospores predominantly in one rank, at least distally. Zoosporangia very long frequently exceeding 0.8 mm. Proliferation internal. Oospore wall thick ($>3\ \mu$m) . . . . **Leptolegnia** **5(6)**

   10'(9') Undischarged zoosporangia with zoospores predominantly in two or more ranks distally. Longest zoosporangia rarely exceeding 0.4–0.8 mm. Proliferation variable.

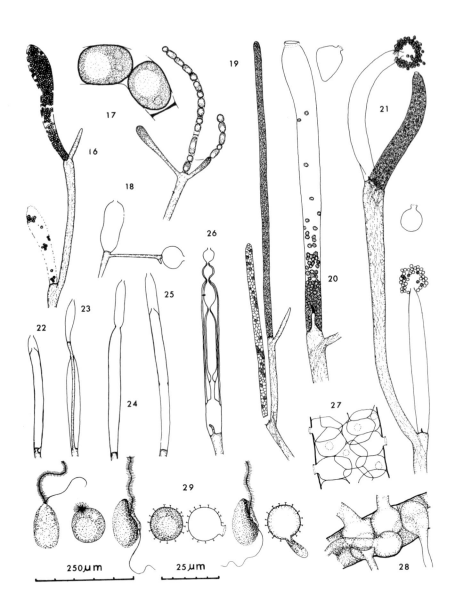

250μm

25μm

PLATE II.   Zoospore details all at the greater magnification. All zoosporangia drawn at the lower magnification.   Fig. 16.   *Thraustotheca clavata.* Zoosporangia in cymose series, showing disintegration of the zoosporangium wall.   Fig. 17.   *Geolegnia septisporangia.* Four zoosporangia in basipetal series with subsequent cymose branching of the hypha. Two multinucleate aplanospores with thick walls.   Fig. 18.   *Pythiopsis cymosa.* Cymose arrangement of inflated zoosporangia.   Fig. 19.   *Dictyuchus monosporus.* Cymose arrangement of filiform zoosporangia. The first-formed zoosporangium is deciduous and secondary zoospores have been released from most of the cysts.   Fig. 20.   *Calyptralegnia achlyoides.* Operculate zoosporangium with angular zoospore cysts. The zoosporangium has a prominent spur to the basal septum.   Fig. 21.

132

Oospore wall thin (<2 μm) . . . . **Scoliolegnia 14(4)**

11(4′) Distal part of zoosporangium filamentous. Undischarged zoosporangia with zoospores in one rank distally 12

11′(4′) Zoosporangium more or less long-clavate, filamentous, or naviculate. Undischarged zoosporangia with zoospores in several ranks distally 15

12(11) Zoosporangia with few large spores (mean dimensions 11 × 20 μm), aplanetic with a wall thicker than the zoosporangium wall . . . .**Geolegnia 9(3)**

12′(11) Zoosporangia with many zoospores (mean dimensions 8–12 μm diameter). planetic, cyst membranes not thicker than the zoosporangium wall 39

13(5′) Zoospores swimming actively from the mouth of the zoosporangia on discharge, proliferation of zoosporangia commonly internal. Oogonia usually smooth 14

13′(5′) At least some zoospores encysting at the mouth or within the zoosporangia on discharge. Proliferation of zoosporangia commonly cymose. Oogonia smooth or papillate 19

14(8,13) Oogonia basipetal or cymose as well as terminal on the main hyphae. Oogonia frequently deciduous. Antheridia absent in most taxa, usually monoclinous when present. Zoosporangia frequently inflated from the basal septum. Cymose proliferation common in some taxa. Mean oospore diameter of the majority of taxa >28 μm
. . . . **"Isoachlya" 1b(3–10)**

14′(8,13) Oogonia terminal or lateral on the main hyphae, rarely basipetal, but never basipetal and deciduous. Antheridia present in most taxa, monoclinous or diclinous. Zoosporangia cylindrical or naviculate. Proliferation of zoosporangia usually internal. Mean oospore diameter of the majority of taxa <28 μm
. . . . **Saprolegnia** sensu stricto **1a(12)**

15(11′) Zoosporangia operculate. Mean oospore diameter >45 μm
. . . . **Calyptralegnia 11(2)**

15′(11′) Zoosporangia indehiscent or with an apical papilla. Mean oospore diameter <40 μm except in *A. megasperma* 16

16(15′) Secondary zoosporangia indehiscent (zoosporangia of young cultures may behave as below). Aplanetic development of zoospores from within zoosporangia rare except in staled cultures 17

16′(15′) Majority of secondary zoosporangia with zoospores emergent and encysting at the mouth of the zoosporangium. In a few taxa zoosporangia very rare with zoospores

Achlya flagellata. Three zoosporangia in cymose series, 2 discharged and with hollow balls of encysted zoospores. Secondary zoospores have been released from that of the lower, naviculate zoosporangium. Figs. 22–25. *Saprolegnia parasitica* zoosporangia with internal proliferation. Fig. 22: nonemergent; Fig. 23: emergent; Fig. 24: partly emergent; Fig. 25: basipetal and nonemergent. Fig. 26. Diagram of a complex arrangement of 4 zoosporangia. The second zoosporangium partly emergent, the third nonemergent, and the fourth basipetal and emergent. Fig. 27. *Dictyuchus monosporus.* Dictyosporangium with empty cysts of primary zoospores, showing papillae emergent through the zoosporangium wall. Fig. 28. *Dictyuchus monosporus.* Aplanosporangium with primary zoospore cysts germinating to give hyphae. Fig. 29. Diagram to show dimorphism of zoospores and zoospore cysts, and diplanetism of the secondary zoospore. (In part after Meier and Webster, 1954).

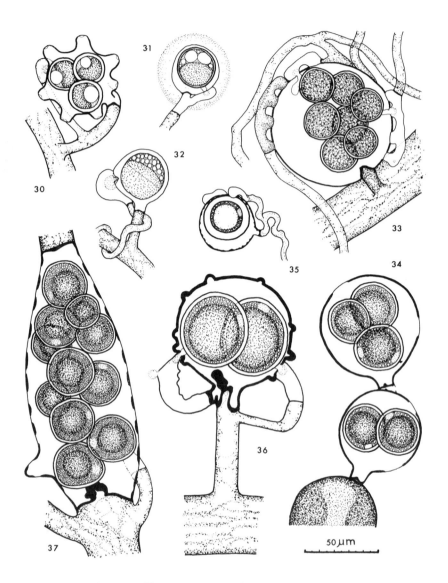

PLATE III.   Fig. 30.   *Achlya recurva (sensu Latham)*. Oogonium with truncate papillae and several eccentric oospores filling the oogonium. The closely monoclinous antheridial branch bears an antheridium laterally applied to the oogonium between the papillae. Note the recurved oogonial hypha.   Fig. 31.   *Pythiopsis cymosa*. Oogonium with a single, aplerotic, subeccentric oospore. Note the mucilage layer around the oogonium. Semihypogynous antheridium with fertilization tube.   Fig. 32.   *Leptolegnia caudata*. Oogonium with a single, plerotic, subeccentric oospore. Note the receptive papilla or bulge at the point of contact with the antheridium. The antheridial hypha is coiled about the oogonial hypha.   Fig. 33.   *Achlya flagellata*. Oogonium with several oospores not filling the oogonium. Several diclinous antheridial hyphae surround

showing aplanetic development from within the zoosporangium

. . . . **Achlya** sensu lato   **2(56)** 19

17(16) Wall of secondary zoosporangia persistent, though perforated after the emergence of zoospores. Zoosporangia frequently deciduous . . . . **Dictyuchus   4(6)**

17'(16) Wall of secondary zoosporangia evanescent, zoospores released by its disintegration. Zoosporangia never deciduous, but they may appear to be so due to disarticulation of blocks of zoospores 18

    18(17') Mycelium stout, zoosporangia broadly cylindrical or short clavate, oogonia pluriovulate . . . . **Thraustotheca   7(2)**

    18'(17') Mycelium slender, zoosporangia long-clavate, naviculate or short and depauperate oogonia uniovulate . . . . **Brevilegnia   10(13)**

19(13') Oospores eccentric 20

19'(13') Oospores centric or subcentric . . . . '. . . 23

    20(19) Oogonia with truncate papillae, or many irregularities 21

    20'(19) Oogonia smooth, or rarely with occasional papillae

. . . . **Achlya** (–eccentric)   **2a(18–21)**

21(20) Oogonia with truncate papillae . . . . **Achlya** (–tuberculate)   **2g(4)**

21'(20) Oogonia irregular, unpitted . . . . **Achlya** (–glomerulate)   **2f(3)**

    22(19') Zoosporangia infrequent, even in young cultures. Oogonia frequently only intercalary at intervals along the same hypha, but also terminal, rarely lateral, and in these cases usually apiculate. Antheridial branches usually closely monogynous and difficult to see . . . . **Achlya** (–aplanes)   **2d(3)**

    22'(19') Zoosporangia freely produced in young cultures. Oogonia usually terminal or lateral on main hyphae, rarely intercalary but then accompanied by obvious antheridia, or lateral oogonia abundant in the vicinity 23

23(22') Antheridial branches present, closely monoclinous or epigynous, antheridia applied by their tips at a wide angle to the oogonium. Oogonia usually in a racemose arrangement about the main hyphae . . . . **Achlya** (-racemose)   **2b(7)**

23'(22') Antheridial branches monoclinous or absent, antheridia applied to oogonia laterally, or antheridia hypogynous. Oogonia scattered on short lateral hyphae about the main hyphae, or terminal on long secondary hyphae 24

the oogonium and have produced large antheridial cells attached to the oogonium by pegs. Note the thin regions of the oogonial wall beneath the antheridial pegs (pits?). Fig. 34. *Saprolegnia* (*Isoachlya*) *toruloides*. Disarticulating catenulate chain of parthenogenetic oogonia. Oospore structure centric to subcentric. Note the granular ooplast and pellucid nuclear spots. The wall of the oogonium is pitted. Fig. 35. *Aphanomyces cladogamus*. Oogonium with roughened internal surface. The single oospore has a subcentric structure with an ooplast possessing very few granules. The antheridium is monoclinous. (Dick, 1971b). Fig. 36. *Achlya colorata*. Papillate oogonium with a few large centric to subcentric oospores. Antheridia are closely monoclinous and exigynous. The antheridial cells are applied to the oogonium by their tips. Fig. 37. *Achlya treleaseana*. Intercalary oogonium possessing both pits and papillae. Oospore structure is subcentric. The antheridium is closely monoclinous.

24(9,23′) Modal length of zoosporangia <about 160–180 μm. Mycelium limited, delicate Zoosporangial proliferation sometimes internal . . . .**Achlya** (-proto) **2e(7)**

24′(9,23) Modal length of zoosporangia >about 200 μm. Mycelium extensive, stout. Zoosporangial proliferation always cymose . . . . **Achlya** (subcentric) **2c(11)**

25(2′) Oogonia with truncate papillae . . . . **Achlya** (-tuberculate) **2g(4)**

25′(2′) Oogonia smooth, with or without papillae, with or without pits, but never with papillae always terminated with a small apical pit 26

26(25′) Oogonia smooth, modal number of oospores 3 or more. Oospore structure eccentric . . . . **Thraustotheca** **7(2)**
Achlya (-eccentric) **2a(18–21)**
S. anisospora

26′(25′) Not with the above combination of characters 27

27(26′) Oogonial wall papillate but not pitted, modal number of oospores 2 or less. Oospore structure *not* eccentric, oospores aplerotic 28

27′(26′) Not with the above combination of characters 32

28(27) Mean oospore diameter <25 μm, except for 2 parasitic species of Aphanomyces with nongranular ooplasts 29

28′(27) Mean oospore diameter > 27 μm and ooplast granular 31

29(28) Immature oogonia lacking a conspicuous mucilage layer . . . **Aphanomyces** subgenus **Asperomyces** of Scott **3(10)**

29′(28) Immature oogonia possessing a conspicuous mucilage layer 30

30(29′) Oospore structure subcentric or nearly so . . . . **Aplanopsis** **13(2)**

30′(29′) Oospore structure markedly subeccentric . . . . **Pythiopsis** **6(2)**

31(28′) Mycelium limited, flaccid and flexuous . . . . **Scoliolegnia** **14(4)**

31′(28′) Mycelium moderately stout and straight or flexuous and extensive 38

32(27′) Oogonial wall unpitted, modal number of oospores three or less, oospore structure subeccentric or eccentric 33

32′(27′) Not with the above combination of characters 34

33(32) Mean oospore diameter >45 μm . . . . **Calyptralegnia** **11(2)**

33′(32) Mean oospore diameter normally <35 μm 38

34(32′) Oogonia normally smooth, occasionally papillate, but never apiculate, modal number of oospores 3 or less. Mean oospore diameter >30 μm, oospore centric or subcentric, ooplast densely granular . . . . **"Cladolegnia"** in part **1c(5)**

34′(32′) Not with the above combination of characters 35

35(34′) Modal number of oospores 1, oospore structure subcentric, often with very few granules in the ooplast 39

35′(34′) Not with the above combination of characters 36

36(35′) Oospores eccentric or subeccentric . . . . **D. pseudodictyon Saprolegnia eccentrica**

36′(35′) Oospores centric or subcentric 37

37(36′) Antheridia present, closely monogynous, and attached to the oogonium by their tips . . . . **Achlya** (−racemose) **2b(7)**

37'(36') Not as above . . . . **Saprolegnia** sensu lato **1(22)**

Note: Separation of the supraspecific groups on oogonial characters alone may be difficult. In general the presence of pits and the absence of papillae or an apiculus indicate a *Saprolegnia*, while the converse applies to the *Achlya*-groups.

Achlya (–proto) **2e (7)**
Achlya (–aplanes) **2d(3)**
Achlya (–subcentric) **2c(11)**

38(31', 33') Oospore plerotic, or if aplerotic then endospore wall with complex layers
. . . . **Leptolegnia** **5(6)**

38'(31', 33') Oospores aplerotic with only a slightly stratified endospore wall
. . . . **Dictyuchus** **4(6)**

Note: Separation of these taxa requires zoosporangial criteria which should be readily obtained on culture. See earlier part of key.

Brevilegnia **10(13)**
Geolegnia **9(3)**
Achlya (–glomerulate) **2f(3)**
Saprolegnia eccentrica
Saprolegnia richteri
Achlya caroliniana
Achlya rodrigueziana
Achlya spinosa
Achlya stellata

39(3, 12', 35) Zoosporangia with an inflated basal portion . . . . **Plectospira** **12(2)**

39'(3, 12', 35) Zoosporangia filamentous throughout their length 40

40(39') Specialized parasites of rotifers by means of modified mycelial tips
. . . . **Sommerstorffia** **8(1)**
(Little-known fungi, which on further study may be more closely related to the Leptolegniellaceae.)

40'(39') No special modifications for catching rotifers . . . . . **Aphanomyces** **3(25)**

B. Check List of Genera and Species of Saprolegniaceae in Chronological Sequence and in Accordance with the Divisions Given in the Key

NOTES: In certain groups of species (2a, 2b, 2c, 9, 10) the taxa are separated by only a very few, probably minor characters. Some may need to be brought to synonymy, but this cannot be done until we understand the bases of morphological variation in the family.

A dagger indicates that the generic status requires reassessment. The subsections of *Achlya* and *Saprolegnia* represent my opinion of likely affinities but no opinion is expressed or implied as to whether these groups are of generic or subgeneric significance. In section 2c oospore structure may be centric or subcentric.

An asterisk indicates that the specific status of a taxon requires reassessment. The absence of an asterisk does not imply that the species is well documented or established.

1a *Saprolegnia ferax* (Gruithuysen) Thuret 1850
    Saprolegnia hypogyna (Pringsheim) de Bary 1883
    Saprolegnia anisospora de Bary 1888
    Saprolegnia diclina Humphrey 1893
    Saprolegnia furcata Maurizio 1899
    Saprolegnia turfosa (Minden) Gaumann 1918
    Saprolegnia litoralis Coker 1923
    Saprolegnia parasitica Coker 1923
    Saprolegnia glomerata (Tiesenhausen) Lund 1934
    Saprolegnia terrestris Cookson ex Seymour 1937
    Saprolegnia uliginosa Johannes 1950
    Saprolegnia australis Elliott 1968
1b *Saprolegnia* ("Isoachlya") *torulosa* de Bary 1881
    Saprolegnia ("Isoachlya") *monilifera* de Bary 1883
  * *Saprolegnia* ("Isoachlya") *toruloides* Kauffman 1921
1c * *Saprolegnia* ("Isoachlya-Cladolegnia")! *Pythiopsis humphreyana*
    Coker 1914
    Saprolegnia ("Isoachlya-Cladolegnia") *eccentrica* Coker 1923
    Saprolegnia ("Isoachlya-Cladolegnia") *megasperma* Coker 1923
    Saprolegnia ("Isoachlya-Cladolegnia") *unisperma* Coker 1923
    Saprolegnia ("Isoachlya-Cladolegnia") *subterranea* Dissmann 1925
    Saprolegnia ("Isoachlya-Cladolegnia") *richteri* Richter ex Seymour
    1937
    Saprolegnia ("Isoachlya-Cladolegnia") *luxurians* (Bhargava &
    Srivastava) Seymour 1970
2a *Achlya* (−eccentric) *prolifera* Nees von Esenbeck 1823
    Achlya (−eccentric) *americana* Humphrey 1893
    Achlya (−eccentric) *debaryana* Humphrey 1893
    Achlya (−eccentric) *caroliniana* Coker 1910
    Achlya (−eccentric) *klebsiana* Pieters 1915
    Achlya (−eccentric) *conspicua* Coker 1923
    Achlya (−eccentric) *dubia* Coker 1923
    Achlya (−eccentric) *flagellata* Coker 1923
    Achlya (−eccentric) *proliferoides* Coker 1923
    Achlya (−eccentric) *orion* Coker and Couch 1923
    Achlya (−eccentric) *subterranea* Coker & Braxton 1926
    Achlya (−eccentric) *inflata* Coker 1927
    Achlya (−eccentric) *rodrigueziana* Wolf 1941
    Achlya (−eccentric) *diffusa* Harvey ex Johnson 1942
    Achlya (−eccentric) *intricata* Beneke 1948
    Achlya (−eccentric) *cambrica* (Trow) Johnson 1956
    Achlya (−eccentric) *oviparvula* Rogers & Beneke 1962

Achlya (–eccentric) *aquatica* Dayal & Thakur Ji 1969
2aa *Achlya* (–heterothallic) *bisexualis* Coker & Couch 1927
Achlya (–heterothallic) *ambisexualis* Raper 1939
Achlya (–heterothallic) *heterosexualis* Whiffen-Barksdale 1965
2b *Achlya* (–racemose) *racemosa* Hildebrand 1867
Achlya (–racemose) *colorata* Pringsheim 1882
Achlya (–racemose) *radiosa* Maurizio 1899
Achlya (–racemose) *sparrowii* Reischer 1949
Achlya (–racemose) *turfosa* Johannes 1949
Achlya (–racemose) *pseudoradiosa* Rogers & Beneke 1962
* *Achlya* (–racemose) *echinulata* Beroqui 1969
2c *Achlya* (–subcentric) *polyandra* Hildebrand 1867
Achlya (–subcentric) *recurva* Cornu 1872
Achlya (–subcentric) *apiculata* de Bary 1888
Achlya (–subcentric) *oblongata* de Bary 1888
Achlya (–subcentric) *oligacantha* de Bary 1888
Achlya (–subcentric) *stellata* de Bary 1888
Achlya (–subcentric) *megasperma* Humphrey 1893
Achlya (–subcentric) *papillosa* Humphrey 1893
Achlya (–subcentric) *spiracaulis* Johnson 1949
* *Achlya* (–subcentric) *braziliensis* Milanez 1965
* *Achlya* (–subcentric) *curvicollis* Beroqui 1969
2d *Achlya* (–aplanes) *spinosa* de Bary 1881
† *Achlya* (–aplanes)! *Aplanes androgynus* (Archer) Humphrey 1893
Achlya (–aplanes) *treleaseana* (Humphrey) Kauffman 1906
2e *Achlya* (–proto) *hypogyna* Coker & Pemberton 1908
† *Achlya* (–proto)! *Protoachlya paradoxa* Coker 1914
†* *Achlya* (–proto)! *Protoachlya polysporus* (Lindstedt) Apinis 1930
†* *Achlya* (–proto)! *Protoachlya hypogyna* Shanor & Conover 1942
Achlya (–proto) *mucronulata* Ziegler 1956
Achlya (–proto) *benekei* Furtado 1965
* *Achlya* (–proto) *bonariensis* Beroqui 1969
2f *Achlya* (–glomerulate) *glomerata* Coker 1912
Achlya (–glomerulate) *abortiva* Coker & Braxton 1926
Achlya (–glomerulate) *lobata* Ziegler & Gilpin 1954
2g † *Achlya* (–tuberculate)! *Thraustotheca primoachlya* Coker & Couch 1924
†*Achlya (–tuberculate)! *Saprolegnia lativica* Apinis 1930
Achlya (–tuberculate) *recurva sensu* Latham 1935
Achlya (–tuberculate) *crenulata* Ziegler 1948
3 *Aphanomyces stellatus* de Bary 1860
Aphanomyces scaber de Bary 1860

Aphanomyces laevis de Bary 1860
Aphanomyces phycophilus de Bary 1860
Aphanomyces norvegicus Wille 1899
Aphanomyces astaci Schikora 1906
Aphanomyces helicoides Minden 1915
Aphanomyces parasiticus Coker 1923
Aphanomyces ovidestruens Gicklhorn 1923
Aphanomyces euteiches Drechsler 1925
Aphanomyces exoparasiticus Coker & Couch 1926
Aphanomyces raphani Kendrick 1927
Aphanomyces cochlioides Drechsler 1929
Aphanomyces cladogamus Drechsler 1929
Aphanomyces camptostylus Drechsler 1929
* *Aphanomyces hydatinae* Valkanov 1931
* *Aphanomyces acinetophagus* Bartsh & Wolf 1938
* *Aphanomyces americanus* (Bartsch & Wolf) Scott 1938
Aphanomyces amphigynus Cutter 1941
Aphanomyces sparrowii Cutter 1941
Aphanomyces apophysii Lacy 1949
Aphanomyces daphniae Prowse 1954
Aphanomyces patersonii Scott 1956
Aphanomyces bosminae Scott 1961
Aphanomyces irregulare Scott 1961
4 *Dictyuchus monosporus* Leitgeb 1869
* *Dictyuchus magnusii* Lindstedt 1872
Dictyuchus achlyoides Coker & Alexander 1927
Dictyuchus missouriensis Couch 1931
Dictyuchus pseudodictyon Coker and Braxton 1931
Dictyuchus pseudoachlyoides Beneke 1948
5 *Leptolegnia caudata* de Bary 1888
* *Leptolegnia bandoniensis* Swan
Leptolegnia subterranea Coker & Harvey 1925
Leptolegnia eccentrica Coker & Matthews 1927
Leptolegnia baltica Höhnk & Vallin 1953
Leptolegnia pontica Artemczuk 1968
6 *Pythiopsis cymosa* de Bary 1888
†* *Pythiopsis papillata* Ookubo & Kobayasi 1955
7 † *Thraustotheca clavata* (de Bary) Humphrey 1893
†* *Thraustotheca irregularis* Coker & Ward 1939
8 *Sommerstorffia spinosa* Arnaudow 1923
9 *Geolegnia inflata* Coker & Harvey 1925
Geolegnia septisporangia Coker & Harvey 1925

* *Geolegnia intermedia* Höhnk 1952
10   *Brevilegnia subclavata* Couch 1927
   Brevilegnia unisperma (Coker & Braxton) Coker 1927
   Brevilegnia linearis Coker & Braxton 1927
   Brevilegnia bispora Couch 1927
   Brevilegnia diclina Harvey 1927
   Brevilegnia megasperma Harvey 1930
   Brevilegnia variabilis Indoh 1941
   Brevilegnia longicaulis Johnson 1950
   Brevilegnia parvispora Höhnk 1952
   Brevilegnia minutandra Höhnk 1952
   Brevilegnia crassa Rossi Valderrama 1956
   Brevilegnia irregularis Rossi Valderrama 1956
   Brevilegnia globosa Ziegler 1958
11   *Calyptralegnia achlyoides* (Coker & Couch) Coker 1927
   Calyptralegnia ripariensis Höhnk 1953
12   *Plectospira myriandra* Drechsler 1927
   Plectospira gemmifera Drechsler 1929
13   *Aplanopsis terrestris* Höhnk 1952
   Aplanopsis spinosa Dick 1960
14   *Scoliolegnia asterophora* (de Bary) Dick 1969
   Scoliolegnia subeccentrica Dick 1969
   Scoliolegnia blelhamensis Dick 1969
   Scoliolegnia depauperata Dick 1971 (*noman rudum*)

VII. IMPORTANT LITERATURE

Cejp, 1959 (monograph); Coker, 1923 (monograph); Coker and Matthews 1937 (monograph); Dick, 1969a (morphology); Dick, 1971a (Leptolegniellaceae); Johnson, 1956 (*Achlya*); Scott, 1961 (*Aphanomyces*); Seymour, 1970 (*Saprolegnia*); Sparrow, 1960 (monograph, particularly Saprolegniales other than Saprolegniaceae).

REFERENCES

Atkins, D. (1954). A marine fungus *Plectospira dubia* n. sp. (Saprolegniaceae), infecting crustacean eggs and small crustacea. *J. Mar. Biol. Ass. U. K.* **33**:721–732.
Barksdale, A. W. (1960). Interthallic sexual reactions in *Achlya*, a genus of the aquatic fungi. *Amer. J. Bot.* **47**:14–23.
Barksdale, A. W. (1966). Segregation of sex in the progeny of a selfed heterozygote of *Achlya bisexualis. Mycologia* **58**:802–804.
Barksdale, A. W. (1968). Meiosis in the antheridium of *Achlya ambisexualis* E. 87. *J. Elisha Mitchell Sci. Soc.* **84**:187–194.

Bessey, E. A. (1950). "Morphology and Taxonomy of Fungi," McGraw-Hill (Blakiston), New York.

Booth, T., and C. E. Miller. (1968). Comparative morphologic and taxonomic studies in the genus *Thraustochytrium*. *Mycologia* **60**:480–495.

Bryant, T. R., and K. L. Howard. (1969). Meiosis in the Oomycetes. 1. A microspectrophotometric analysis of nuclear deoxyribonucleic acid in *Saprolegnia terrestris*. *Amer. J. Bot.* **56**:1075–1083.

Cantino, E. C. (1966). Morphogenesis in aquatic fungi. *In* "The Fungi" (G. C. Ainsworth and A. S. Sussman, eds.), Vol. 2, pp. 283–337. Academic Press, New York.

Cejp, K. (1959). "Oomycetes I. (Flora ČSR)". Cesko. Akad., Prague.

Coker, W. C. (1914). An Achlya of hybrid (?) origin. *Science* **40**:386.

Coker, W. C. (1923). "The Saprolegniaceae with Notes on Other Water Molds." Univ. of North Carolina Press, Chapel Hill.

Coker, W. C., and V. D. Matthews (1937). Saprolegniales. *In N. Amer. Flora* 2, Part **1**:15–76.

Darley, W. M., and M. S. Fuller. (1970). Cell wall chemistry and taxonomic position of *Schizochytrium*. *Amer. J. Bot.* **57**:761.

de Bary, A. (1852). Beitrag zur Kenntniss der *Achlya prolifera*. *Bot. Ztg.* **10**:473.

de Bary, A. (1881). Untersuchuingen über Peronosporeen und Saprolegnieen. *In* "De Bary and Woronin's Beiträge zur Morphologie und Physiologie der Pilze," 4th Ser., pp. 1–145. Frankfurt a. M.

de Bary, A. (1888). Species der Saprolegnieen. *Bot. Ztg.* **46**:599–610, 613–621, 629–636, and 645–653.

Dick, M. W. (1961). *Brevilegniella keratinophila* gen. nov. sp. nov. *Pap. Mich. Acad. Sci., Arts Let.* **46**:195–204.

Dick, M. W. (1969a). Morphology and taxonomy of the Oomycetes, with special reference to Saprolegniaceae, Leptomitaceae and Pythiacea. I. Sexual reproduction. *New Phytol.* **68**:751–775.

Dick, M. W. (1969b). The *Scoliolegnia asterophora* aggregate, formerly *Saprolegnia asterophora* DeBary (Oomyotes) *J. Linn. Soc. London Bot.* **62**:255–266.

Dick, M. W. (1971a). Leptolegniellaceae, fam. nov. *Trans. Brit. Mycol. Soc.* **57**:417–625.

Dick, M. W. (1971b). Oospore structure in *Aphanomyces*. *Mycologia* **63**:686–688.

Dick, M. W. (1972). Morphology and taxonomy of the Oomycetes, with special reference to Saprolegniaceae, Leptomitaceae and Pythiaceae. II. Cytogenetic systems. *New Phytol.* **71**:1151–1159.

Dick, M. W. and Win-Tin (1973). The development of cytological theory in the Oomycetes. *Biol. Rev.* **48**:133–158.

Emerson, R. (1950). Current trends of experimental research on the aquatic phycomycetes. *Annu. Rev. Microbiol.* **44**:169–200.

Fischer, A. (1892). Die Pilze Deutschlands, Oesterreichs und der Schweig. IV. Phycomycetes. *Kryptogamenflora* **1**:310–367.

Fuller, M. S., B. E. Fowles, and D. J. McLaughlin. (1964). Isolation and pure culture study of marine phycomycetes. *Mycologia* **56**:745–756.

Gaertner, A. (1972). Characters used in the classification of thraustochytriaceous fungi. *Veröff. Inst. Meeresforsch. Bremerhaven Sonderb.* **13**:138–194

Gay, J. L., and A. D. Greenwood. (1966). Structural aspects of zoospore production in *Saprolegnia ferax* with particular reference to the cell and vacuolar membranes. *In* "The Fungus Spore" (M. F., Madelin, ed.), Colston Pap. No. 18, pp. 95–110. Butterworth, London.

Gleason, F. H., Stuart, T. D., Price, J. S. and Nelbach, E. T. (1970). Growth of certain aquatic oomycetes on amino acids. II. *Apodachlya, Aphanomyces,*, and *Pythium*. *Physiol. Plant.* **23**:769–774.

Goldstein, S. (1963a). Morphological variation and nutrition of a new monocentric marine fungus. *Arch. Mikrobiol.* **45**:101–110.

Goldstein, S. (1963b). Studies of a new species of *Thraustochytrium* that displays light stimulated growth. *Mycologia* **55**:799–811.

Howard, K. L. (1971). Oospore types in the Saprolegniaceae. *Mycologia* **63**:679–686.

Humphrey, J. E. (1893). The Saprolegniaceae of the United States, with notes on other species. *Trans. Amer. Phil. Soc.* [N. S.] **17**:63–148.

Huneycutt, M. B. (1948). Keratinophilic phycomycetes. I. A new genus of the Saprolegniaceae. *J. Elisha Mitchell Sci. Soc.* **64**:277–285.

Huneycutt, M. B. (1952). A new water mold on keratinized materials. *J. Elisha Mitchell Sci. Soc.* **68**:109–112.

Johnson, T. W. (1956). "The Genus *Achlya*: Morphology and Taxonomy." Univ. of Michigan Press, Ann Arbor.

Johnson, T. W. (1966). *Ectrogella* in marine species of Licmophora. *J. Elisha Mitchell Sci. Soc.* **82**:25–29.

Kanouse, B. B. (1927). A monographic study of special groups of the water molds. II. Leptomitaceae and Pythiomorphaceae. *Amer. J. Bot.* **14**:335–357.

Kazama, F. (1972). Ultrastructure of *Thraustochytrium* sp. zoospores. I. Kinetosome. *Arch. Mikrobiol.* **83**:179–188.

Kützing, F. T. (1843). "Phycologia Generalis." Lipsiae.

Lechmere, A. E. (1911). Further investigations of methods of reproduction in the Saprolegniaceae. *New Phytol.* **10**:167–203.

Machlis, L. (1966). Sex hormones in fungi. *In* "The Fungi" (G. C. Ainsworth and A. S. Sussman, eds.), Vol. 2, pp. 415–433. Academic Press, New York.

Meier, H. and Webster, J. (1954). An electron microscope study of cysts in the Saprolegniaceae. *J. Exp. Bot,* **5**:401–409.

Migula, W. (1903). Myxomycetes, Phycomycetes, Basidiomycetes... *In Flora Deut., Oesterr. Schweitz* **3**:1–510.

Nees von Esenbeck, C. G. D. (1824). Addendum to C. G. Carus, Beitrag zur Geschichte der unter Wasser an verwesenden Thierkörpern sich erzeugenden Schimmel- oder Algengattungen. *Nova Acta Leopold-Carol.* **11**:507–522.

Perkins, F. O. (1972). The ultrastructure of holdfasts, "rhizoids", and "slime tracks" in thraustochytriaceous fungi and *Labyrinthula* spp. *Arch. Mikrobiol.* **84**:95–118.

Pringsheim, N. (1858). Beiträge zur Morphologie und Systematik der Algen. II. Die Saprolegnieen. *Jahrb. Wiss. Bot.* **1**:284–306.

Pringsheim, N. (1882). Neue Beobachtungen über den Befruchtungsact der Gattungen Achlya und Saprolegnia. *Sitzungs-ber. Akad. Berlin* pp. 855–890.

Sansome, E. (1965). Meiosis in diploid and polyploid sex organs of Phytophthora and Achlya. *Cytologia* **30**:103–117.

Schroeter, J. (1893). Phycomycetes. *In* "Engler and Prantl, Natürlichen Pflanzenfamilien," Vol. I, pp. 63–141.

Scott, W. W. (1961). A monograph of the genus Aphanomyces. *Va., Agr. Exp. Sta., Tech. Bull.* **151**:1–95.

Scott, W. W., R. Seymour, and C. Warren. (1963). Some new and unusual fungi from Virginia. I. Lower phycomycetes. *V. J. Sci.* **14**:11–15.

Seymour, R. (1970). The genus *Saprolegnia. Nova Hedwigia* **19**:1–124.

Sparrow, F. K. (1936). Biological observations on the marine fungi of Woods Hole waters. *Biol. Bull.* **70**:236–263.

Sparrow, F. K. (1950). Some Cuban phycomycetes. J. Wash. Acad. Sci. **40**:50–55.

Sparrow, F. K. (1960). "Aquatic Phycomycetes," 2nd rev. ed. Univ. of Michigan Press, Ann Arbor.

Sparrow, F. K. (1968). Remarks on the Thraustochytriaceae. *Veroeff. Inst. Meeresforsch. Bremerhaven, Sonderb.* **3**:7–18.

Sparrow, F. K., and D. Gotelli. (1969). Is *Atkinsiella* holocarpic? *Mycologia* **61**:199–201.

Thomas, D. des S., and J. T. Mullins. (1969). Cellulose induction and wall extension in the water mold *Achlya ambisexualis. Physiol. Plant.* **22**:347–353.

Trow, A. H. (1905). Fertilization in the Saprolegniales. *Bot. Gaz.* **39**:300.

Von Minden, M. (1912). Phycomycetes. IV. Reihe: Saprolegniineae. *Kryptogamenflora Mark Brandenburg* **5**:479–608.

Webster, J., and C. Dennis. (1967). The mechanism of sporangial discharge in *Pythium middletonii. New Phytol.* **66**:307–313.

Ziegler, A. W. (1948). A comparative study of zygote germination in the Saprolegniaceae. *J. Elisha Mitchell Sci. Soc.* **64**:13–40.

Zopf, W. (1884). Zur Kenntnis der Phycomyceten. I. Zur Morphologie und Biologie der Ancylisteen und Chytridiaceen, zugleich ein Beitrag zur Phytopathologie. *Nova Acta Ksl. Leopold-Carol. Deut. Akad. Naturforsch.* **47**:141–236.

CHAPTER 8

Leptomitales

M. W. Dick

Department of Botany
University of Reading
Reading, England

I. INTRODUCTION

The Leptomitales is a small order of about 20 species distinguished from other Oomycetes by the thallus, which is constricted at more or less regular intervals. In other respects it shows similarities to the Saprolegniales and Peronosporales and may be presumed to share with these orders the basic oomycete characters—a zoospore possessing two flagella heterokont in morphology and probably also in length; a cell wall of which the principal structural component is a glucan polymer; mitochondria with microtubular cristae, and an oogamous sexual reproduction without discrete male gametes. Very few Leptomitales have been unequivocally shown to possess all these features. Thus *Leptomitus lacteus* has no known sexual reproduction and zoosporangia are unknown in *Apodachlyella completa*. Mitochondrial form is known only in *L. lacteus* and *Aqualinderella fermentans*. However, in the latter species there are biochemical and physiological grounds for expecting the mitochondria to be aberrant. Data on cell walls rely largely on the chlor-zinc iodide reaction. Recent reports (Sietsma *et al.*, 1969; Lin and Aronson, 1970) differ in their conclusions regarding wall structure in *Apodachlya*. The presence of chitin is reported by Lin and Aronson (1970) and in this *Apodachlya* differs from all other oomycetes investigated, but shows some resemblance to the Hyphochytridiomycetes. In view of the reports of a second vestigial blepharoplast in this class and the current opinion that the bacillariophyte sperm (with a single tinsel flagellum) indicates a derivation from a heterokont algal stock, it might be worth investigating the relationship between the Hyphochytridiomycetes and the Oomycetes.

II. GENERAL MORPHOLOGICAL ACCOUNT

The thallus of Leptomitales may be composed either of a basal cell from which elongate vegetative or reproductive segments are produced, or alternatively it may be a mycelium formed by new segments budding from preexisting segments. Emerson and Weston (1967) have suggested that there is a graded series connecting these two types of thallus construction, linked primarily by the thallus form of *Sapromyces* and *Leptomitus*. However, descriptions of the thallus of *Leptomitus* are conflicting and it may well be that different isolates of this species differ in their initial thallus form. In this connection, it should be recalled that *Leptomitus lacteus* is known only as an asexual taxon and thus any clonal variation could be perpetuated. There are reports that *Leptomitus* possesses a distinct basal cell and a mycelium with dichotomous branching, but there are also strains which have monopodial branching and no apparent basal cell. In *Sapromyces* the basal cell is identifiable, but little differentiated from subsequently formed segments. In this genus new segments develop terminally or subterminally from the last formed segment. There is no evidence for this type of segment polarity in *Leptomitus*. Neither *Apodachlya* nor *Apodachlyella* has a basal cell; any segment may initiate a new branch and the branch origin is not always subterminal. In the remaining genera the basal cell is well developed even though elongate vegetative segments may be absent. When present, these segments may be terminated by reproductive structures or they may show lateral branching without constrictions below the zoosporangia. The basal cell may become inflated and lobed. This expansion is accompanied by very thick secondary wall deposits and there are indications that the outer wall layers may be lost by exfoliation.

The cytoplasm of the vegetative thallus frequently contains conspicuous refractive granules referred to as cellulin granules. The chemistry of these granules has not been satisfactorily determined, nor has it been definitely established that they are equivalent to the material deposited in the constrictions of the thallus which separate the reproductive organs from the subtending vegetative segments. The granules are most conspicuous in *Apodachlya* and *Leptomitus* and are usually located near the constriction. They are not disturbed by protoplasmic streaming, but may respond to mechanical damage or possibly osmotic shock by becoming lodged in the constriction. This "stopcock" explanation is not entirely satisfactory since it presumes a plugging action due to turgor of the intact or damaged segments. If this were so, the hypothesis needs to account for the fact that in *Leptomitus* or *Apodachlya* only one cellulin granule is usually present in each segment, and that this granule is normally located in each segment close to the constriction proximal to the hyphal tip. In these genera the cellulin granule is already conspicuous in the penultimate, complete segment. In *Apodachlyella* there is a number of cellulin granules formed, usually in a group. Individually their diameter is not normally such as to completely occlude the constriction. It should also be noted that the nature of the constriction differs between *Leptomitus, Apodachlya,* and *Apodachlyella,* and those other genera with a well-defined basal cell. In the latter group the constriction is extended giving a pedicellate appearance to the segments and the thallus wall is thicker in the region of the constriction. In this they recall the constrictions found in certain Siphonaceous green algae. Constrictions are not found in any branches of small diameter. They are not found in rhizoids or at the point of initiation of rhizoids, nor are they found on antheridial branches or at the point of initiation of antheridial branches. Thus, even in *Leptomitus,* which may not have rhizoids developed from a basal cell, fine vegetative branches near a particulate substrate may be of even diameter throughout their length. Rhizoids developed from the basal cells of monocentric thalli are characteristically blunt-tipped. Johnson (1951) has described the rhizoids of *Mindeniella assymetrica* as sharp-pointed, but his figure does not indicate any marked deviation from the above generalization. These intramatrical rhizoids may thus be distinguished from the finely tapering rhizoids of Chytridiomycetes. Complete septation occurs only at the formation of reproductive structures and apparently always involves the deposition of cellulinlike material in the lumen of the constriction. The antheridial septum is not formed in a constriction.

Asexual reproduction is by means of zoospores produced in zoosporangia. On rare occasions fragmentation of the thallus may occur, as may rejuvenation from the basal cells of monocentric thalli. Chlamydospores are not known unless the spiny sporangia of *Mindeniella* and *Araiospora* (see below) are regarded as such. The formation of endogenous asexual spores in vegetative segments has been reported for *Apodachlyella.*

Zoosporangia are usually terminal segments of the principal filaments of the thallus. In *Apodachlya minima* they are found terminally on lateral branch systems and in *Mindeniella, Aqualinderella,* and rarely in *Araiospora* they are formed directly from the basal cell. In *Leptomitus* and *Apodachlya* a basipetal series of zoosporangia may develop. The zoosporangia are cylindrical, ovoid, or pyriform and usually distinct from vegetative segments or oogonia. A detailed discussion of zoospore morphology and terminology is given for the Saprolegniales (see p. 113) and will not be repeated here. Sparrow (1960) has reported that the zoospores of *Araiospora, Rhipidium,* and *Mindeniella* may be distinguished by their cell inclusions. In the Leptomitales a distinct difference in zoosporogenesis is found between *Leptomitus* and *Apodachlya* and the other genera. In these two genera the principal zoospore form is only developed after a preceding zoosporic phase which may merely serve to place the zoospore at the mouth of the zoosporangium before its encystment. Information on polyplanetism of the principal zoospore form is for the most part lacking, although it has been reported for *Mindeniella.*

In all monocentric genera the principal zoospore form develops within the zoosporangium and zoosporangium dehiscence is probably accompanied by the formation of an evanescent vesicle or at least a momentary restraint to the dissemination of the first-released zoospores. However, there are divergent views on this latter feature, and a state of affairs similar to that known in *Phytophthora* may prevail. Spiny zoosporangia are known only in *Mindeniella* and *Araiospora*. These zoosporangia have thicker walls and their production appears to be linked to environmental conditions, in particular to low oxygen concentration. Their dehiscence has been insufficiently studied but is apparently the same as for normal zoosporangia. Zoospore germination in *Leptomitus* and *Apodachlya* and endogenous spore germination in *Apodachlyella* is by the production of one to three buds which become the first segments of the mycelium. There is no evidence of polarity. In *Mindeniella*, *Sapromyces*, *Araiospora*, and *Aqualinderella* the zoospore germinates by producing, without constrictions, the rhizoidal primordium. This is accompanied by some swelling of the zoospore cyst succeeded by a bulge (more or less lateral in *Mindeniella* and *Aqualinderella*) from which the basal cell develops. This polar development may be presumed in *Rhipidium*.

The oogonium is smooth walled, though there may be some exfoliation of the outer wall layers in *Sapromyces* and, under natural conditions, *Aqualinderella* may be found to have an investing layer of very fine spines. In the absence of any critical studies of the protoplasmic contents of the spiny bodies of *Mindeniella spinospora*, I think it preferable to consider the structures described as oogonia to be resting zoosporangia. The genus is therefore treated here as having no known sexual reproduction.

Sexual reproduction is homothallic where known, except for one species of *Sapromyces* (Bishop, 1940). Antheridia are normally present. The oogonia of the Leptomitales are characteristically uniovulate except for *Apodachlyella completa* which has four to five oospheres. *Aqualinderella* is known to produce pluriovulate oogonia in culture. The antheridial branch frequently arises close to the oogonium and the antheridial cell penetrates the oogonial wall to produce the fertilization tube. The oosphere is plerotic in *Apodachlya*, aplerotic and formed without periplasm in *Apodachlyella* and aplerotic and periplasmic in the remaining genera. Periplasm is least persistent in *Sapromyces* and *Aqualinderella* and most persistent, forming a unique ornamentation to the developing oospore, in *Araiospora*. In periplasmic genera the exospore is more or less reticulate, probably from a condensation of periplasmic remains, but in *Araiospora* it is likely that the periplasm forms carbohydrate walls before the epispore is formed. In all genera the endospore is thick and stratified. In *Apodachlya punctata* this layer is reputed to be punctate but no photographic evidence has been published. The oospore structure is apparently the same as in the Peronosporales except that in *Apodachlya* and *Apodachlyella* two types of cell inclusion become manifest by aggregation and redistribution in the mature oospore. A good account of the older literature on nuclear cytology is given by Sparrow (1960). Recent work (Howard and Bryant, 1972; Win-Tin, 1972, unpublished) suggests that meiosis occurs in the gametangia of *Apodachlyella*, but conclusive cytological evidence (i.e., the recognition of bivalents) has yet to be provided for any member of the order.

Species in which the basal cell is conspicuous tend to have very large oospores.

Oospore germination is known only for *Apodachlya* and in this genus the information is incomplete. The description of the changes in oospore structure accompanying germination is not readily reconciled with those for *Phytophthora*. In view of the apparent similarities in the structure of the dormant oospores similar germination changes might have been expected.

III.   DEVELOPMENT OF TAXONOMIC THEORY

Leptomitus lacteus was one of the earliest known water molds that was figured by Dillwyn (1809) and which he referred to the fungus allegedly described by Roth in 1789. The genus *Leptomitus* was established by Agardh (1824) for all water molds, which up to that time had been classed with filamentous algae under the generic name *Conferva*. The questionable status of the water molds as aberrant algae remained in the literature for several decades, notably upheld by Pringsheim. He regarded differences in thallus construction as of little importance, as did others, up to the time of Hartog (1887).

Kützing (1843) was the first to use the family name Leptomitaceae; he narrowed the diagnosis of *Leptomitus* by excluding the genera *Achlya* and *Saprolegnia* established by Nees von Esenbeck (1824). At this time no sex organs in any water mold were recognized. It was not until 1852, when de Bary reiterated the generic distinction between *Achlya* and *Saprolegnia* and utilized sexual characters at the specific level, that the confusion between the generic names in use in the aquatic phycomycetes was resolved. No other recognizable member of the Leptomitales was described until 1867 when Hildebrand described *Leptomitus* (*Apodachlya*) *brachynema* and 1872 when Cornu published his descriptions of *Rhipidium* and introduced a new generic name, *Apodya*, for *L. lacteus*. In 1872 Lindstedt established a subfamily of the Saprolegniaceae, the Leptomiteae, on the basis of the constricted thallus for *Leptomitus* (including *Apodachlya*) and *Rhipidium* (including *Sapromyces* and *Araiospora* as now recognized). Thus, the group has always been distinguished on the basis of thallus morphology. It is also noteworthy that Lindstedt used the site of zoospore formation to separate these two genera. Fischer (1892) established the Apodyeae based on Cornu's genus and Humphrey (1893) followed Lindstedt in placing the Leptomitales in a subfamily of the Saprolegniaceae. At about the same time Schroeter (1893) established the Leptomitaceae as a family and emended the diagnosis to include the character of the single oospore. It should be remembered that Schroeter placed the Pythiaceae in his group Saprolegniineae, and so by implication regarded the family as intermediate between the Saprolegniaceae and Pythiaceae. Thaxter (1896) was the first explicitly to raise the question of relationship with the Saprolegniaceae or Pythiaceae. His student, King (1903), on the basis of a study of *Araiospora*, dismissed the resemblance (based on habitat and zoosporangium formation) to the cohort Saprolegniineae in favor of the resemblance to the Peronosporineae in sexual structures, principally using the characters of the uniovulate oogonium and the periplasmic oosporogenesis. De Bary had also earlier suggested this affinity between the Leptomiteae and *Pythium*. The similarity of zoosporogenesis in *Leptomitus* and *Apodachlya* to the Saprolegniaceae on the one hand, and the similarity of oosporogenesis in *Araiospora* and *Rhipidium* with the Peronosporales on the other hand, together with the unique thallus construction, lead Kanouse (1927) to raise the group to ordinal level as the Leptomitales.

Unfortunately the early emphasis on asexual reproduction in the saprophytic oomycetes and the subsequent debate on the significance of the zoospore in the phylogeny of all flagellate fungi have overshadowed the relationship of sexual reproduction between the primary parasitic Peronosporales and the saprophytic Leptomitales. Sparrow (1960) has stressed the occurrence of the auxiliary form zoospore in some taxa of the group and the aquatic saprophytic habit. Thus it has become customary to bracket together the Saprolegniales and Leptomitales. Only recently has Dick (1969) demonstrated the basic similarity of oospore structure between the Leptomitales and Peronosporales and emphasized its dissimilarity to the Saprolegniaceae (sensu stricto).

Indoh (1939) suggested that the type of thallus and method of oosporogenesis were sufficiently different between the genera *Leptomitus*, *Apodachlya*, and *Apodachlyella* and the genera *Sapromyces*, *Araiospora*, and *Rhipidium* to justify their separation. Sparrow (1943, 1960) has added to these characters that of the zoospore form at zoosporogenesis and erected (but failed to validate) the family Rhipidiaceae. A Latin diagnosis for the family, which may be accepted as a validation, was given by Cejp (1959).

Cejp erected a third family, the Mindeniellaceae, on the absence of vegetative filamentous segments (the reproductive structures developing directly from the basal cell) and the lack of periplasm in the oospore. In my opinion it is probable that the second character is based on a misconception of the description of so-called oogonia in *Mindeniella*. The first character is probably also untenable since *Araiospora* may exceptionally develop sporangia from the basal cell. Furthermore, *Aqualinderella* has no filamentous vegetative segments but has periplasmic oosporogenesis and would thus be impossible to place were this family accepted.

Finally, Indoh (1939) has suggested that by including details of sexual reproduction in the diagnosis of the Leptomitaceae, Schroeter was not accepting *L. lacteus* as the type of the family. The formal taxonomic status of the families of the Leptomitales has thus been short of ideal.

A closer understanding of the interrelations of the Leptomitales with other aquatic fungi has come from a number of physiological investigations (Cantino, 1966; Gleason, 1968; Gleason *et al.*, 1970; Golueke, 1957). On the whole these have confirmed the conclusions based on critical morphological study, contrary to Burnett's (1968) recent generalization.

Thus it has been shown that the cytochrome system and lysine synthesis of the Leptomitales and Saprolegniales are similar, but differ from those of the Chytridiomycetes. Both orders are apparently basically autotrophic for essential vitamins, although *Sapromyces* has been shown to require thiamine. The Leptomitales differ from the Saprolegniales in that they have retained the ability to utilize the sulfate ion while they have lost the ability to utilize nitrogen supplied as the ammonium or nitrate ion (except *Mindeniella* which can utilize ammonium ion). On this basis, Cantino (1966) has postulated an earlier evolutionary divergence of Saprolegniales from Leptomitales and Peronosporales, the latter two orders differing in their autotrophy for vitamins.

Recent studies (Gleason and Unestam, 1968a, b; Held, 1970) have shown that within the Leptomitales a trend to fermentative metabolism can be traced. Thus, under reduced oxygen tension *Leptomitus* and *Apodachlya* are unable to produce acid while *Sapromyces* and *Mindeniella* will do so. *Rhipidium* is strongly fermentative and *Aqualinderella* is obligately so, having lost its cytochrome system. These four acid-producing genera may also be united by their possession of a rarely encountered NAD-dependent D-lactate dehydrogenase.

In carbon metabolism *Leptomitus* and *Apodachlya* may be united by their ability to utilize both amino acids and carboxylic acids, and share with *Araiospora* an inability or restricted ability to utilize sugars. It is suggested that the ancestral type possessed the abilities to metabolize amino acids and carboxylic acids and to ferment sugars. The former has been lost by the *Rhipidium*-group genera, the latter, at least partially, by the *Apodachlya*-group genera.

Lund (1934), on ecological grounds, has suggested that *Apodachlya* and *Sapromyces* require a well-oxygenated environment, while *Rhipidium* thrives on only a small supply of oxygen. Emerson's studies of *Aqua-linderella* would support this ecological dichotomy. It thus appears that evolution in the Leptomitales is proceeding in two diametrically opposed directions. It is clear that our knowledge of the order will be enhanced by extensive ecological studies as these impinge upon both the physiological and the morphological characters of the species.

IV. CONCLUSION

In view of the imperfect nature of the genus *Leptomitus* and the uncertainty regarding its zoospore germination, it would be unwise at the present time to accept the arguments of Emerson and Weston (1967) for retaining only one family in the Leptomitales based on the range of diversity of thallus form. Sparrow (1960) has pointed out that a resemblance in body plan of various fungi has given rise to faulty ideas of relationships, and even though one may divide the Leptomitaceae and Rhipidiaceae, on this character, I do not regard it as preeminent. Differences in asexual and sexual reproduction are sufficient.

The difference in the form of the zoospore at zoosporogenesis sets apart the genera *Leptomitus* and *Apodachlya*. Lindstedt's original separation based on zoospore discharge, though difficult to apply and of doubtful use under certain environmental conditions, is probably valid and may have an ultrastructural basis even when a manifest vesicle is not apparent. Furthermore, there is no evidence of polarity at germination in the zoospore of *Apodachlya*, as is found in *Sapromyces*, *Araiospora*, *Mindeniella*, and *Aqualinderella*.

In oosporogenesis *Apodachlya* and *Apodachlyella* have no periplasm. A gradation from no periplasm, through evenescent periplasm to persistent periplasm can be traced in the Leptomitales just as it can in the Perono-sporales, starting with plerotic species of Pythium. This character reenforces the diverging evolutionary lines represented by *Apodachlya* and *Rhipidium*. Oospore structure in *Apodachlya* is clearly derived and distinct from the other genera. *Apodachlyella* is exceptional in having pluriovulate oogonia. Oospore diameters in the Rhipidiaceae are on the whole larger than those of the Leptomitaceae (Dick, 1969).

If the Leptomitales contained a plexus of species with intermediate characters one could support Emerson and Weston's contention. But with the only partial exception of the monotypic imperfect *Leptomitus*, the species fall into two discrete groups, with speciation apparently centered on *Apodachlya* and *Rhipidium*. Therefore, two families are recognized in the Leptomitales, the Leptomitaceae probably showing convergent evolution to the Saprolegniaceae, and the Rhipidiaceae showing extreme morphological and physiological adaptation to a highly specialized ecological niche.

V. DIAGNOSTIC CHARACTERS IN THE LEPTOMITALES

The families are separated by the presence in Leptomitaceae of an auxiliary zoospore form at zoosporogenesis and the absence of a vesicle at zoosporangial discharge; the apparently more complex oospore structure in the Leptomitaceae in that two types of reserve globule show aggregation patterns; the presence of periplasm in Rhipidiaceae; the basically monocentric thallus in the Rhipidiaceae, and the myceliar thallus in the Leptomitaceae. Segments tend to be pedicillate and the constrictions thickened in the Rhipidiaceae.

Generic distinctions in the Leptomitaceae are based on the dimensions of the hyphal segments and the degree of differentiation of zoosporangial segments from vegetative segments. There are several characters of diagnostic value associated with sexual reproduction of which the most convenient are oospore number and the plerotic or aplerotic oospore. The degree of differentiation of antheridial branches may also be used in diagnosis.

Generic distinctions in the Rhipidiaceae are based on the degree of differentiation of vegetative segments from the basal cell and the development of lobes from the basal cell initial. These characters are supported by oospore dimensions, the presence of spiny resting sporangia, and zoosporangial dimensions.

Leptomitus, *Apodachlyella*, and *Aqualinderella* are monotypic. In the remaining genera the species may be distinguished by the dimensions of reproductive structures, of which the most valuable is the arithmetic mean of the oospore diameter. The relative origins of the antheridia and oogonia are also specific characters. In *Araiospora* the morphology of the resting sporangium is used.

VI. KEY

Order Leptomitales

This order consists of aquatic, saprobic fungi with a constricted, segmented thallus. The zoosporangia are cylindrical to ovoid or pyriform and the zoospores are delimited within the zoosporangia; zoosporangial discharge is various. The zoospores are biflagellate. The oogonia are usually uniovulate. Antheridia are usually present. The oospores are large with a central reserve globule, as in the Peronosporales, and their wall is thick.

KEY TO FAMILIES AND GENERA OF THE LEPTOMITALES

1. Thallus composed of extensive hyphae with nonpedicellate segments and nonthickened constrictions. Principal form zoospores not developed in the zoosporangium. Oosporogenesis not periplasmic . . . . **Leptomitaceae**   2

1′ Thallus usually with a conspicuous basal cell, development bipolar, vegetative segments pedicellate, constrictions with thicker walls. Principal form zoospores developed within the zoosporangium. Oosporogenesis periplasmic . . . . **Rhipidiaceae**   4

   2(1) Vegetative segments indistinct and irregular in length. Zoosporangia unknown. Oogonia present, pluriovulate. Oospores aplerotic . . . . **Apodachlyella**

   2′(1) Vegetative segments well-defined and regular in length. Zoosporangia usually present. Oogonia present or absent, uniovulate. Oospores plerotic   3

3(2′) Vegetative segments long and stout (up to 400 μm $\times$ 45 μm). Zoosporangia little different from vegetative segments, formed in basipetal series, zoospores not encysting at mouth of zoosporangium. Oogonia unknown . . . . **Leptomitus**

3′(2′) Vegetative segments slender (up to 500 μm $\times$ 20 μm). Zoosporangia more or less ovoid and distinct from vegetative segments, rarely in basipetal series, zoospore encystment variable. Oogonia present . . . . **Apodachlya**

   4(1′) Basal cell little differentiated from extensive vegetative segments. Oospores small ($<$30 μm) or absent . . . . **Sapromyces**

   4′(1′) Basal cell well developed, vegetative segments present or absent. Oospores large ($>$35 μm) or absent   5

5(4′) Vegetative segments absent. Spiny zoosporangia present or absent. Sexual reproduction parthenogenetic or absent   6

5′(4′) Vegetative segments normally present. Spiny zoosporangia present or absent. Sexual reproduction usually with oogonia and antheridia   7

   6(5) Basal cell lobed, reproductive structures sessile. Spiny zoosporangia absent (probably tropical and warm temperate) . . . . **Aqualinderella**

   6′(5) Basal cell clavate, reproductive structures pedicellate. Spiny zoosporangia present . . . . **Mindeniella**

7(5′) Basal cell frequently lobed. Zoosporangia frequently sympodial, spiny zoosporangia absent. Oospores with a reticulate exospore . . . . **Rhipidium**

7′(5′) Basal cell stout, truncate but not lobed. Zoosporangia commonly whorled, spiny zoosporangia present. Oospores with an elaborate periplasmic envelope, exospore absent . . . . **Araiospora**

VII.   IMPORTANT LITERATURE

Cejp, 1959 (monograph); Coker and Matthews, 1937 (monograph); Dick, 1964 (*Apodachlyella*); Dick, 1969 (morphology); Emerson and Weston, 1967 (*Aqualinderella*); Gleason, 1968 (physiology); Kanouse, 1927 (monograph); Sparrow, 1960 (monograph); Thaxter, 1896 (Rhipidiaceae); Tiesenhausen, 1912 (*Apodachlya*).

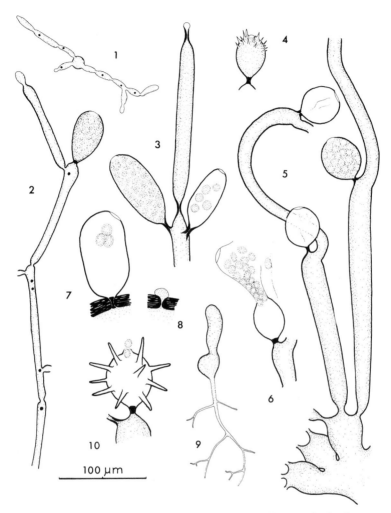

PLATE I. FIG. 1. Germinating zoospore of *Apodachlya*, showing nonpolar development of the mycelium by repeated budding, and the early formation of cellulin granules. Fig. 2 Hypha of *Apodachlya* bearing a terminal zoosporangium. Note the nonpolar branching from one of the segments. Fig. 3. *Sapromyces* with pedicellate zoosporangia and vegetative segment. Contrast tip growth with *Apodachlya*. Fig. 4. *Mindeniella*, spiny zoosporangium. Fig. 5. Small portion of a basal cell and vegetative filaments of *Rhipidium*. Note sympodial renewal without constriction of the vegetative filaments. Fig. 6. *Rhipidium* zoosporangium with evanescent vesicle and operculum. Fig. 7. *Aqualinderella*, sessile zoosporangium. Fig. 8. *Aqualinderella*, zoosporangial initial developing through the thick wall of the basal cell. Fig. 9. *Aqualinderella*, polar germination of the zoospore. Note that the zoospore cyst had swollen appreciably before the rhizoidal and basal cell systems had started development. Fig. 10. *Araiospora*, spiny zoosporangium. [Fig. 3 after Indoh (1953) and Emerson (1958); 4 after Johnson (1951); 6, 10 after Thaxter (1896).]

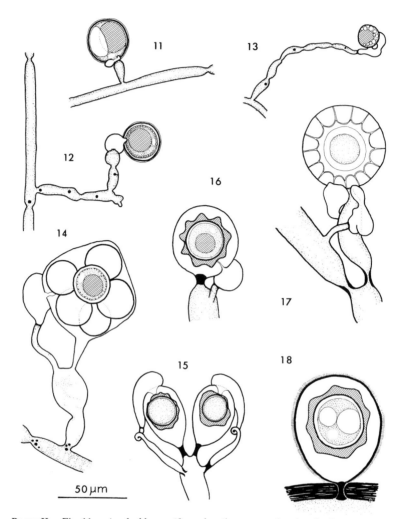

PLATE II.   Fig. 11.   *Apodachlya pyrifera*, plerotic oospore showing the 2 reserve aggregates, the nuclear spot, and the laterally displaced hypogynous antheridium. Fig. 12.   *Apodachlya brachynema*, plerotic oospore showing the different distribution of one of the reserve aggregates Fig. 13.   *Apodachlya minima*, oogonium with immature oospore showing the central reserve globule partially surrounded by globules aggregating to give a second reserve globule in a configuration similar to Fig. 11. Note the semihypogynous antheridium. Fig. 14.   *Apodachlyella*, oogonium and antheridium. Note the absence of the oogonial septum, and the absence of the constriction at the base of the antheridial branch. Fig. 15.   *Sapromyces*, oogonia and antheridia. Details of oospore structure not known. Epispore (hatched) irregular in outline. Fig. 16. *Rhipidium*, oogonium and antheridium. Note the absence of constrictions to the antheridial branch. Oospore structure is presumed. Epispore reticulate. Fig. 17.   *Araiospora*, oogonium and antheridium. Note the walled layer developed from the periplasm. The wall material between this layer and the endospore is presumed to be exospore. Fig. 18.   *Aqualinderella*, sessile parthenogenetic oogonium with reticulate epispore. Oospore contents incompletely described. [Fig. 15 after Thaxter (1896) and Cejp (1959); 16 after Thaxter (1896) and Indoh (1953); 17 after Shanor and Olive (1942), and King (1903); 18 after Emerson and Weston (1967).]

REFERENCES

Agardh, C. A. (1824) "Systema Algarum." Vol. 1. Lund.
Bishop, H. (1940). A study of sexuality in *Sapromyces reinschii. Mycologia* **32**: 505–529.
Burnett, J. H. (1968). "Fundamentals of Mycology." Arnold, London.
Cantino, E. C. (1966). Morphogenesis in aquatic fungi. *In* "The Fungi" (G. C. Ainsworth and A. S. Sussman, eds.), Vol. 2, pp. 283–337. Academic Press, New York.
Cejp, K. (1959). "Oomycetes I. (Flora ČSR)." Cesko, Acad., Prague.
Coker, W. C., and V. D. Matthews. (1937). Saprolegniales, *N. Amer. Flora* **2**, Part 1:15–76.
Cornu, M. (1872). Monographie des Saprolégniées; étude physiologique et Systematique, *Ann. Sci. Natur.:Bot.* [5] **15**:1–198.
de Bary A. (1852). Beitrag zur Kenntniss der *Achlya prolifera. Bot. Ztg.* **10**:473–479, 489–496 and 505–511.
Dick, M. W. (1964). *Apodachlyella completa* (Humphrey) Indoh. *J. Linn. Soc. London, Bot.* **59**:57–62.
Dick, M. W. (1969). Morphology and taxonomy of the Oomycetes, with special reference to Saprolegniaceae, Leptomitaceae and Pythiaceae I. Sexual reproduction. *New Phytol.* **68**:751–775.
Dillwyn, L. W. (1809). "British Confervae." London.
Emerson, R. (1958). Mycological organisation. *Mycologia* **50**:589–621.
Emerson, R., and W. H. Weston. (1967). *Aqualinderella fermentans* gen. et sp. nov., a phycomycete adapted to stagnant waters. I. Morphology and occurrence in nature. *Amer. J. Bot.* **54**:702–719.
Fischer, A. (1892). Die Pilze Deutschlands, Oesterreichs und der Schweiz. IV. Phycomycetes. *Kryptogamenflora* **1**:310–367.
Gleason, F. H. (1968). Nutritional comparisons in the Leptomitales. *Amer. J. Bot.* **55**:1003–1010.
Gleason, F. H., and T. Unestam. (1968a). Comparative physiology of respiration in aquatic fungi. I. The Leptomitales. *Physiol. Plant* **21**:556–572.
Gleason, F. H., and T. Unestam. (1968b). Cytochromes of aquatic fungi. *J. Bacteriol.* **95**: 1599–1603.
Gleason, F. H., T. D. Stuart, J. S. Price, and E. T. Nelbach. (1970). Growth of certain aquatic Oomycetes on amino acids. *Physiol. Plant.* **23**:769–774.
Golueke, C. G. (1957). Comparative studies on the physiology of *Sapromyces* and related genera. *J. Bacteriol.* **74**:337–343.
Hartog, M. M. (1887). On the formation and liberation of the zoospores of the Saprolegnieae. *Quart. J. Microsc. Sci.* **27**:427–438.
Held, A. A. (1970). Nutrition and fermentative energy metabolism of the water mold *Aqualinderella fermentans. Mycologia* **57**: 339–358.
Hildebrand, F. (1867). Mykologische Beiträge. 1. Ueber einige neue Saprolegnieen. *Jahrb. Wiss. Bot.* **6**:249–269.
Howard, K. L., and T. R. Bryant. (1972). Meiosis in the Oomycetes. II. A microspectrophotometric analysis of DNA in *Apodachlya brachynema. Mycologia* **63**:58–68.
Humphrey, J. E. (1893). The Saprolegniaceae of the United States, with notes on other species. *Trans. Amer. Phil. Soc.* [N. S.] **17**:63–148.
Indoh, H. (1939). Studies on the Japanese aquatic fungi. I. On *Apodachlyella completa* sp. nov. with revision of the Leptomitaceae. *Sci. Rep. Tokyo Bunrika Daigaku, Sect. B* **4**:43–50.
Indoh, H. (1953). Observations on Japanese aquatic fungi. I. *Apodachlya, Sapromyces* and *Rhipidium. Nagaoa* **3**:25–35.
Johnson, T. W. (1951). A new *Mindeniella* from submerged, rosaceous fruits. *Amer. J. Bot.* **38**:74–78.
Kanouse, B. B. (1927). A monographic study of special groups of the water molds I. Blastocladiaceae. II. Leptomitaceae and Pythiomorphaceae. *Amer. J. Bot.* **14**:287–306 and 335–357.

King, C. A. (1903). Observations on the cytology of *Araiospora pulchra* Thaxter. *Proc. Boston Soc. Natur. Hist.* **31**:211–245.

Kützing, F. T. (1843). "Phycologia Generalis." Lipsiae.

Lin, C. C., and J. M. Aronson. (1970). Chitin and cellulose in the cell walls of the oomycete, *Apodachlya* sp. *Arch. Mikrobiol.* **72**:111–114.

Lindstedt. K. (1872). "Synopsis der Saprolegniaceen und Beobachtungen über einige Arten." Berlin.

Lund, A. (1934). Studies on Danish freshwater Phycomycetes and notes on their occurrence particularly relative to the hydrogen ion concentration of the water. *Kgl. Dans. Vidensk., Selsk. Skr., Naturv. Math. Afd.*, **9** *Raekke* **6**:1–97.

Nees von Esenbeck, C. G. D. (1824). Addendum to: C. G. Carus, Beiträge zur Geschichte der unter Wasser an verwesenden Thierkörpern sich erzengenden Schimmel- oder Algengattungen. *Nova Acta Leop.-Carol.* **11**:507–522.

Schroeter, J. (1893). Phycomycetes. *In* "Engler and Prantl; Natürichen Pflanzenfamilien," Vol. I, pp. 63–141.

Shanor, L., and L. S. Olive. (1942). Notes on *Araiospora streptandra*. *Mycologia* **34**:536–542.

Sietsma, J. H., D. E. Eveleigh, and R. H. Haskins. (1969). Cell wall composition and protoplast formation of some oomycete species. *Biochim. Biophys. Acta.* **184**:306–317.

Sparrow, F. K. (1943). "The Aquatic Phycomycetes, exclusive of the Saprolegniaceae and Pythium." Univ. of Michigan Press, Ann Arbor.

Sparrow, F. K. (1960). "Aquatic Phycomycetes," 2nd rev. ed. Univ. of Michigan Press, Ann Arbor.

Thaxter, R. (1896). New or peculiar aquatic fungi. 4. *Rhipidium, Sapromyces,* and *Araiospora,* nov. gen. *Bot. Gaz.* **21**:317–330.

Tiesenhausen, M. (1912). Beiträge zur Kenntnis der Wasserpilze der Schweiz. *Arch. Hydrobiol. Planktonk.* **7**:261–308.

List of Taxa in the Leptomitales

Leptomitus lacteus (Roth) Agardh 1824

Apodachlya brachynema (Hildebrand) Pringsheim 1883

Apodachlya pyrifera Zopf 1888

Apodachlya punctata Minden 1912

Apodachlya minima Coker 1938

Apodachlya seriata Lund 1934

Apodachlyella completa (Humphrey) Indoh 1893

Rhipidium interruptum Cornu 1872

Rhipidium thaxteri Minden 1912

Rhipidium americanum Thaxter 1896

Rhipidium parthenosporum Kanouse 1927

Sapromyces elongatus (Cornu) Coker 1872

Sapromyces androgynus Thaxter 1896

Araiospora spinosa (Cornu) Thaxter 1872

Araiospora pulchra Thaxter 1896

Araiospora coronata Linder 1926

Araiospora septandra Kevorkian 1934

Mindeniella spinospora Kanouse 1927

Mindeniella assymetrica Johnson 1951

Aqualinderella fermentans Emerson & Weston 1967

CHAPTER 9

Lagenidiales

F. K. SPARROW

Department of Botany
University of Michigan
Ann Arbor, Michigan

I. ORDER LAGENIDIALES

This order consists of microscopic, holocarpic, endobiotic, primarily parasitic fungi. Their body plan is simple, consisting of one-cell, sometimes there is a septate mycelial development of slight extent or it can be profuse and occasionally fragment into subthalli, forming one or more reproductive organs. The walls generally give a cellulose reaction. The zoospores are laterally biflagellate with an anterior tinsel flagellum. The posterior flagellum is of the whiplash type. The spores are formed within or outside the sporangium and often in a vesicle. The resting spores are asexually formed by encystment of a thallus or parts of it, or sexually after conjugation of simple thalli, or oogamously by antheridial and oogonial formation; the oosphere often lacks periplasm.

Lagenidiales are parasites of freshwater and marine algae, other Phycomycetes and microscopic animals, and in roots of cereals.

A.   General Characteristics

This heterogeneous group consists primarily of endoparasites of algae, other aquatic fungi, and both microscopic and larger animals and their eggs. They are found in both freshwater and marine hosts. A few are saprophytic or attack only moribund hosts and many appear to be capable of living in artificial culture.

The thallus may be one- (Plate I, A–C, F) or few-celled, often linklike (Plate I, D, I) and monophagous, or may be segmented and hyphalike (Plate I, E, L) and extend through many host cells. In the marine Sirolpidiaceae the mycelial unbranched or branched elements tend to separate into more or less independent units ("subthalli") (Plate I, G–H) which

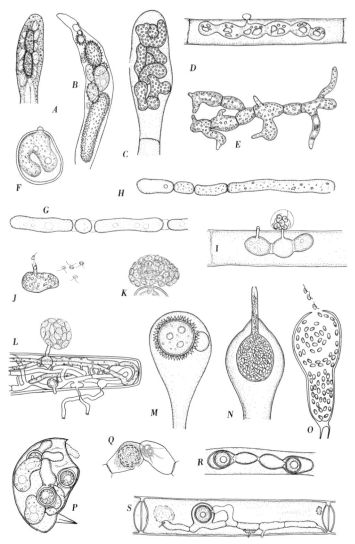

PLATE I.   Lagenidiales.   A.   Two delicately membranated l-celled thalli of *Olpidiopsis achlyae* parasitic in a hyphal tip of *Achlya*. They are centrally suspended by cytoplasmic strands of host. B. Eight 1-celled thalli of *Olpidiopsis varians* parasitic in *Achlya*. They have been converted into tuberculate, smooth- and spiny-walled zoosporangia, and (top) a spiny resting spore and its companion cell. C. Contorted, tubular 1-celled thallus of *Petersenia irregulare* parasitic in *Achlya* hypha. D. One-celled thallus of *Myzocytium proliferum*, parasitic in *Mougeotia*, about to become segmented at constrictions. Epibiotic cyst of infecting zoospore persistent. E. Segmented thallus of *Lagenidium humanum* on human skin. F. One-celled thallus of *Lagenidium oophilum* parasitic in a rotifer egg. Discharge papilla of future zoosporangium already formed. G. Tubular thallus of *Sirolpidium bryopsidis* dividing into subthalli. H. Same.

after varying times become converted into zoosporangia (Plate I, J). A sort of holocarpy may be said to prevail, although not comparable with that found among chytrids. Thus, in *Sirolpidium* and *Haliphthorus* parts of the thallus may be used up in zoospore formation whereas other parts may remain vegetative (Plate I, G, H).

Zoospores are always of the laterally biflagellate type, and in the Lagenidiaceae tend to be relatively large, few, and formed as in *Pythium* outside sporangium in an external vesicle (Plate I, I, L). In the other families they are relatively small and produced in great numbers and completely within the sporangium (Plate I, J, N, O).

Resting spore production is preceded by processes which in the Lagenidiaceae can be termed sexual in nature, involving formation of male and female gametangia of varying degrees of differentiation into oogonia and antheridia (Plate I, P-S). In the Olpidiopsidaceae (primarily parasites of other aquatic fungi) an appearance of conjugation of thalli of two sizes is found (Plate I, B, M). Whether there is karyogamy in the receptive thallus which becomes the resting spore is questionable. No true resting structures or sexuality has been found in the marine Sirolpidiaceae and from analogy with autotrophic marine algae alternation of heteromorphic generations is a distinct possibility.

The left-hand subthallus has become converted into a sporangium and has discharged its zoospores. I. Three-celled thallus of *Myzocytium proliferum* parasitic in *Mougeotia*. The middle segment has become converted into a sporangium whose few large zoospores are completing maturation in a vesicle formed at tip of discharge tube. J. Subthallus of *Sirolpidium* which has been converted into a zoosporangium and is discharging numerous small zoospores. K. Zoospores of *Lagenidium oophilum* parasitic in rotifer egg (same as Fig. F) discharging its individual zoospores into vesicle at tip of discharge tube where flagella are beginning to form. L. *Lagenidium closterii* parasitic in *Closterium*, showing mycelial nature of segmented thallus 1 section of which is maturing its relatively large zoospores in a vesicle at tip of a discharge tube. M. Spiny resting spore and companion cell of *Olpidiopsis saprolegniae* in swollen tip of hypha of *Saprolegnia*. N. Mature zoosporangium of *Olpidiopsis saprolegniae* in swollen apex of *Saprolegnia* hypha. The very numerous small biflagellate zoospores are about to emerge from open tip of discharge tube of sporangium. O. Numerous small zoospores of *Rozella inflata* emerging from parasitized sporangium of *Phytophthora* which was completely occupied by parasite. P. *Myzocytium proliferum* parasitic in rotifer. Some segments of thallus have been converted into sex organs. the receptive segments bearing "oospores." Note the antheridium is undifferentiated but bears a fertilization beak protruding into receptive thallus. Q. Two segments of a thallus of *Lagenidium humanum*, one of which is contributing its contents to the other which is forming a resting spore. Only a pore is formed in mutual wall. R. Thallus of *Myzocytium proliferum* parasitic in *Mougeotia*. The 2 middle links have functioned as contributing cells to their respective end cells which have formed "oospores" in them. S. Segmented thallus of *Lagenidium rabenhorstii* in *Spirogyra*. The second and third segments from left have formed sex organs with oospore and well-defined antheridium with fertilization tube. The other segments have formed zoosporangia which have already discharged their zoospores. (From various authors; in Sparrow, 1960.)

B. Key

KEY TO FAMILIES AND IMPORTANT GENERA OF LAGENIDIALES[1]

1. Zoospores relatively small or numerous, formed within the sporangium 2

1′. Zoospores relatively large and few, formed or completing their maturation in a vesicle at the orifice of a discharge tube. **(Lagenidiaceae)** 3

    2(1) Thallus saccate always 1-celled, resting spores lying free in host not loosely within a sac
. . . . **(Olpidiopsidaceae)** 5

    2′(1) Thallus tubular unbranched, or branched, predominantly multicellular at maturity; marine . . . .·(Sirolpidiaceae) 8

3(1′) Thallus 1-celled, attached to inner host wall by a thick collar of callus; parasitic in roots of cereals . . . . **Lagena**
      Lagena radicicola, parasitic in roots of wheat and other cereals; Vanterpool and Ledingham (1930).

3′(1′) Thallus uni- or multicellular, lying free in host cell or adherent to persistent penetration tube of zoospore 4

    4(3′) Thallus when multicellular unbranched, occupying 1 cell of host, strongly constricted at cross walls, the segments beadlike; antheridial cell poorly if at all differentiated
. . . . **Myzocytium** p. 973
      Myzocytium proliferum; parasite of Conjugatae; *M. vermicola* parasite of eel worms.

    4′(3′) Thallus unbranched or branched, occupying 1 or more cells of host, little or not at all constricted at crosswalls; antheridial cell usually well differentiated
. . . . **Lagenidium** p. 982
      Lagenidium rabenhorstii in *Spirogyra*; *L. pygmaeum* in pollen; *L. giganteum* in mosquito larvae; *L. callinectes* in crab eggs.

5(2) Parasitic in Euglenophyceae or Cryptophyceae . . . . **Pseudosphaerita** p. 961
    Pseudosphaerita euglenae in cytoplasm of *Euglena spp.*

5′(2) Parasitic in aquatic Phycomycetes or in freshwater or marine algae 6

    6(5′) Thallus completely filling the host cell, sporangia at maturity single or in linear series
. . . . **Rozellopsis** p. 922
      Rozellopsis septigena in *Saprolegnia*; *R. waterhousii* in *Phytophthora*.

    6′(5′) Thallus not completely filling host cell 7

7(6′) Sporangia unlobed, not tubular, with 1 (rarely more) discharge tube; resting spores formed by conjugation of walled thalli . . . . **Olpidiopsis** p. 926
    Olpidiopsis saprolegniae in *Saprolegnia*; *O. schenkiana* in Conjugatae.

7′(6′) Sporangia predominantly irregularly lobed or tubular, with more than 1 discharge tube
. . . . **Petersenia** p. 957
    Petersenia lobata in *Callithamnion, Ceramium; P. irregulare* in *Saprolegnia*.

    8(2′) Thallus richly branched, 1-celled at maturity with a single discharge tube
. . . . **Lagenisma**
      Lagenisma coscinodisci, parasitic in marine species of *Coscinodiscus*; Drebes (1966, 1968).

[1]The page numbers following generic names in this key refer to Sparrow (1960) where descriptions of genera and species up to 1957 may be found.

8′(2′) Thallus unbranched or branched, septate, or divided into disarticulating subthalli at maturity 9

9(8′) Thallus with rudimentary branches, not disarticulating into subthalli
. . . . **Pontisma** p. 970
Pontisma lagenidioides in marine Red Algae; not recognized as distinct from *Sirolpidium* by some authors.

9′(8′) Thallus unbranched or branched profusely, disarticulating into numerous subthalli which function nonsynchronously as zoosporangia 10

10(9′) Thallus composed of irregular elements; heterotrichous on solid media
. . . . **Sirolpidium** p. 965
Sirolpidium bryopsidis in marine Green Algae; *S. zoophthorum,* parasitic in larvae of *Venus mercenaria, V. mortoni.*

10′(9′) Thallus composed of $\pm$ isodiametric elements; homotrichous on solid media
. . . . **Haliphthorus**
Haliphthorus milfordensis in Vishniac (1958). Considered by Dick (p. 127) to be in Saprolegniales; placed by Vishniac in its own family, Haliphthoraceae, along with *Atkinsiella* (Saprolegniales).

REFERENCES

Drebes, G. (1966). Ein parasitischer Phycomycet (Lagenidiales) in Coscinodiscus. *Helgolaender Wiss. Meeresunters.* **13**:426–435.

Drebes, G. (1968). *Lagenisma coscinodisci* gen. nov., spec. nov., ein Vertreter der Lagenidiales in der marinen Diatomee Coscinodiscus. *Veroeff. Inst. Meeresforsch. Bremerhaven* **3**: 67–70.

Gotelli. D. (1971). *Lagenisma coscinodisci,* a parasite of the marine diatom Coscinodiscus occurring in the Puget Sound, Washington. *Mycologia* **63**:171–174.

Sparrow, F. K. (1960). "Aquatic Phycomycetes," 2nd rev. ed. Univ. of Michigan Press, Ann Arbor.

Vanterpool, T. C., and G. A. Ledingham. (1930). Studies on "browning" root rot of cereals. I. The association of *Lagena radicicola,* n. gen., n. sp., with root injury of wheat. *Can. J. Res.* **2**:171–194.

Vishniac, H. (1958). A new marine phycomycete. *Mycologia* **50**:66–79.

CHAPTER 10

Peronosporales

GRACE M. WATERHOUSE

Formerly of the Commonwealth Mycological Institute
Kew, Surrey, England

I. CHARACTERS OF THE ORDER

Peronsporales are coenocytic mycelial fungi which are parasitic in plants or saprobic in water or soil. The hyphae are usually colorless, 3–15 μm wide with walls composed mainly of a glucan-cellulose complex plus protein containing hydroxyproline. There are cross partitions, sometimes pluglike, which seal off empty parts or delimit the reproductive organs. The parasites form haustoria.

Vegetative reproduction occurs by zoosporangia or conidia, whether or not they are deciduous. The sporangia have one or few pores with or without a papilla and produce laterally biflagellate zoospores or a plasma, either in a vesicle or naked, or a germ tube. The anterior flagellum is of the tinsel type and the posterior is of the whiplash type. Perennation occurs by thick-walled nondeciduous chlamydospores or by sealed-off portions of mycelium.

The oogonia are thin- or thick-walled, smooth or ornamented, hyaline or brownish. One (rarely up to 6) oosphere is fertilized by one or more smaller antheridia; the oospore is usually single, filling (plerotic) or not filling (aplerotic) the oogonium. The wall is hyaline or brownish, smooth or ornamented with a thick wall up to 10 (usually 2–5) μm wide, and it germinates by a hypha or by a zoosporangium.

Saprobic members can utilize sulfate but are partially or completely heterotrophic for thiamine.

II. KEYS

KEY TO THE FAMILIES OF PERONOSPORALES

Obligate parasites of plants with unbranched, clavate sporangiophores, each bearing a basipetal chain of deciduous sporangia in dense subepidermal clusters, forming on the host white or creamish sori, eventually erupting with the shedding of the sporangia; oogonial periplasm persistent and conspicuous; haustoria knoblike . . . . . **Albuginaceae (Albugo)**

White blister: *A. candida*, worldwide on Cruciferae; *A. tragopogonis*, common on Compositae; *A. bliti* on Amaranthaceae and *A. ipomoea-panduratae* on Convolvulaceae, more common in the tropics.

Obligate parasites of plants with branched treelike sporangiophores or conidiophores of determinate growth (no subsporangial regrowth), differentiated from the mycelium, emerging singly or in tufts from the epidermis usually via the stomata, producing sporangia or conidia singly at the branch tips; periplasm persistent and conspicuous; haustoria varied, usually branched . . . . **Peronosporaceae**

Nonobligate parasites or saprobes; sporangiophores or conidiophores usually undifferentiated from the mycelium, branched, indeterminate, resuming growth after the production of a sporangium or conidium either from below or within the previous empty sporangium; periplasm a thin layer or absent; haustoria absent or branched . . . . **Pythiaceae**

KEY TO THE PERONOSPORACEAE (DOWNY MILDEWS)

1. Primary aerial sporophore emerging from host surface a stout trunk 10 μm or more broad, usually 15–25 μm; oogonial wall thick and rough or ornamented 2

1'. Primary emergent sporophore narrow, not more than 15 μm, usually 8–10 μm broad; oogonial wall, except in *Bremiella*, unornamented 3

    2(1) Sporophore unbranched, apex with short sterigmata bearing papillate sporangia germinating by zoospores; oospore aplerotic . . . . **Basidiophora**

    2'(1) Sporophore repeatedly (2–3) times dichotomously branched in the upper part, spores usually nonpapillate germinating by zoospores or germ tube; oospore plerotic . . . . **Sclerospora**
        All on Gramineae: *S. graminicola* on sorghums, millets, maize, and grasses; *S. sacchari* and *S. spontanea* on sugar cane; *S. maydis*, *S. philippinensis*, and *S. sorghi* on maize and sorghum.

3(1') Spore wall uniformly thick (nonporoid), germination only by a germ tube emerging at any point on the surface . . . . **Peronospora**
        P. parasitica (on crucifers); *P. tabacina* (or *P. hyoscyami*) on tobacco (blue mold); *P. viciae* on peas, beans, lucerne, and vetches; *P. farinosa* on spinach and beet; *P. destructor* on leeks and other alliums.

3'(1') Spore wall poroid, germination by zoospores or plasma emerging through an apical pore with or without a papilla 4

    4(3') Secondary and later branches of the sporophore at right angles to the axis or nearly so, tips blunt . . . . **Plasmopara**
    Plasmopara viticola widespread on grape vine, *P. nivea* widespread on Umbelliferae, *P. australis* on cucurbits in warm areas. *P. halstedii* on Compositae (sunflower).

    4'(3') Branches arising at acute angles 5

5(4') Tips of branches acute . . . . **Pseudoperonospora**
    Pseudoperonospora cubensis on cucurbits, *P. humuli* on hops.

5'(4') Tips of branches much enlarged and bearing 3 or 4 peripheral sterigmata; oogonial wall and oospore wall rather thin and unornamented . . . . **Bremia**
    Bremia lactucae on lettuce.

5"(4') Tips of branches blunt and slightly enlarged; oogonial wall thick and ornamented . . . . **Bremiella**
    Bremiella megasperma on *Viola*.

KEY TO THE PYTHIACEAE

1. Asexual spores spiny, spherical (av. 35 mm diam.), apical on ordinary hyphae, germinating by tubes; antheridia amphigynous, oogonial wall bullate . . . . **Trachysphaera**
Trachysphaera fructigena, pod rot of cacao and other tropical fruits.

1'. Asexual spores smooth, elongated transversely to the growing axis, on verticillate sporophores, germinating by zoospores or hyphae . . . . **Diasporangium**

1". Neither of these combinations of characters 2

2(1") Sporangia usually ovoid or obpyriform with a distinct apical emission zone (thickening or papilla); protoplasm fashioned into zoospores before emission or nearly so; zoospores dispersing quickly, vesicle, if formed, quickly evanescent 3

2'(1") Sporangia filamentous, inflated hyphal, spherical, ovoid or occasionally obpyriform; protoplasm emerging almost or completely undifferentiated into a spherical vesicle usually at the tip of a short discharge tube and remaining at the orifice for some minutes; zoospores formed in the vesicle; antheridia always paragynous; oogonia smooth or spiny; oospore plerotic or aplerotic . . . . **Pythium**
Pythium ultimum and *P. irregulare* (common causes of damping-off and root rots), *P. aphanidermatum*, *P. graminicola* and *P. arrhenomanes* (commoner in warmer areas as root rot organisms); true *P. debaryanum* is rare. Most species are soil inhabiting, e.g., *P. intermedium*, *P. oligandrum*, or aquatic, e.g., *P. middletonii*, *P. undulatum*.

2"(1") Sporangia usually elongated transversely; protoplasm emerging via a long discharge tube into an elongated vesicle or naked; zoospores fashioned in the external medium; proliferating through the empty sporangium; antheridia paragynous; oogonia smooth; aquatic . . . . **Pythiogeton**

3(2) Antheridia amphigynous or paragynous; oogonia smooth or occasionally ornamented (never spiny); oospore aplerotic . . . . **Phytophthora**
Phytophthora infestans (late blight of potato and tomato and other Solananceae); *P. palmivora* (cacao black pod and diseases of rubber and other tropical crops); *P. cactorum*, *P. cinnamomi*, *P. megasperma*, and *P. nicotianae* and its var. *parasitica* (all widespread on many hosts). All species are parasitic, attacking most world crop plants, often with more than one species on a crop.

3'(2) Antheridia always paragynous; oogonial wall fused with the oospore and ornamented; oospore plerotic . . . . **Sclerophthora**
All on Gramineae. *S. macrospora* widespread on cereals.

III. DISCUSSION

Recent classifications of this group reflect very divergent opinions. Moreau (1953), taking up a suggestion of Fitzpatrick (1923), placed *Phytophthora* as a subgenus of *Pythium* and made the order Pythiales (as distinct from Peronosporales). In marked contrast Chadefaud and Emberger (1960), Yurova (1962) and others leave *Pythium* in the Pythiaceae but group *Phytophthora* in the Peronosporaceae or Phytophthoraceae. This is going back to Pethybridge (1913), who favored separate families, and Wilson (1914), who considered *Phytophthora* so different that he erected the Phytophthorales. However, the general consensus is that all the genera should be placed in one major group, be it Peronosporales (Fitzpatrick, 1930) or Peronosporaceae (Gäumann, 1964; Skalický, 1964; von Arx, 1967) with subdivision into (usually) three families (Fitzpatrick) or three tribes for, respectively (1) *Pythium* and related genera, (2) *Peronospora* and related genera, and (3) *Albugo*.

This divergence in the use of orders, families, and lower taxa and the different positions of well-known genera serves to emphasize the state of flux in which the taxonomy of this group moves and the state of doubt concerning the relative meaning of order and family. The classification used here is the most widely accepted one at the present time and it does appear to be in line with the characters and relationships of the genera; these will be discussed under each family. But much more information about all the genera is needed before the status and composition of groups can be accepted with confidence and parallelled with those of other fungus groups.

A. Albuginaceae

The distinctiveness of *Albugo* is clear and generally accepted, but whether the sum of its divergent features merits a separate family is a debatable matter not to be resolved at present. The asexual state with its pustules of subepidermal erect sporangiophores budding off sporangia continuously in basipetal succession (Fig. 1a) is unique, but the sex organs and oospores (Fig. 1c) do not differ markedly from those of *Peronospora* (Fig. 3b,c) except that the oogonial wall is thin and colorless and nonpersistent. The oospore is aplerotic. Knoblike haustoria are present in all species so far investigated.

Albugo

Two groups have been recognized on the form of the sporangium: *Aequales* Fischer, with a uniformly thin wall, and *Annulati* Fischer with an internal equatorial thickening on the wall (Fig. 1b). Otherwise the sporangia are extraordinarily similar throughout, being smooth, colorless or nearly so, and spherical or shortly ellipsoid-cylindrical, usually in the range 12–20 μm but in a few species up to 25 μm. Săvulescu (1946), however, used small differences in size (2–3 μm) and shape to make many varieties, forms, and other infraspecific taxa, but Biga (1955) reduced these to nine varieties, two in *candida*, two in *ipomoea-panduratae*, and five in *trago-pogonis*. The pros and cons of this type of splitting are discussed under *Peronospora*. Other sporangial features used to characterize species are the color of the pustule (white to shades of cream) and the size and fertility or sterility of the terminal sporangium.

Oospore ornamentation (Fig. 2) is distinctive between species on different host families (Wilson, 1907) and confirms that specialization to one family

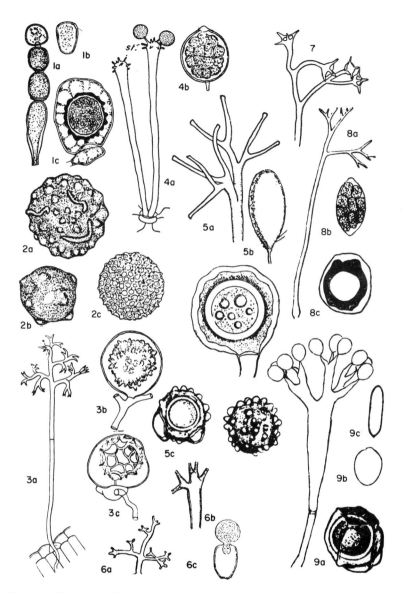

PLATE I. Figs. 1, 2. *Albugo*. 1a, *A. tragopogonis* sporangiophore, × 340; b, *A. bliti* spore, × 332; c, *A. candida* sex organs, × 340 (after de Bary). Fig. 2. Oospores. a, *A. candida*; b, *A. tropica*, c, *A. swertiae* (all × 425, after Wilson). Fig. 3. *Peronospora*. a, *P. destructor* conidiophore, × 76; b, *P. arenariae* and c, *P. myosotidis* sex organs, × 332 (after de Bary). Fig. 4. *Basidiophora*. a, *B. entospora* sporangiophore, × 20; b, sporangium, × 468 (after Roze and Cornu); c, *B. butleri* sex organs, × 595 (after Weston). Fig. 5. *Bremiella megasperma*. a, Sporangiophore apex, × 382; b, sporangium × 255; c, sex organs, × 425 (after Berlese). Fig. 6. *Plasmopara*. a, *P. australis* sporangiophore tips, × 170 (after Berlese); b, *P. pygmaea* sporangiophore tips, × 459, and c, sporangium emitting plasma, × 340 (after de Bary). Fig. 7. *Bremia lactucae* sporangiophore branches, × 170 (after Gäumann). Fig. 8. *Pseudoperonospora celtidis*. a, sporangiophore, × 190; b, sporangium, × 544; c, sex organs, × 544 (after Waite). Fig. 9. *Sclerospora*. a, *S. sorghi* conidiophore and sex organs, × 183 (after Weston); b, *S. graminicola* sporangium, × 510; c, *S. spontanea* conidium, × 340.

is a guide to species delimitation. Unusually, the internal structure of the oospore, including the number of nuclei fusing initially and present in the mature spore, differs from species to species. Biga (1955) gives a key to 30 species based on host, oospores, and conidia but without using features of the sorus or sporangial wall.

B.   Peronosporaceae

In this family the sporangiophore and its sporangia have been the basis for generic delimitation and also the cause of much controversy, resulting in the suggested merging of genera, e.g., *Pseudoperonospora*, and much movement of species from genus to genus. The sex organs are not markedly different between genera except in a few instances to be mentioned, hence the reliance placed on the asexual state, but Skalický (1966) gives a key based on oogonia and oospores.

The sporangiophores emerge through the stomata and grow to a height of 100–750 μm depending on the species and on the conditions. In *Basidiophora* and *Sclerospora* the initial trunk is broad (Figs. 4a, 9a) and in the rest of the genera narrow (Fig. 3a) and on this depends the difference in the mode of spore liberation (Ingold, 1971); in the former group the spores are ejected from the tips of the sporophores by a propulsive mechanism while in the latter the whole sporophore twists convulsively on drying and the spores are shaken off. They are always deciduous. The initial branching of the trunk may be irregular. Subsequent branching may be monopodial [*Plasmopara, Pseudoperonospora* (Fig. 8a)], reduced to sterigmatalike tips (*Basidiophora* Fig. 4a), or dichotomous (the remaining genera). In *Plasmopara* the branches emerge at nearly a right angle (Fig. 6a), in the rest they are at an acute angle; the tips may be blunt or slightly expanded [*Plasmopara, Bremiella* (Fig. 5a)], broadened to form a disk with sterigmata at the edge (*Bremia*) (Fig. 7), or pointed (the rest).

Little difference was recognized in the form of the spores until Shaw (1950) pointed out, as Fischer (1892) had done, that the wall in *Peronospora* and most species of *Sclerospora* is completely uniform and germination can only be by a tube, whereas in all the other genera it is apically poroid and germination may be by zoospores or a plasma as well as hyphal.

The oogonia are larger and the antheridia broader in comparison with the Pythiaceae; the antheridium makes broader application to the oogonium, so that it looks hemispherical, in contrast to a narrow meeting in Pythiaceae. The oogonial wall is thin or not much thickened and unornamented except in one section of *Peronospora, Basidiophora* (Fig. 4c), and *Sclerospora* (Fig. 9a) where it is thick and ornamented though Safeeulla *et al.* (1963) maintain that it remains thin and that it is the residual protoplasm which contributes the thickening. The residual periplasm, as in *Albugo*, envelops the oospore, forming a thick, dark, often sculptured layer [epispore or exosporangium (Shaw, 1970)]. This feature was used by de Bary to make sections and subsections in *Peronospora* but in the other genera the sculpturing is not sufficiently characteristic to form a diagnostic feature. The oospore is very much smaller than the oogonium except in *Sclerospora* where it is plerotic and the epispore is fused with the oogonial wall.

The status of each genus and reasons for maintaining its position are discussed under the individual genera. Skalický (1966) reviewed the morphological features separately and in detail and commented on their diagnostic value. As a result, he merged *Pseudoperonospora* partly in *Peronospora* and partly in *Plasmopara*, emended the diagnosis of *Peronospora* to admit species with poroid sporangia and monopodial sporangiophores, and accepted *Dicksonomyces* (in which he incorporated *Basidiophora* (*Sclerospora*) *butleri*). *Dicksonomyces* has been shown to be based partly on a hyphomycete (Waterhouse, 1968a). The reasons for rejecting his other proposals are discussed below.

Now that some obligate downy mildews are being grown in tissue culture (e.g., *Peronospora parasitica*, *Pseudoperonospora humuli*, *Plasmopara viticola*, *Sclerospora*, etc.) additional characters of diagnostic value may emerge.

Peronospora

This is by far the largest genus, recorded on many Dicotyledon families, but on few Monocotyledons. The conidiophore is fairly uniform throughout, though there are characteristic differences in its morphology between host families as regards the stiffness (Divaricatae Fischer) or curvature of the branches (Undulatae Fischer), density of branching, and features of the ultimate branchlets. However, though Gäumann (1923) figured these and commented on the differences, he made no use of them in his biometric studies. Similarly, Gustavsson (1959a) dismissed them completely. Gäumann's use of small differences in conidial dimensions as a basis for erecting species set the pattern of almost one host, one species for many years. Gustavsson (1959a) corrected some of Gäumann's nomenclature and measurements and deprecated the use of single collections on which to base species, as well as the lack of recognition of natural variations. Nevertheless, he followed the biometric line. Yerkes and Shaw (1959) deemed these small differences to be insufficient to establish species and pointed out that when strains on all hosts in one family were considered there was a continuous overlapping series. Consequently they merged all species on Cruciferae under *P. parasitica* and all those on Chenopodiaceae under *P. farinosa*. This does not mean that there may not be more than one species per family: on Leguminosae there are probably at least two distinguishable by their oospores.

The point at issue is whether these biometric differences are constant and always distinguishable and if so whether they rank at specific or only at an infraspecific level. McMeekin (1969) found that conidia of *P. parasitica* on cabbage and radish grown in a growth chamber varied from 22 × 16 to 26 × 18.5 μm irrespective of host. Differences in shape were regarded as chance variations from a single population mean but they did not occur in the field. Smith (1970), in carefully controlled studies, found that the size and shape of conidia of *Peronospora* on tobacco did not vary with different light intensities, host species, or inoculum (single conidia or many), but there were significant differences with temperature (20.6 × 17.4 μm at 10–15°C, against 18.2 × 14.9 μm at 25–30°C), with locality and day of collection (probably due to weather), and also between pathogenic strains (a difference of 2 × 1.5 μm). Humidity was not tested, though Gustavsson (1959a) thought this to be a factor contributing to variation. Therefore, before such small differences can be accepted as meaningful, they must be shown to be nonvariable; even so, they do not seem to be worthy of species status. Indeed, Skalický (1964) and Shepherd (1970) made forma speciales for *Peronospora* on tobacco, *Nicotiana* spp., and *Hyoscyamus*.

Between host families there are also differences in the color of the "down" (white, gray, or mauvish) and in the general shape of the conidium. The biometric forms lie within this pattern and there seems to be a case for aggregate species on host families, forma where biometric differences are adequate, and forma speciales for host specific strains morphologically indistinguishable. Races have been detected in some species.

Probably because sex organs and oospores are rarely found (absent because of heterothallism or seasonal formation, or present but missed) their use in speciation has been lost sight of since Gäumann placed heavy emphasis on conidia. Skalický (1966) revived de Bary's sections (1863) and subsections based on oogonial and oospore wall ornamentation: Leiothecae (unornamented oospore wall) divided into parasiticae (thick oogonial wall) and effusae (thin), and Calothecae (ornamented) divided into verrucosae (Fig. 3b) and reticulatae (Fig. 3c). Gustavsson (1959b), having investigated the oospores of several species, rejected their diagnostic value because they were too variable. It is surprising that the sexual state which is considered to be of such prime importance in other groups should be thought valueless here.

Bremia

The genus is easily recognized by the dead white "down," the dichotomous sporangiophores branching at an acute angle (Fig. 7), the enlarged tips bearing a ring of three to six sterigmata, and the papillate sporangia, but these nearly always germinate by a tube; they are nearly spherical, being slightly longer than broad.

Until 1949 the genus was known only on Compositae. Then *B. graminicola* on *Arthraxon* was described, also *Plasmopara oplismeni* on *Oplismenus* and *P.* sp. on pearl millet, both in Africa. It is very unusual to have a downy mildew other than *Sclerospora* sp. on Gramineae and an investigation is needed to make sure that two additional genera are really involved.

On Compositae many hosts are attacked but usually only one species, *B. lactucae*, is recognized, though about ten species have been erected (Săvulescu, 1962, and others) on slight differences (1–3 μm) in dimensions of sporangia and sporangiophores, for example, sporangia on *Lampsana* and *Senecio* are slightly smaller and those on *Sonchus* slightly larger than the type. The same arguments apply to the advisability of using such differences as a basis for species differentiation as are mentioned under *Peronospora*. There are differences in host specificity and there are races on *Lactuca*.

There is also a *Plasmopara* sp. on Compositae which has been confused with *Bremia*.

The sex organs and oospores are similar to those of *Peronospora*: the oogonium wall is thin and the aplerotic oospore slightly rugulose. The haustoria are globose.

Basidiophora

The three species described are characterized by a short stout sporangiophore, the reduced trunk failing to branch and producing only sterigmata each of which bears one sporangium. Germination is by zoospores (Fig. 4a,b).

The sex organs differ from those of related genera in that the oogonial wall is ornamented (bullate) and the oospore smooth (Fig. 4c).

Speciation is on host, dimensions of sporangiophore and sporangia, and oogonial ornamentation and size.

Sclerospora

The type species is in a somewhat anomalous position as the only one that has a sporangium germinating by zoospores of the thirteen or so "good" species (Waterhouse, 1964), if the transfers of *S. butleri* to *Basidiophora* and of *S. macrospora* to *Sclerophthora* or to *Phytophthora* are accepted. The other species, except for four in which no conidia have been described, have nonporoid spores germinating only by tubes. The asexual organs develop very early in the morning when dew is heaviest and because they are very ephemeral, they are often missed.

The species can be distinguished by their conidiophores which have a basal "cell" differing in length and the bulbosity of the base (Fig. 9a); the length of the sterigmata; and the size and shape of the conidia. The conidia in all species are smooth and colorless, but range from almost spherical to cylindric-oval (Fig. 9b,c).

The sex organs differ from those of other genera (except *Bremiella*) in that the oogonial wall appears to be ornamented and the oospore smooth (Fig. 9a), though Safeeulla *et al.* (1963) maintain that the oogonial wall is very thin and the ornamentations formed from the periplasm—thus it would be epispore which is fused to the oogonial wall. The ornamentations differ from species to species as does the size of the oogonium which is usually deeply colored some shade of brown.

The genus is confined to Gramineae (mostly Panicoideae); some species are host specific but those on the maize-sorghum-sugarcane-millet group exhibit cross infectivity. Indeed, there is some doubt as to whether they are distinct species, but they can be separated on both conidial and oogonial characters (Waterhouse, 1964). Until more work is done on collections from single-species inoculations on a variety of hosts under different conditions the names serve as distinguishing epithets. The fact that one species can infect different hosts and more than one species can be found on a single host has led to confusing descriptions of what may be mixed infections. The haustoria are digitate.

Plasmopara

This genus has retained its status quite firmly in a family in which there has been some merging and suggested merging. Also the tendency has been to move species into the genus rather than out of it.

The noticeable feature is the branching of the sporangiophore which is rather random (not dichotomous); the later branches emerge at right angles or nearly so. They are stiff and straight in contrast to the curved or willowy nature of some peronosporas, *Bremia* and *Bremiella*. The ultimate branchlets are very short and truncate. The sporangiophore is, however, much more varied than in *Peronospora*, ranging from the short (100–300 μm), stout, scarcely branched (Fig. 6b) one of *Plasmopara pygmaea* (Supinae Fischer) to the slender much branched, almost *Peronospora*-like one (Fig. 6a) of *P. viticola* (Altae Fischer).

Before *Plasmopara* was founded, de Bary (1863) placed the species now in the genus in two sections: Plasmatoparae, sporangia emitting a plasma from which the zoospores are fashioned externally (Fig. 6c), and Zoosporiparae, emitting zoospores. Wilson (1914) thought this difference sufficient to make two genera, with *Rhysotheca* for the second, but this has not been accepted. The species do not fit neatly into either the branching or the zoospore divisions. Skalický (1966) made two subgenera: Plasmopara for the *pygmaea* group and Hyalodendron for the *viticola* group, again on the branching of the sporangiophore. Certainly these two groups have very different characters.

The oogonial wall is persistent and somewhat thickened but not sculptured, the oospore is aplerotic but not markedly so, and the epispore is thick and dark, and sculptured in some species. The haustoria are globose. The "down" is white.

There has been no monograph on the genus since Berlese (1902), though Săvulescu and Săvulescu (1952) covered the European species and Wilson (1914) described and keyed twelve zoospore-producing species (*Rhysotheca*).

Pseudoperonospora

The name and status of this genus have been the subject of much discussion. The claim for priority of *Peronoplasmopara* cannot be accepted as it was based on Berlese's subgenus which has no priority outside its own rank and as a genus is antedated by *Pseudoperonospora*.

When it was thought that the conidia of *Peronospora* could act as zoosporangia, as well as germinate by tubes, it was argued that there was no reason for separating the two genera.

However, Shaw (1950) showed that the conidial wall in *Peronospora* is nonporoid (uniformly thick) and germination can only be by a tube, whereas the conidia of *Pseudoperonospora* are true sporangia with a poroid apex. This is a good diagnostic character since it is one of fundamental wall structure.

Skalický (1966) would dissolve the genus on the grounds that it is heterogeneous. If all the names included under the genus were accepted it undoubtedly would be because of misdeterminations, but, if the wrongly named species are removed, a compact definable group is left (*P. cubensis, P. humuli, P. urticae, P. cannabina,* and *P. celtidis*). The last four are morphologically close and are on closely related host families (Urticaceae, Cannabinaceae, and Ulmaceae) and might prove to be forms of one species.

The sporangiophores (Fig. 8a) are, at first glance, like those of *Peronospora,* but they are more delicate; branching is not dichotomous and is much less profuse, and the branches are straight. The sporangia are ellipsoid, poroid and papillate, and nearly always germinate by zoospores (Fig. 8b); they are faintly colored brownish or mauvish-gray.

The oogonium has a thin wall and the aplerotic oospore a thick, dark, slightly rough wall (Fig. 8c). The haustoria are knoblike to digitate (Fraymouth, 1956). There is no monograph but one is in preparation (G. M. Waterhouse, unpublished).

Bremiella

The single species, known only in North America, placed first in *Peronospora* and then in *Plasmopara*, forms a white "down" on *Viola* spp. and has been confused with *Peronospora violae* which is rightly placed.

The sporangiophore is very graceful with long curved dichotomous branches (Fig. 5a). The ultimate branchlets are also long, truncate at the tips, and slightly inflated (termed an apophysis by Berlese). The sporangia (Fig. 5b) are very large compared with those in the rest of the family. Berlese (1902) gives 80–95 × 40–60 μm; they are papillate and germinate by zoospores.

The oogonial wall is thick, sculptured, and persistent and the smooth oospore aplerotic (Fig. 5c).

With this combination of characters the species cannot be placed elsewhere.

C. Pythiaceae

All the genera of this family except, in some respects, *Phytophthora*, are markedly different from those of the Peronosporaceae. The members are either saprophytes or facultative parasites growing readily in artificial media; the sporangiophores are undifferentiated from the other hyphae and renew growth after the production of a sporangium (Fig. 11a,b,c). This is in contrast to the obligate parasitism of the Peronosporaceae and the differentiated sporangiophores whose growth is terminated by the production of a sporangium. This segregation is accepted in all recent classifications for all genera except *Phytophthora*. The reason why the place of the latter is in dispute is because some taxonomists consider the sporangiophores of the type species to be *Peronospora*-like. The resemblance is purely superficial; in development they are quite different after the initial branching. In *Phytophthora* (*all* species) the branch forms a terminal sporangium and then growth is renewed either from the hypha below the sporangium in a sympodial manner—with short growth in some conditions (aerial growth) or longer when growth is more active (in water)—or from within the empty sporangium (Fig. 11b).

In the Peronosporaceae the primary branches branch again and again but eventually growth is terminated by a sporangium with no growth renewal. Thus there must be a fundamental difference of some importance based on inherent genetic and physiological qualities as great as those at the basis of sympodial and cymose branching in other groups. On these grounds alone *Phytophthora* is rightly placed with the Pythiaceae, but other characteristics also place it apart from the Peronosporaceae. For example, when the oosphere is delimited there is very little periplasm remaining outside the primary membrane of the oospore and this either gradually disappears or remains as a very thin barely visible layer adhering to the outer oospore wall. In the other two families the periplasm is very much thicker and constitutes the epispore.

Middleton (1952) gave a detailed and satisfactory account of the status of the genera of the Pythiaceae. All except two have few species which are rather rare.

Pythium

This is the largest genus (92 spp., Waterhouse, 1968b–d). Several species have been encountered only in aquatic situations and are accepted as saprobic, though it is possible that some may be weak parasites of aquatic plants or animals but not yet recognized as such. The majority are soil inhabitants and many of these are implicated in root rots (particularly of Gramineae), some as primary attackers, others as secondary invaders enhancing the pathogenic process; a few are mycorrhizal. Very few venture above ground and the aerial parts attacked are at or near soil level: prominent among these are cucurbit fruits lying on the soil. Yet names appear only rarely in lists of soil fungi because isolation methods necessary for their detection have not been used. Rarely are they host specific (Rangaswami, 1962).

The distinguishing features are: the delicate hyphae (av. 5 μm, max. 10 μm); sporangial shapes—usually hyphal (swollen or not) or spherical, much less commonly ellipsoid or ovoid, rarely papillate, but often with a discharge tube left after dehiscence; the emission of the protoplasm with little differentiation into a vesicle where the zoospores are formed (Fig. 10 a–c), a process which takes several minutes; the smooth or spiny (Fig. 10d), and colorless oogonial wall; the paragynous or hypogynous antheridium; and the usually colorless and unornamented, and often plerotic oospore.

In the species so far tested, the chemical composition of the walls, some enzyme complexes, and nutrient requirements differ from those of *Phytophthora* (see below).

The characters used in speciation (Waterhouse, 1968b) are the morphology of the sporangium, dimensions, and whether it proliferates internally; oogonial dimensions and the morphology of the spines; the plerotic or aplerotic character of the oospore and (rarely) the nature of its wall; the point of origin of the antheridium in relation to the oogonium and the number per oogonium; and maximum growth temperature.

Phytophthora

The position of the genus has stabilized in spite of attempts to merge it with *Pythium* or remove it to another family.

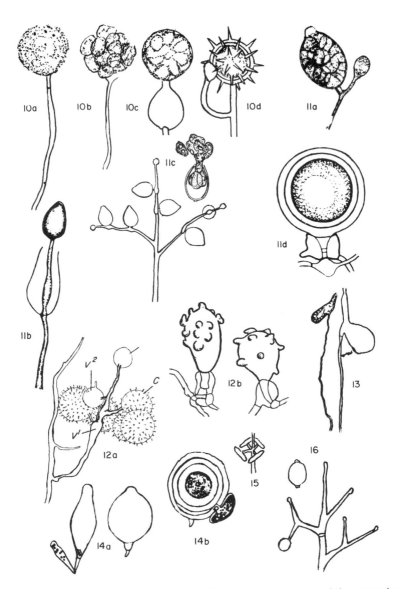

PLATE II. Fig. 10. *Pythium.* a, *P. afertile* filamentous sporangium emitting protoplasm, and, b, with zoospores formed (× 340); c, *P. echinocarpum* sporangium, and d, sex organs (× 361, after Tokunaga). Fig. 11. *Phytophthora.* a, *P. nicotianae* sporangiophore and sporangium Zeiss 4 × (after Sawada); b, *P. cinnamomi* sporangiophore and sporangium, × 294 (after Rands); c, *P. infestans* sporangiophore, × 170 and sporangium, × 332 (after de Bary); d, *P. erythroseptica* sex organs, × 765. Fig. 12. *Trachysphaera fructigena.* a, conidia, × 636, b, sex organs, × 510 (after Tabor and Bunting). Fig. 13. *Pythiogeton utriforme* sporangia, × 200 (after von Minden). Fig. 14. *Sclerophthora macrospora.* a, sporangia, × 340; b, sex organs, × 255 (after Ito and Tokunaga). Fig. 15. *Diasporangium* sporophore. Fig. 16. *Peronophythora* sporophore and sporangium, × 255.

178

The morphological features distinguishing it from *Pythium* are: the rather wider hyphae (av. 6 μm, max. 14 μm), the limoniform, obpyriform, or ovoid sporangia (shapes much less common in *Pythium*), the fashioning of the zoospores in the sporangium (Fig. 11a,c) and their emission fully flagellated, the smooth, rough, or warted (never spiny) oogonial wall which may thicken and become brownish, the amphigynous (Fig. 11d) antheridium (very few species are entirely paragynous), and the more parasitic nature. It is possible that haustoria are always present (and absent from *Pythium*) but this has been little investigated (Blackwell, 1953). The separation of the two genera is supported by the different protein and oxidase patterns tested to date (Clare *et al.*, 1968), the greater protein content of the hyphal wall in *Pythium* (Bartnicki-Garcia, 1968; Novaes-Ledieu and Jiménez-Martínez, 1969), and the thiamine requirement of *Phytophthora* (Cantino, 1966).

Except for a few apparently saprophytic water molds, which may prove to be only casually so, all the species are parasites; a few have not been brought into culture, but most are able to grow in senescing or moribund tissues or in agar culture. Many have been isolated from soils and attack underground parts of plants but there is little evidence that they are able to grow and reproduce in soil, though it is possible that mycelium in plant detritus could produce sporangia and zoospores since this occurs under laboratory conditions.

Many species are able to attack the aerial parts of plants causing collar rots, trunk cankers, twig blights, leaf fall, leaf blights, and fruit rots, often of economically important crops (Rangaswami, 1962).

The species fall into three groups based on sporangial features: one with little or no apical thickening ("nonpapillate"), proliferating through the empty sporangium (Fig. 11b), and nondeciduous; the other two have conspicuous apical thickening, are usually papillate (Fig. 11a), are non-proliferating, and usually deciduous; in one of these the apical thickening is shallow (up to 3.5 μm), the sporangium scarcely papillate and usually nondeciduous, while the other has a deep thickening (4 μm or more), is conspicuously papillate, and usually deciduous. In the first group are the aquatic types that cause root and tuber rots; these are rare above ground in contrast to the majority of the last group which are common on aerial parts.

A computer analysis (J. C. Gower and G. M. Waterhouse, unpublished), using 33 diagnostic morphological characters, gave a very definite segregation into two groups—group one above as distinct from the other two groups, with a lesser segregation between groups two and three.

Sporangia are much more variable in size and shape than in the Peronosporaceae; in consequence, the tendency has been to discount sporangial dimensions as a diagnostic criterion. However, under uniform and near natural conditions there are significant differences, particularly of shape, between species (Waterhouse, 1970a,b).

A key to 48 species (Waterhouse, 1963) uses sporangial papillation, proliferation, dimensions, and, to a lesser extent, shape, antheridial form and size, oogonial size, ornamentation and shape, oospore production in dual cultures and wall thickness, culture patterns, maximum and minimum growth temperatures, and host specialization as diagnostic features. These cannot often be used in isolation, as groups of characteristics, rather than single ones, indicate specific differences. The use of ultrastructure, serology, electrophoresis of proteins (particularly enzymes), and microchemical analyses to differentiate between species has been reviewed by Waterhouse (1970b).

Pythiogeton

This genus is close to *Pythium* and *Phytophthora* but all species, of which there are six, all aquatic, according to Sparrow (1960), have unusual, somewhat irregular, large sporangia (often up to 100 μm and can be up to 300 μm), elongated transversely to the growing axis, emitting the contents as undifferentiated protoplasm through a long discharge tube (up to 70 μm) into an elongated vesicle or naked, and proliferating through the base of the empty sporangium (Fig. 13). The distortion from the usually neat appearance of the Pythiaceae sporangium suggests something like a virus infection.

The sex organs have not been studied in detail and have not been seen in four species. They are *Pythium*-like with paragynous antheridia and a plerotic oospore having a very thick wall.

Those so far tested are able to grow under conditions of low oxygen, in contrast to *Pythium* and *Phytophthora*, are heterotrophic for thiamine, like *Phytophthora*, and are able to utilize cellulose (Cantino, 1949). Sparrow gives a key (1960) based on sporangial shape and size and length of discharge tube.

Trachysphaera

The single species is well known because of the devastation caused to cacao pods and other tropical fruits. Its spiny conidia (Fig. 12a) are reminiscent of those of *Pythium oligandrum*, but the amphigynous antheridia and bullate oogonial wall (Fig. 12b) place it near *Phytophthora*. No zoospores are produced.

Zoophagus

There are two species both capturing rotifers, but the second described species is doubtful because no sporangia or zoospores were found. According to the evidence so far, inclusion in the Pythiaceae is indicated (Middleton, 1952; Sparrow, 1960), but this is still a doubtful placing until a full description of the sex organs is available. The sporangia are like those of a filamentous *Pythium* (Prowse, 1954).

Diasporangium

The single species is known only from Höhnk's original description. Its verticillate sporangiophores (Fig. 15) would place it outside the family were it not for the report of the emission of the contents as zoospores or a plasma. The oospores are like those of Pythiaceae, but no antheridia were described. It is a soil fungus.

Sclerophthora

The genus is based on *Sclerospora macrospora* which has also been placed in *Phytophthora*. The unbranched or sympodial sporangiophores with their citriform or obpyriform sporangia are *Phytophthora*-like (Fig.14a). The sex organs would fit either genus but the mature oospore has a thick epispore which fuses with the oogonial wall to form a thick brown covering over the oospore (Fig. 14b) proper which can easily be pressed out.

There are four species, all on Gramineae (mostly Pooideae), where they cause a downy mildew type of infection (Raychaudhuri, 1970). Skalický (1966) places the genus in the Phytophthoraceae on the grounds of the multinucleate oospore, the hyphalike sporangiophore, the lemon-shaped sporangia, and the wide host range. Thirumalachar (1969) adds the type of disease caused.

Peronophythora

This genus, known only from the original description (Chen, 1961), was so named because of some resemblances to both genera. The only *Peronospora*-like feature is the sporangiophore which appears to be determinate in its growth. The ease with which the fungus is grown in artificial culture, the narrow hyphae (2–5 μm) and sporangiophores (5–7 μm), the papillate sporangia (Fig. 16), the chlamydospores, and the amphigynous antheridia would place it in the Pythiaceae. An unusual feature is the large size of the zoospores (33 × 18 μm). If the development of the sporangiophores were followed perhaps this too would place it in this family.

REFERENCES

Bartnicki-Garcia, S. (1968). Cell wall chemistry, morphogenesis, and taxonomy of fungi. *Annu. Rev. Microbiol.* **22**:87–108.

Berlese, A. N. (1902). Saggio di una monographia della Peronosporaceae. *Riv. Patol. Veg.* [1] **9**:1–126.

Biga, M. L. B. (1955). Riesaminazione delle specie del genere *Albugo* in base alla morfologia dei conidi. *Sydowia* **9**:339–358.

Blackwell, E. M. (1953). Haustoria of *Phytophthora infestans* and some other species. *Brit. Mycol. Soc. Trans.* **36**:138–158.

Cantino, E. C. (1949). The growth and nutrition of *Pythiogeton. Amer. J. Bot.* **36**:747–756.

Cantino, E. C. (1966). Morphogenesis in aquatic fungi. *In* "The Fungi" (G. C. Ainsworth and A. S. Sussman, eds.), Vol. 2, pp. 283–337. Academic Press, New York.

Chadefaud, M., and L. Emberger. (1960). Les Phycomycètes dimastiguiés = Leptomitales, Saprolégniales, Péronosporales. *In* "Traité de Botanique—Systématique," Vol. I, pp. 844–858. Masson, Paris.

Chen, C.-C. (1961). A species of *Peronophythora* gen. nov. parasitic on lichi fruit in Taiwan. *Coll. Agr., Nat. Taiwan Univ., Spec. Publ.* **10**:1–37.

Clare, B. G., N. T. Flentje, and M. R. Atkinson. (1968). Electrophoretic patterns of oxidoreductases and other proteins as criteria in fungal taxonomy. *Aust. J. Biol. Sci.* **21**:275–295.

de Bary, A. (1863). Recherches sur le développement de quelques champignons parasites. *Ann. Sci. Natur. Bot. IV* **29**:5–148.

Fischer, A. (1892). "Die Pilze. IV. Phycomycetes," pp. 383–490. Kummer, Leipzig.

Fitzpatrick, H. M. (1923). Generic concepts in the Pythiaceae and Blastocladiaceae. *Mycologia* **15**:166–173.

Fitzpatrick, H. M. (1930). "The Lower Fungi. Phycomycetes." McGraw-Hill, New York.

Fraymouth, J. (1956). Haustoria of the Peronosporales. *Brit. Mycol. Soc. Trans.* **39**:79–107.

Gäumann, E. (1923). Beiträge zu einer Monographie der Gattung *Peronospora. Beitr. Kryptogamenflora Schweiz* **5**:1–360.

Gäumann, E. (1964). Peronosporaceae. *In* "Die Pilze," pp. 66–76. Birkhaeuser, Basel.

Gustavsson, A. (1959a). Studies on Nordic Peronosporas. I. Taxonomic revision. II. General account. *Opera Bot.* **3**:1–271; 1–61.

Gustavsson, A. (1959b). Studies on the oospore development in *Peronospora. Bot. Notis.* **112**:1–16.

Ingold, C. T. (1971). "Fungal Spores. Their Liberation and Dispersal." Oxford Univ. Press (Clarendon), London and New York.

McMeekin, D. (1969). Other hosts for *Peronospora parasitica* from cabbage and radish. *Phytopathology* **59**:693–696.

Middleton, J. T. (1952). Generic concepts in the Pythiaceae. *Tijdschr. Plantenziekten* **58**:226–235.

Moreau, F. (1953). Mastigomycètes à spores seules mobiles: Péronosporales, Leptomitales, Pythiales. *In* "Les Champignons. II. Systématique," pp. 1094–1163. Lechevalier, Paris.

Novaes-Ledieu, M., and A. Jiménez-Martínez. (1969). The structure of cell walls of Phycomycetes. *J. Gen. Microbiol.* **54**:407–415.

Pethybridge, G. (1913). On the rotting of potato tubers by a new species of *Phytophthora* having a method of sexual reproduction hitherto undescribed. *Roy. Dublin Soc. Sci. Proc.* **13**:529–565.

Prowse, G. A. (1954). *Sommerstorffia spinosa* and *Zoophagus insidians* predacious on rotifers, and *Rozellopsis inflata* endoparasite of *Zoophagus. Brit. Mycol. Soc. Trans.* **37**:134–150.

Rangaswami, G. (1962). "Pythiaceous Fungi (A Review)." Indian Council of Agricultural Research, New Delhi.

Raychaudhuri, S. P. (1970). Workshop on the downy mildew diseases of maize (corn) and sorghum (21–25 September, 1969). *Indian Phytopathol.* **23**:169–435.

Safeeulla, K. M., M. J. Thirumalachar, and C. G. Shaw. (1963). Gametogenesis and oospore formation in *Sclerophthora cryophila* on *Digitaria marginata. Mycologia* **55**:819–823.

Săvulescu, O. (1946). Studiul speciilor de *Cystopus* Lév. din Europa cu privire specială asupra speciilor din România. Thesis 213, Faculty of Science, University of Bucharest.

Săvulescu, O. (1962). A systematic study of the genera *Bremia* Regel and *Bremiella* Wilson. *Rev. Biol. Acad. Rep. Pop. Rom.* **7**:43–62.

Săvulescu, T., and O. Săvulescu. (1952). Studiul *Sclerospora, Basidiophora, Plasmopara* si *Peronoplasmopara*. *Bull. Stiint. Acad. Rep. Rom.* pp. 327–457.

Shaw, C. G. (1950). The genera of the Peronosporaceae. *Phytopathology* **40**:25.

Shaw, C. G. (1970). Morphology and physiology of downy mildews—significance in taxonomy and pathology. *Indian Phytopathol.* **23**:364–370.

Shepherd, C. J. (1970). Nomenclature of the tobacco blue mould fungus. *Brit. Mycol. Soc. Trans.* **55**:253–256.

Skalický, V. (1964). Beitrag zur infraspezifischen Taxonomie der obligat parasitischen Pilze. *Acta Univ. Carol., Biol., Suppl.* pp. 25–90.

Skalický, V. (1966). Taxonomie der Gattungen der Familie Peronosporaceae. *Preslia* **38**: 117–129.

Smith, A. (1970). Biometric studies on conidia of *Peronospora tabacina*. *Brit. Mycol. Soc. Trans.* **55**:59–66.

Sparrow, F. K. (1960). "Aquatic Phycomycetes," 2nd rev. ed. Univ. of Michigan Press, Ann Arbor.

Thirumalachar, M. J. (1969). Morphological basis for the characterisation and separation of the genera *Phytophthora, Sclerophthora,* and *Sclerospora*. *Indian Phytopathol* **22**:155.

von Arx, J. A. (1967). "Pilzekunde." Cramer, Lehre.

Waterhouse, G. M. (1963). Key to the species of *Phytophthora* de Bary. *Mycol. Pap.* **92**:1–22.

Waterhouse, G. M. (1964). The genus *Sclerospora*. *Misc. Publ. C. M. I.* **17**:1–30.

Waterhouse, G. M. (1968a). *Dicksonomyces* based on a misdetermination. *Mycologia* **60**: 976–978.

Waterhouse, G. M. (1968b). Key to *Pythium* Pringsheim. *Mycol. Pap.* **109**:1–15.

Waterhouse, G. M. (1968c). The genus *Pythium* Pringsheim. *Mycol. Pap.* **110**:1–71.

Waterhouse, G. M. (1968d). Nomina conservanda proposita. *Pythium* Pringsheim. *Taxon* **17**:88.

Waterhouse, G. M. (1970a). The genus *Phytophthora*. Diagnoses and descriptions. Revised edition. *Mycol. Pap.* **122**:1–59.

Waterhouse, G. M. (1970b). Taxonomy in *Phytophthora*. *Phytopathology* **60**:1141–1143.

Wilson, G. W. (1907). Studies in the North American Peronosporales. I. The genus *Albugo*. *Bull. Torrey Bot. Club* **34**:61–84.

Wilson, G. W. (1914). Studies in the North American Peronosporales. V. A review of the genus *Phytophthora*. *Mycologia* **6**:54–83.

Yerkes, W. D., and C. G. Shaw. (1959). Taxonomy of *Peronospora* species on Cruciferae and Chenopodiaceae. *Phytopathology* **49**:499–507.

Yurova, N. F. (1962). On the systematic position of the genus *Phytophthora*. *Bot. Zh. (Leningrad)* **47**:1499–1503.

Zygomycotina
Zygomycetes

CHAPTER 11

Mucorales

C. W. Hesseltine and J. J. Ellis

Northern Regional Research Laboratory
Agricultural Research Service
U. S. Department of Agriculture
Peoria, Illinois

I. GENERAL DISCUSSION OF THE CLASS ZYGOMYCETES

The Zygomycetes are fungi which reproduce asexually by aplanospores (nonmotile sporangiospores), by modified sporangial units functioning as conidia, or by true conidia. The sexual state is represented by zygospores. Gametangia are often morphologically similar to each other but sometimes vary greatly in size. The Zygomycetes are represented by three orders: the Mucorales, Entomophthorales, and Zoopagales. Possibly a fourth order, Endogonales, should be recognized, but this matter will be discussed under the Endogonaceae (p. 201). The Entomophthorales are fungi in which the sporangium has been reduced to function as a single conidium that is forcibly discharged at maturity. In the Zoopagales, conidia are borne singly or in chains but are not forcibly discharged; its members are parasitic on invertebrates. In both orders the gametangial vesicle does not form a part of the zygospore wall, and conidia are the only type of asexual spore produced. Both orders are typically parasitic on insects, invertebrates, and other animals, although some may be saprobic. The third order, Mucorales, consists of fungi which reproduce asexually by means of nonmotile but sometimes appendaged sporangiospores borne in sporangia, merosporangia, sporangiola, or as one-spored sporangia or conidia. The outer zygospore wall is formed by modification of the gametangia walls. Although most members are saprobic, some genera attack other fungi and may, in some instances, attack animals and plants.

The Zygomycetes, as a class, have long been considered a rather natural group because of the production of aplanospores as asexual means of reproduction and, second, because in its members sexual reproduction is the result of the fusion of gametangia to produce the zygospore. A further indication of the unity of these fungi is the composition of the cell wall which is chitosan-chitin in nature where species have been investigated by modern methods. Chitin has been found in the cell walls of three families of Mucorales (*Mucor, Rhizopus, Phycomyces, Absidia, Cunninghamella,* and *Mortierella*) and in the Entomophthorales (*Entomophthora* and *Basidiobolus*), but it has not been investigated in the Zoopagales.

Asexual evolution appears to have gone from multispored sporangia, to reduced number of sporangiospores, to one-spored sporangiola where the sporangial wall is distinct from the sporangiospore wall, to forms on which it is difficult to demonstrate the two walls except with special techniques.

Evolution also appears to have progressed from a purely saprobic type of nutrition to that of dependence upon other fungi in the Mucorales and to parasitism of more highly evolved animals and plants in the other two orders. Members of the Entomophthorales are often insect pathogens and, in a few species, are plant and fungal pathogens as well. In the Zoopagales, which seem to have paralleled the Entomophthorales in evolution, their members are parasitic on soil- and aquatic-inhabiting animals. Consequently, extensive development of mycelium is no longer a necessity and, as a result, has been greatly reduced as it also has been in the more advanced families of the Mucorales. Nutrition of the Zygomycetes has also followed the same direction of evolution. In the primitive Mucorales, no vitamins or growth factors are required; they can grow readily on an inorganic nitrogen source with minerals and a sugar. In more specialized forms, such as *Pilobolus*, all species have an absolute requirement for such growth factors as ferrichrome (coprogen). Members of the Entomophthorales are capable of growing saprobically in pure culture, but the nutrient medium is a complex one. In Endogonaceae, no true pure cultures can be maintained, probably because of the complexity of their growth requirements which they obtain in nature from their association with the roots of higher plants. Associated with the development in the Entomophthorales and Zoopagales of a parasitic existence on insects and other animals has been the development of very potent proteolytic enzyme systems, especially in the Entomophthorales. On the other hand, a variety of secondary metabolites does not appear to have evolved, based upon our current knowledge.

A further specialization in the Entomophthorales is the evolution of the forcible discharge of spores. If the spore does not alight on a favorable substrate, it, in turn, produces a second and smaller conidium which is again shot away; this may be repeated again.

We believe that there is little or no evidence now available to support even a tentative conclusion on the mysterious origin of the Zygomycetes.

II. ORDER MUCORALES

A. General Characteristics

The thallus is typically coenocytic and eucarpic with extensive mycelium. Reproduction occurs by one to many sporangiospores (single-celled aplanospores) formed in terminal sporangia, or less often by conidia. The conidia are produced singly (with one exception). Mucorales are zygosporic and heterothallic or homothallic; the outer wall of the zygospore is formed from the modified wall of the gametangium. They are typically saprobic or, less often, parasitic on other fungi and rarely parasitic on insects or vertebrates, including man.

The order Mucorales consists of fourteen families. It is a group of filamentous, typically saprobic fungi. Members of the Mucorales are among the most common fungi encountered when isolating microorganisms from soil, air, dung, or decaying plant material. Because of the rapid growth of the mycelium, the Mucorales are conspicuous in isolation plates, often overgrowing the slower growing fungi, bacteria, and actinomycetes. Its members are nearly all terrestrial. Some forms are obligate parasites on other fungi, especially on Mucorales and mushrooms, and a few are weak parasites on higher plants and animals. Members of one family occur frequently in a mycorrhizal association with many higher plants, both annuals and perennials.

When a spore of one of the Mucorales germinates, it forms one or more germ tubes which repeatedly divide to form a multibranched mycelium. Since members of this order grow at such a rapid rate and are often those species first seen in the decay of vegetable matter, they have been referred to as the "sugar fungi." This is an apt name for they utilize most simple carbohydrates efficiently, leaving more complex material for other microorganisms to attack. The mycelium typically shows large radiating hyphae with shorter, thinner hyphae attached to it. Generally, the mycelium is devoid of cross walls or septa at an early age although septations are often seen in the mycelium of older colonies. In advanced families septa are formed in hyphae, and these often have a median plug. Asexual reproduction usually occurs abundantly after the mycelium has grown extensively. Fertile hyphae are formed aerially and may be positively phototrophic as in species of *Pilobolus* and *Phycomyces*. The fertile hyphae are either sporangiophores or conidiophores. The sporangiophores may or may not branch and are terminated with a sporangium which consists of an outer membrane containing, typically, many spores. Columellae often form in the center base of sporangia. Columellae are vesicles or central sterile regions continuous with the sporangiophores inside the sporangium. In some genera the spores are borne linearly within a sporangial wall. Such sporangia lack columellae and are referred to as merosporangia. In others, the spores are greatly reduced in number, and such sporangia are referred to as sporangioles. When sporangioles are present, they are much smaller than the larger, multispored sporangia. Sometimes the sporangioles differ in shape and are borne differently from the sporangia. In some genera the spores are produced singly without a clearly defined sporangial membrane and can be called sporangiola. In addition to these asexual means of reproduction, the hyphae of some species produce cells, which have thick walls and contain a dense cytoplasm. These are referred to as chlamydospores and occur both in the aerial and subtrate mycelium. In some species, oidia are produced which are thin-walled and usually globose in shape. They tend to bud as in yeasts and break off as individual units, especially in liquid media. They may be terminal or intercalary, single or in chains. All spores produced in the Mucorales germinate to form mycelium and never zoospores. Trophocysts form in two genera of the Pilobolaceae. They are large, variously shaped, swollen cells from which the sporangiophores grow and are located in or on the substrate.

The sexual state is always a zygospore which is the result of fusion of multinucleate gametangia. Zygospores are thick-walled, nearly always heavily pigmented, and range from yellow to black. Their walls may be very roughened with peaks or nearly smooth and punctate. The outer wall of the zygospore is developed from the two fused gametangia. The zygospores are formed between two sections of hyphae (suspensors) which may be in a straight line on opposite sides of the zygospore and equal in diameter, or the suspensors may be twisted about each other and then separated like the jaws of a pair of tongs with the zygospore held by their tips. In some species the suspensors may be very unequal in diameter and length. In a few genera such as *Absidia* and *Phycomyces* numerous simple or branched outgrowths develop from the suspensors. In the more advanced mucors, the gametangia may be undifferentiated from the adjacent hyphae even when the zygospore is mature. Usually the zygospore contains a large globule. When the zygospore germinates, the germ tube develops either as a sporangiophore and sporangium or as mycelium. Germination of zygospores has been infrequently seen. In those species where two strains of opposite mating reaction must be brought together in order to produce zygospores, the species is said to be heterothallic; in those forms which carry both mating types in the same mycelium and where two strains are not required for zygospore formation, the condition is referred to as being homothallic. In some instances, zygospore-appearing structures are formed without the fusion of gametangia, and they are referred to as azygospores. The (+) and (−) signs of heterothallic species can be traced back to Blakeslee's designations since imperfect mating reactions often occur between species of the same genus or even different genera.

The source of cultures from nature is often a useful character in aiding one in the identification of Mucorales. For example, all members of the family Pilobolaceae are dung inhabiting, especially on dung of herbivorous animals. Members of the family Choanephoraceae are found growing primarily on the flowers and fruits of certain flowering plants, although occasionally isolates may be made from litter under the plants. *Choanephora cucurbitarum*, a member of this family, can usually be found growing on members of the family Cucurbitaceae. Species of *Syncephalis* and *Piptocephalis* are found parasitizing other Mucorales and are obligate parasites, although a few species have been reported parasitizing other fungi. All species belonging to the genera *Spinellus*, *Syzygites*, and *Dicranophora* are found only on fleshy basidiomycetes. Species such as *Gongronella butleri*, *Absidia spinosa*, *A. glauca*, *Mucor genevensis*, *M. hiemalis*, *M. ramannianus*, most species of *Mortierella*, and all species of *Zygorhynchus* are found in soil or soil-contaminated material. On the other hand, many species of *Mucor* and *Rhizopus* can be encountered almost anywhere in nature where nutrient is available and moisture is sufficient to allow growth and reproduction.

B. Economic Importance

Economically Mucorales are of importance because they cause a number of storage decays. Species of *Rhizopus* are important agents in the soft rot of sweet potatoes and in the so-called "leak" of strawberries, raspberries, peaches, and other fruit. Some members of *Rhizopus*, *Mucor*, and *Absidia* are always found in stored grains. In the past *Rhizopus* has been such a common cause of rapid spoilage of bread that one of its species has been referred to as the "black bread mold." Species of *Absidia*, *Mucor*, and *Rhizopus* have been found as causes of mucormycosis in man and animals. A recent survey in Great Britain of fungus diseases of domestic animals showed that next to *Aspergillus fumigatus* the most common fungus pathogens were *Absidia corymbifera* and *A. ramosa* which are nearly indistinguishable species.

Various species of *Rhizopus* form large amounts of fumaric and lactic acids. Many Mucorales, under appropriate fermentation conditions, are capable of rearranging steroid molecules, sometimes into more useful compounds. Species of *Mucor* and *Rhizopus* are used in the amylo process for making alcohol. *Rhizopus* is used in the fermenting of certain Oriental foods such as tempeh, and *Actinomucor* is used in making sufu or Chinese cheese from soybeans. Other genera such as *Blakeslea* make rather large amounts of beta-carotene.

C. Culture

Studies on members of the Mucorales should be made in pure culture. In those forms which are obligate parasites, the host as well as the parasite must be grown together. Good success in maintaining *Syncephalis* and *Piptocephalis* can be obtained by inoculating plates with a low and slow-growing *Mucor* such as *M. ramannianus*. Spores can be picked carefully from the parasite's sporangia and inoculated at the same places inoculated with *Mucor*. Usually the parasite will grow beyond the host as well as above it and, therefore, can be studied with relative ease. There are, however, some parasitic forms which do not attack *Mucor*, and, hence, another host must be used, such as *Cokeromyces*.

Media commonly used for the study of Mucorales are: Potato dextrose agar (PDA), Malt agar, Synthetic *Mucor* medium, Czapek's solution agar, Corn steep agar, YPSS medium (for details, see Booth, 1971), and Tomato paste-Oatmeal agar.

Solution I

Heinz Baby oatmeal food . . . . 20.0 gm

Tomato paste . . . . 20.0 gm

    Add to 500 ml of boiling water.

Solution II

Difco agar . . . .   15.0 gm

Tap water . . . . 500.0 ml

Melt agar by steaming, then mix two solutions and autoclave.

Ordinarily the strain being studied is inoculated onto two plates of the first four media above. Inoculation is made on one plate at a single place in the center of the dish and at three equally spaced places on the second plate. Potato dextrose agar is an excellent sporulation medium and can be used for carrying nearly all stock cultures. Malt agar is used for the same purpose and, in addition, it was used to describe many different species before 1945. Czapek's solution agar usually allows only sparse growth except in the case of *Actinomucor* and *Cunninghamella* and a few other species. It is often used to detect the presence of slower-growing contaminants. Czapek's solution agar is very useful to demonstrate branching of individual sporangiophores which are often too tangled and twisted to be seen clearly on the richer media given above. All descriptive work should, however, be done from cultures growing on Synthetic *Mucor* agar, since it is a simple defined medium which can be readily prepared and reproduced. Most Mucorales will grow readily at 25°C, but some forms such as *Mucor miehei* must be incubated at a higher temperature. Others such as *Dicranophora* and *Chaetocladium jonesii* must be incubated at a lower temperature in order to get normal growth and sporulation.

For the production of zygospores in heterothallic forms, matings are made on malt agar and potato dextrose agar. Each of the strains to be mated is inoculated a short distance from the other on the agar. Ordinarily two strains are contrasted on a single plate, although sometimes it is convenient to place three or four strains on a single plate. To supplement these two media, tomato paste-oatmeal, corn steep agar, and YPSS are commonly used. In a number of cases, temperatures below or above 25°C are required for zygospore formation even though asexual reproduction occurs readily at 25°C. *Thamnidium elegans, Pirella circinans*, and *Phycomyces nitens* form zygospores at low temperatures, whereas 34°C is optimum for zygospore production in *Absidia ramosa* and *A. corymbifera*.

One should study all forms under low magnifications to determine the general habits of growth. Microscopic mounts should be made by removing a small amount of the fruiting hyphae, wetting it in 70% alcohol, and adding distilled water. Usually this will show the material in excellent condition for study. Most mounting fluids are unsatisfactory because of the distortion and collapse of the very delicate microscopic structures. The substrate mycelium should also be studied, and the cultures should be examined again when they are about 1 month old for the development of chlamydospores, oidia, and other changes that occur in the mycelium with age.

D. Families

Fourteen families are recognized. Separation is based on the nature of the asexual means of reproduction including the nature of the sporangium and the sporangiophores. To a lesser degree, the key is based on the habitat and the means of sexual reproduction. The latter characteristic would be given more importance except for two reasons: The sexual state often is not known, for instance, in *Actinomucor* and in many species of *Mortierella*; and secondly, the sexual state is hard to induce and, therefore, is not a very practical characteristic for the average person who wants to identify his isolates. With only a few exceptions, the authors have seen and studied every family and genus in the key. The keys are constructed in such a manner that, in most cases at least, two or more characteristics are included so that if the reader is confused by one, he will have at least one other characteristic to consider.

It becomes obvious in the key to the families and genera of Mucorales that at least two new families must be recognized; namely, Saksenaeaceae and Radiomycetaceae, each containing two genera. They will be described elsewhere.

Choanephoraceae

Three genera are known (*Choanephora*, *Blakeslea*, and *Gilbertella*). All members are characterized by possessing large columellate sporangia with persistent sporangial walls which break open as two halves. Sporangiospores are often striate, brown to purple, and possess long, stiff, hairlike spines at their ends. In addition to sporangia, all but two species possess distinct fruiting stalks which bear conidia (one-spored sporangiola) or sporangiola. These, as well as the sporangia, are dark colored. The sporangia are typically borne circinately at least when young. Zygospores are known in all genera and in *Choanephora* and *Blakeslea* are lightly roughened and borne between tong-shaped suspensors. In *Gilbertella* they are roughened and borne between opposed suspensors, a characteristic indicating a relationship to the Mucoraceae. All members are heterothallic. All members grow on the fruit or flowers of higher plants and in *Blakeslea* and *Choanephora* they are tropical, except for *C. cucurbitarum*.

Cunninghamellaceae

The family consists of four genera: *Cunninghamella*, *Mycotypha*, *Phascolomyces*, and *Thamnocephalis*. All members of this family produce conidia with no other asexual reproductive structures on their conidiophores, although chlamydospores are formed in the substrate mycelium. Zygospores are known in two genera: *Cunninghamella* and *Mycotypha*; in each they are *Mucor*-like. Only one species in *Mycotypha* and one in *Cunninghamella* are known to be homothallic. Consequently, this family represents a very homogeneous group of genera with respect to both sexual and asexual means of reproduction. They are separated on the basis of the shape of vesicles on which the conidia are borne. In *Mycotypha* the vesicles are elongate and appear like miniature cattails. In *Phascolomyces* vesicles are sessile and the single species possess large, conspicuous, dark-colored chlamydospores in the substrate mycelium. The least known of the genera is *Thamnocephalis* which has been reported only twice. One was found in the United States and the other in India, and each was considered as a distinct species. Only *Cunninghamella* is commonly encountered, and it occurs frequently in soil all over the world. Its members are also quite common on nuts from the Tropics. Many strains of species of *Cunninghamella* can grow luxuriantly on a simple medium with inorganic nitrogen.

Dimargaritaceae

This family consists of four genera: *Spinalia, Dimargaris, Dispira,* and *Tieghemiomyces.* Our information on this family is almost entirely based on the recent studies of Dr. R. K. Benjamin who also proposed the family. The four genera represent a natural group of forms all characterized by two-spored sporangia with the sporangiospores arranged in a linear fashion and, hence, are reduced merosporangia. The genera are separated on the basis of morphology of the sporangiophores, which range from simple to very complex structures. Zygospores are globose, hyaline, without pronounced roughening, punctate, and borne between undifferentiated hyphae. The sexual state is known for all genera except *Spinalia* which was described from slides. The members of this family are either parasitic or are facultative parasites on other Mucorales. All members of the family have regular septa with a central plug and these septa are formed from the beginning of the growth of hyphae. The least known genus, *Spinalia,* has not been grown in culture. Because of all the characteristics noted above, this family parallels the Kickxellaceae and is a very homogeneous assemblage far advanced in the evolution of Mucorales. These two families also may represent geographical isolation in arid regions since most of the species have been found in the Southern California area growing on the dung of rodents, frogs, and in soil.

Endogonaceae

As considered here, the family consists of three genera: *Endogone, Sclerocystis,* and *Glaziella,* as arranged by Zycha (1935). Of all the families of the Mucorales, this is the least known and most in need of monographing. Classification is based upon the nature of the sporocarps which may contain sporangia, chlamydospores, or zygospores. In *Endogone,* many species form vesicular-arbuscular mycorrhiza on the roots of both annual and perennial plants, but either they do not form sporocarps or they have not yet been found. The family, because of its importance as a common mycorrhizal form, has received a great deal of attention but, as yet, no pure cultures have been made. This has handicapped an accurate determination of the number of genera and species. The knowledge we have now clearly indicates the placement of the group in the Zygomycetes because of the zygospores and the nonmotile sporangiospores, but it is not clear whether the group belongs in the Mucorales or whether it should be recognized as a fourth order in the Zygomycetes. The most recent account of the group by Gerdemann (1971) follows that of Zycha and is also the one we have followed. Undoubtedly, the nearest relatives of the Endogonaceae are to be found in the Mortierellaceae. This conclusion is based on the following facts. In the Mortierellaceae, reduced sporangia without columellae are found; its members are almost all confined to the soil; cultures tend to fruit rather poorly in culture; reduced sporangia and terminal chlamydospores are found; and in some instances, the zygospores are surrounded by a weft of mycelium. All these seem to be characteristics which are forerunners of the condition found in the Endogonaceae.

If one were to consider all the Zygomycetes with sporocarps as an order character, then *Sclerocystis*, *Glaziella*, and part of *Endogone*, with perhaps some species of *Mortierella*, might be placed in it. However, this does not seem to be a wise move until much more is known about the Endogonaceae and their life histories. Moreau (1953) recognized the order Endogonales with one family containing *Endogone*, *Glaziella*, and *Sclerocystis*.

Two genera of very uncertain position in the Mucorales which may belong in the Endogonaceae are *Azygozygum* and *Massartia*. In *Azygozygum* no sporangia or conidia were found, but it does possess chlamydospores and zygosporelike structures. It was isolated as an organism causing a collar or root rot of *Antirrhinum majus* and, in Chesters' opinion (1933), it is closely related to some species of *Mortierella*. The chlamydospores may be intercalary or terminal and possess short spines as seen in some members of the genus *Endogone*. The zygospores are similar to those illustrated by Nicolson and Gerdemann (1968). These workers showed slender suspensors slightly or not at all swollen just as illustrated by Chesters for *Azygozygum*. *Massartia* does not possess sporangia but does possess zygospores formed between tonglike suspensors with nearly smooth walls.

Helicocephalidaceae

This family was proposed by Boedijn (1958) for *Rhopalomyces* and *Helicocephalum*. At first, these genera seem unrelated. In *Rhopalomyces* the large deciduous conidia are produced on vesicles at the tip of long, stout conidiophores. Conidia produced in *Helicocephalum* are in a single coiled chain at the end of a conidiophore. All the spores are formed simultaneously and, at maturity, they collect to form a viscous droplet. Unfortunately, we do not know the nature of the zygospores in either genus. On the other hand, there are some characteristics which are alike and show relationships between them: (1) Both genera produce conidia only at ends of unbranched conidiophores. (2) Conidia are brown or blackish-brown and very large. (3) The vegetative mycelium is hyaline and small in diameter, similar to that of *Syncephalis* but not with the type of septa seen in *Syncephalis*. (4) Rhizoids are found at the base of the conidiophores. (5) Both genera are found in habitats of rotting and decaying material often contaminated with animal feces and typically supporting a large nematode population. (6) Species in both genera attack and destroy nematode eggs. (7) In some species of *Helicocephalum* the tip of the conidiophore enlarges to appear as the initiation of a vesicle. (8) Conidia germinate as a single germ tube from the end of the spore. In *Rhopalomyces* the germ tube develops from the apiculus. *Rhopalomyces* requires a special type of medium for growth (Ellis, 1963). While *Helicocephalum* has never been grown in pure culture, but one suspects it will grow on the same special medium as required by *Rhopalomyces*. This consists of liver and animal fat incorporated into a nutrient medium.

Kickxellaceae

The family is represented by *Spirodactylon, Spiromyces, Kickxella, Martensiomyces, Linderina, Dipsacomyces, Martensella,* and *Coemansia*. The family is a homogeneous assemblage of forms all possessing one-spored sporangia and with zygospores formed from progametangia that are undifferentiated from the hyphae. The zygospore walls are smooth, or nearly so, and thick walled. Vegetative mycelium is septate from the beginning and the septa have median plugs. The colonies are often in colors of orange to yellow and typically restricted in growth. The fertile branches which are septate or nonseptate (sporocladia) bear one-spored sporangiola acrogenously on ovoid or elongate-ellipsoid cells (pseudophialides). The members of the family are found in soil, dung, dead insects, and organic debris and a few species are parasites or facultative parasites on other fungi. The family is obviously very closely related to the Dimargaritaceae because of the manner of formation of zygospores and the reduced number of spores in a sporangium. The genera are differentiated on the basis of the nature of the sporangiophores and the sporocladia. Most of our information is based on the studies of Linder (1943) and R. K. Benjamin (1958, 1959, 1966).

Mortierellaceae

Four genera are placed in this family, namely: *Haplosporangium, Mortierella, Dissophora,* and *Aquamortierella*. Recently Linnemann in Zycha *et al.* (1969) monographed the family and included the above genera except *Aquamortierella* but added the genus *Echinosporangium*. She further concluded that *Herpocladium* probably belonged here, but it is considered as a *nomen dubium*. The family is composed of species which produce from one- to many-spored sporangia with columellae lacking or only vestigal. In the species with sporangia that contain only one to two spores, the sporangial wall can readily be seen as a separate structure from the wall of the sporangiospore. *Mortierella* and *Haplosporangium* are rather common genera usually found in soil. Although the colonies grow rapidly and are typically white with wavy zonations, they tend to sporulate sparsely. Chlamydospores with spiny and rough thick walls are formed at the ends of branches in the aerial mycelium. These are often encountered in the genus *Mortierella*. Zygospores are unknown except in *Mortierella* and even here they have been reported for only a few species. They are formed between tong-shaped suspensors, are surrounded by a weft of mycelium, and appear to anticipate the sporocarps found in the Endogonaceae. *Haplosporangium* is a genus which has sporangia with only one or two spores. *Mortierella* possesses sporangia with a few to many sporangiospores, although one section has only stylospores and completely lacks sporangia. In *Dissophora*, many-spored sporangia are present but are borne on simple sporangiophores arising from hyphae of indeterminate growth. In *Aquamortierella* the sporangia contain many spores which have two vermiform appendages at either end, a condition not encountered in any other genus in the Mucorales.

Mucoraceae

This is the family from which all the other families in the Mucorales developed. It also contains the largest number of genera (20). Although a number of genera are monotypic, three genera, *Absidia*, *Mucor*, and *Rhizopus*, each contains large numbers of species. Members of the family possess a number of characteristics in common. All have large multispored sporangia which contain well-defined columellae. The exception to this is the curious mold, *Azygozygum*, which does not produce sporangia. In the family, no sporangioles are present except in *Backusella* which has conidia as well as sporangioles.

Evolution in the Mucoraceae went in several directions and at different rates from the basic *Mucor*-type organism. In *Mucor*, hyaline sporangiophores that are variously branched are formed from substrate mycelium and each bears a multispored sporangium at its tip. In some cases the immature sporangium may be circinate, but eventually all are borne in an upright position. The globose to spherical sporangia have thin sporangial walls which may persist or deliquesce. Within the sporangium is a sterile, variously shaped region (columella) delimited by a wall from the spore-filled area. The columella is enlarged from the basal region of the sporangium. In some instances, chlamydospores are formed in the aerial and the substrate mycelium in great profusion.

The zygospore in *Mucor* is formed as a result of the fusion of two gametangia delimited from the suspensor by a septum. The suspensors are typically enlarged, equal, and opposite to one another. The two gametangia fuse, and this results in the zygospore which is always brown to black, heavy-walled, and covered with rough blunt conical projections or warts. Zygospores in *Mucor* are formed in the aerial mycelium.

Advanced characteristics in the family are (1) pyriform sporangia;

(2) persistent thick sporangial walls; (3) simple pigmented sporangiophores; (4) production of rhizoids and stolons; (5) reduction of the sporangial state; (6) production of subsporangial swellings; (7) simple to complex nutrition including parasitism; (8) suspensors with appendages; (9) zygospores borne on or in the substrate; (10) zygospores with thin, nearly smooth walls; (11) heterogametic.

As noted above, the sporangial wall of *Mucor* is either persistent or deliquescent and its shape is spherical or globose. From this basic type has developed the condition in *Pirella* in which the terminal sporangia are globose but the branches below the terminal sporangia produce large sporangia that are pyriform and borne circinately at maturity. Further evolution caused all the sporangia to be pyriform, the condition found in *Absidia*. Furthermore, the aerial hyphae have been modified to produce stolons and rhizoids, but the fertile branches or sporangiophores are never opposite the rhizoids. From *Absidia* two genera, *Gongronella* and *Chlamydoabsidia*, are segregated: the first because of a swelling below the sporangium and *Chlamydoabsidia* because of many large blackish chlamydospores found regularly in the aerial mycelium. Related to *Mucor* is a separate line in which the globose sporangia possess persistent walls and all the sporangia are borne circinately. This condition is found in *Circinella* except in one species, *C. linderi*, which has upright terminal sporangia, suggesting that it may be the ancestral type from which Thamnidiaceae arose through the genus *Helicostylum*.

Another evolutionary development in the sporangium was differentiation of a portion of the sporangial wall at its attachment to the sporangiophore into an apophysis. This evolved through *Pirella* and *Absidia* and also in the genus *Rhizopus*; however, *Rhizopus* retained globose sporangia. Like *Absidia*, *Rhizopus* developed rhizoids and stolons, but the sporangiophores developed opposite the rhizoids. *Rhizopus* also possesses large and blackish sporangia quite unlike those seen in *Absidia*. Its sporangiospores are pigmented and typically marked by striations. A further evolution in the *Rhizopus* line which began with the genus *Rhizopodopsis* was the development of branches in umbels bearing sporangia and sterile spines, leading up to the condition seen in *Rhizopus*. Thus, the sporangiophores evolved into stolons, and the umbels were the clusters of sporangiophores seen in *Rhizopus*. Undoubtedly, because of degeneration of the ability to form sporangia and sporangiospores, the genus *Chlamydomucor* evolved with reproduction dependent upon the enormous number of chlamydospores produced in both the aerial and substrate mycelium.

An entirely different development was the reduction in the branching of sporangiophores to simple, thick-walled, colored sporangiophores which bore large columellate sporangia with thick, persistent walls at their ends. This is represented by the *Phycomyces-Spinellus* line. *Spinellus* evolved further since it is restricted to a parasitic existence on mushrooms. Parallel with this evolution was the development of zygospores borne between tong-shaped suspensors that possess thornlike outgrowths. In *Spinellus*, outgrowths of the suspensor are less specialized; in both *Phycomyces* and *Spinellus* the zygospore wall is not so distinctly warted as that found in *Mucor*, perhaps because of the protection offered by the appendages from the suspensors surrounding the zygospore.

Several other genera represent only slight specialization over the *Mucor* condition. *Zygorhynchus* sporangia are like those in *Mucor* and all the species are homothallic (some species of *Mucor* are also); however, *Zygorhynchus* is distinct because of the pronounced heterogametic condition. The genus is always restricted to a soil habitat. In *Actinomucor*, rhizoids and stolons are produced but they are not so pronounced as in *Rhizopus* or *Absidia*. *Actinomucor* sporangia are hyaline and globose in shape, but are not dark in color. Its sexual state is not known. In *Syzygites* the sporangiophores are regularly dichotomously divided, and it grows in nature only on mushrooms. The zygospores are like *Mucor* except for sterile mycelium surrounding the zygospore which arises at the base of the opposed suspensors. *Sporodiniella* is related to *Syzygites*, with sporangia on dichotomous branches. In addition it has long, slightly bent spines and is, presumably, heterothallic. It was found in Indonesia growing on remnants of insects. *Parasitella* appears almost identical in its asexual state with *Mucor hiemalis* but has simple projections on its suspensors and is parasitic on mucors. Like *Mucor* it grows luxuriantly as a saprobe. The curious genus *Backusella* appears to be like *Mucor* but, in addition, possesses conidia and sporangioles all on the same sporangiophore and is undoubtedly evolved from *M. dispersus*.

The genus *Rhizomucor* contains two or three species with dark-colored sporangia which lack an apophysis and have poorly developed stolons and rhizoids. *Rhizomucor* is the only thermophilic genus in the whole order.

The remaining genus, *Azygozygum*, is perhaps the most difficult to place in the order and, therefore, for the time being it is left in the Mucoraceae. It completely lacks sporangia, sporangioles, or conidia. As its name implies, it forms azygospores in abundance. Unlike the Endogonaceae, it grows readily in culture; however, it may eventually be removed to this family. Chlamydospores are formed both terminally and intercalary and some possess short, blunt spines, and are similar to the stylospores of some species of *Mortierella*. Chesters (1933) suggests it may belong in the family Mortierellaceae. The suspensors may be tong-shaped or they may be more or less opposed. The gametangia are swollen and the zygospores, if they are zygospores, appear to be smooth walled.

As noted above, some family members are exclusively inhabitants of mushrooms (*Syzygites*, *Dicranophora*, and *Spinellus*). Others are strictly soil inhabitants (*Zygorhynchus* and many species of *Mucor* and *Absidia*). Some are dung inhabiting (*Pirella* and some species of *Mucor*). On the other hand, members of *Rhizopus* and *Absidia* are widely distributed on stored grain, decaying fruits and vegetables, in the air, and in compost. Species of *Mucor*, *Rhizopus*, and *Actinomucor* are of economic importance because of their use in fermented foods (tempeh, sufu, and lao-chao) and in steroid transformations; they also cause rapid decay of ripe and harvested fruit and vegetables.

Pilobolaceae

Three genera, *Pilobolus*, *Pilaira*, and *Utharomyces*, are known, and these constitute a natural group always found growing on dung. All species possess sporangia with dark-colored, persistent walls containing many spores and with well-defined columellae that are often applanate. Sporangiophores are usually large and elongate and often phototrophic. In *Pilobolus* and *Utharomyces* special swollen areas separated from the vegetative mycelium by a septum are found at the base of the sporangiophores. These swellings are on, or more often in, the substrate and are called trophocysts. Zygospores are known for some species of *Pilobolus* and *Pilaira*. These are located on the substrate and are formed between tong-shaped suspensors. The zygospore wall is thick and smooth or nearly so and light brown to black. So far as is known, all members of the family are heterothallic. The nearest relative of this family is *Phycomyces*. Evolution in the family is quite clear, with *Pilaira* being the most primitive since its sporangia are not discharged; it does not have trophocysts and it grows readily on the usual culture media used for Mucorales. *Utharomyces* is very rarely found and apparently is restricted to the Tropics. It is intermediate between *Pilaira* and *Pilobolus* in that it has trophocysts and subsporangial swellings; however, the sporangia are not shot away at maturity. *Pilobolus* represents the end of the evolutionary development. Its sporangia are always shot off the sporangiophore, it has trophocysts, and there is a large swelling below the sporangium. Also, all species of *Pilobolus* require a special growth factor, coprogen (ferrichrome), that is not required by *Pilaira* or *Utharomyces*.

Piptocephalidaceae

This family consists of two genera, *Piptocephalis* and *Syncephalis*. Both genera are parasitic on other fungi and, in almost all cases, on other Mucorales. Both are worldwide in their distribution and are often encountered in material in which other Mucorales are growing. The vegetative mycelium is delicate and fine. Asexual reproduction is exclusively by means of one- to many-spored merosporangia and these are usually borne on vesicles. In *Syncephalis* the vesicles are not deciduous; in *Piptocephalis* they usually are. The genera are also separated easily by whether or not the sporangio-phores are branched. In *Piptocephalis* they are always branched dichoto-mously while in *Syncephalis* they are unbranched except in three instances. Zygospores in both are produced between opposed or tong-shaped and coiled suspensors. Their walls are thick and somewhat roughened and/or reticulated.

Radiomycetaceae[1]

This new family contains two genera: *Radiomyces* and *Hesseltinella*. The family is made up of genera which possess only sporangioles borne on secondary vesicles which are, in turn, borne on branches arising from an enlarged primary vesicle. Some sporangioles bear long spinose processes. Both genera possess stolons and rhizoids.

The zygospores are known in the two species of *Radiomyces*. The game-tangia are opposite one another and are enlarged as in the gametangia of *Mucor*. However, adjacent to the zygospore long, flexuous, branched appendages grow out and sheath the zygospore. The zygospores are smooth or nearly so.

The genera are distinguished by the fact that in *Hesseltinella*, a single many-spored sporangiole is borne on the secondary vesicles on a short apical stalk while in *Radiomyces* numerous one- to many-spored spor-angioles are borne on the secondary vesicle.

Hesseltinella was isolated from paddy soil in Brazil; the two species of *Radiomyces* were isolated from mouse and lizard dung in California.

Saksenaeaceae[1]

This new family is admittedly an unnatural one since the zygospores in neither *Saksenaea* nor *Echinosporangium* are known. For this reason, it is not certain whether these genera are closely related or not. On the basis of asexual reproduction, they have in common sporangia which are unlike any other known in the Mucorales. The sporangia of *Echinosporangium* are borne aerially and are transversely elongate, cylindrical to sausage-shaped, with one to five spines at the apices. In *Saksenaea* the sporangio-phore arises from above short rhizoids and forms a long-necked, flask-shaped sporangium with a distinct columella in the basal venter. Spores are released through the long neck of the sporangium. Both genera are soil inhabitants, *Saksenaea* having been reported in India and the United States and *Echinosporangium* in the United States.

[1] This new family is to be published in *Mycologia* **66** (1974).

The two genera are alike because each has: (1) neither globose nor pyriform sporangia, (2) soil habitat, (3) rapid-growing mycelium, and (4) no other means of reproduction except for the many-spored sporangia.

Syncephalastraceae

The family is monogeneric. The single species, *Syncephalastrum race-mosum*, reproduces by merosporangia borne deciduously on vesicles. The mycelium grows rapidly and the branched sporangiophores may be modified into stolonlike structures which produce rhizoids. Septa are not produced except to delimit reproductive structures or they are produced elsewhere only when the culture is old. This heterothallic species readily produces zygospores formed between equal gametangia in the aerial hyphae. The zygospores are dark and roughened and are similar to *Mucor* zygospores in formation and appearance. It is a common saprobic form worldwide in distribution. It grows rapidly and well on simple media and is found in soil, on dung, and in grain in storage. Undoubtedly, this is the ancestor of *Syncephalis* and *Piptocephalis*.

Thamnidiaceae

This family consists of seven genera (*Thamnostylum*, *Cokeromyces*, *Chaetostylum*, *Chaetocladium*, *Thamnidium*, *Helicostylum*, and *Dicrano-phora*). All but *Dicranophora* are undoubtedly members of the family. All species of the family have sporangiola which are distinct from the large terminal *Mucor*-like sporangia found in some genera, such as *Thamnidium* and *Helicostylum*. In others, only sporangiola are produced and the branches end in thornlike sterile spines. Zygospores are known in *Cokeromyces*, *Thamnidium*, *Thamnostylum*, *Dicranophora*, and *Chaetocladium*, but unknown in *Chaetostylum* and *Helicostylum*. Zygospores are known in *H. piriforme*, but this species was transferred to *Thamnostylum* by von Arx and Upadhyay in von Arx (1970). In *Thamnidium*, *Chaetostylum*, and *Cokeromyces*, the zygospores are formed between enlarged, nearly equal suspensors and the zygospores are dark, heavy-walled, and *Mucor*-like. In *Chaetocladium brefeldii*, zygospores are formed on zygophores near the surface of the agar and one suspensor is enlarged to the size of the zygospores. Zygospores are brown in color with rough walls. The genus *Chaetocladium* is free living, but it also can be an active parasite on other Mucorales and, when parasitic, produces gall-like structures on the host. Information on *Dicranophora* is mostly obtained from the study of Dobbs (1938). Zygospores are creviced but never warty with the tong-shaped suspensors very unequal in size—one the size of the hyphae, the other swollen. The single species is homothallic and is found growing on mushrooms. Dobbs was of the opinion that this genus belongs near *Syzygites* and *Spinellus* rather than *Thamnidium*. It also has a sexual state that closely resembles oomycetes because of the manner in which copulation of the gametangia occurs. Sporangioles are produced on dichotomously branched sporangiophores which also possess a large columellate sporangium. Columellae are also found in the sporangioles.

None of the members of the family Thamnidiaceae is commonly found, although many of the species are worldwide in distribution. They occur on dung, mushrooms, soil, and decaying vegetation. *Thamnidium* and *Chaetostylum* have been reported often from meat products stored at low temperature.

The genera are separated rather easily by the manner of the branching of the sporangiophores. In *Cokeromyces* and *Thamnostylum*, the sporangioles are always borne circinately and the sporangioles are pyriform. *Cokeromyces* lacks a large *Mucor* sporangium and its known species are all homothallic. On the other hand, *Thamnostylum* is heterothallic and possesses a terminal sporangium. The *Mucor*-like zygospores are similar for these genera. In the other five genera, the sporangioles are borne on straight branches and are globose. In *Chaetostylum* and *Chaetocladium*, the ends of fertile branches end in long, stiff spines. In *Chaetostylum*, the sporangioles are many spored while in *Chaetocladium* they are always single-spored. In *Thamnidium*, the branches of the sporangiophore that bear the sporangioles are repeatedly dichotomously divided. *Dicranophora* also has more or less irregular dichotomously divided branches but, unlike *Thamnidium*, grows on mushrooms, is homothallic, and has zygospores not formed like *Mucor*. Its zygospore wall is not covered with blunt projections. *Helicostylum* has its sporangioles on branches which are never dichotomously divided but are simple or irregularly branched.

Most authorities on the order have come to approximately the same point of view regarding primitive and advanced morphological and physiological characters. These may be summarized as follows: Primitive—(1) Mycelium nonseptate, septate only in old cultures or to delimit reproductive structures; (2) sporangia large and columellate; (3) progametangia opposed and highly differentiated; (4) zygospores on aerial hyphae; (5) zygospore wall thick, dark brown or black, roughened with projections; (6) nutrition simple; (7) free living. Advanced—(1) Vegetative and fruiting hyphae septate from beginning with septa highly modified; (2) sporangia greatly reduced to one or two spores; (3) progametangia little differentiated, often with the suspensors tong-shaped or parallel to each other; (4) zygospores on or in the substrate; (5) zygospore light colored, smooth or punctate; (6) growing as a facultative or obligate parasite on other fungi or in a mycorrhizal association with plants; (7) nutrition complex, not growing on simple media but requiring growth factors as well as organic nitrogen.

Of course, it is understood that evolution did not progress in all these characters at the same rate. For example, in the Cunninghamellaceae, the sexual state is a primitive type with dark, heavy-walled zygospores having very rough walls. The progametangia are opposed to each other. Asexually, all sporangia have been reduced to one-spored sporangia. Other advanced characters found in some families are sporangia forcibly being discharged and zygospores being surrounded by mycelium to form sporocarps.

III. KEY TO FAMILIES AND GENERA

KEY TO FAMILIES AND GENERA OF MUCORALES

1. Sporocarps produced containing mostly noncolumellate sporangia, zygospores, or chlamydospores or, sometimes, combinations of these. Often growing as vesicular-arbuscular mycorrhiza on roots of higher plants and these sometimes without sporocarps (*Endogone*) . . . . **Endogonaceae** (see p. 195)   2

1'. Sporocarps missing; not forming vesicular-arbuscular mycorrhiza on roots of higher plants 4

2(1) Sporocarps naked or surrounded by a variably developed peridium or sometimes covered with tomentum of hyphal tufts. Sporocarps compact or loosely coherent or absent, with chlamydospores or zygospores produced singly in soil or in root tissue . . . . **Endogone**   (Plate II, f)

Found associated with roots of higher plants; many species. Nicolson and Gerdemann, 1968.

2'(1) Sporocarps with sporangia, zygospores, or chlamydospores organized in a definite region 3

3(2') Sporangia, zygospores, or chlamydospores organized in a single layer; sporocarp not hollow . . . . **Sclerocystis**

Three species. Zycha *et al.*, 1969.

3'(2') Sporocarp with sporangia, zygospores, or chlamydospores in the wall of the hollow sporocarp . . . . **Glaziella**

One species. Zycha *et al.*, 1969.

4(1') Vegetative and fruiting hyphae regularly septate from the beginning; septa with median plugs; many-spored sporangia absent 5

4'(1') Vegetative and fruiting hyphae nonseptate, often irregularly septate in age; septa without median plugs; many-spored sporangia present or absent 16

5(4) Imperfect state represented by 1-spored sporangia functioning as conidia . . . . **Kickxellaceae** (see p.197)   6

5'(4) Imperfect state represented by 2-spored sporangiola (merosporangia) . . . . **Dimargaritaceae** (see p.195)   13

6(5) Fertile branches or fertile sporangiophore segments distinctly coiled 7

6'(5) Fertile branches or fertile sporangiophore segments relatively straight or slightly arched 8

7(6) Sporocladium septate, tapering to a pointed apex; sporangiola arising from pseudophialides arranged transversely on inner surface of sporocladium . . . . **Spirodactylon**
   On mouse or rat dung, California; Benjamin, 1959; 1 species, *S. aureum.*

7'(6) Sporocladium aseptate, constricted terminally; sporangiola pleurogenous on terminal swelling, arising successively by budding . . . . **Spiromyces**
   On mouse dung, California; Benjamin, 1963; 1 species, *S. minutus.*

8(6') Sporangiophores determinate; sporocladia arranged in a terminal whorl or umbel
   9

8'(6') Sporangiophores indeterminate; sporocladia formed laterally or apically but appear pleurogenous through continued growth of the main sporangiophore axis 10

9(8) Sporocladia sessile, forming simultaneously in an apical whorl . . . . **Kickxella**
   Syn: *Coronella, Coemansiella*; Linder, 1943; Benjamin, 1958; 1 species, *K. alabastrina.*

9'(8) Sporocladia stalked, forming successively in apical umbels on recurved branches
   . . . . **Martensiomyces**
   From forest soil, Belgian Congo; Meyer, 1957; Benjamin, 1959; 1 species, *M. pterosporus.*

10(8') Sporangiola arising from pseudophialides clustered distally on an aseptate, ovoid or globose vesicle . . . . **Linderina**
   From soil, worldwide; Raper and Fennell, 1952; Benjamin, 1959; 2 species.

10'(8') Sporangiola arising from pseudophialides arranged more or less transversely on elongate, septate sporocladia 11

11(10') Sporocladia arising pleurogenously as lateral outgrowths; sporangiospores elliptic-fusiform with a long terminal spinous protuberance . . . . **Dipsacomyces**
   From soil; Benjamin, 1961; 1 species, *D. acuminosporus.*

11'(10') Sporocladia arising acrogenously but later appearing pleurogenous through continued growth of the main sporangiophore axis; sporangiospores elongate-elliptical to elongate-ovoid 12

12(11') Sporangiola arising from pseudophialides arranged transversely on the upper surface of septate sporocladia . . . . **Martensella**
   Presumably parasitic on other fungi; Linder, 1943; Benjamin, 1959; 2 species.

12'(11') Sporangiola arising from pseudophialides arranged transversely on the lower surface of septate sporocladia . . . . **Coemansia** (Plate I, e, f)
   Widespread; Linder, 1943; Benjamin, 1958, 1959; over 14 species.

13(5') Sporangiophores simple or sparsely branched in age; merosporangia arising directly from terminal vesicles . . . . **Spinalia**
   Presumably parasitic on *Mucor*; Benjamin, 1959, 1966; 1 species, *S. radians.*

13'(5') Sporangiophores abundantly branched but may be simple at first; merosporangia produced from sporiferous branchlets 14

14(13') Sporangiophore branches all ending in a terminal cluster of sporiferous branchlets; spores at maturity either moist or dry . . . . **Dimargaris** (Plate I, a–c)
   Parasitic on other fungi (mostly Mucorales); Benjamin, 1965; 7 species.

14'(13') Sporangiophore branches both sterile and fertile; spores at maturity dry . . 15

15(14') Sporangiophore branches coiled or recurved; sporiferous branchlets borne in terminal clusters . . . . **Dispira**
   Parasitic on other fungi (especially Mucorales), Benjamin, 1959, 1963, 1966; 3 species.

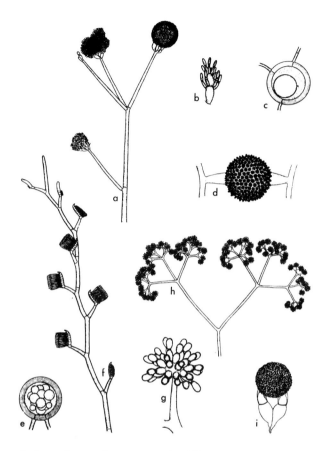

PLATE I.   a, b, Sporophore and merosporangia of *Dimargaris*, × 75, × 490; c, Zygospore of *Dimargaris*, × 176; d, Zygospore of *Syncephalastrum*, × 95; e, Zygospore of *Coemansia*, × 235 f, Sporophore and sporangiola *Coemansia*, × 136; g, h, Sporophore and merosporangia of *Piptocephalis*, × 600, × 62; i, Zygospore of *Piptocephalis*, × 170.

a, b, c, g, h (From R. K. Benjamin, 1959); d, i, (From R. K. Benjamin, 1966); e, f (From R. K. Benjamin, 1958).

15′(14′) Sporangiophore branches straight or only slightly arched; sporangiferous branchlets borne laterally . . . . **Tieghemiomyces**
        Found in California and Illinois; parasitic on Mucorales; Benjamin, 1959, 1961, 1966;
        2 species.

    16(4′) Producing merosporangia only 17

    16′(4′) Not producing merosporangia but sporangia, sporangiola, or conidia may be present. 19

17(16) Saprobic: sporangiophores irregularly racemosely branched; zygospores borne between opposed suspensors . . . . **Syncephalastraceae**   (see p.203)

One genus . . . . **Syncephalastrum** (Plate I, d)
    Widespread: saprobic in soil, decaying vegetation, and dung; 1 species, *S. racemosum.*
    Benjamin, 1959, 1966.

17'(16) Obligate parasites usually on other Mucorales; sporangiophores unbranched or, if
branched, then branched dichotomously; zygospores borne between 2 tong-shaped suspensors
. . . . **Piptocephalidaceae** (see p. 201) 18

  18(17') Sporangiophores typically dichotomously branched and cylindrical; merosporangiferous vesicles comparatively small and usually deciduous
  . . . . **Piptocephalis** (Plate I, g–i)
    Syn: *Mucoricola*; all species appear to be obligate parasites on other fungi (mostly
    Mucorales in nature); Benjamin, 1959, 1966; 22 species.

  18'(17') Sporangiophores simple or once or twice branched and usually tapered either
  apically or basally; merosporangiferous vesicles usually relatively large and not decidous
  . . . . **Syncephalis**
    Syn: *Monocephalis, Calvocephalis, Microcephalis*; common in cultures of soil and
    dung, typically parasitic on other Mucorales; Benjamin, 1959, 1966; 35 species.

19(16') Sporangiospores dark-colored with tufts of bristles at their ends. Sporangial wall
splitting into two halves. . . . . **Choanephoraceae** (see p.194) 20

19'(16') Sporangiospores never with tufts of bristles at their ends and usually not dark-colored.
Sporangial wall not splitting into 2 halves 22

  20(19') With sporangia only; zygospores formed as in *Mucor* and with large projections on
  zygospore wall. (One species of *Choanephora* has only sporangia) . . . . . **Gilbertella**
    On fruit; 2 species, *G. persicaria* is fairly common. Hesseltine, 1960b.

  20'(19') With sporangia and sporangiola or conidia. Zygospores formed between tong-
  shaped suspensors; zygospore wall without large projections 21

21(20') With sporangia and conidia (conidial stage missing in 1 species, *C. circinans*)
. . . . **Choanephora** (Plate III, 3, 4)
    On blossoms or in soil, generally in Tropics; *C. cucurbitarum* on cucurbit flowers,
    also in temperate regions; 4 species. Hesseltine, 1953; Hesseltine and Benjamin, 1957.

21'(20') With sporangia and sporangiola . . . . **Blakeslea**
    On plants and in soil; 2 species, the common one is *B. trispora.* Hesseltine, 1953.

  22(19') Sporangia flask-shaped with a distinct spherical venter and a long neck or irregular
  in shape, never pyriform, globose or spherical . . . . **Saksenaeaceae** (see p.202) 23

  22'(19') Sporangia always globose, spherical, or pyriform, or somewhat applanate . . 24

23(22) Sporangia flask-shaped with a spherical center and a long neck; columella present
. . . . **Saksenaea**
    In soil; 1 species, *S. vasiformis.* Saksena, 1953.

23'(22) Sporangia cylindrical to sausage-shaped with 1 to 5 spines at apices; no columella
. . . . **Echinosporangium**
    In arid soils; 1 species, *E. transversalis*; Malloch, 1967. Genus placed in the Mor-
    tierellaceae by Zycha *et al.*, 1969.

  24(22') Sporangial walls densely cutinized above; sporangia violently discharged or
  passively separated as a whole from the sporangiophores; suspensors tong-shaped
  . . . . **Pilobolaceae** (see p. 201) 25

  24'(22') Sporangial wall thin, not densely cutinized (see possible exception in *Phycomyces*
  in family Mucoraceae); sporangia not violently discharged or passively separated as a

whole; suspensors opposite or sometimes tong-shaped 27

25(24) Discharging sporangia violently . . . . **Pilobolus** (Plate II, a, b)
A number of species, all on dung; Buller, 1934.

25'(24) Not discharging sporangia violently 26

26(25') Without a subsporangial swelling . . . . **Pilaira**
Several species which are all dung inhabiting, especially common on rabbit dung;
Buller, 1934.

26'(25') With a subsporangial swelling and with trophocysts . . . . **Utharomyces**
On dung in Tropics; 1 species, *U. epallocaulus.* Boedijn, 1958.

27(24') Conidia present only and these borne upon a vesicle or in a coiled fashion 28

27'(24') Sporangia, sporangiola, or both present. If conidia then not borne directly on vesicles
or in coils 33

28(27) Conidiophores branched or if unbranched then with vesicles that are cattail-like.
Growing on ordinary nutrient media except *Thamnocephalis*
. . . . **Cunninghamellaceae** (see p.194) 29

28'(27) Conidiophores unbranched; not growing on typical mold media; often parasitic
on nematodes . . . . **Helicocephalidaceae** (see p. 196) 32

29(28) Conidia borne on globose or oval vesicles 30

29'(28) Conidia borne on greatly elongate vesicles which resemble cattails . . . . **Mycotypha**
Air and soil; saprobic; 2 species. Novak and Backus, 1963.

30(29) With many large swollen dark-colored chlamydospores in the substrate mycelium;
conidiophores with sessile vesicles on which conidia are borne along the conidiophores;
capitate vesicles are also present . . . . **Phascolomyces**
In soil and dung; 1 species, *P. articulosus.* Boedijn, 1958.

30'(29) Without greatly enlarged dark-colored chlamydospores in the substrate mycelium;
conidia borne on vesicles at the ends of conidiophores or on elongate branches; conidia
more or less roughened with spines 31

31(30') Conidiophores unbranched except in the upper quarter; globose vesicles borne rigidly at
right angles to the branches; ultimate ends of branches sterile and curved
. . . . **Thamnocephalis**
From dung; 2 species, very uncommon. Blakeslee, 1905.

31'(30') Conidiophores indefinite in length with branches ending in fertile vesicles; vesicles
never borne rigidly at right angles to the branch; ultimate ends of branches not sterile and
curved . . . . **Cunninghamella** (Plate II, g, h)
Saprobic; soil, nuts, decaying vegetation; many species. Cutter (1946) recognized
5 species.

32(28') Conidia large, brown, formed over the surface of a vesicle; never in chains
. . . . **Rhopalomyces**
[non *Rhopalomyces* Harder et Sorgel, Blastocladeaceae.] Found in very foul environ-
ments with bacteria and nematodes; 13 species. Ellis, 1963.

32'(28') Conidia formed as a single chain usually coiled at the tip of a single fruiting stalk;
vesicles never present or vestigial . . . . **Helicocephalum**
On decaying plant material; 3 species. Drechsler, 1943.

33(27') Sporangia without distinct columella (reduced to 1- or 2-spored sporangia in *Haplosporangium*); zygospores borne between tong-shaped suspensors often in a weft of mycelium;

stylospores often encountered in *Mortierella* . . . . **Mortierellaceae** (see p.197)   34

33′(27′) Sporangia with a distinct columella; often with sporangiola; zygospores formed variously 37

    34(33) Sporangia with 1 or 2 sporangiospores . . . . **Haplosporangium**
        On dung and soil; 6 species. Linnemann in Zycha *et al.*, 1969.

    34′(33) Sporangia with several or many sporangiospores 35

35(34′) Sporangiospores with 2 vermiform appendages. Found in aquatic habitats
. . . . **Aquamortierella**

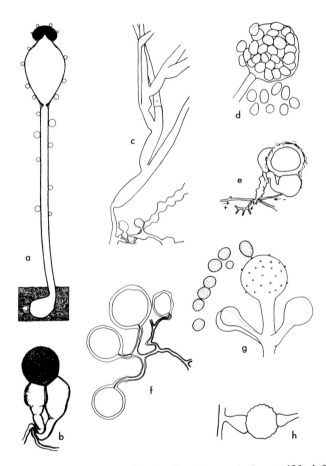

PLATE II.   a, b, *Pilobolus*, × 8; c–e, *Mortierella*. c, Sporangiophores × 125; d, Sporangium and sporangiospores, × 280; e, Zygospore, × 150; f, *Endogone* chlamydospores, × 80. g–h, *Cunninghamella* g, Conidiophore and conidia, × 300; h, zygospore, × 125.
    a, b (From Buller, 1934); e (From Gams and Williams 1963); f (From Gerdemann, 1965).

On pupae of net-winged midge in New Zealand; 1 species, *A. elegans.* Embree and Indoh, 1967.

35′(34′) Sporangiospores without vermiform appendages. Soil, dung, or slime-flux-inhabiting
. . . . 36

    36(35′) Sporangiophores arising from the ordinary vegetative mycelium; almost always soil-inhabiting . . . . **Mortierella** (Plate II, c–e)
    Common, worldwide in soil; 83 species recognized by Linnemann in Zycha. *et al.,* 1969.

    36′(35′) Sporangiophores arising in progression on special fertile hyphae of indeterminate growth . . . . **Dissophora**
    From dung or slime flux of trees; 2 species. Linnemann in Zycha *et al.,* 1969.

37(33′) With sporangiola only. Branches of fruiting mycelium never with sterile spines. Fertile branches always from vesicles . . . . **Radiomycetaceae** (see p.202) 38

37′(33′) With sporangia and sporangiola; if with sporangiola always with large sporangia (in *Chaetocladium* with 1-spored sporangiola but always with sterile spines) 39

    38(37) Secondary vesicles bearing many sporangiola on short stalks. Primary vesicle without a sterile short process above; homothallic . . . . **Radiomyces**
    On mouse dung and lizard dung, California; 2 species. Embree, 1959; Benjamin, 1960.

    38′(37) Secondary vesicles bearing but a single-stalked sporangiolum. Primary vesicle extended as a sterile short blunted process. Heterothallic . . . . **Hesseltinella**
    In paddy fields in Brazil; 1 species, *H. vesiculosa.* Upadhyay, 1970.

39(37′) With sporangiola always present, but in 1 genus single-spored. Most species with large *Mucor*-like terminal sporangia . . . . **Thamnidiaceae** (see p.203) 40

39′(37′) Without sporangiola (an exception is *Backusella* which also has conidia). Never with sporangiophores or their branches ending in sterile spines . . **Mucoraceae** (see p.198) 46

    40(39) Sporangiola borne circinately and pyriform in shape 41

    40′(39) Sporangiola not borne circinately and globose in shape 42

41(40) With a large terminal sporangium; heterothallic. . . . . **Thamnostylum**
    Saprobic and common; in dung and soil. Species of *Helicostylum* which have sporangiola borne circinately and pyriform in shape. Two species, of which the most common is *T. piriforme (H. piriforme).* von Arx and Updhyay in von Arx, 1970.

41′(40) Without a large terminal sporangium; homothallic . . . . **Cokeromyces**
    Saprobic on dung; 2 species. Benjamin, 1960.

    42(40′) Ends of fertile branches ending in stiff sterile spines 43

    42′(40′) Ends of fertile branches not ending in sterile spines 44

43(42) Several sporangiospores to a sporangium . . . . **Chaetostylum**
    Growing at low temperature on meat and dung; 1 species, *C. fresenii.* Hesseltine and Anderson, 1957.

43′(42) Single-spored sporangiola only . . . . **Chaetocladium**
    Soil and dung; often parasitic on other Mucorales; 8 species. Hesseltine and Anderson, 1957.

    44(42′) Homothallic; zygospores nearly smooth-walled with suspensors unequal; parasitic on higher fungi . . . . **Dicranophora**
    On mushrooms; 1 species, *D. fulva.* Dobbs, 1938.

    44′(42′) With none of these characteristics 45

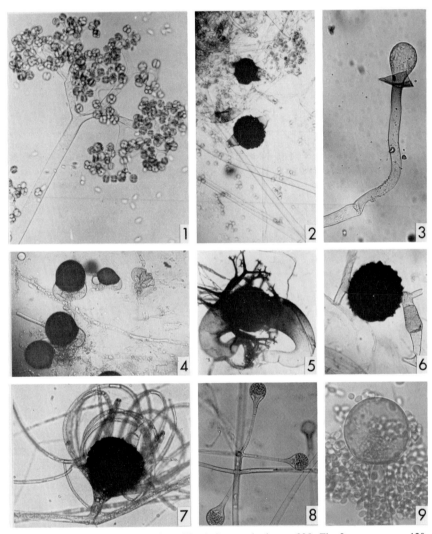

PLATE III.   Figs. 1, 2. *Thamnidium*. Fig. 1, Sporangiophore × 225; Fig. 2. zygospores × 120. Figs. 3, 4. *Choanephora*. Fig. 3, sporangiophore × 112; Fig. 4, zygospores × 112. Fig. 5, *Phycomyces* zygospore × 22. Fig. 6, *Mucor* zygospore × 146. Figs. 7, 8. *Absidia*. Fig. 7, zygospore × 128; Fig. 8, sporangia × 180. Fig. 9. *Mucor* deliquesced sporangium × 146.

Figs. 1, 2 (From Hesseltine and Anderson, 1956); Figs. 3, 4 (From Hesseltine and Benjamin, 1957); Figs. 6, 9 (From Hesseltine, 1954); Fig. 7 (From Ellis and Hesseltine, 1965); Fig. 8 (From Hesseltine and Ellis, 1964).

45(44′) With all the branches bearing sporangiola dichotomously divided

. . . . **Thamnidium** (Plate III, 1–2)
    On dung, leaf mold, soil, and cold stored meat; 4 species. Hesseltine and Anderson, 1956.

45′(44′) With the branches bearing sporangiola not dichotomously divided

. . . . **Helicostylum**

46(39′) All sporangia pyriform or at least part of the sporangia pyriform 47

46′(39′) Sporangia globose or at least not pyriform . . . . 50·

47(46) All sporangia pyriform and with an apophysis 48

47′(46) Terminal sporangia globose while the secondary sporangia are pyriform and borne circinately . . . . **Pirella**
    On dung; 2 species, *P. circinans*. Hesseltine, 1960a.

48(47) Regularly with a swelling below the sporangium; sporangia small

. . . . **Gongronella**
    Worldwide, in soil; 2 species, the common species is *G. butleri*. Hesseltine and Ellis, 1964.

48′(47) Without a swelling below the sporangium 49

49(48′) With large dark aerial chlamydospores . . . . **Chlamydoabsidia**
    From pea roots: 1 species, *C. padeni*. Hesseltine and Ellis, 1966.

49′(48′) Without large dark aerial chlamydospores . . . . **Absidia** (Plate III, 7, 8)
    Worldwide, in soil, decaying vegetation, grain, sometimes growing in animals and man; at least 28 species and varieties. Hesseltine and Ellis, 1961, 1964, 1966; Ellis and Hesseltine, 1966.

50(46′) Sporangia always abortive or not formed; always with many chlamydospores throughout the colony 51

50′(46′) Sporangia not abortive and usually formed in great abundance 52

51(50) With abortive sporangia constantly formed which resemble sporangia of *Rhizopus*

. . . . **Chlamydomucor**
    In Oriental fermented foods; 5 species. Brefeld, 1889.

51′(50) Without abortive sporangia but with many zygospores and chlamydospores

. . . . **Azygozygum**
    From diseased *Antirrhinum majus* plants in Great Britain; 1 species, *A. chlamydosporum*. Chesters, 1933.

52(50′) With all sporangia borne circinately and often in umbels (an exception is *C. linderi* which has upright terminal sporangia) . . . . **Circinella**
    On dung, nuts, especially Brazil nuts, soil; common and worldwide; 8 species. Hesseltine and Fennell, 1955.

52′(50′) With all or at least part of the sporangia not borne circinately at maturity . . 53

53(52′) Sporangia always with a distinct apophysis 54

53′(52′) Sporangia without an apophysis 56

54(53) Sporangiophores formed mostly on stolons and often with rhizoids 55

54′(53) Sporangiophores not on stolons or in umbels; growing on mushrooms

. . . . **Spinellus**
    On mushrooms in the cooler regions of world, often on *Mycena*; 7 species. Zycha *et al.*, 1969.

55(54) Sporangia borne in umbels with short blunt sterile projections in the center of the umbel . . . . **Rhizopodopsis**

On fallen fruits of *Elaeagnus* in Java; 1 species, *R. javensis*. Boedijn, 1958.

55'(54) Sporangiophores usually borne opposite rhizoids and typically unbranched; sporangia not in umbels on a sporangiophore . . . . **Rhizopus**

Very common, worldwide, in soil, decaying fruit, and vegetation; many species. Inui *et al.* (1965) recognized 14 species but at least 120 species and varieties have been described.

56(53') Homothallic; with the suspensors heterogamous; parasitic on mushrooms or soil inhabitants 57

56'(53') Generally heterothallic, but if homothallic with suspensors equal or nearly so 58

57(56) With *Mucor*-like zygospores; soil-inhabiting; sporangia all of one type . . . . **Zygorhynchus**

Always in soil, worldwide and common; 7 species. Hesseltine *et al.*, 1959.

57'(56) With nearly smooth-walled zygospores; growing on mushrooms; sporangia of 2 types . . . . **Dicranophora**

(see under Thamnidiaceae).

58(56') Sporangiophores simple, very large, over 80 mm in height, with a metallic luster; suspensors tonglike and with branched, fingerlike projections growing from the suspensors . . . . **Phycomyces**   (Plate III, 5)

Worldwide but not common; on dung, wood, and materials containing fatty material; 3 species. C. R. Benjamin and Hesseltine, 1959.

58'(56') Some sporangiophores in any colony divided and without a metallic luster   . 59

59(58') With rhizoids and stolons 60

59'(58') Without rhizoids and stolons 61

60(59) Thermophilic, growing above 40°C; colonies gray or blackish . . . . . **Rhizomucor**

Common in heating vegetation and grain, also soil; 2 species. This genus is usually reported as species of *Mucor* with globose sporangiospores. Lucet and Costantin, 1899.

60'(59) Not thermophilic, that is not growing at 40°C. Colonies white to light tan . . . .**Actinomucor**

Common and worldwide in air, dung, and all sorts of plant material; 1 species, *A. elegans*; used industrially to produce Chinese cheese. C. R. Benjamin and Hesseltine, 1957.

61(59') With sporangiophores divided regularly dichotomously, or ending in umbels . . . 62

61'(59') With sporangiophores divided variously but not dichotomously divided many times or in regular umbels 63

62(61) Sporangiophores bearing sporangia in umbels with each ultimate branch dichotomously divided with 1 arm bearing a sporangium while the other arm ends in a long sterile spine . . . . **Sporodiniella**

On dead insects, Indonesia; 1 species, *S. umbellata*. Boedijn, 1958.

62'(61) Sporangiophores dichotomously divided with all ultimate branches ending in sporangia; growing on mushrooms . . . . **Syzygites**

On mushrooms, worldwide; 1 species, *S. megalocarpus*. Hesseltine, 1957. *Sporadinia* is the generic name commonly used.

63(61') With sporangia, sporangiolalike sporangia, and dark-colored conidia

· **Backusella**
  Soil in the United States; 1 species *B. circina*. Ellis and Hesseltine, 1969.

63'(61') Not with sporangia, sporangiola, and conidia in one culture · · · · · · · · · · 64

64(63') Not attacking other Mucorales and forming galls; suspensors not with fingerlike
  projections · **Mucor**   (Plate III, 6, 9)
  Very common in soil, dung, decaying vegetation, stored grains; over 600 species
  reported. Hesseltine, 1950; Zycha *et al.*, 1969; Schipper, 1970.

64'(63') *Mucor*-like, free-living but often attacking other *mucors*; suspensors with fingerlike
  projections (appears like *M. hiemalis* in pure culture) · · · · · · · · · · · **Parasitella**
  In soil or on other Mucorales; 1 species, *P. simplex*. Bainier, 1903.

REFERENCES

Asterisk (*) indicates some important publications on the Mucorales not mentioned in the text.
*Ainsworth, G. C. (1971). "Ainsworth and Bisby's Dictionary of the Fungi," 6th ed. Common-
  wealth Mycol. Inst. Kew, Surrey, England.
*Alexopoulos, C. J. (1962). "Introductory Mycology," 2nd ed., pp. 184–210. Wiley, New York
  (1st ed., 1952).
*Bainier, G. (1882). Étude sur les Mucorinées. 136 pp. Thesis, Paris.
Bainier, G. (1903). Sur quelques espéces de Mucorinées nouvelles ou peu connues. *Bull. Soc.
  Mycol. Fr.* **19**:153–172.
Benjamin, C. R., and C. W. Hesseltine. (1957). The genus *Actinomucor*. *Mycologia* **49**:240–249.
*Benjamin, C. R., and C. W. Hesseltine. (1959). Studies on the genus *Phycomyces*. *Mycologia*
  51:751–771.
Benjamin, R. K. (1958). Sexuality in the Kickxellaceae. *Aliso* **4**:149–169.
Benjamin, R. K. (1959). The merosporangiferous Mucorales. *Aliso* **4**:321–433.
Benjamin, R. K. (1960). Two new members of the Mucorales. *Aliso* **4**:523–530.
Benjamin, R. K. (1961). Addenda to "The Merosporangiferous Mucorales." *Aliso* **5**:11–19.
Benjamin, R. K. (1963). Addenda to "The Merosporangiferous Mucorales." II *Aliso* **5**:273–288.
Benjamin, R. K. (1965). Addenda to "The Merosporangiferous Mucorales," III. *Dimargaris*.
  Aliso **6**:1–10.
Benjamin, R. K. (1966). The merosporangium. *Mycologia* **58**:1–42.
*Berkeley, M. J., and C. E. Broome. (1875). Enumeration of the fungi of Ceylon. Part II. *J. Linn.
  Soc. London, Bot.* **14**:29–140.
*Bessey, E. A. (1950). "Morphology and Taxonomy of Fungi," pp. 150–172. McGraw-Hill
  (Blakiston), New York.
Blakeslee, A. F. (1905). Two conidia-bearing fungi *Cunninghamella* and *Thamnocephalis*. *Bot.
  Gaz. (Chicago)* **40**:161–170.
Boedijn, K. B. (1958). Notes on the Mucorales of Indonesia. *Sydowia* **12**:321–362.
*Bonorden, H. F. (1851). "Handbuch der Allgemeinen Mykologie." pp. 119–129. Stuttgart.
Booth, C. (1971). Fungal culture media. *In* "Methods in Microbiology" (J. R. Norris *et al.*,
  eds..), Vol. 4, pp. 57–91. Academic Press, New York.
Brefeld, O. (1889). "Untersuchungen aus dem Gesammtgebeite der Mykologie," Vol. 8,
  p. 223. Leipzig.
Buller, A. H. R. (1934). "Researches on Fungi," Vol 6, pp. 1–224. Longmans, Green, New York.
Chesters, C. G. C. (1933). *Azygozygum chlamydosporum* nov. gen. et sp. A Phycomycete
  associated with a diseased condition of *Antirrhinum majus*. *Trans. Brit. Mycol. Soc.* **18**:
  119–214.
*Clements, F. E., and C. L. Shear. (1931). "The Genera of Fungi." Wilson, New York.

*Cutter, V., Jr. (1942a). Nuclear behavior in the Mucorales. I. The *Mucor* pattern. *Bull. Torrey Bot. Club* **69**:480–508.

*Cutter, V., Jr. (1942b.) Nuclear behavior in the Mucorales. II. The *Rhizopus, Phycomyces* and *Sporodinia* patterns. *Bull. Torrey Bot. Club* **69**:592–616.

Cutter, V., Jr. (1946). The genus *Cunninghamella* (Mucorales). *Farlowia* **2**:321–343.

Dobbs, C. G. (1938). The life history and morphology of *Dicranophora fulva* Schröt. *Trans. Brit. Mycol. Soc.* **21**:167–192.

Drechsler, C. (1943). A new non-helicoid bisporous *Helicocephalum* parasitizing nematode eggs. *Mycologia* **35**:134–141.

Ellis, J. J. (1963). A study of *Rhopalomyces elegans* in pure culture. *Mycologia* **55**:183–198.

Ellis, J. J., and C. W. Hesseltine. (1965). The genus *Absidia*: globose-spored species. *Mycologia* **57**:222–235.

Ellis, J. J., and C. W. Hesseltine. (1966). Species of *Absidia* with ovoid sporangiospores. II. *Sabouraudia* **5**:59–77.

Ellis, J. J., and C. W. Hesseltine. (1969). Two new members of the Mucorales. *Mycologia* **61**: 863–872.

Embree, R. W. (1959). *Radiomyces*, a new genus in the Mucorales. *Amer. J. Bot.* **46**:25–30.

Embree, R. W., and H. Indoh. (1967). *Aquamortierella*, a new genus in the Mucorales. *Bull. Torrey Bot. Club* **94**:464–467.

*Fischer, A. (1892). Phycomycetes: Mucorinae. *In* "Rabenhorst's Kryptogammen flora von Deutschland, Osterreich und der Schweiz" Vol. 1, Part IV, pp. 161–310. Kummer, Leipzig.

*Fitzpatrick, H. M. (1930). "The Lower Fungi. Phycomycetes," pp. 234–280. McGraw-Hill, New York.

Gams, W., and S. T. Williams. (1963). Heterothallism in *Mortierella parvispora* Linnemann. I. Morphology and development of zygospores and some factors influencing their formation. *Nova Hedwigia* **5**:347–357.

Gerdemann, J. W. (1965). Vesicular-arbuscular mycorrhizae formed on maize and tuliptree by *Endogone fasciculata*. *Mycologia* **57**:562–575.

Gerdemann, J. W. (1971). Fungi that form the vesicular-arbuscular type of endomycorrhiza. *U.S., Dep. Agr., Misc. Publ.* **1189**, 9–18.

*Gilman, J. C. (1957). "A Manual of Soil Fungi," 2nd ed., pp. 13–70. Iowa State Coll. Press, Ames.

Hesseltine, C. W. (1950). A revision of the Mucorales based especially upon a study of the representatives of this order in Wisconsin. Ph. D. Thesis, University of Wisconsin.

Hesseltine, C. W. (1953). A revision of the Choanephoraceae. *Amer. Midl. Natur.* **50**:248–256.

Hesseltine, C. W. (1954). The section Genevensis of the genus *Mucor*. *Mycologia* **46**:358–366.

*Hesseltine, C. W. (1955). Genera of Mucorales with notes on their synonymy. *Mycologia* **47**:344–363.

Hesseltine, C. W. (1957). The genus *Syzygites* (Mucoraceae). *Lloydia* **20**:228–237.

Hesseltine, C. W. (1960a). The zygosporic stage of the genus *Pirella* (Mucoraceae). *Amer. J. Bot.* **47**:225–230.

Hesseltine, C. W. (1960b). *Gilbertella* gen. nov. (Mucorales). *Bull. Torrey Bot. Club* **87**:21–30.

Hesseltine, C. W., and P. Anderson. (1956). The genus *Thamnidium* and a study of the formation of its zygospores. *Amer. J. Bot.* **43**:696–703.

Hesseltine, C. W., and P. Anderson. (1957). Two genera of molds with low temperature growth requirements. *Bull. Torrey Bot. Club* **84**:31–45.

Hesseltine, C. W., and C. R. Benjamin. (1957). Notes on the Choanephoraceae. *Mycologia* **49**:723–733.

Hesseltine, C. W., and J. J. Ellis. (1961). Notes on Mucorales, especially *Absidia*. *Mycologia* **53**:406–426.

Hesseltine, C. W., and J. J. Ellis. (1964). The genus *Absidia: Gongronella* and cylindrical-spored species of *Absidia. Mycologia* **56**:568–601.

Hesseltine, C. W., and J. J. Ellis. (1966). Species of *Absidia* with ovoid sporangiospores. I. *Mycologia* **58**:761–785.

Hesseltine, C. W., and D. I. Fennell. (1955). The genus *Circinella. Mycologia* **47**:193–212.

Hesseltine, C. W., C. R. Benjamin, and B. S. Mehrotra. (1959). The genus *Zygorhynchus. Mycologia* **51**:173–194.

Inui, T., Y. Takeda, and H. Iizuka. (1965). Taxonomical studies on genus *Rhizopus. J. Gen. Appl. Microbiol., Suppl.* **11**:1–121.

*Lendner, A. (1908). Les Mucorinées de la Suisse. *Mater Flore Cryptog. Suisse* **3**(1):1–180.

Linder, D. H. (1943). The genera *Kickxella, Martensella,* and *Coemansia. Farlowia* **1**:49–77.

Lucet, A., and J. Costantin. (1899). Sur une nouvelle mucorinée pathogneé. *Compt. Rendu. Acad. Sci.* **129**:1031–1034.

Malloch, D. (1967). A new genus of Mucorales. *Mycologia* **59**:326–329.

Meyer, J. (1957). *Martensiomyces pterosporus* nov. gen. nov. sp. Nouvelle Kickxellacée isolée du sol. *Bull. Soc. Mycol. Fr.* **73**:189–201.

Moreau, F. (1952–1953). "Les champignons" (*Encycl. Mycol.* **12** and **13**.). Lechevalier, Paris.

*Naumov, N. A. (1939). "Clés des Mucorinées" (*Encycl. Mycol.* **9**). Lechevalier, Paris.

Nicolson, T. H., and J. W. Gerdemann. (1968). Mycorrhizal *Endogone* species. *Mycologia* **50**:313–325.

Novak, R. O., and M. P. Backus. (1963). A new species of *Mycotypha* with a zygosporic stage. *Mycologia* **55**:790–798.

Raper, K. B., and D. I. Fennell. (1952). Two noteworthy fungi from Liberian soil. *Amer. J. Bot.* **39**:79–86.

Saksena, S. B. (1953). A new genus of the Mucorales. *Mycologia* **45**:426–436.

*Samson, R. A. (1969). Revision of the genus *Cunninghamella* (Fungi, Mucorales). *Proc., Kon. Ned. Akad. Wetensch., Ser. C* **72**:322–335.

Schipper, M. A. A. (1970). Two species of *Mucor* with oval- and spherical-spored strains. *Antonie van Leeuwenhoek; J. Microbiol. Serol.* **36**:475–488.

*Schroeter, J. (1886). Die Pilze Schlesiens. *Kryptogamen-Flora Schles.* **3**(2): 198–217.

*Schroeter, J. (1897). Mucorineae. *In* "Engler and Prantl, die Natürlichen Pflanzenfamilien," Vol. I. (1)5 pp. 119–134.

*Thaxter, R. (1914). New or peculiar Zygomycetes. 3: *Blakeslea, Dissophora,* and *Haplosporangium,* nova genera. *Bot. Gaz. (Chicago)* **58**:353–366.

Upadhyay, H. P. (1970). Soil fungi from north-east and north Brazil. VIII. *Persoonia* **6**:111–117.

*Van Tieghem, P. (1875). Nouvelles recherches sur les Mucorinées. *Ann. Sci. Natur. Bot.* [6] **1**: 5–175.

*van Tieghem, P., and G. Le Monnier. (1873). Recherches sur les Mucorinées. *Ann. Sci. Nat. Bot.* [5] **17**:261–399.

von Arx, J. A. (1970). "Genera of Fungi Sporulating in Pure Culture." Cramer, Weinheim.

*Vuillemin, P. (1903). Importance taxinomique de l'appareil zygospore des Mucorinées. *Bull. Soc. Mycol. Fr.* **19**:106–118.

Zycha, H. (1935). Mucorineae. *Kryptogamenflora Mark Brandenburg* **6a**:1–264.

Zycha, H., R. Siepmann, and G. Linnemann. (1969). "Mucorales." Cramer, Weinheim.

CHAPTER 12

Entomophthorales

GRACE M. WATERHOUSE

Formerly of the Commonwealth Mycological Institute
Kew, Surrey, England

I. CHARACTERS OF THE ORDER ENTOMOPHTHORALES

Entomophthorales are coenocytic mycelial fungi, which are either parasitic in animals (rarely plants) or saprophytic in soil or dung. Their mycelium consists of longer or shorter hyphal segments (the latter are termed hyphal bodies) averaging 13 μm or up to 30 μm or more in diameter and with rather thick walls mainly composed of chitin which is colorless or tinted brown.

Vegetative reproduction occurs by (1) the mycelium breaking up into hyphal bodies or the budding of these bodies; (2) primary uni- or multinucleate conidia, their shapes are varied and they usually have a basal papilla, formed singly at the apex of simple or branched, hyaline or colored sporophores; or (3) secondary conidia produced on germination of the primary, with the conidia often being forcibly dicharged. Perennation takes place by thick-walled hyphal bodies or single-walled chlamydospores.

Sexual reproduction occurs by the fusion of equal or unequal multinucleate hyphal enlargements (gametangia) which gives rise, usually by lateral budding, to a zygospore with a thick (up to 6.5 μm) two- to three-layered, hyaline or dark, smooth or ornamented wall. Similar azygospores are produced without fusion.

There is one family, the Entomophthoraceae.

II. LIST OF GENERA

Entomophthoraceae Warming 1884
  Entomophthora Fres. 1856 syn. *Empusa* Cohn 1855 preempted by *Empusa* Lindley 1824
    Type species *Entomophthora muscae* (Cohn) Fres. 1856 (see MacLeod, 1963). Divided by Batko (1964a) into five genera:
      Entomophthora Fres.
      Triplosporium (Thaxt.) Batko type species *T. fresenii* (Nowak.) Batko possibly antedated by *Neozygites* Witlaczil 1885 type species *N. aphidis* Witl. = *T. fresenii* fide Gustafsson, 1965.

219

Culicicola Niewl. 1916 type species *C. culicis* (Braun.) Nieuwl.
Entomophaga Batko type species *E. grylli* (Fres.) Batko
Zoophthora Batko type species *Z. radicans* (Bref.) Batko syn.
    Entomophthora sphaerosperma Fres.
Ancylistes Pfitzer 1872 emend Berdan 1938
    Type species *A. closterii* Pfitzer (see Berdan, 1938; Couch, 1949)
Completoria Lohde 1874
    Type species *C. complens* Lohde
Massospora Peck 1879
    Type species *M. cicadina* (see MacLeod, 1963)
Conidiobolus Bref. 1884
    Type species *C. utriculosus* Bref. (see Drechsler, 1954, 1965; Srinivasan and Thirumalachar, 1967b).
Basidiobolus Eidam 1886
    Type species *B. ranarum* Eidam (see Drechsler, 1958; Benjamin, 1962; Srinivasan and Thirumalachar, 1967a)
Doubtful or rejected genera
Blastocystis Alexeieff 1911 (near *Conidiobolus* or *Basidiobolus* fide Lavier, 1952)
Botryobolus Arnaud 1952 (invalid—no Latin diagnosis—conidial state only described, insufficient characters to place it)
Boudierella Cost. 1897 preempted by *Boudierella* Sacc.
Delacroixia Sacc. & Sydow 1899 a new name for *Boudierella* Cost.
    Type species *D. coronata* (Cost.) Sacc. & Sydow now placed in *Entomophthora* or *Conidiobolus*, though the genus is retained by Drechsler (1952).
Ichthyophonus Plehn & Mulsow 1911. For *Ichthyosporidium* Caullery & Mesnil 1905, thought to have priority, an animal type species has been selected. This leaves *Ichthyophonus* for the fungus species but some are thought to fall into *Basidiobolus* (Léger, 1927; Drechsler, 1955).
Lamia Nowak. 1884 preempted by *Lamia* Endlich. 1841, type species *L. culicis* (Braun) Nowak. replaced by *Culicicola*.
Loboa Ciferri *et al.* 1956, type species *L. loboi* (Fonseca & Leão) Ciferri *et al.* (syn. *Paracoccidioides loboi* (Fons. & Leão) Alm. & Lacaz. Placed by Ciferri *et al.* in Entomophthorales Imperfectae in the Paracoccidioidaceae (see below).
Myiophyton Lebert 1957, type species *M. cohnii* Leb. = *Empusa* fide Sacc. Revived by von Arx (1970) for *E. muscae* and related species separated from *Entomophthora*.
Paracoccidioides Almeida 1930, type species *P. brasiliensis* (Splend.) Alm. Placed by Ciferri *et al.* in Entomophthorales Imperfectae. Carmichael (1962) placed this genus in the Hyphomycetes near *Chrysosporium* Corda.
Sorosporella Sorokin 1888 type species *S. agrotidis* Sorok. Included in the Entomophthorales by Giard in 1889 but is a hyphomycete.
Strongwellsea Batko & Weiser 1965 type species *S. castrans* Batko & Weiser. Conidial state only described. Not sufficiently known or different from *Entomophthora* as usually understood.
Tarichium Cohn 1870 type species *T. megaspermum* Cohn. Based on resting spores of *Entomophthora* (see MacLeod and Müller-Kögler, 1970)
Zygaenobia Weiser 1951 type species *Z. intestinalis* Weiser. No Latin diagnosis, therefore invalid. Conidia only decribed; said to be near *Massospora*.

KEY TO THE ENTOMOPHTHORACEAE

1. Parasitic in algae or ferns 2
1'. Parasitic in animals or saprobic 3

2(1) Parasitic in desmids; mycelium divided by walls into linear multinucleate segments which may break up; gametangia unequal or scalariform . . . . **Ancylistes**
Ancylistes closterii in *Closterium.*

2'(1) Parasitic in fern prothallia; mycelium segments botryose, not breaking; azygospores
. . . . **Completoria**
Completoria complens in ferns.

3(1') Parasitic in arthropods; mycelium wholly or partially breaking up into multinucleate segments; primary conidiophores basidiumlike in *Entomophthora*; gametangia equal, unequal, or absent; zygospores arising as a lateral bud, wall colored, often dark 4

3'(1') Parasitic in vertebrates or saprobic; mycelium septate but not breaking up; gametangia equal or unequal; zygospores colorless or faintly colored 5

4(3) Conidia verrucose or smooth, produced inside the host, not forcibly discharged
. . . . **Massospora**
Two spp. in cicadas.

4'(3) Conidia smooth, produced outside the host, forcibly discharged . . **Entomophthora**
Entomophthora culicis in gnats and mosquitos etc., *E. forficulae* in earwigs, *E. erupta* in the green apple bug, *E. sphaerosperma* in *Pieris* caterpillars and many other hosts; 10 spp. in aphids.

5(3') Conidia uni- or binucleate, forming sporangiospores; nuclei clearly visible under light microscope; secondary conidia on long slender stalks adhesive; gametangia parallel projections at right angles to the fusing hyphae, fusing laterally . . . . **Basidiobolus**
Basidiobolus ranarum common in dung of amphibia and reptiles, *B. meristosporus* and probably other species responsible for phycomycosis in man (Clark, 1968).

5'(3') Conidia not forming sporangiospores; adhesive conidia not formed or rare; gametangia fusing end to end . . . . **Conidiobolus**
Conidiobolus coronatus common in soil.

III. DISCUSSION

In spite of the very different modes of life of the members of the various genera of this order, on morphological grounds they form a closely related group. Outstanding features are the rather wide coarse mycelium usually segmented by cross partitions forming, by loss of protoplasm in neighboring sectors, characteristic hyphal bodies which may become separate; the budding off of single terminal conidia, which are forcibly ejected in most genera; the characteristic shapes of the conidia (Fig. 4) which germinate by the production of secondary conidia; and the thick-walled zygospores or azygospores produced by the hyphal bodies.

The general consensus of opinion is that all the genera fit into one family, but Langeron and Vanbreuseghem (1952) took up the Basidiobolaceae Engler & Gilg 1924 on the grounds that *Basidiobolus* is saprobic and not living in animals, but only passing through and appearing in their excreta; that the mycelium is made up of uninucleate, not multinucleate elements that do not break up as in *Entomophthora*; and that the gametangia are unequal. It is now known that species of *Basidiobolus* are parasitic (in vertebrate animals); that the nuclear condition and tendency to segment are not exclusive to particular genera; and that unequal gametangia are characteristic also of other genera and that those of *Basidiobolus* are not always unequal. In view of these facts, there do not appear to be sufficient grounds for maintaining this family.

Ubrizsy and Vörös (1968), in moving *Ancylistes* into this order, retained in their classification the Ancylistaceae, but this does not appear to be necessary.

A. Entomophthoraceae

Entomophthora

This is by far the largest and most well-known genus. The early history of its divisions and mergings is summarized by MacLeod (1963). von Arx (1970) has suggested that *Myiophyton* Lebert 1857 has priority over *Empusa* because its type species, *M. cohnii*, is obviously the same as *E. muscae* as indicated by Saccardo. Von Arx solves the prospect of a large number of name changes by adopting *Myiophton* for only a few species close to *E. muscae* (cf., Batko below) and leaving the remainder in *Entomophthora*.

The tendency recently has been to include in it all the species parasitic on arthropods (and only those species). In 1963, however, MacLeod resuscitated *Massospora* (comprising possibly only two species, both on cicadas) on account of the binucleate conidia produced within the body of the insect, and not externally, and not forcibly discharged. This seems to be a reasonable separation which was fortified at first by the unique warted character of the conidia (Fig. 4), but a second species described in 1963 has smooth conidia. MacLeod rejects the other two species described.

On the other hand MacLeod and Müller-Kögler (1970) place *Tarichium* (species forming resting spores only) within *Entomophthora*. The resting spores conform with those of the genus, and conidial states may eventually be found; those that have been found place the species in *Entomophthora*. There seem to be a slight taxonomic grounds for maintaining it as a separate genus. MacLeod and Müller-Kögler have retained it for convenience as a section name without legal status.

Batko (1964a,b), in contrast, divided *Entomophthora* drastically, leaving in the genus only those species (nine in all) with multinucleate hyphal bodies, simple conidiophores (the *Empusa* of Cohn, Fresenius, etc.; Fig. 2), single-walled multinucleate conidia, and azygospores, but producing no rhizoids or pseudocystidia. (Most authors use the term cystidia but I prefer to follow Batko because these structures differ functionally and morphologically from basidiomycete cystidia). The remaining species, apart from most of those in *Tarichium* (three species are transferred to *Zoophthora*), and in *Massospora* (both of which he retains as genera), are distributed between four genera—two new ones, *Zoophthora* and *Entomophaga*, *Triplosporium* elevated from Thaxter's subgenus, and Nieuwland's *Culicicola*. Gustafsson (1965) suggests that *Triplosporium* is antedated by *Neozygites*.

Zoophthora is the largest and in it Batko places the bulk of the remaining species, all those with dichotomously or irregularly branched conidiophores (Fig. 1) and pseudocystidia, uninucleate conidia with a double wall which Batko interprets as monosporal sporangia (cf., *Basidiobolus*), and rhizoids (Fig. 3). Batko (1966) divides them among four subgenera: *Zoophthora* (based on *Entomophthora radicans*), *Pandora* (*aphidis*), *Erynia* (*ovispora*), and *Furia* (*virescens*). The status of the type species chosen, *Entomophthora radicans* Bref., is in some dispute. What was undoubtedly the same fungus was described earlier as *E. sphaerosperma* Fres. on resting spores only. Brefeld described *E. radicans* on conidia only, but after inoculations with his conidia he obtained resting spores which he said were the same as those of *E. sphaerosperma* and placed the two in synonymy. Batko (1964a), however, rejects the latter species on the grounds that it is a *nomen confusum*, as the figures given by Fresenius include drawings of structures not referable to the Entomophthorales.

Triplosporium and *Entomophaga* have no rhizoids or pseudocystidia, the main differences between them lying in the multinucleate condition of the hyphal bodies and conidia of the latter genus, in contrast to the four-nucleate condition of the former. Thaxter's subgenus *Triplosporium* included two species with smoky-colored, thick-walled conidia having evenly granular contents and arising from capillary conidiophores, and ellipsoidal zygospores arising from budlike outgrowths on vertical gametangia. Batko keeps these two species and adds a third. *Culicicola* (five species) appears to differ only in that rhizoids are present.

Admitting that it is possible (and perhaps convenient) to divide *Entomophthora* into groups of species with aggregations of common characters, one wonders why these groups should be genera rather than subgenera or some other infrageneric taxon. Hutchison (1963) used, for convenience of identification, twelve conidial types, following earlier workers. Lakon (1963) named eight types. Gustafsson (1965) suggested eight groups which are rather similar to those of Hutchison though he does not refer to the latter's work or give a key to the eight groups or a basis for the grouping. But Batko's divisions cut right across these and appear to have no similarity at all except for *Triplosporium*. Similarly a tentative segregation following a preliminary determination of fatty acids by Tyrrell (1967) does not fit in with any of the other groupings. Clearly, division

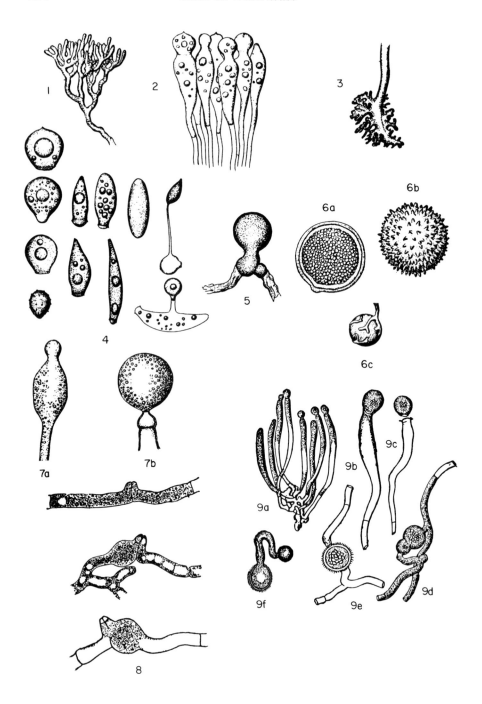

of the genus needs much substantiation based on more morphological and cytological work.

There has not been a world monograph on or key to *Entomophthora* during this century though regional studies have been published (Hutchison, 1963; Gustafsson, 1965). There are over 100 epithets attached to *Entomophthora* (MacLeod, 1963) but probably not more than two-thirds will turn out to be "good" species when the genus is monographed. MacLeod and Müller-Kögler (1970) give a key to 24 "Tarichium" species. Many of the species are host specific while others (e.g., *E. sphaerosperma, E. grylli*) attack widely different hosts. There are many features which appear to be without excessive variation under various conditions and these, together with the host group or species, enable distinctions to be made. Some distinguishing characters have been mentioned in the discussion on the divisions of the genus (above). For example, the presence of rhizoids does constitute a constant feature of certain species.

The primary conidia have diagnostic shapes (twelve are depicted by Hutchison, 1963; Fig. 4) and sizes and, though apparently hyaline, are of different colors in the mass; in some species a regular contour may be disturbed by a characteristic apiculus or by a papillate base of a variety of shapes. The zygospores (or azygospores) characterize species by their size and ornamentation as in most "Tarichium" species (Fig. 6). In addition, certain features of the hyphal bodies (shape, wall-thickness) may become more useful as more species are grown in culture and some field characters, particularly the color of the infected insect, are useful starting points. When further data have been amassed computerization should give firmer bases on which to make groupings within the genus.

Massospora

As indicated above, this genus differs from the foregoing in that the binucleate conidia are produced inside the host and are not released until the host body breaks up; they then emerge passively, in contrast to the typical emergence of the conidiophores from the host and the forceful discharge of the conidia in *Entomophthora*. The resting spores are

PLATE I. Fig. 1, Branched conidiophores, *Entomophthora sphaerosperma*. Fig. 2, Unbranched conidiophores, *E. muscae*. Fig. 3, Rhizoid, *E. echinospora*. Fig. 4, Types of conidia and secondary conidia. Fig. 5, Formation of zygospore, *E. occidentalis*. Fig. 6, Types of zygospores, a, smooth, b, spiny, c, hyphal (after Thaxter, 1888; all × 304 except Figs. 1 and 6c × 161). Fig. 7, *Basidiobolus ranarum* conidiophore, a, conidium forming, b, mature. Fig. 8, conjugation (after Thaxter, 1888, × 244). Fig. 9, *Conidiobolus utriculosus*, a, b, c, conidiophores, with conidia, d, conjugation, e, f, zygospores mature and germinating (after Brefeld: a, × 56; d, e, × 105; the rest, × 140).

azygospores with a sculptured wall. The conidia of the type species are unique in that they are verrucose.

Conidiobolus

Srinivasan and Thirumalachar (1967b) give a key to 31 species of *Conidiobolus*. This genus is very near *Entomophthora*, but as the species are mainly saprobic and can be grown readily in culture, the mycelium is much more in evidence and of the normal hyphal type. Hyphal bodies are produced on some media under certain conditions but without the tendency to break up. The conidia are forcibly ejected (Fig. 9c) but the mechanism is different from that of *Entomophthora* (Ingold, 1953). The gametangia form on short or long hyphal branches which meet end on (Fig. 9d), in contrast to *Basidiobolus* and some species of *Entomophthora* where fusion is lateral. The two layers of the zygospore wall tend to separate.

Species separation is on zygospore position and size and the thickness and sculpturing of its wall; size of conidia and the formation of microconidia on radial sterigmata; the size of hyphae and their form and ability to mass in definite aggregates; and the formation of chlamydospores (azygospores).

One species that has been moved from genus to genus, thus illustrating the narrow separations between genera, is *C. coronatus* (= *C. villosus*) = *Delacroixia coronata* = *Entomophthora coronata*. It was placed in *Conidiobolus* because of its saprobic habit, being originally found in cultures of mushroom spores and it is common in soil. When it was found as a parasite of insects there appeared to be no good characters to separate it from *Entomophthora*. It was made the type of *Delacroixia* because of the production of microspores. Later this genus was dropped but was recently revived by Drechsler (1952) on the grounds that the conidium, by producing microspores, even though externally, was acting as a sporangium and was therefore unique. This condition, however, also occurs in *Basidiobolus* (Benjamin, 1962) and the grounds for retaining *Delacroixia* appear to be removed, though Tyrrell and MacLeod (1972) use it as a subgenus.

Basidiobolus

Until the last decade or so species of this genus had been reported only from saprobic habitats, if rather specialized ones (dung and gut contents of amphibia and reptiles; Hutchison and Nickerson, 1970). Drechsler (1955) included parasites of fish transferred from *Ichthyosporidium* following Léger (1927) and others. From 1950, isolates from human patients suffering from phycomycosis (Clark, 1968; Srinivasan and Thirumalachar, 1967a; Cutler and Swatek, 1969; Emmons *et al.*, 1970) were identified as belonging to this genus and new species were added.

Isolates grow well in culture where the mycelium tends to form hyphal bodies but does not break up. Vertical light-sensitive conidiophores have a characteristic apical form; an initial cylindrical swelling produces a spherical conidium at its tip. At maturity the apical part of the swelling erupts and is shot off propelling the conidium (Fig. 7) in front of it (Ingold, 1953). Germinating conidia may produce (1) similar secondary conidia, (2) microspores on radiating sterigmata (as in *E. coronata*), (3) sporangiospores by division of the contents, a feature not occurring in any other genus of the order, or (4) slender pedicels (1–2 μm wide) on which elongated obclavate conidia with adhesive tips are borne. Sporangiospores may also be produced by the secondary conidia or by the hyphal segments. Gametangia develop as protuberances at right angles to the hyphal long axis. The contents of one gametangium pass into the other. The two walls of the zygospore are often partially or completely separated as in *Conidiobolus*. The fatty acids differ from those in *Conidiobolus* (Tyrrell, 1967).

Speciation within the genus (6 spp.; Benjamin, 1962; Srinivasan and Thirumalachar, 1967a) is on the size of the zygospores and the nature of the wall, the mycelium in culture, and the odor of the colony. Certain physiological characters, e.g., maximum growth temperature, production of pigment on tyrosine agar (Cutler and Swatek, 1969), and antigenic differences, appear to vary as much between isolates as between species, and isolates from mycoses are often different in these respects from nonpathogenic isolates.

Completoria

The two remaining genera are very specialized and, although very different from other Entomophthorales in their parasitism in that they are parasites of plants, nevertheless exhibit the usual features of the order.

Completoria is monotypic and the only species is very rare, being found only in fern prothallia and recorded only a few times (in two countries). The mycelium forms a compact cluster of hyphal bodies which, while vegetative, seem to be confined to each cell (Fig. 11a); the hyphae do not ramify the host as in insects though an infection hypha can penetrate an adjacent cell and develop into another cluster. Conidiophores penetrate the surface and produce conidia typical of the order; they are forcibly ejected and germinate to give secondary conidia if no host is near. Otherwise the germ tube penetrates the host. Azygospores develop in the hyphal cluster in the cell (Fig. 11b).

Ancylistes

This has been recorded only from desmids. The mycelium in the unicellular alga is restricted, because of the small size of the host, to sparingly branched wide septate hyphae, somewhat constricted at the septa. True hyphal bodies do not develop. Infection is able to proceed to another desmid cell

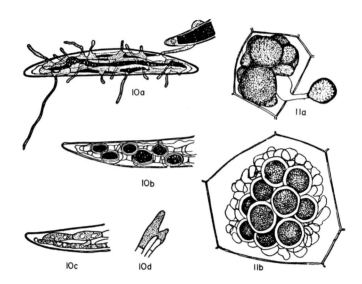

PLATE II. Fig. 10, *Ancylistes closterii*, a, young thallus with conidiophores and infection hypha to another desmid; b, zygospores; c, d, conjugation (after Pfitzer, × 240). Fig. 11, *Completoria complens*, a, hyphal bodies and a conidiophore, b, azygospores (after Atkinson, 1895, × 336).

when it comes in contact by means of an infection hypha (Fig. 10). The genus was placed in the Lagenidiales until Berdan (1938) showed that the conidia were of the Entomophthorales type forming only when the emerging hyphae are able to reach the air, and that hyphal conjugation preceded the formation of resting spores—thus they were truly zygospores (Fig. 10 c–d). Couch (1949) added another species to the two investigated by Berdan.

The species are differentiated on the nature of the wall of the cell enclosing the zygote, the sizes of the zygospores and conidia, and the nature of the exit hyphae.

REFERENCES

Atkinson, G. F. (1895). Damping off. *N. Y. Agr. Expt. Sta., Ithaca, Bull.* **94**:233–272.
Batko, A. (1964a). On the new genera: *Zoophthora* gen. nov., *Triplosporium* (Thaxter) gen. nov. and *Entomophaga* gen. nov. (Phycomycetes: Entomophthoraceae). *Bull. Acad. Pol. Sci., Ser. Sci. Biol.* **12**:323–326.
Batko, A. (1964b). Remarks on the genus *Lamia* Nowakowski 1883 vs. *Culicicola* Nieuwland 1916 (Phycomycetes: Entomophthoraceae). *Bull. Acad. Pol. Sci., Ser. Sci. Biol.* **12**:399–402.
Batko, A. (1966). On the subgenera of the fungus genus *Zoophthora* Batko 1964 (Entomophthoraceae). *Acta Mycol.* **2**:15–21.
Benjamin, R. K. (1962). A new *Basidiobolus* that forms microspores. *Aliso* **5**:223–233.
Berdan, H. (1938). Revision of the genus *Ancylistes*. *Mycologia* **30**:396–415.

Carmichael, J. W. (1962). *Chrysosporium* and some other aleuriosporic hyphomycetes. *Can. J. Bot.* **40**:1137–1173.

Clark, B. M. (1968). Epidemiology of phycomycosis. *In* "Systemic Mycoses" (G. E. W. Wolstenholme and R. Porter, eds.), pp. 179–192. Churchill, London.

Couch, J. N. (1949). A new species of *Ancylistes* on a saccoderm desmid. *J. Elisha Mitchell Sci. Soc.* **65**:131–136.

Cutler, J. E., and F. E. Swatek. (1969). Pigment production by *Basidiobolus* in the presence of tyrosine. *Mycologia* **61**:130–135.

Drechsler, C. (1952). Widespread occurrence of *Delacroixia coronata* and other saprophytic Entomophthoraceae in plant detritus. *Science* **115**:575–576.

Drechsler, C. (1954). Two species of *Conidiobolus* with minutely ridged zygospores. *Amer. J. Bot.* **41**:567–575.

Drechsler, C. (1955). A southern *Basidiobolus* forming many sporangia from globose and from elongated adhesive conidia. *J. Wash. Acad. Sci.* **45**:49–56.

Drechsler, C. (1958). Formation of sporangia from conidia and hyphal segments in an Indonesian *Basidiobolus*. *Amer. J. Bot.* **45**:632–638.

Drechsler, C. (1965). A robust *Conidiobolus* with zygospores containing granular parietal protoplasm. *Mycologia* **57**:913–926.

Emmons, C. W., C. H. Binford, and J. P. Utz. (1970). "Medical Mycology," 2nd ed., pp. 230–255. Lea & Febiger, Philadelphia.

Gustafsson, M. (1965). On the species of the genus *Entomophthora* Fres. in Sweden. I. Classification and distribution. *Lantbruks Hoegsk. Anna.* **31**:103–212.

Hutchison, J. A. (1963). The genus *Entomophthora* in the western hemisphere. *Trans. Kans. Acad. Sci.* **66**:237–254.

Hutchison, J. A., and M. A. Nickerson. (1970). Comments on the distribution of *Basidiobolus ranarum*. *Mycologia* **62**:585–587.

Ingold, C. T. (1953). "Dispersal in Fungi." Oxford Univ. Press (Clarendon Press), London and New York.

Lakon, G. (1963). Entomophthoraceae. *Nova Hedwigia* **5**:7–26.

Langeron, M., and R. Vanbreuseghem. (1952). "Précis de Mycologie." Masson, Paris.

Léger, L. (1927). Sur la nature et l'evolution des sphérules décrites chez les Ichthyophones, phycomycètes parasites de la Truite. *C. R. Acad. Sci.* **184**:1268–1271.

MacLeod, D. M. (1963). Entomophthorales infections. *In* "Insect Pathology" (E. A. Steinhaus, ed.), Vol. 2, pp. 180–231. Academic Press, New York.

MacLeod, D. M., and E. Müller-Kögler. (1970). Insect pathogens: Species originally described from their resting spores mostly as *Tarichium* species (Entomophthorales: Entomophthoraceae). *Mycologia* **62**:33–66.

Srinivasan, M. C., and M. J. Thirumalachar. (1967a). Studies on *Basidiobolus* species from India with discussion on some of the characters used in the speciation of the genus. *Mycopathol. Mycol. Appl.* **33**:56–64.

Srinivasan, M. C., and M. J. Thirumalachar. (1967b). Evaluation of taxonomic characters in the genus *Conidiobolus*, with a key to known species. *Mycologia* **59**:698–713.

Thaxter, R. (1888). The Entomophthoreae of the United States. *Mem. Boston Soc. Natur. Hist.* **4**:133–201.

Tyrrell, D. (1967). The fatty acid composition of 17 *Entomophthora* isolates. *Can. J. Microbiol.* **13**:755–760.

Tyrrell, D., and D. M. MacLeod. (1972). A taxononomic proposal regarding *Delacroixia coronata* (Entomophthoraceae). *J. Invert. Pathol.* **20**:11–13.

Ubrizsy, G., and J. Vörös. (1968). "Mezögazdasàgi Mykologia". Akadémiai Kiadó, Budapest.

von Arx, J. A. (1970). "The Genera of fungi Sporulating in Pure culture." Cramer, Lehre.

CHAPTER 13

Zoopagales

C. L. DUDDINGTON

Formerly of the Department of Life Sciences
The Polytechnic of Central London
London, England

I. GENERAL CHARACTERISTICS

The mycelium consists of branched, nonseptate hyphae, or a thallus which is usually a coiled filament, branched or unbranched. The Zoopagales are usually predacious on or parasitic in Protozoa, Nematoda, or other small animals. Asexual reproduction occurs by conidia, sexual reproduction by the conjugation of two similar or dissimilar gametangia. The zygospores are spherical and covered with hemispherical warts. (Figs. 1–8).

The Zoopagales consist of microscopic fungi occurring in soil, in rotting vegetation of many kinds, and in water. The great majority are either predacious on rhizopod protozoa or on free-living nematodes, or are endoparasitic in them. Two genera, *Bdellospora* and *Amoebophilus*, are ectoparasitic on amoebae.

The predacious species of Zoopagales possess a mycelium of branched, nonseptate hyphae to which the prey, consisting chiefly of amoebae or other small rhizopods, but sometimes of free-living nematodes, adheres owing to the secretion of a sticky substance by the hyphae. After capture, the protozoa are invaded by fine, branched haustoria arising from the hyphae at the point of contact. Species that capture nematodes intrude a system of trophic hyphae into the body of the victim. In either case, the body contents of the prey are consumed by the fungus.

In the endoparasitic species of Zoopagales the mycelium is replaced by a thallus of variable shape, the commonest form being a broad, filamentous structure coiled in a helix of about one-and-a-half turns; such thalli may be simple or branched, sometimes many times. In *Aplectosoma* the thallus is more or less cushion-shaped. In *Euryancale* a mycelium of non-septate hyphae is developed inside the body of the host, a nematode; this is the only instance where an endozoic mycelium is developed. In the

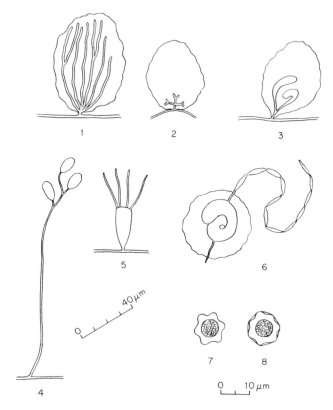

PLATE I.   Characteristics of the Zoopagales. Fig. 1, Bushlike haustoria of *Stylopage rhic-nacra* in the body of an amoeba. Fig. 2, Pedicellate haustoria of *Stylopage haploe*. Fig. 3, Inflated haustoria of *Zoopage thamnospira*. Fig. 4, Three conidia of *Stylopage cymosa* on their erect conidiophore. Fig. 5, Conidium of *Acaulopage tetraceros*, with its empty distal appendages, borne directly on a vegetative hypha. Fig. 6, Coiled thallus of *Cochlonema verrucosum* inside an amoeba, bearing a chain of spindle-shaped conidia. Fig. 7, Zygospore of *Stylopage rhabdoides*. Fig. 8, Zygospore of *Acaulopage ischnospora*, surrounded by the collapsed, persistent zygosporangium.

ectoparasitic genus *Bdellospora*, the infecting conidium swells to form a lemon-shaped thallus external to the host which draws nourishment from it by means of a branched system of haustoria. In *Amoebophilus*, which is also ectoparasitic on amoebae, the infecting conidium puts out a system of haustoria into the host and then functions as a thallus without enlargement.

Asexual reproduction in the Zoopagales is by conidia, the mode of production of which is variable. These are of great taxonomic importance, the genera and species being defined according to their shape, size, and manner of production. Sexual reproduction has not been recorded for all members of the group; where known, it takes place by the fusion of a pair of gametangia which usually consists of undifferentiated hyphal branches in the mycelial species; the parasitic species tend to have specially differentiated gametangia. Fusion between gametangia produced by the mycelium on the one hand, and from a germinating spore on the other, is common in some species. Sexual reproduction may be isogamous or anisogamous, one gametangium frequently being distinctly longer than its fellow. The zygospores are extremely characteristic, consisting of spherical, thick-walled bodies, their walls being ornamented with regularly arranged, hemispherical, wartlike excrescences. In some species the collapsed wall of the zygosporangium persists as a covering to the zygospore; the presence or absence of a persistent zygosporangium is of taxonomic importance in distinguishing between species, but not between genera.

Nothing is known about nuclear phenomena in the group. They are usually regarded as being obligate predators or parasites, but this point has not been sufficiently tested. Some of them also show a certain amount of host-specificity, but this again has not been tested by experiment.

All the Zoopagales have formerly been placed in one family, the Zoopagaceae, but there would seem to be sufficient difference between the predacious and the parasitic genera to justify their separation into different families. Accordingly, in this work the Zoopagaceae is here emended[1] to include the predacious genera only, and a new family, the Cochlonemaceae, is proposed[2] to accommodate the parasitic genera.

II. IMPORTANT LITERATURE

C. Drechsler, Some conidial phycomycetes destructive to terricolous amoebae, *Mycologia* **27**;6–40, 1935 *et seq.*; Predaceous fungi, *Biol. Rev.* **16**:265–90, 1941; Duddington, Further records of British predacious fungi, *Trans. Brit. Mycol. Soc.* **33**:209–15, 1951 *et seq.*; The predacious fungi: Zoopagales and Moniliales, *Biol. Rev.* **31**:152–96, 1956.

[1]Zoopagaceae, fam. emend. Mycelium of nonseptate hyphae, capturing small animals or occasionally saprobic; asexual reproduction by hyaline conidia; sexual reproduction by the fusion of two gametangia; zygospores small, spherical, ornamented with hemispherical warts.

[2]**Cochlonemaceae**, fam. nov. *Soma thallosa aut aliquando mucelica, intra corpus aut aliquando thallo extra corpus, in parvis animalibus; gignens sine sexu hyalinis conidiis; gignens sexu coniunctione duorum gametangiorum; zygosporae parvae, globosae, et teretibus verrucis.*

Typus: *Cochlonema.*

III.   KEY

KEY TO FAMILIES AND IMPORTANT GENERA OF ZOOPAGALES

1. Habit predacious; mycelium of nonseptate hyphae produced outside the host
. . . . **Zoopagaceae**   2

1'. Habit ecto- or endoparasitic; only fertile hyphae appearing outside the host
. . . . **Cochlonemaceae**   5

    2(1) Conidia absent. Asexual reproduction by lateral or intercalary chlamydospores
. . . . **Cystopage**
        7 species

    2'(1) Conidia present   3

3(2') Conidia borne in chains on short projections from the vegetative hyphae
. . . . **Zoopage** (Fig. 3)
        8 species

3'(2') Conidia not borne in chains   4

    4(3') Conidia borne singly on short projections from the vegetative hyphae, often with an
    empty appendage or appendages . . . . **Acaulopage** (Figs. 5, 8)
        28 species

    4'(3') Conidia borne singly or in groups on erect conidiophores
. . . . **Stylopage** (Figs. 1, 2, 7)
        13 species

5(1') Ectoparasitic; thallus derived from an infecting conidium   6

5'(1') Endoparasitic; thallus separate from conidium   7

    6(5) Infecting conidium enlarging to form a lemon-shaped thallus; conidia formed in
    long chains which may be branched . . . . **Bdellospora**
    Monotypic: *B. helicioides.*

    6'(5) Infecting conidium not enlarging; conidia budded off either singly or in short chains
. . . . **Amoebophilus**
    Monotypic: *A. sicyosporus.*

7(5') Conidia borne in chains   8

7'(5') Conidia not in chains   9

    8(7) Thallus coiled, branched, or unbranched . . . . **Cochlonema** (Fig. 6)
        18 species

    8'(7) Thallus flattened or cushion-shaped . . . . **Aplectosoma**
    Monotypic: *A. microsporum.*

9(7') Conidia sessile or nearly so, borne singly at intervals on prostrate fertile hyphae
. . . . **Endocochlus**
    4 species

9'(7') Conidia on long or short stalks which may bear successive conidia . . . .   **Euryancale**
    3 species

Zygomycotina
Trichomycetes

CHAPTER 14

Trichomycetes

ROBERT W. LICHTWARDT

Department of Botany
University of Kansas
Lawrence, Kansas

I. TRICHOMYCETES

A. General Characteristics

Trichomycetes is a class of obligate symbionts of arthropods which are attached to the chitinous gut linings or externally to the exoskeleton. Their thalli are unbranched and coenocytic or branched and septate. Asexual reproduction occurs by exogenous spores with appendages (trichospores), sporangiospores, arthrospores, or amoeboid cells. Biconical zygospores are produced in one order (Harpellales).

Trichomycetes can be distinguished from other fungi both by their habitat and morphological characteristics. In most instances it is necessary to dissect the host to find them. Occasionally, the internal species may be seen protruding from the anus. *Amoebidium parasiticum* is an exception, in that it occurs attached to external parts of minute crustacea and immature insects.

The hosts of trichomycetes are widely distributed geographically and include a variety of marine, freshwater and terrestrial arthropods. The most common ones are insects (aquatic larvae of Diptera, mayfly nymphs, beetles), Crustacea (crabs, isopods, amphipods), and millipeds. The degree of infection (or infestation) of a given population may range from uninfected to virtually 100 %. Some host individuals, such as black fly larvae, may harbor three or four distinct genera at one time.

The type of symbiotic relationship between trichomycetes and their hosts is uncertain; the fungi are sometimes referred to as commensals. Species of only two genera have been cultured to date: *Amoebidium* and *Smittium*. The nutritional requirements of these particular species do not differ from those of other common saprobic fungi.

Thalli of all trichomycetes are more or less firmly attached to their hosts

237

by means of a secreted holdfast or specialized holdfast cell. Species of most genera are found in the hindgut, a few in the peritrophic membrane of the midgut, and some in the foregut or stomach. Unbranched forms are coenocytic and form septa only to delimit reproductive cells. Branched genera normally produce septa in nonsporulating thalli, but in at least several genera the septa are known to be perforate.

Asexual reproduction includes the production of trichospores (Fig. 1) (Harpellales) and sporangiospores (Fig. 2) (Eccrinales). Trichospores form exogenously with one or more hairlike appendages (not flagella) produced in the "generative cell" from which the spore develops. Sporangia typically produce but one spore. The development of both sporangia and trichospores is basipetalous. Arthrospores are produced in the Asellariales, and amoeboid cells in the Amoebidiales. The Amoebidiales is rather distinct from the other orders in several respects, and there is some question about its phylogenetic relationship to other trichomycetes.

Sexual reproduction has not been confirmed in three of the orders. The fourth, the Harpellales, contains several genera that produce zygospores. These are characteristically biconical and thus differ from those of other Zygomycotina. Some of the zygospores have attached appendages similar to those of the trichospores.

B. Important Literature

R. W. Lichtwardt. "The Trichomycetes: What are Their Relationships?," *Mycologia* 65:1–20, 1973, J.–F. Manier, "Trichomycètes de France," *Ann. Sci. Nat. Bot.,* Ser. XII, 10:565–672, 1969; J.–F. Manier and R. W. Lichtwardt, "Révision de la Systématique des Trichomycètes," *Ann. Sci. Nat. Bot.,* Ser. XII, 9:519–532, 1968.

KEY TO ORDERS OF TRICHOMYCETES

1. Spores (trichospores) produced exogenously, usually bearing one or more long, fine appendages; zygospores produced in many genera . . . . **Harpellales,** p. 241

1′. Spores (sporangiospores) produced endogenously, or thallus breaking up into arthrospores; zygospores not present 2

    2(1′) Thallus branched and septate, producing arthrospores . . . . **Asellariales,** p. 242

    2′(1′) Thallus unbranched or branched only at base; nonseptate; producing sporangiospores. 3

3(2′) Sporangiospores usually produced singly in series of terminal sporangia that release spores in sequence; no amoeboid cells . . . . **Eccrinales,** p. 242

3′(2′) Entire thallus functioning as a sporangium, releasing spores more or less simultaneously; amoeboid cells produced at some stage . . . . **Amoebidiales,** p. 243

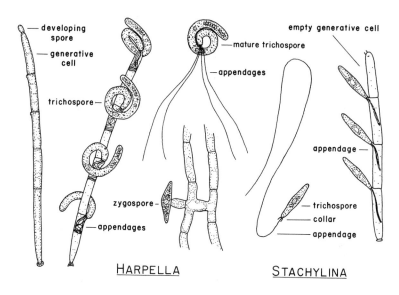

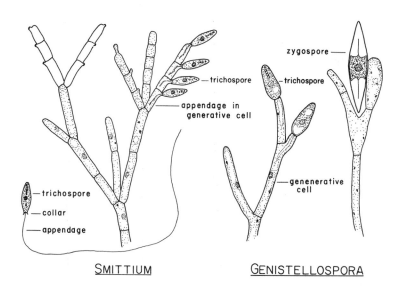

Fig. 1. Representative genera of Trichomycetes. Harpellaceae: *Harpella*. *Stachylina*. Genistellaceae: *Smittium, Genistellospora*.

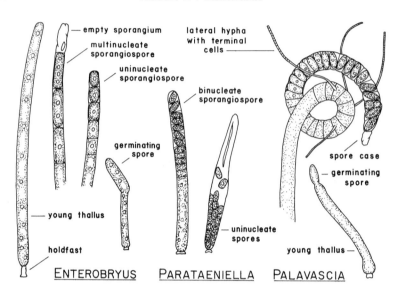

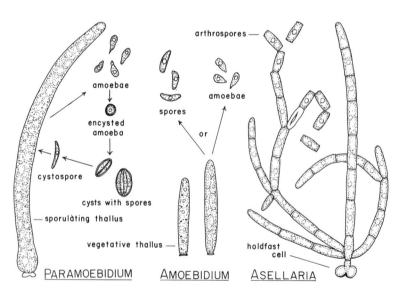

FIG. 2. Representative genera of Trichomycetes. Eccrinaceae: *Enterobryus*. Parataeniella-ceae: *Parataeniella*. Palavasciaceae: *Palavascia*. Amoebidiaceae: *Paramoebidium, Amoebidium*. Asellariaceae: *Asellaria*.

II. HARPELLALES

This order contains branched or unbranched thalli attached to the hindgut lining or peritrophic membrane of immature insects in aquatic habitats. The trichospores usually have one to many appendages attached to their basal end. Biconical zygospores are known in many genera.

KEY TO FAMILIES AND GENERA OF HARPELLALES

1. Thallus unbranched, attached to the peritrophic membrane of the midgut
. . . . **Harpellaceae** 2

1'. Thallus branched, attached to the hindgut lining . . . . **Genistellaceae** 4

2(1) Trichospores cylindrical and slightly curved to coiled, with 4 appendages
. . . . **Harpella** (Fig. 1)

2'(1) Trichospores ellipsoidal with 1 or no appendages 3

3(2') Trichospores with 1 appendage . . . . **Stachylina** (Fig. 1)

3'(2') Trichospores with no appendages, remaining attached to disarticulated generative cells
. . . . **Carouxella**

4(1') Trichospores allantoid; appendages unknown . . . . **Orphella**

4'(1') Trichospores straight with 1 or more appendages 5

5(4') Trichospores with 1 appendage 6

5'(4') Trichospores with more than 1 appendage 9

6(5) Trichospores ellipsoidal, with a short to long collar . . . . . **Smittium** (Fig. 1)

6'(5) Trichospores cylindrical or pyriform, without a collar 7

7(6') Trichospores cylindrical . . . . **Pteromaktron**

7'(6') Trichospores pyriform 8

8(6') Holdfast cell lobed at base . . . . **Spartiella**

8'(6') Holdfast cell swollen but not lobed . . . . **Graminella**

9(5') Trichospores with 2–4 appendages 10

9'(5') Trichospores with 6 or more appendages 13

10(9) Trichospores with 1 long and 2 short appendages . . . . **Glotzia**

10'(9) Trichospores with 2–4 appendages of equal length 11

11(10') Base of thallus embedded in a mucilaginous substance; trichospores cylindrical
. . . . **Stipella**

11'(10') Base of thallus with a distinct holdfast; trichospores ovoid or ellipsoidal 12

12(11') Trichospores ovoid; zygospore attachment oblique . . . . **Genistella**

12'(11') Trichospores ellipsoidal; zygospore attachment perpendicular to zygosporophore
. . . . **Simuliomyces**

13(9') Trichospores with a distinct collar; zygospore attachment oblique
. . . . **Trichozygospora**

13'(9') Trichospores without a collar; zygospore attachment parallel to axis of zygosporophore
. . . . 14

14(13′) Base of thallus embedded in a mucilaginous substance; zygospores formed subsequent to conjugation between hyphae . . . . **Pennella**

14′(13′) Holdfast distinct; zygospores formed without conjugations
. . . . **Genistellospora** (Fig. 1)

III. ASELLARIALES

Branched thalli are attached to the hindgut lining of isopods or insects. Reproduction occurs by arthrospores. Sexual reproduction is unknown. This order consists of a single family, the Asellariaceae.

KEY TO GENERA OF ASELLARIACEAE

1. Thallus in a dipteran larva; basal cell relatively unmodified with a simple holdfast
. . . . **Trichoceridium**

1′. Thallus not in a dipteran larva; basal cell noticeably modified 2

2(1′) Thallus is an isopod; basal cell swollen, lobed, rhizoidal, or with toothlike outgrowths; basal cell unbranched or branched only at distal end . . . . . . **Asellaria** (Fig. 2)

2′(1′) Thallus in a Collembola; basal cell forming pectinate branches, with initial spore case persistent . . . . **Orchesellaria**

IV. ECCRINALES

In this order, the thalli are unbranched or branched only at the base, and are coenocytic and attached to the chitinous lining of the hindgut and/or foregut. It is found in millipeds, isopods, amphipods, decapods, and insects. Sporangiospores are usually produced singly in a series of terminal sporangia; they are uni- or multinucleate and sometimes thick-walled. Sexual reproduction is unknown.

KEY TO FAMILIES AND GENERA OF ECCRINALES

1. Thalli producing directly but one type of spore; sporangia sometimes germinating *in situ*
. . . . **Palavasciaceae**
Consisting of a single genus . . . . **Palavascia** (Fig. 2)

1′. Thalli producing at least 2 types of spores; sporangia do not germinate *in situ* 2

2(1′) Uninucleate spores produced in thalli which become converted entirely or partly into multispored sporangia . . . . **Parataeniellaceae** 3

2′(1′) All spores produced singly in series of terminal sporangia . . . . . **Eccrinaceae** 4

3(2) Thalli in an isopod, producing uninucleate or binucleate spores
. . . . **Parataeniella** (Fig. 2)

3′(2) Thalli in a beetle larva, producing uninucleate spores or multinucleate cells which cleave into uninucleate bodies . . . . **Lajassiella**

4(2′) Thalli in an amphipod 5

4′(2′) Thalli not in an amphipod 7

5(4) Mature thalli branched at base . . . . **Ramacrinella**

5'(4) Mature thalli unbranched at base 6

    6(5') Ovoid spores with more than 1 nucleus; larger thalli usually curved at base
    . . . . **Astreptonema**

    6'(5') Ovoid spores 1-nucleate; thalli usually not curved at base . . . . . **Taeniellopsis**

7'(4') Thalli with swollen, lobed bases; in an isopod . . . . **Alacrinella**

7'(4') Thallus bases not swollen or lobed; not in an isopod 8

    8(7') Typically growing in tufts of individual thalli attached to a common holdfast system;
    in stomach of decapods . . . . **Enteromyces**

    8'(7') Thalli not growing in tufts 9

9(8') Producing thin-walled uninucleate and multinucleate spores and occasionally other kinds,
but not thick-walled spores; most species in millipeds, some in crabs or beetles
. . . . **Enterobryus** (Fig. 2)

9'(8') Producing thick-walled resting spores, often at the time of host molting, as well as thin-
walled spores 10

    10(9') Thalli in a milliped 11

    10'(9') Thalli in a decapod 12

11(10) Thick-walled spores unilocular at maturity . . . . **Eccrinoides**

11'(10) Thick-walled spores bilocular at maturity . . . . **Eccrinidus**

    12(10') Persistent spore case at apex oval; in stomach or hindgut of hermit crabs or crayfish
    . . . . **Arundinula**

    12'(10') Persistent spore case at apex cylindrical; in hindguts of true crabs
    . . . . **Taeniella**

V. AMOEBIDIALES

Amoebidiales, which has unbranched thalli, occurs in the hindgut or on the external parts of aquatic crustacea and insects. The entire thallus functions as a sporangium releasing amoeboid, motile cells, or spores with rigid walls. The amoebae encyst and produce cystospores which develop into new thalli. Sexual reproduction is unknown.

The order consists of a single family, the Amoebidiaceae.

KEY TO GENERA OF AMOEBIDIACEAE

1. Thallus attached to outer surface of host, producing either spores with rigid walls or amoeboid
cells . . . . **Amoebidium** (Fig. 2)

1'. Thallus attached to hindgut lining of host, producing only amoeboid cells
. . . . **Paramoebidium** (Fig. 2)

Basidiomycotina
Teliomycetes
Uredinales, p. 247
Ustilaginales, p. 281

CHAPTER 15

Uridinales

G. F. Laundon

Plant Diagnostic Service
Ministry of Agriculture and Fisheries
Levin, New Zealand

The rust fungi total about 4000 species distributed among 100 genera in use today. In nature all are obligate parasites, although recently there has been a breakthrough in artificial culture (Scott and Maclean, 1969). They parasitize green plants from the ferns and conifers to most of the more highly evolved families of both dicotyledons and monocotyledons. They grow internally with intercellular mycelium which rarely bears clamp connections. Sporulation is by means of sporogenous hyphae emerging through the stomata, by breaking through the surface of the host (erumpent), or in some cases by internal production of spores, released only by decay of the host.

I. LIFE CYCLES

The rusts have a pleomorphic life cycle. Sometimes this life cycle is quite simple but frequently it achieves a complexity not attained by any other fungi. Moreover, imposed on the full life cycle and its shorter versions are variations in pattern which greatly add to the complexity and which have caused difficulties and dispute in terminology.

The full life cycle in its "basic" form contains five distinctive spore states differing in morphologic and cytologic ways. These are the pycnia, aecia, uredinia, telia, and basidia. In most rusts these occur on one host (or group of related collateral hosts): they are autoecious. But there are a substantial number of rusts which are heteroecious, that is they have a life cycle which is split into two parts alternating between two groups of plants of divergent relationship (alternate hosts), the pycnia and aecia occurring on one group and the uredinia and telia on the other. Some rusts have added complexity by possessing two types of urediniospores (normal and amphispores) or two types of teliospores, one type soon germinating and the other a resistant type resting for a period before germination.

As explained later, the complex five-state life cycle (including heteroecism) was probably developed very early on in the phylogeny of the rusts and all variations of this which we know today are derived by reduction, that is by elimination or modification of parts of it. In extreme cases of reduction, only one or two states occur, as for example in microcyclic species of *Coleosporium* where the pycnia, aecia, and uredinia are eliminated and even the so-called teliospores are perhaps no more than immature basidia.

II. TERMINOLOGY

It is in the reduced life cycles that the difficulties and dispute in terminology arise. An awareness of these is essential in using the rust literature and the key included herein. A somewhat lengthy explanation seems unavoidable for a proper understanding.

Morphology, ontogeny, and homology are the bases of the traditional or "common sense" terminology. But the lack of precision in this, the variations and intergradations of morphology, and the uncertainty of homologies has led some to reject this basis and to adopt an "exact" terminology based solely on cytologic and ontogenic events in the life cycle.

Despite considerable morphologic variation there is, in most rusts, very little difficulty in dividing up the life cycle by morphology and designating the various states. The pycnia with their minute gametic spores never pose any difficulty despite variation from a well-defined flask shape to an effused and indefinite one. The aecia typically bear catenulate spores with a characteristic "verrucose" ornamentation; the uredinia, pedicellate spores with characteristic "echinulate" ornamentation. The telia and teliospores vary greatly in structure but can be denoted by their position in the life cycle, by their bearing the basidia, by the absence of aecial or uredinial features, and by morphologic comparison with the telia of related species whose life cycle is known. Traditionally there has been one important exception to this scheme. It occurs in the heteroecious life cycles of some rusts such as *Coleosporium* and *Chrysomyxa*. In these, sori bearing catenulate and verrucose ornamented spores occur not only in the normal position following the pycnia (and therefore are true aecia), but also on the alternate host associated with the telia (in the uredinial position). These aeciosporelike spores in the uredinial position have traditionally been known as "uredospores." Thus their position in the life cycle has overruled their morphology.

In the cytologically based system the terms are applied as follows: the pycnia represent the gametic stage, the aecia represent the stage in which plasmogamy (diploidization) occurs, the uredinia represent the conidial or repeating asexual stage, and the telia and basidia represent the stages in which syngamy and meiosis occur, respectively.

It will be noted that the terms given here are identical for both systems and only the basis by which they are applied is different. Not surprisingly the terms are commonly used throughout the rust literature and in speech without any indication as to the basis of their application. In such a situation it is obvious that confusion and errors will arise. To add to the confusion the terms themselves appear in the literature in variant forms (e.g., uredosori, uredinia, uredia) which might be expected to indicate the basis of their application but do not. For example, the terms aecidium, uredosorus, and teleutosorus, traditionally applied on a morphologic basis, have also been applied on a cytologic basis (Wilson and Henderson, 1966); the term uredium, first used in the cytologic system (Arthur 1932), has often since been used on a morphologic basis (e.g., Laundon, 1964) (and moreover is etymologically unsound; Savile 1968); and the term uredinium, originally used (Arthur, 1905) and still used (Savile, 1968) on a morphologic basis, has also been used on a cytologic basis (e.g., R. S. Peterson's papers).

Some examples of actual rusts will greatly assist an understanding of terminology. Endocyclic forms such as *Endophyllum* and *Kunkelia* are thought to be reduced forms which have lost the (morphological) uredinial and telial states of their ancestors; their basidia are now borne on what were called the aeciospores. Morphologically these are still typical aeciospores and are termed aeciospores in traditional terminology. Cytologically, however, they are the spores in which syngamy occurs and hence are teliospores in cytologic terminology (or better, they are aecidioid teliospores because both plasmogamy and syngamy occur together). Thus, in the literature, these spores are called aeciospores by some authors and teliospores by others (or, of course, variants of these such as aecidiospores or teleutospores) usually without any explanation as to the basis of the terminology.

Other important examples occur in the large number or rusts, in various genera, which have lost the aecial state (e.g., *Puccinia punctiformis*), or the uredinial state (e.g., *P. lagenophorae*) of their ancestors. In the former type of life cycle, naturally plasmogamy occurs in the first-formed uredinia; hence these first-formed sori called uredinia (often referred to as "primary uredosori") in traditional terminology are called aecia (often "uredinioid aecia") in cytologic terminology. In the latter type of life cycle, the aecia have often evolved a repeating or conidial function. Thus the first-formed sori in which plasmogamy occurs are termed aecia in either terminology but subsequent sori which are termed aecia (or "secondary aecidia") in traditional terminology are, because of their conidial nature, termed uredinia (or "aecidioid uredinia") in cytologic terminology. Here again confusion may arise in the literature. Statements such as "all spores bearing basidia are teliospores," "aecia never repeat," "urediniospores are always pedicellate" are true only in one system of terminology. Often the reader is not aware of such limitations and cannot understand statements which seem to go against the facts as he knows them.

Some of the advantages and disadvantages of each system of terminology have already been alluded to. The precision of the cytologic system might seem a real advantage; in practice such precision does not exist since cytology is not readily observed, and it is not normal to perform cytologic studies before applying terms to sori or spores. Thus, usually the only way in which a term can be used with any hope of reliability is by morphologic equation, or ontogenic equation, or both, with other spore states of known or presumed cytology. One must say "presumed" because the vast majority of rusts have not been cytologically investigated. Thus, the terms which are defined by cytologic events alone are normally applied by morphologic comparisons alone. That few cytologic errors result only confirms the reliability of morphologic features.

Although in traditional terminology it is impossible to provide foolproof and concise definitions, in practice this seems to cause very little difficulty: the use of traditional terminology tends to be intuitive with little regard for formal definition. It is for this reason that I have dubbed it the "common sense" system. The point about the traditional terms is that they are based not on definitions but on usage; they are natural, like natural taxa, instead of artificial and mechanistic. Although the use of clearly defined terms may seem essential in science, their application for multiple purposes in highly complex biological systems may prove unworkable. This, as explained above, seems the probable fate of cytologic terminology. But practicability apart, the morphologic terms carry so much useful and meaningful information that it seems shameful to abandon them.

Instead I have proposed elsewhere (Laundon, 1967) an adapted form of the traditional system which I hoped would be a compromise solution to the terminology problem. This system, which is employed in this text and in the key, is basically morphologic but is able to carry cytologic information appended to the basic terms. As a result it embodies maximum information content and does not necessarily imply knowledge which does not exist. It departs from the traditional system in only one important respect: aecia and uredinia are more strictly defined than is traditional, aeciospores being invariably catenulate and urediniospores being invariably pedicellate. This means that in rusts such as *Coleosporium* and *Chrysomyxa* the catenulate spores associated with the telia are termed aeciospores instead of uredospores (or some variant of this) as they are known elsewhere. The justification for choosing 'catenulate' as the only essential aecial character and 'pedicellate' as the only essential uredinial character is published elsewhere (Laundon, 1972).

TABLE I
TERMINOLOGY OF SPORE STATES

Traditional [a]		Arthur, 1934 [b]	Laundon, 1967 [c]	
O	Spermagonia [d,e]	Pycnia [f]	O	Pycnia
I(I[I])	(Primary) aecidia [d]	Aecia	I(I[I])	(Aecial) aecia
I(I[II])	(Secondary) aecidia	(Aecidioid) uredia	I[II]	Uredinial aecia
I	Aecidia (endo-form) [g,]	(Aecidioid) telia	I[III]	Telial aecia
II[I]	(Primary) uredosori	(Uredinoid) aecia [h]	II[I]	Aeciàl uredinia
II(II[II])	(Secondary) uredosori	Uredıa	II(II[II])	(Uredinial) uredinia
III	Teleutosori	Telia	III	Telia
IV	Basidia	Basidia	IV	Basidia

a The basic terms were proposed by de Bary (1865).

b Arthur used these terms in his well-known "Manual" (1934) but had previously published them in a German paper (1932).

c Laundon (1967) used the spelling "uredia" but now accepts "uredinia" as correct following Savile (1968).

d Cunningham (1930, 1931) used the terms "pycniosori" and "aecidiosori."

e Some modern-day adherents of the traditional terminology use "pycnia."

f There is a tendency to use the term "spermagonia" nowadays especially among the followers of Arthur's terminology. Others dislike this term because of lack of homology with the spermagonia of the Ascomycetes. Rust pycnia are hermaphroditic while true spermagonia are "male" sex organs.

g Some authors have referred to the endo-type aecidia as "teleutosori" (e.g., Saccardo, 1888).

h Arthur also used the term "stylosporic aecia" as an alternative.

Tables I and II compare terminology, the former correlating spore states and the latter correlating life cycles.

III. CORRELATED SPECIES AND TRANZSCHEL'S LAW

The theory that rusts with shorter life cycles are derived from those with the more complex has been mentioned. For many rusts it is possible, by means of morphologic similarity and occurrence on the same or related hosts, to surmise that one or more species derive from another more complex species. Species in groups such as this are termed "correlated species" (Arthur *et al.*, 1929).

Correlated species can also arise by reduction of morphology alone rather than of life cycle. Thus, a species having one-celled teliospores may be thought to be a reduced or correlated species of one having two-celled teliospores.

What is known as Tranzschel's law arises from a characteristic of

TABLE II
TERMINOLOGY OF LIFE CYCLES

Example	Traditional	Arthur, 1934 [a]	Laundon, 1967
Puccinia graminis	O I II III IV eu-form	O I II III IV macrocyclic	O I II III IV macrocyclic
P. punctiformis	O II III IV brachy-form	O I II III IV macrocyclic	O II^{I} II^{II} III IV brachycyclic
Gymnosporangium (most species)	O I III IV opsis-form	O I III IV demicyclic	O I III IV demicyclic
Coleosporium	O I II III IV eu-form	O I II III IV macrocyclic	O I^{I} I^{II} III IV demicyclic
P. lagenophorae	(O) I III IV opsis-form	(O I) II III IV macrocyclic	(O I^{I}) I^{II} III IV demicyclic

(The uredinial aecia of such rusts as this have not generally been recognized on account of their being morphologically indistinguishable from ordinary aecia. The pycnia and true aecia are rarely found.)

P. heterospora	(O) III IV micro-form	(O) III IV microcyclic	(O) III IV microcyclic

(Rusts like *P. malvacearum* with teliospores which germinate immediately have been called lepto-forms).

Endophyllum	O I IV endo-form	O III IV microcyclic	O I^{III} IV endocyclic
P. chrysanthemi	II III IV hemi-form		

[a] Arthur did not use roman numerals for the spore states; they are used here for convenience.

correlated species. This law states that in reductions from the parent heteroecious macrocyclic type to the descendent microcyclic the telia of microcyclic species occur on the aecial host of the macrocyclic species. Moreover they often adopt the habit of the aecia of the parent macrocyclic species. Thus, a heteroecious species with grouped aecia and scattered telia will reduce to a microcyclic species with grouped telia while a heteroecious species with systemic aecia (e.g., *Tranzschelia discolor*) will reduce to a microcyclic species with systemic telia (e.g., *T. anemones*). It appears to be universally true that Tranzschel's law is applicable in respect of microcyclic telia occurring on the aecial host of the parent heteroecious species, the only exception which had been firmly suspected (*Chrysomyxa arctostaphyli*) having been disproved (Peterson, 1961). It should be noted that

Tranzschel's law does not necessarily apply to autoecious macro- or demicyclic rusts derived from heteroecious species. Such rusts may occur on either the aecial host (e.g., *Puccinia lagenophorae*) or the telial host (e.g., *P. graminella*) of the parent heteroecious species. However, the law does appear to apply to endocyclic rusts.

Arthur (1934) used the concept of correlated species on an extensive scale. An example is *Puccinia extensicola* (now known as *P. dioicae*), with aecia on Compositae and telia on Gramineae, for which Arthur gives eight correlated species. One of these is merely a morphologically reduced form with one-celled teliospores (*Uromyces perigynius*). The other seven are microcyclic species demonstrating Tranzschel's law: they all occur on the Compositae. Another excellent example, cited by Cummins (1959), is of *P. interveniens* (heteroecious, demicyclic); *P. graminella* (autoecious, demicyclic); *P. sherardiana* (microcyclic) and *Endophyllum tuberculatum* (*P. neotuberculata*) (endocyclic). The first of these species has aecia on Malvaceae and telia on Gramineae. The second is autoecious on Gramineae. The other species occur on Malvaceae in agreement with Tranzschel's law.

IV. MORPHOLOGY

A. Pycnia

The pycnia (Fig. 1 A, B) are produced on a haploid thallus resulting from basidiospore infection. They contain a palisade of sporogenous cells which produce one-celled pycniospores in a sweetish nectar. This spore-laden nectar is exuded from the pycnia and carried by insects. The spores are sperm cells and effect fertilization by fusion with flexuous hyphae of pycnia of the opposite mating type.

Hiratsuka and Cummins (1963) defined eleven types of pycnia based on various combinations of four features: growth (determinate or indeterminate), bounding structures (present or absent), shape (flat or convex, i.e., conical or globose), and position (subepidermal, intraepidermal, subcuticular, or intracortical). These eleven types can be reduced to three basic forms: globose, conical and indeterminate (i.e., spreading indefinitely and not limited to a well-defined shape). However, these three types are not always clear cut; some species of the genera *Uredinopsis*, *Chrysomyxa*, and *Melampsora* show lens-shaped to hemispheric intergradations between the truly conical type, as in other species of these genera, and some other genera and the globose type as in *Milesina* (= *Milesia*).

The work of Hiratsuka and Cummins has much improved our understanding of the significance of pycnia in taxonomy. Previously many authors had tended to regard pycnial position, for example, subepidermal, intra-

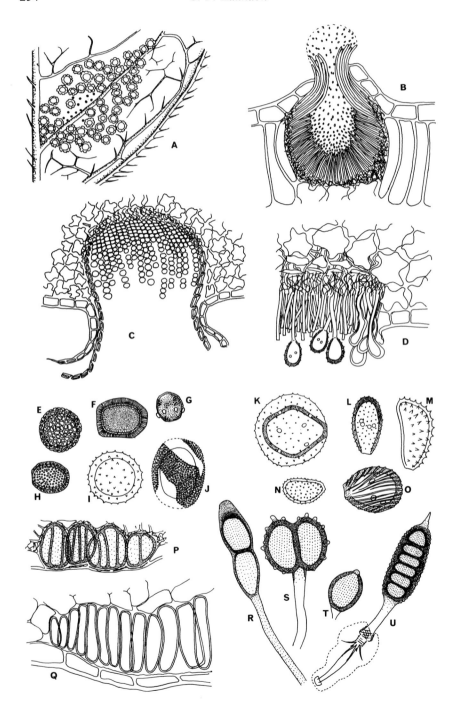

epidermal, or subcuticular, as the only pycnial feature of importance in differentiating genera (e.g., Arthur, 1934; Cummins, 1959; Thirumalachar, 1961). Now it is clear that structure rather than position is significant since a number of rusts which are clearly related have pycnia in different positions but of basically similar structure (e.g., *Melampsora* and *Hapalophragmium* in each of which all species have pycnia but some are subcuticular while others are subepidermal).

Hiratsuka and Cummins discuss the relationship between the types of pycnia and give diagrams of suggested phylogenetic development of pycnia as well as possible lines of evolution of the genera as suggested by the pycnia. The three basic types of pycnia all occur in both primitive and advanced rusts suggesting that there may be several lines of evolution.

B. Aecia

The aecia are sori containing a palisade of sporogenous cells which invariably bear the aeciospores in chains (as aecia are defined herein) (Fig. 1 A, C). Aeciospores are invariably single celled and usually have hyaline walls, without visible pores and with characteristically "verrucose" ornamentation (Fig. 1 E–J). Aecia usually result from a fertilization process and thus are associated with pycnia, but they may also be a repeating spore form (uredinial aecia). Aeciospores normally germinate to produce a diploid thallus bearing uredinia or telia, but alternatively they may produce basidia (telial aecia).

Some aecia are nonperidiate (naked or caeomatoid) although paraphyses may be present. This type is characteristic of some genera such as *Melampsora* and some phragmidious genera; it also occurs occasionally in some genera such as *Puccinia*, probably as a result of degeneracy since a few peridial cells are often present.

The aecia with peridia are often segregated into three types. The cupulate ("aecidioid") aecium is the best-known type. It occurs in all or nearly all

FIG. 1. A, A group of pycnia and aecia, × 5; B, a globose subepidermal pycnium, × 125; C, a cupulate aecium, × 63; D, a paraphysate uredinium, × 125; E, *Phragmidium bulbosum* aeciospore, × 250; F, *Puccinia stenandri* aeciospore, × 250; G, *Puccinia dioicae* aeciospore, × 250; H, *Pucciniastrum americanum* aeciospore, × 250; I, *Phragmidium violaceum* aeciospore, × 250; J, *Puccinia tetragoniae* aeciospore, × 250; K, *Puccinia tetragoniae* urediniospore, × 250; L, *Puccinia graminis* urediniospore, × 250; M, *Hemileia vastatrix* urediniospore, × 250; N, *Pucciniastrum americanum* urediniospore, × 250; O, *Puccinia macropoda* urediniospore, × 250; P, subepidermal multicellular teliospores of *Pucciniastum americanum*, × 250; Q, subepidermal single-celled teliospores of *Melampsora populnea*, × 250; R, *Puccinia graminis* teliospore, × 250; S, *Diorchidium woodiae* teliospore (after Cummins, 1959), × 250; T, *Uromyces minor* teliospore, × 250; U, *Phragmidium tuberculatum* teliospore, from a water mount to demonstrate the hygroscopic pedicel, × 125.

species of *Puccinia* and in many other genera, especially those with globose pycnia. The aecial peridium is a distinctive structure quite unlike the peridia which enclose some uredinia. It consists of rhomboid thick-walled cells with characteristic labyrinthiform sculpturing. The peridium may be short and truly cupulate in shape or long and tubular. On breaking open, the sides curve back ("revolute"), sometimes arching back considerably and tearing in the process to give an irregular frill.

The peridermioid aecium is an elongated, large structure often flattened on one side and tonguelike or blisterlike. This type occurs in a number of the more primitive genera such as *Milesina, Pucciniastrum, Coleosporium,* and *Cronartium.* The peridial cells are generally longer and narrower than in cupulate aecia but they are otherwise rather similar. In some species the peridium is multi-layered or ribbed. Typically the peridium opens irregularly or is circumscissile.

The roestelioid aecium is elongated and hornlike ("cornute"). The peridial cells tend to be greatly elongated and their walls are often ribbed. The peridium opens by splitting along the sides becoming torn into shreds ("lacerate"). This form is typical of the genus *Gymnosporangium* although some species of this genus have ordinary cupulate aecia.

The four types of aecia have usually been segregated in separate genera: *Caeoma, Aecidium, Peridermium,* and *Roestelia,* respectively. However, they do intergrade and are, perhaps, only retained for convenience. In any case most rusts with peridermioid aecia occur on gymnosperms and by common consent *Peridermium* is limited to rusts on such hosts. Similarly the genus *Roestelia* is limited to likely aecial stages of *Gymnosporangium.*

The characteristic "verrucose" ornamentation shown by most aeciospores is probably formed by splitting of the outer wall layer into mosaiclike pieces. The mosaic pieces can vary greatly as to depth and width. In some species they are broad plaques with almost no apparent depth. At the other extreme they are long and needlelike. Occasionally aeciospores have a different type of ornamentation altogether as in the echinulate aeciospores of some *Phragmidium* species.

One feature of aeciospores which should gain more attention is the presence in some species of refractive bodies or pore plugs (Dodge, 1924; Cunningham, 1931). The origin and function of these objects is still obscure. They often occur in a regular arrangement around the aeciospores but easily become detached from their original position and then adhere irregularly. They sometimes seem to intergrade with the normal ornamentation of the spores and Holm (1966) has suggested that they are merely enlarged warts. They are characteristic of certain species, for instance; they occur in most *Carex* rusts but not in any grass rusts that were examined (Savile, 1965). They also occur in the autoecious demicyclic *Puccinia lagenophorae* which Holm (1966) suggests may be derived from a heteroecious *Carex* rust. These objects are rarely mentioned in technical descriptions of rusts, an unfortunate omission because they can constitute a useful diagnostic feature.

C. Uredinia

The uredinia (Fig. 1 D) are sori bearing one-celled urediniospores singly on pedicels (as defined herein). Usually the spores (Fig. 1 K–O) act as conidia but in some rusts which have lost their aecia they also replace the aeciospores (aecial urediniospores) and in one rust, *Hemileia vastatrix*, it has been claimed that they produce basidia on germination (telial urediniospores) (Rajendren, 1967a).

In many rusts the uredinia do not possess any bounding structures or other special features, but many others have paraphyses associated with the uredinia. In most cases these are marginal and incurved, clavate to capitate and with a somewhat thickened dorsal wall, as in, for example, many *Puccinia* species and in *Phragmidium* and *Tranzschelia*. In some cases the paraphyses are basally united and septate; this type seems to be characteristic of a number of genera such as *Phakopsora*, *Physopella*, *Crossopsora*, *Hamaspora*, *Dicheirinia* (Cummins, 1959). *Melampsora* is unique in possessing erect capitate hyaline paraphyses scattered throughout the sorus; the genus can be identified by this feature alone. A few rusts, especially the primitive genera (e.g., *Milesina*, *Melampsorella*, *Pucciniastrum*, and *Cronartium*) have peridiate uredinia but the peridium is quite unlike the aecial peridium.

Most rusts have urediniospores with colored walls perforated by more or less conspicuous pores and with characteristic echinulate ornamentation. There are, however, many species with hyaline or very pale walls and in these the pores are usually rather obscure. The arrangement and number of pores is an important taxonomic feature. Cummins (1936) discusses these characteristics of pores in a large number of genera and considers their phylogenetic significance. In addition to arrangement and number of pores he gives the following three additional features: size, presence of cuticular caps, and presence of ornamentation over pores. He also gives a technique for improving the visibility of inconspicuous pores.

The characteristic echinulate ornamentation of urediniospores is quite different from that of verrucose aeciospores. It takes the form of pointed conical spines composed of a continuous layer, which Savile (1954) says is the middle or inner layer of the wall with a very thin coating of shrunken exospore over it. Some urediniospores are described as verrucose but probably in most cases this type of ornamentation is simply blunt spines and in no way resembles the true verrucose ornamentation of aeciospores. Other forms of ornamentation do occasionally occur: reticulate, striate or smooth, but even these may arise in the same basic fashion.

The spores of aecial uredinia may be identical to those of the normal uredinia or they may differ slightly in morphology or position. For example, in *Augusia* the aecial uredinia are subcuticular like the pycnia whilst the true uredinia are subepidermal; their spores differ slightly in size and wall thickness.

Some rusts, notably the primitive genera *Uredinopsis* and *Hyalopsora* and a few *Puccinia* species, have a resistant type of urediniospore (amphispore) in addition to the normal ones. These are often thicker walled, darker colored, and germinate only after a period of rest.

D. Telia

Telia and teliospores (Fig. 1 P–U) vary enormously in their morphology. Only a few basic features will be noted here; many other variations are indicated in the key. The distinction between sessile and pedicellate teliospores has long been regarded as of great significance and segregation of the rusts into families has usually been based principally on this feature. There is now some doubt, however, as to whether those rusts with sessile teliospores or those with pedicellate teliospores do represent monophyletic lines (Hiratsuka and Cummins, 1963).

There are two features commonly shown by telia and teliospores which are never shown by aecia or uredinia and their spores: sessile spores embedded in the host, and spores with two or more cells. Many teliospores are smooth, unlike most aeciospores and urediniospores. Among those which are ornamented, their ornamentation never resembles that of aeciospores and very rarely resembles that of urediniospores; usually it takes the form of very blunt warts, wrinkling, or reticulation; sometimes there are large processes like fingers or spines.

In some genera teliospores develop in clusters on basal cells and this characteristic has been used as one of major value in delimiting certain genera. The significance of these basal cells is reviewed by Thirumalachar and Cummins (1949) who point out that their value is reduced owing to intergradation between prominent basal cells in some species and obscure component cells of the cellular hymenium (which may or may not be regarded as basal cells) in other species.

Many species of *Puccinia* have stromatic telia. This is a feature that has often been misinterpreted and wrongly labelled "paraphysate," "closed," "long covered," or, more accurately, called "loculate," or "pseudoparaphysate." They do in fact have a stroma composed of vertical tightly packed pseudoparaphyses, often divided into locules containing the spores, and remaining closed until the host decays. The term "corbiculae" has also been applied to these structures (Kuhnholz-Lordat, 1941).

V. PHYLOGENY AND CLASSIFICATION

A basic assumption which has been made by many authors concerning the phylogeny of rusts is that host taxonomy and rust taxonomy generally go hand in hand. Thus, the rusts found today on the ferns and conifers are descendants of early evolved rusts little changed from their ancestors, while the rusts on more advanced hosts, such as the Rosaceae, Labiatae, Compositae, and Monocotyledons, are new forms which have developed along with their hosts. Certain trends in life cycle and morphology accompanying evolution of hosts tend to support this hypothesis, as follows:

a. Life Cycles. Almost all the rusts occurring on the ferns are both heteroecious and macrocyclic; so also are most of those occurring on the conifers. On the more advanced hosts autoecism tends to predominate and shortened life cycles become common. This trend may seem to run contrary to the expected direction: is it possible that the most primitive rusts could possess the most complex life cycles? There is little doubt that this is, indeed, the case since it is inconceivable that the extraordinarily complex heteroecious macrocyclic life cycle could have evolved many times. On the other hand, reductions can easily occur at any stage of evolution.

b. Sessile and Pedicellate Teliospores. Almost all the rusts occurring on the ferns have sessile teliospores; so also do all those occurring on the conifers (except *Gymnosporangium*). On more advanced hosts most rusts have pedicellate teliospores in a single layer. Those rusts with sessile teliospores which occur on more advanced hosts tend predominantly to have catenulate teliospores.

c. Sorus and Spore Morphology. Correlated with the evolution of the host is an increase in complexity of morphology especially in the telial stage: septation and arrangement of cells, ornamentation, pigmentation, wall layering, and pores.

Apart from these trends there is one other aspect which supports the theory that host and rust taxonomies are linked. This is the occurrence of tightly knit groups of rusts each associated with a similar group of hosts. The rusts occurring on the ferns represent one such group, but there are a number of other examples such as those on the Rosaceae (the phragmoteliospore rusts) and those on the legumes including *Uromycladium, Dicheirinia, Ravenelia,* and *Hapalophragmium.*

In classifying rust genera into major divisions, one feature above all others has been used by most authors. This is the sessile or pedicellate nature of the teliospores. It is the one "universal" feature which seems to segregate the primitive rusts from the more advanced. Dietel (1900) was the first to use this feature for the primary division in the rusts and he further divided the "sessile" group into those with teliospores in single layers and solitary or in crusts (Melampsoraceae), those with teliospores in waxy crusts of one to two layers (Coleosporiaceae), and those with teliospores in chains (Cronartiaceae). The "pedicellate" group was called the Pucciniaceae.

Most later authors have followed this basic scheme while making minor variations and modifications of their own, such as changing the rank of the divisions and adding or subtracting subdivisions (e.g., Dietel, 1928; Cunningham, 1931; Arthur, 1934). In general they accept two families, the Melampsoraceae with sessile teliospores, and the Pucciniaceae with pedicellate teliospores, and they divide these up into a varying number of tribes.

The classifications all imply that each of these two families is monophyletic. However, Hiratsuka and Cummins (1963) show that pycnial structure suggests three main lines of development. These lines cut across the two previously recognized families so that if new families were recognized based on the these three lines, they would each take genera from both of the two original families. Of course such a novel classification could not be accepted without further supporting evidence. Some such evidence is indicated by Hiratsuka and Cummins (1963) who are, however, undoubtedly aware that more is needed before a formal classification along these lines can be proposed. Because of the present uncertainties in classifying rusts, I prefer to avoid presenting any particular system herein.

Finally, it is necessary to say a word about the significance of life cycles in rust taxonomy. There was a period when life cycles were given considerable significance by some authors, notably by Arthur (1906, 1907) and the Sydows (e.g., Sydow, 1921). A large number of genera were erected or resurrected for the purpose of segregating genera on life cycle alone. Certainly there was some practical value in such a scheme but it was too artificial to gain wide acceptance and was later abandoned even by its most determined initiators. Nowadays it is recognized that very closely related species can have differing life cycles (as in "correlated species") and even that variations in life cycle can occur within a single species, for instance *Uropyxis petalostemonis* with life cycles of O II^{I} II^{II} III or O II^{I} III or O III (Baxter, 1959, p. 223) and *Gymnoconia nitens* which possesses both demicyclic and endocyclic forms (Laundon, 1967).

VI. NOMENCLATURE

Rust nomenclature is fraught with certain difficulties which have not occurred to such an extent (or even at all) in other fungi. These difficulties have occurred because the imperfect states of rusts have not really been recognized as such; they have never been placed in the fungi imperfecti and in some cases they are so distinctive that they can readily be placed in the correct perfect genus. Three factors have contributed to the problem:

1. Prior to 1930 and to a considerable extent continuing up to 1940 it was usual to treat epithets based on imperfect states as equally applicable to the perfect states and as transferable to "perfect" genera. Even today, some uninformed or dissenting authors continue this practice.

2. Uredinial states were regarded as part of the perfect state by Arthur (1934) and he regarded epithets applied to the uredinial state as equally applicable to the telial state and as transferable to "perfect" genera.

3. Because of very characteristic morphology in some aecial and uredinial states, such states have often been placed in "perfect" genera.

Hylander, Jørstad, and Nannfeldt (1953), Wilson and Bisby (1954), and Cummins and Stevenson (1956) made a determined attempt to put rust nomenclature in line with the International Code of Botanical Nomenclature and brought the requirements of the Code to the attention of other uredino-logists (some of whom preferred to ignore them!), but they failed to appreciate the full effect of the Code and did not realize that the Code as it then stood did not deal adequately with all the problems. Deighton (1960) drew attention to these problems and the deficiencies in the Code and proposed modifications to deal with them. His proposals were adopted into the 1966 Code.

The main points to bear in mind in relation to the rusts and the present requirements of the Code are covered by Article 59 and the examples thereunder [see also Deighton (1960) for numerous helpful examples but ignore his Example 8 which was wrongly included]. Note that names published in contravention of paragraphs 1–3 of this Article are "illegiti-mate" (though validly published) but that this illegitimate status does not affect the illegitimacy of later homonyms (Article 64) (this means that some well-known rusts such as *Chrysomyxa piperiana*, *Coleosporium eupatoriae* and *Gymnosporangium hyalinum* have no legitimate names). Combinations published in contravention of paragraph 4 are not validly published as new combinations but can be valid as new names, provided other conditions of the Code are fulfilled. Often, these conditions are not fulfilled until sometime after the "combination" has first been made and often by another author. In such cases the author citation appended to the name of the rust is that of the later author only.

There is a problem in the nomenclature of rust genera caused by the type being wrongly cited; for example, a genus described for a perfect state being given a type based on an imperfect state. Proposals have been put forward which would deal with such cases, but none have yet been accepted and incorporated in the Code. For the present the status of these generic names must remain unsettled.

VII.   IMPORTANT LITERATURE

Arthur, J. C. (1934). "Manual of the Rusts in United States and Canada." 1962 ed., Hafner, New York.

Cummins, G. B. (1959). "Illustrated Genera of Rust Fungi." Burgess, Minneapolis, Minnesota.

Cummins, G. B. (1971). "The Rust Fungi of Cereals, Grasses and Bamboos." Springer-Verlag, Berlin and New York.

Cunningham, G. H. (1931). "The Rust Fungi of New Zealand." Privately printed, Palmerston North, New Zealand.

Gäumann, E. (1959). "Die Rostpilze Mitteleuropas." Büchler, Bern.

Hiratsuka, N. (1958). "Revision of taxonomy of the Pucciniastreae." Kasai Publ. and Printing Co., Tokyo.

Hylander, N., I. Jørstad, and J. Nannfeldt. (1953). Enumeratio Uredinearum Scandinavicarum. *Opera Bot.* **1**:1–102.

Laundon, G. F. (1965). The generic names of Uredinales. *Mycol. Pap.* **99**:1–24.

Sydow, P., and H. von Sydow (1902–1924). "Monographia Uredinearum." Vols 1–4. Borntraeger, Leipzig.

Thirumalachar, M. J., and B. B. Mundkur (1949). Genera of rusts. I and II. *Indian Phytopathol.* **2**:65–101, 193–244.

Thirumalachar, M. J., and B. B. Mundkur. (1950). Genera of rusts. III and Appendix. *Indian Phytopathol.* **3**:4–42, 203–204.

Wilson, M., and D. M. Henderson. (1966). "British Rust Fungi." Cambridge Univ. Press, London and New York.

VIII.   USE OF KEY

Although rust genera are classified chiefly by the features of the telial state, their other states have been utilized, and to some extent must be used, in a generic key to separate certain genera. It is as yet impossible to devise a key which will run out to the correct genus with any spore state. Indeed a considerable number of rusts (including some genera) are still incompletely known and their classification remains uncertain. Often a specimen to be identified may have only one or two of the possible states present and very often not the telial state. Rarely does one have the time or the means to search for or to carry out investigations into the full life cycle and all the spore states before making an identification. Therefore it very often happens that a generic key is of little help in identifying a routine rust specimen.

On the other hand, the host specialization of the rusts means that most host species or even genera are attacked by only a few specific rusts. Thus, a relatively easy method to identify a rust specimen is by way of the host, especially if a host index is available. By reference to such an index only a few rust names will be obtained (sometimes only one) which need to be checked one by one to see which of them matches. For some host groups or geographical areas there are host keys to aid this process [see, for example, Laundon (1965) and Cummins (1971)]. This method has an added advantage in that, even if the species has previously been wrongly classified, there is still a good chance that the material at hand can be matched with it. It is, however, dependent on correct host identification and this is usually absolutely essential in rust identification, especially for such large genera as *Puccinia* and *Uromyces*.

The purpose of a generic key to the rusts is, then, not so much for routine identification use but for research purposes or critical studies, as for example, research into the affinites of difficult taxa, the identification of species new to science, or in establishing the need for a new genus.

The key which follows is novel in that it deals separately with the imperfect states (aecia and uredinia), and the perfect state (telia). Each state is keyed out as far as possible independent of the other states, and although one cannot go very far with the aecial or uredinial states alone (for the reasons indicated above) it is hoped that this key will assist more than previous keys have done in dealing with specimens bearing only parts of their pathogens' life cycle.

It will be noted that the key also gives "imperfect" generic names for the aecial and uredinial states where such names are available. This has resulted in the use of many more such generic names than the few "form genera" (e.g. *Aecidium* and *Uredo*) which have been recognized by most authors. This is because, in my opinion, a taxonomy of imperfect forms can fulfil a useful purpose and is not obsolete as the trend followed by other authors would indicate.

KEY TO GENERA OF UREDINALES

1. Aecia 2

1'. Uredinia 13

1''. Telia 21

    2(1) On gymnosperms 3

    2'(1) On angiosperms 7

3(2) Caeomatoid . . . . **Caeoma**
    See 10'

3'(2) Peridermioid . . . . **Peridermium**
    Aecial states of *Chrysomyxa, Coleosporium, Cronartium, Milesina, Pucciniastrum, Uredinopsis*, etc. About 20 species still remain here. For possible subdivisions, see 4.

    4(3') Pycnia spreading, not well-defined 5

    4'(3') Pycnia discrete, clearly defined 6

5(4) On needles . . . . no generic name available
　　　Aecial states of *Coleosporium*.

5′(4) On stems and cones . . . . **Peridermium** s. str.
　　　Aecial states of *Cronartium*. Two species which have been treated as endocyclic because
　　　they produce basidial-like germ tubes have been segregated in the genus *Endo-
　　　cronartium*, which, however, is illegitimate because it contains the type species of
　　　Peridermium. Reference: Peterson, 1967.

　　6(4′) Pycnia conical . . . . **Pomatomyces**
　　　　Aecial states of *Chrysomyxa, Pucciniastrum, Uredinopsis*, etc.

　　6′(4′) Pycnia globose. . . . . no generic name available
　　　　Aecial states of *Milesina*.

7(2′) Pycnia spreading, not well-defined . . . . **Epitea**
　　　Aecial states of *Frommea, Kuehneola, Phragmidium, Xenodochus,* and *Triphragmium*.
　　　These would probably be placed in *Caeoma* by most authors.

7′(2′) Pycnia discrete, clearly defined 8

　　8(7′) Aecia roestelioid and on Rosaceae . . . . **Roestelia**
　　　　Aecial states of *Gymnosporangium*. There are about 9 species which probably should
　　　　be placed here but which have been illegitimately placed in *Gymnosporangium*.

　　8′(7′) Aecia not roestelioid or not on Rosaceae 9

9(8′) Aecia caeomatoid 10

9′(8′) Aecia cupulate . . . . **Aecidium**
　　　Used by some authors for all aecial states except *Peridermium*, but more strictly
　　　could be limited to those of *Puccinia, Uromyces, Cumminsiella, Maravalia, Miyagia,
　　　Nyssopsora, Physopella, Spumula, Tranzschelia,* etc. Several hundred species still
　　　placed here. If pycnia or basidia present, see 12.

　　10(9) Spores intermingled with a network of hyphal threads ("elater" hyphae)
　　　. . . . . **Elateraecium**
　　　　Two species in Africa, India, and the Philippines.

　　10′(9) Spore mass without hyphal threads . . . . **Caeoma**
　　　　Aecial states of *Arthuria, Crossopsora, Melampsora,* etc. Uredinial aecia of *Arthuria,
　　　　Chrysomyxa,* and *Coleosporium* should be placed here. About 15 species still remain
　　　　here and a few more could be added but are wrongly placed elsewhere. Endoforms
　　　　can be placed in *Kunkelia*. If pycnia are present, see 11.

11(10′) Pycnia stilboid, raised high above the epidermis . . . . **Kunkelia**
　　　Aecial states of *Gymnoconia*. Actually based on an endoform but could be expanded
　　　to include any aecial state as characterized herein.

11′(10′) Pycnia conical . . . . **Caeoma** s. str.
　　　Aecial states of *Crossopsora, Melampsora, Olivea, Phragmopyxis,* and *Pucciniostele*.

11″(10′) Pycnia globose . . . . no generic name available
　　　Aecial states of *Chrysocelis* and *Polioma*.

　　12(9′) Pycnia conical . . . . **Monosporidium**
　　　　Aecial states of *Cerotelium, Dasturella, Leucotelium, Masseeella, Ochropsora, Phy-
　　　　sopella,* and *Tranzschelia*. Thirumalachar (1961) interprets *Monosporidium* as endo-
　　　　cyclic but this appears not to have been proven. If it is not, then *Kulkarniella* is available
　　　　for endoforms; Kulkarni, 1968.

　　12′(9′) Pycnia globose . . . . **Aecidium** s. str.

Aecial states of *Puccinia, Uromyces, Cumminsiella, Maravalia, Miyagia,* and *Zaghouania.* Endoforms can be placed in *Endophyllum.*

13(1′) Uredinia peridiate, peridium of fused paraphyses . . . . . no generic name available
Uredinial states of *Miyagia.*

13′(1′) Uredinia peridiate with cellular peridium 14

13″(1′) Uredinia not peridiate but may be paraphysate 18

14(13′) Ostiolar cells of peridium bearing spines . . . . . . no generic name available
Uredinial states of *Melampsoridium.*

14′(13′) Not so 15

15(14′) Urediniospores colorless, even when fresh 16

15′(14′) Urediniospores colored, at least when fresh 17

16(15) Urediniospores beaked . . . . no generic name available
Uredinial states of *Uredinopsis.* At least 1 species (*Uredo investita*) could be placed here.

16′(15) Urediniospores not beaked but sometimes pointed . . . . **Milesia**
Uredinial states of *Milesina.* Several species at present in *Uredo* could be segregated here.

17(15′) Urediniospores minutely warty . . . . no generic name available
Uredinial states of *Hyalopsora.* At least 1 species (*Uredo obovata*) could be segregated here.

17′(15′) Urediniospores echinulate . . . . **Peridiopsora**
Uredinial states of *Cronartium, Melampsorella,* and *Pucciniastrum.* Several species at present in *Uredo* could be segregated here. Aecial uredinia belonging to *Uraecium* s. str. key out here but most authors regard this name as available for aecial uredinia of any type.

18(13″) Paraphyses intermixed, broadly capitate, hyaline and erect . . . . **Rubigo**
Uredinial states of *Melampsora.* There are a few species at present in *Uredo* which could be segregated here.

18′(13″) Paraphyses absent or marginal, colored or incurved, usually not broadly capitate 19

19(18′) Urediniospore pedicels bearing thin-walled branches in a whorl . . . . **Mapea**
Uredinial states of *Telomapea.*

19′(18′) Not so 20

20(19′) Urediniospores reniform with large spines and hyaline walls
. . . . no generic name available
Uredinial states of *Hemileia.* Several species at present incorporated in *Uredo* could be segregated here.

20′(19′) Urediniospores not reniform or without large spines or colored . . . . **Uredo**
Uredinial states of *Puccinia, Uromyces, Phakopsora, Phragmidium, Physopella, Pileolaria, Prospodium, Ravenelia,* etc., but used by most authors for all uredinial states. Several hundred species still placed here. Especially in warm countries many rusts persist in the uredinial state and form no other states.

21(1″) Teliospores sessile 22

21′(1″) Teliospores pedicellate 69

22(21) Teliospores scattered either in the epidermal cells or subepidermally 23

22′(21) Teliospores crustose or erumpent, never in the epidermal cells 30

23(22) Teliospores subepidermal 24

23′(22) Teliospores in the epidermal cells 26

24(23) All spore forms devoid of pigment. On ferns . . . . **Uredinopsis**
About 27 species, probably all heteroecious with aecia on needles of *Abies*. References: Faull, 1938; Hiratsuka, 1958.

24′(23) At least the uredinia colored yellowish to orange when fresh. Not on ferns . . 25

25(24′) Uredinia with ostiolar cells each drawn out into a long spine. Teliospores nonseptate but laterally united . . . . **Melampsoridium**
Three species are well-known and are heteroecious with aecia on needles of *Larix*; 2 other species have been described. Reference: Hiratsuka, 1958.

25′(24′) Uredinia with rounded ostiolar cells, sometimes spiny but not drawn out into a single spine. Teliospores laterally septate . . . . **Pucciniastrum** s. str.
About 23 species are included here and a further 14 species which key out at 26′ are included in the genus in most recent works. In addition, a number of species known only as uredinia may belong here. Where known, aecia occur on needles of *Abies* and *Picea*. Reference: Hiratsuka, 1958.

26(23′) Pycnia subepidermal or, if not, then all spore states devoid of pigment. On ferns
. . . . 27

26′(23′) Pycnia subcuticular. Never on ferns 28

27(26) All spore forms devoid of pigment . . . . **Milesina**
The name *Milesia* has often been used for this genus but really applies only to the uredinial state. About 34 species probably all of which are heteroecious with aecia on needles of *Abies*. Another 22 species which are known only in the uredinial state probably belong here. Reference: Faull, 1932; Hiratsuka, 1958.

27′(26) At least the uredinia colored yellowish to orange when fresh . . . . . . **Hyalopsora**
Eight species probably all of which are heteroecious with aecia on needles of *Abies*. Another 5 species which are known only in the uredinial state probably belong here. Reference: Hiratsuka, 1958.

28(26′) Teliospores laterally united, usually pale or colorless . . . . **Melampsorella**
Only 3 species, all are heteroecious with aecia on *Abies* and *Picea* causing "witches" brooms. Reference: Hiratsuka, 1958.

28′(26′) Teliospores vertically septate, usually brownish . . . . **Pucciniastrum**
Pucciniastrum in the strict sense keys out to 25′. The other part keys out here. This is divided into 2 segregates by some authors; see 29.

29(28) Uredinia absent. Telia causing swelling of the host tissues and forming "witches" brooms. . . . . **Calyptospora**
One species *C. goeppertiana*, on *Abies* and *Vaccinium*.

29′(28) Uredinia present. Not causing much hypertrophy or "witches" brooms
. . . . **Thekopsora**
About 13 species belong here.

30(22′) Teliospores 1-celled 31

30′(22′) Teliospores 2-celled or more 64

31(30) Teliospores in a single layer 32

31′(30) Teliospores in 2 or more layers or irregularly arranged 42

33(32) True urediniospores occurring in the life cycle . . . . **Goplana**
Seven species all occurring in tropical regions of the world.

33'(32) True urediniospores (as defined herein) absent from the life cycle; replaced by uredinial aecia or the species microcyclic . . . . **Coleosporium**
About 86 species still recognized but probably many of these do not differ significantly in morphology. Another 5 or more species known only as uredinial aecia probably belong here. An important genus with aecia on *Pinus* and on various flowering plants. In many species the aecia are still not known; some other species are microcyclic (*Gallowaya*) and 1 species has true aecia but not uredinial aecia (*Synomyces*). A few species have catenulate teliospores and key out at 42. More temperate in distribution than *Goplana*. Important species are *C. asterum* and *C. tussilaginis*.

34(32') Telia remaining covered by host tissue. With erect capitate papaphyses interspersed in the uredinia . . . . **Melampsora**
Syn. *Chnoopsora*. About 74 species still recognized but probably many of these do not differ significantly in morphology. Some species are heteroecious with telia on *Populus* and *Salix* and aecia on various conifers and angiosperms. The autoecious species occur on various families. The most economically important species is *M. lini*, autoecious on flax.

34'(32') Telia becoming erumpent. Marginal paraphyses often present around the uredinia but no erect capitate interspersed ones present. 35

35(34') Teliospores becoming 3-septate (internal basidia) . . . . **Ochropsora**
Three species. *O. ariae* can infect apple and pear.

35'(34') Teliospores remaining nonseptate (external basidia) 36

36(35') Teliospores borne in groups on basal cells 37

36'(35') Teliospores borne singly 39

37(36) Telia subcuticular. Brachycyclic . . . . **Tegillum**
One species, *T. fimbriatum* on *Vitex*, C. America.

37'(36) Telia subepidermal (but may be raised on a thick cellular base above the epidermis). Macro-, brachy-, or microcyclic 38

38(37') Macrocyclic. Telia raised on a thick cellular base . . . . **Olivea**
Four species in tropical regions.

38'(37') Brachy-, or microcyclic. Telia not raised on a cellular base . . . . . **Chaconia**
Eight species referred here. They occur in warm regions. Reference: Mains, 1938a.

39(36') Basidia large and robust 40

39'(36') Basidia narrow 41

40(39) Uredinia present in the life cycle. On angiosperms . . . . **Mikronegeria**
One species, *M. fagi* on *Fagus*, S. America.

40'(39) Uredinia absent. On conifers . . . . **Ceropsora**
One species, *C. piceae* on *Picea*, India.

41(39') Uredinia present in the life cycle . . . . **Aplopsora**
Two species, probably heteroecious, but the pycnia and aecia are unknown.

41'(39') Uredinia absent . . . . **Chrysocelis**
Four species in tropical regions. Autoecious, demicyclic.

42(31′) Telia with a peridium, like aecia 43

42′(31′) Telia without such a peridium 44

43(42) Teliospores with intercalary cells . . . . **Endophylloides**
    One species, *E. portoricensis* on *Mikania*, tropical America.

43′(42) Teliospores without intercalary cells . . . . **Dietelia**
    One species, *D. verruciformis* on *Sida*, Argentina.

44(42′) Telia cushionlike, flat, or $\pm$ globose 45

44′(42) Telia tall and columnar, often hairlike 57

45(44) Telia gelatinous 46

45′(44) Telia not gelatinous 47

46(45) Teliospores becoming 3-septate (internal basidia) . . . . **Coleosporium**
    A few *C.* spp with catenulate teliospores (*Gallowaya* and *Stichopsora*) may key out
    here. The main part of the genus keys out at 33.

46′(45) Teliospores remaining nonseptate (external basidia) . . . . **Coleopuccinia**
    See 98′.

47(45′) Telia remaining covered by host tissue 48

47′(45′) Telia becoming erumpent 50

48(47) Teliospores in regular chains 49

48′(47) Teliospores disarranged . . . . **Phakopsora**
    About 50 species throughout warm regions of the world. Autoecious species with
    nonparaphysate uredinia havé been placed in *Bubakia*. *Phakopsora gossypii* can be
    important on cotton.

49(48) Aecia cupulate. Only 1 type of teliospore occurring . . . . **Physopella**
    Syn. *Angiopsora*. About 18 species throughout warm regions of the world. *Physopella*
    ampelopsidis on *Vitis* and *P. zeae* on *Zea* are important species. Reference: Cummins
    and Ramachar, 1958

49′(48) Aecia caeomatoid. Two types of teliospores occur . . . . **Pucciniostele**
    Three species belong here and all occur on *Astilbe*. A fourth species, *P. hashiokai*
    on *Ampelopsis*, does not belong. The primary teliospores are 4-celled tetradlike
    discs. The secondary teliospores which develop in separate sori consist of single
    cells fused in a mass but regularly arranged. Reference: Cummins and Thirumalachar,
    1953.

50(47′) Teliospores regularly arranged in either horizontal or vertical direction (or both)
    51

50′(47′) Teliospores disarranged . . . . **Uredopeltis**
    Two species, Africa and India. Reference: Laundon, 1963.

51(50) Teliospores separating in horizontal plates. Microcyclic . . . . **Alveolaria**
    Three species in tropical America.

51′(50) Teliospores adhering vertically, not separating in horizontal plates 52

52(51′) Uredinia occur in the life cycle 53

52′(51′) Uredinia absent from the life cycle 55

53(52) Telia black and hard with cells firmly fused together . . . . **Dasturella**
    Four species all occurring in India. One is known to be heteroecious.

53′(52) Telia paler, not forming a hard-fused mass 54

54(53′) Macrocyclic. Teliospore chains laterally ± adherent . . . . **Cerotelium**
About 24 species mostly throughout warm regions of the world. *C. fici* on fig is of some importance.

54′(53′) Brachycyclic. Teliospore chains laterally free . . . . **Phragmidiella**
Four species of tropical Africa and India.

55(52′) Pycnia conical, subcuticular or subepidermal. Teliospores pale 56

55′(52′) Pycnia globose, subepidermal. Teliospores brownish . . . . **Baeodromus**
Five species all occurring in America and all microcyclic.

56(55) Pycnia subcuticular . . . . **Arthuria**
Three species occurring in S. America and India.

56′(55) Pycnia subepidermal . . . . **Chrysomyxa**
About 20 species mostly heteroecious with aecia on *Picea* and telia on Ericaceae and related families. *Hiratsukaia* has been segregated on the basis of its ramicolous telia and very long teliospore chains.

57(44′) Teliospores with long gelatinizing intercalary cells . . . . **Trichopsora**
One species, *T. tournefortiae* on *Tournefortia*, S. America. Microcyclic.

57′(44′) Teliospores without intercalary cells 58

58(57′) Teliospores regularly arranged in horizontal plates. Microcyclic . . . **Alveolaria**
See 51.

58′(57′) Teliospores regularly arranged in vertical chains 59

58″(57′) Teliospores disarranged 60

59(58′) Aecia caeomatoid. Uredinia paraphysate . . . . **Crossopsora**
About 18 species throughout warm regions of the world.

59′(58′) Aecia peridiate. Uredinia peridiate . . . . **Cronartium**
Fifteen species of which at least 2 are doubtfully distinct. In addition two *Peridermium* species are known to be distinct and to possess telia but have not yet received *Cronartium* names and there are 2 forms which survive in the aecial stage alone on *Pinus* (*Endocronartium*, see 5′). Otherwise, *C.* spp. are heteroecious with aecia on *Pinus* and telia on various dicotyledons. Several species are economically important especially *C. ribicola*. Reference: Peterson, 1967.

60(58″) Telia gelatinous 61

60′(58″) Telia not gelatinous 63

61(60) Demicyclic (as defined herein) or microcyclic. Teliospores becoming 3-septate (internal basidia). . . . . **Coleosporium**
A few *C.* spp. with catenulate teliospores (*Stichopsora*) may key out here. The main part of the genus keys out at 33′.

61′(60) Macrocyclic. Teliospores remaining nonseptate (external basidia) 62

62(61′) Aecia cupulate. Uredinia paraphysate . . . . **Masseeella**
Five species mostly autoecious, macrocyclic, and occurring on Euphorbiaceae. They occur in Asia.

62′(61′) Aecia caeomatoid. Uredinia without paraphyses but like pycnidia and opening by an ostiole . . . . **Kamatomyces**
One species, *K. narasimhanii* on *Flueggea*, India, segregated from *Masseeella*.

63(60′) Teliospores with a long beak . . . . **Skierka**
Nine species throughout tropical regions of the world.

63′(60′) Teliospores rounded or pointed but not with a long beak . . . . **Cionothrix**
Microcyclic and all occurring in S. America except for 1 in Australia.

64(30′) Teliospores catenulate 65

64′(30′) Teliospores not caténulate 68

65(64) Telia in hairlike columns . . . . **Gambleola**
One species: *G. cornuta* on *Berberis*, Asia.

65′(64) Telia not hairlike . . . . ´66

66(65′) Telia with a peridium, like cupulate aecia . . . . **Pucciniosira**
Eleven species throughout warmer regions of the world.

66′(65′) Telia not peridiate, not like cupulate aecia 67

67(66) Pycnia not occurring. Telia ± gelatinous . . . . **Coleopuccinia**
See 98′.

67′(66) Pycnia occur in the life cycle. Telia not gelatinous . . . . **Didymopsora**
Six species all microcyclic and occurring in Africa and S. America.

68(64′) Teliospores in discoid spore heads . . . . **Nothoravenelia**
Two species both on Euphorbiaceae. The species might be confused with *Kernkampella* (see 137).

68′(64′) Teliospores not in discoid heads but usually grouped on basal cells . . **Polioma**
Four species, in America, 3 are microcyclic on *Salvia*, one is macrocyclic on *Geranium*. Reference: Baxter and Cummins, 1951.

69(21′) Teliospores 1-celled 70

69′(21′) Teliospores 2-celled 95

69″(21′) Teliospores 3-celled or more 117

70(69) Teliospore wall not colored, not thickened, not ornamented and pore(s) obscure
. . . . 71

70′(69) Teliospore wall either colored, thickened, ornamented, or with visible pore(s)
. . . . 82

71(70) Teliospores becoming 3-septate (internal basidia) 72

71′(70) Teliospores remaining nonseptate (external basidia) 73

72(71) Pycnia globose, subepidermal. Teliospores borne singly . . . . **Chrysella**
One species: *C. mikaniae* on *Mikania*, C. America.

72′(71) Pycnia conical, subcuticular. Teliospores borne in groups on basal cells
. . . . **Achrotelium**
Four species throughout warm regions of the world.

73(71′) Telia ± gelatinous . . . . **Coleopuccinia**
See 98′.

73′(71′) Telia not gelatinous 74

74(73′) Urediniospore pedicels bearing thin-walled branches in a whorl . . . **Telomapea**
One species: *T. inocarpi* on *Inocarpus*, Asia.

74′(73′) Urediniospore pedicels not bearing branches 75

75(74′) Telia superstomatal 76

75′(74′) Telia intraepidermal 78

75″(74′) Telia subepidermal 79

76(75) Urediniospores reniform, basidia slender and symmetrical . . . . . . **Hemileia**
About 40 species belong here and although some are known only in the uredinial state, this state is so characteristic that they have often been given *Hemileia* names. Pycnia and aecia are unknown, but 1 species is said to have telial uredinia (Rajendren, 1967a). On both monocotyledons and dicotyledons (mostly on Rubiaceae) in tropical regions, mostly Africa and Asia. *Hemileia vastrarix* is of major importance on coffee. Reference: Gopalkrishnan, 1951.

76'(75) Urediniospores symmetrical 77

77(76) Uredinia intraepidermal. Basidia slender, curved 78

77'(76) Uredinia superstomatal. Basidia robust, straight and thick . . . . **Blastospora**
Two species, on *Smilax*, Japan. Reference: Mains, 1938b.

78(75' and 77) On Rosaceae . . . . **Gerwasia**
About 17 species (including *Mainsia*) autoecious on *Rosa* and *Rubus* throughout warmer regions of America and Asia. Reference: Jackson, 1931.

78' (75' and 77) On Leguminosae . . . . **Scopellopsis**
One species: *S. dalbergiae* on *Dalbergia*, India. Reference: Rajendren, 1967b.

79(75'') Pycnia unknown. Teliospores ellipsoid . . . . **Botryorhiza**
One species: *B. hippocrateae* on *Hippocratea*, W. Indies.

79'(75'') Pycnia occur in the autoecious life cycle. Teliospores cylindric 80

80(79') Pycnia globose, subepidermal . . . . **Maravalia**
About 10 species mostly on legumes and occurring in warmer regions of the world. Reference: Mains, 1939.

80'(79') Pycnia conical, subcuticular 81

81(80') Aecial uredinia subcuticular, paraphysate . . . . **Angusia**
One species: *A. lonchocarpi* on *Lonchocarpus*, Zambia.

81'(80') Aecial uredinia subepidermal . . . . **Scopella**
See 94.

82(70') Telia subcuticular . . . . **Lipocystis**
One species: *L. caesalpiniae* on *Mimosa*, W. Indies.

82'(70') Telia subepidermal 83

83(82') Pedicels each bearing 2 or more teliospores or vesicles 84

83'(82') Pedicels each bearing a single teliospore 85

84(83) Teliospores and vesicles attached quite separately to the pedicels
. . . . **Uromycladium**
Seven species all on *Acacia* and *Albizzia* in Australia and New Zealand. All autoecious and some forming large galls.

84'(83) Teliospores borne in pairs on common apical cells, usually partially joined together
. . . . **Diabole**
See 114.

85(83') Urediniospore pedicels bearing thin-walled branches in a whorl . . . . . **Telomapea**
See 74.

85'(83') Urediniospore pedicels not bearing branches 86

86(85') Teliospore pedicels 1-septate . . . . **Trachyspora**
About 5 species all on *Alchemilla* except 1 on *Sapium*.

86'(85') Teliospore pedicels not septate 87

87(86') Uredinia and telia peridiate . . . . **Miyagia**
See 104.

87'(86') Uredinia and telia not peridiate 88

88(87') Telia gelatinous 89

88'(87') Telia not gelatinous 91

89(88) Teliospores in hairlike columns and with thickened apical walls . . . . **Chardoniella**
One species: *C. gynoxidis* on *Gynoxis*, S. America.

89'(88) Teliospores not combining these features 90

90(89') Telia on Cupressaceae . . . . **Gymnosporangium**
See 98.

90'(89') Telia on Rosaceae . . . . **Coleopuccinia**
See 98'.

91(88') Teliospores with 1 germ pore or none visible 92

91'(88') Teliospores with 2–3 germ pores . . . . **Phragmidium**
See 126.

92(91) Pycnia globose, subepidermal 93

92'(91) Pycnia conical, subcuticular 94

93(92) Teliospores hyaline, verrucose and thick-walled . . . . **Zaghouania**
Two species (including *Cystopsora*) both on *Oleaceae*.

93'(92) Teliospore walls colored or smooth or thicker above than at the sides . . . **Uromyces**
Several hundred species on both monocotyledons and dicotyledons throughout the
world. Especially common on Leguminosae. Intergrades with *Puccinia* (since there are
a number of species with various proportions of 1- and 2-celled spores) and only
maintained for convenience. There are several economically important species, e.g.,
U. betae on beet, *U. dianthi* on carnations, *U. appendiculatus* and *U. viciae-fabae*
on beans, *U. pisi* on peas.

94(92') Teliospores smooth, cylindric . . . . **Scopella**
About 14 species, mostly on the Sapotaceae and occurring in warmer regions of the
world. Reference: Cummins, 1950.

94'(92') Teliospores usually ornamented, globose, or ellipsoid . . . . **Pileolaria**
About 20 species mostly on Anacardiaceae throughout the world. The few species
occurring on legumes may not be related.

95(69') Only 1 cell of the teliospore attached to the pedicel; usually with 1 cell ± vertically
above the other 96

95'(69') Both cells of the teliospore (or sterile apical cells beneath them) attached to the pedicel;
the cells arranged horizontally side by side 113

96(95) Teliospore cells each becoming 3-septate (internal basidia) . . . . . **Chrysopsora**
One species: *C. gynoxidis* on *Gynoxis*, S. America.

96'(95) Teliospores remaining nonseptate (external basidia) 97

97(96') Telia gelatinous 98

97'(96') Telia not gelatinous 99

98(97) Telia on Cupressaceae . . . . **Gymnosporangium**

Fifty-seven species throughout temperate regions of the world. In addition there are about 11 species known only in the aecial stage. All species but one are heteroecious with aecia mostly on Rosaceae and telia on *Juniperus*. Only 2 species possess uredinia, the remainder are demicyclic. Several species, especially *G. clavipes* and *G. juniperi-virginianae*, are important on apples and pears. Reference: Kern, 1964.

98'(97) Telia on Rosaceae . . . . **Coleopuccinia**
Four species, including *Coleopucciniella,* which is interpreted as possessing pedicellate phragmospores. All on Rosaceae and microcyclic, perhaps derived from *Gymnosporangium.* Occurring in China and Japan.

99(97') Teliospore wall not colored, not thickened, not ornamented, and pore(s) obscure
. . . . 100

99'(97') Teliospore wall either colored, thickened, ornamented, or with visible pore(s) . . 102

100(99) Basidia formed by elongation of the spore apex . . . . **Chrysocyclus**
Three species, all microcyclic and on Compositae and Solanaceae in S. America.

100'(99) Basidia formed through an obscure pore 101

101(100') Macrocyclic . . . . **Leucotelium**
Three species with telia on *Prunus* and aecia on Ranunculaceae.

101'(100') Brachycyclic . . . . **Sorataea**
Several species belong here, including *Allopuccinia* and *Mimema* and perhaps some species still retained in *Puccinia,* but which possess subcuticular pycnia.

102(99') Teliospores with 4 or more germ pores per cell. Pycnia globose, subepidermal
. . . . **Cleptomyces**
Two species both occurring in S. America.

102'(99') Teliospores with 3 germ pores per cell. Pycnia conical, subcuticular . . **Dipyxis**
Two species both on Bignoniaceae in America. *Stereostratum* with one species, *S. corticioides* on Gramineae in China and Japan also keys out here, but its affinities are uncertain.

102"(99') Teliospores with 2 germ pores per cell 103

102'"(99') Teliospores with 1 germ pore per cell 104

103(102") Macrocyclic. Pycnia globose, subepidermal . . . . **Cumminsiella**
Six species all autoecious on Berberidaceae. Reference: Baxter, 1957.

103'(102") Brachycylic or microcyclic. Pycnia conical, subcuticular . . . . **Uropyxis**
Thirteen species all autoecious; 10 occur on Leguminosae, 1 on Cucurbitaceae, and 2 which occur on Bignoniaceae and which differ from the others in showing no distinct lamination of the teliospore wall are included with some uncertainty. Reference: Baxter, 1959.

104(102'") Uredinia and telia peridiate . . . . **Miyagia**
Syn.: *Peristemma.* Six species all autoecious on Compositae. Three have 1-celled teliospores (*Corbulopsora*) and the other three have 2-celled teliospores. Reference: Hiratsuka, 1969.

104'(102'") Uredinia and telia not peridiate 105

105(104') Telia superstomatal 106

105'(104') Telia subepidermal 107

106(105) On ferns or *Berberis* . . . . **Desmella**
Syn.: *Edythea.* Eight species, all in S. America. Perhaps this genus should be united with *Prospodium.*

106'(105) On monocotyledons . . . . **Desmellopsis**
One species: *D. afromomicola*, Africa. Perhaps this genus should be united with *Prospodium*.

106"(105) On Bignoniaceae and Verbenaceae . . . . **Prospodium**
About 40 species, all occurring in warm regions of America. Reference: Cummins, 1940.

107(105') Telia with hygroscopic paraphyses . . . . **Didymopsorella**
Two species (including *Gymnopuccinia*) both on *Toddalia*.

107'(105') Telia without hygroscopic paraphyses 108

108(107') Teliospores adhering in long hairlike columns . . . . **Kernella**
One species, *K. lauricola* on *Litsea*, India.

108'(107') Teliospores not adhering in long hairlike columns 109

109(108') Teliospores in fascicles . . . . **Tranzschelia**
Ten species either heteroecious with aecia on Ranunculaceae and telia on Rosaceae or autoecious (macro-, demi-, or microcyclic) on Ranunculaceae. *Tranzschelia discolor* is important on peach and other *Prunus* spp.

109'(108') Teliospores not in fascicles 110

110(109') Pycnia conical, subcuticular 111

110'(109') Pycnia conical, subepidermal . . . . **Dasyspora**
One species: *D. gregaria* on *Xylopia*, tropical America.

110"(109') Pycnia globose, subepidermal 112

111(110) Brachy- or microcyclic . . . . **Prospodium**
See 106".

111'(110) Demi- or endocyclic . . . . **Gymnoconia**
One species: *G. nitens* on *Rubus* which occurs in both demi- and endocyclic forms. Some authors segregate the endocyclic form in *Kunkelia*.

112(110") Teliospores sessile on short basal cells which may be mistaken for pedicels. Often more than 1 teliospore attached to each basal cell . . . . **Polioma**
See 68'.

112'(110") Teliospores truly pedicellate . . . . **Puccinia**
About 3000 species on both monocotyledons and dicotyledons throughout the world. There are many important species, e.g., *P. graminis* on wheat (black stem rust), *P. sorghi* and *P. polysora* on maize (*Zea*), *P. allii* on onions, *P. menthae* on mint, *P. helianthi* on sunflower.

113(95') With sterile apical cell(s) between pedicel and teliospore cells 114

113'(95') Teliospore cells borne directly on the pedicels 116

114(113) Teliospore cells 2 on each apical cell . . . . **Diabole**
One species: *D. cubensis* on *Mimosa*, C. America and W. Indies.

114'(113) Teliospore cells with an apical cell each 115

115(114') Teliospores with 1 germ pore per cell . . . . **Dicheirinia**
Nine species on legumes throughout warmer regions of America except one in the Canary Islands and one in Mauritius. Reference: Cummins, 1935.

115'(114') Teliospores with 2 germ pores in each cell . . . . **Diorchidiella**
One species: *D. australis* on *Mimosa*, S. America.

116(113') Teliospores colorless, in compact waxy sori . . . . **Sphenospora**
Nine species all on monocotyledons in tropical areas of America.

116'(113') Teliospores colored, not in compact waxy sori but pulverulent

. . . . **Diorchidium**
Syn: *Allotelium*. About 12 species, mostly on legumes in tropical regions of the world. *Diphragmium* probably belongs here..

117(69'') Teliospore cells arranged as in phragmospores, that is, serially with pedicel attached to the lower one only 118

117'(69'') Teliospore cells arranged triquetrously (triangularly 3-celled) 127

117''(69'') Teliospore cells in a muriform arrangement 129

117'''(69'') Teliospore cells arranged radially in a discoid head 130

118(117) Telia $\pm$ gelatinous, on Cupressaceae . . . . **Gymnosporangium**
See 98.

118'(117) Telia not gelatinous, on angiosperms, mostly Rosaceae 119

119(118') Teliospore wall hyaline 120

119'(118') Teliospore wall colored 123

120(119) Teliospores lanceolate, cells firmly joined . . . . **Hamaspora**
Eleven species, all on *Rubus* in Africa, Asia, Australia, and New Zealand. Reference: Monoson, 1969.

120'(119) Teliospores not lanceolate 121

121(120') Teliospores usually 4- or more-celled, the cells rather easily separating

. . . . **Kuehneola**
About 10 species, several on *Rubus*. *Kuehneola uredinis* is important on brambles (*Rubus*) especially the European blackberry.

121'(120') Teliospores 2- or 3-celled, rarely more; not easily separating 122

122(121') Pycnia conical and subcuticular . . . . **Sorataea**
See 101'.

122'(121') Pycnia globose and subepidermal . . . . **Puccinia**
See 112'.

123(119') Teliospores with 1 germ pore in each cell 124

123'(119') Teliospores with more than 1 germ pore in each cell (except perhaps the apical cell)
. . . . 125

124(123) Pycnia conical and subcuticular . . . . **Frommea**
Three species on *Potentilla* and related genera.

124'(123) Pycnia globose and subepidermal . . . . **Puccinia**
See 112'.

125(123') Teliospores with 1 germ pore in the apical cell, 2 in the other cells. Pedicels very short
. . . . **Xenodochus**
Two species, one demicyclic and the other microcyclic; both on *Sanguisorba*.

125'(123') Teliospores with 2 or more germ pores in all cells. Pedicels usually long 126

126(125') Teliospores without conspicuous outer hygroscopic layer . . . **Phragmidium**
About 60 species, all autoecious on Rosaceae (but see Peterson and Cronin, 1967). About 10 brachycyclic species with smooth-walled teliospores germinating without dormancy on nonhygroscopic basally 1-septate pedicels are sometimes segregated

as *Phragmotelium*. *Phragmidium mucronatum* and *P. tuberculatum* are important on ornamental roses and *P. rubi-idaei* is important on raspberries.

126′(125′) Teliospores with conspicuous outer hygroscopic layer . . . . . **Phragmopyxis**
Four species, all autoecious on legumes, in America and Africa.

127(117′) Teliospore cells arranged in an erect triangle . . . . **Hapalophragmium**
About 10 species, all autoecious on legumes. I include *Hapalophragmiopsis* here, but see Thirmulachar (1961) and Hiratsuka and Cummins (1963, p. 502).

127′(117′) Teliospore cells arranged in an inverted triangle 128

128(127′) Teliospores with only 1 germ pore in each cell . . . . **Triphragmium**
Four species are autoecious on Rosaceae; one other is recorded on Compositae. All occur in the northern hemisphere.

128′(127′) Teliospores with 2 or more germ pores in each cell . . . . **Nyssopsora**
Ten species (including *Triphragmiopsis*) autoecious on various families.

129(117″) Teliospores considerably elongated with the cells arranged in laterally united chains
. . . . **Cumminsina**
One species: *C. clavispora* on *Grewia*, Southern Africa.

129′(117″) Teliospores ± globose . . . . **Sphaerophragmium**
About 15 species mostly on legumes in tropical regions of the world.

130(117‴) Teliospore heads on a cellular base surrounded by paraphyses
. . . . **Nothoravenelia**
See 68.

130′(117‴) Teliospore heads truly pedicellate, mostly nonparaphysate 131

131(130′) Teliospore pedicels single 132

131′(130′) Teliospore pedicels several per head of spores and fused together 136

132(131) Teliospore heads with hygroscopic cysts 133

132′(131) Teliospore heads with nonhygroscopic basal cells 134

133(132) Pedicels attached to the fused cysts . . . . **Cystomyces**
One species: *C. costaricensis*.

133′(132) Pedicels attached to the teliospores, cysts separate . . . . **Spumula**
Four species all autoecious on legumes.

134(132′) Teliospores with 1 basal cell to each fertile cell 135

134′(132′) Teliospores with a pair of fertile cells to each basal cell . . . . **Diabole**
See 114.

135(134) Teliospores dark brown, ornamented . . . . **Dicheirinia**
See 115.

135′(134) Teliospores pale brown, smooth . . . . **Anthomyces**
One species: *A. brasiliensis*.

136(131′) Teliospore heads with hygroscopic cysts 137

136′(131′) Teliospore heads with nonhygroscopic basal cells . . . . . . **Anthomycetella**
One species: *A. canarii* on *Canarium*, Philippines.

137(136) Teliospore heads with a layer of sterile "epipatella" cells between the fertile teliospore cells and the cysts. Marginal epipatella cells each ornamented with one conspicuous projection. Fertile teliospore cells with conspicuous pores . . . . **Kernkampella**
Two species already assigned here but six more, at present placed in *Ravenelia*, should be added; all on Euphorbiaceae.

137′(136) Teliospore heads without a layer of "epipatella" cells between the fertile teliospore cells and the cysts or, if such cells are present, the heads either smooth or ornamented with several projections to each marginal cell. Fertile teliospore cells without visible pores
· **Ravenelia**
About 150 species, autoecious on legumes. Most, if not all of those described on Euphorbiaceae belong in *Kernkampella*. None of the species is of much importance.

REFERENCES

Arthur, J. C. (1905). Terminology of the spore structures in the Uredinales. *Bot. Gaz. (Chicago)* **39**:219–222.

Arthur, J. C. (1906). *In* "Résultats Scientifiques du Congrès International de Botanique, Vienne, 1905," pp. 331–348. Fischer, Jena.

Arthur, J. C. (1907). Uredinales. *N. Amer. Flora* **7**:85–1151.

Arthur, J. C. (1932). Terminologie der Uredinales. *Ber. Deut. Bot. Ges.* **50**a:24–27.

Arthur, J. C. (1934). "Manual of the Rusts in United States and Canada." 1962 ed., Hafner, New York.

Arthur, J. C., *et al.* (1929). "The Plant Rusts (Uredinales)." Wiley, New York.

Baxter, J. W. (1957). The genus *Cumminsiella. Mycologia* **49**:864–873.

Baxter, J. W. (1959). A monograph of the genus *Uropyxis. Mycologia* **51**:210–226.

Baxter, J. W., and G. B. Cummins. (1951). *Polioma* Arth., a valid genus of the Uredinales. *Bull. Torrey Bot. Club* **78**:51–55.

Cummins, G. B. (1935). The genus *Dicheirinia. Mycologia* **27**:151–159.

Cummins, G. B. (1936). Phylogenetic significance of the pores in urediospores. *Mycologia* **28**: 103–132.

Cummins, G. B. (1940). The genus *Prospodium* (Uredinales). *Lloydia* **3**:1–78.

Cummins, G. B. (1950). The genus *Scopella* of the Uredinales. *Bull. Torrey Bot. Club* **77**: 204–213.

Cummins, G. B. (1959). "Illustrated Genera of Rust Fungi." Burgess, Minneapolis, Minnesota.

Cummins, G. B. (1971). "The Rust Fungi of Cereals, Grasses and Bamboos." Springer-Verlag, Berlin and New York.

Cummins, G. B., and P. Ramachar. (1958). The genus *Physopella* (Uredinales) replaces *Angiopsora. Mycologia* **50**:741–744.

Cummins, G. B., and J. A. Stevenson. (1956). A check list of North American rust fungi (Uredinales). *Plant Dis. Rep., Suppl.* **240**.

Cummins, G. B., and M. J. Thirumalachar. (1953). *Pucciniostele*, a genus of the rust fungi. *Mycologia* **45**:572–578.

Cunningham, G. H. (1930). Terminology of the spore forms and associated structures of the rust fungi. *N. Z. J. Sci. Technol.* **12**:123–128.

Cunningham, G. H. (1931). "The Rust Fungi of New Zealand." Privately printed, Palmerston North, New Zealand.

de Bary, A. (1865). Neue Untersuchungen ueber die Uredineen, insbersondere die Entwicklung der *Puccinia graminis* und der Zuzammenhang derselben mit *Aecidium berberidis*. I. *Monatsber. Adad. Wiss. Berlin* pp. 15–49.

Deighton, F. C. (1960). Article 59. *Taxon* **9**:231–241.

Dietel, P. (1900). Uredinales. *In* "Engler and Prantl, Natürlichen Pflanzen Familien," Vol. I. Part 1, pp. 24–81.

Dietel, P. (1928). Uredinales. *In* "Engler and Prantl, Natürlichen Pflanzen Familien," 2nd ed., Vol. 6, pp. 24–98.

Dodge, B. O. (1924). Aecidiospore discharge as related to the character of the spore wall. *J. Agr. Res.* **27**:749–756.

Faull, J. H. (1932). Taxonomy and geographical distribution of the genus *Milesia*. *Contrib. Arnold. Arbor.* **2**:1–138.

Faull, J. H. (1938). Taxonomy and geographical distribution of the genus *Uredinopsis*. *Contrib. Arnold Arbor.* **1**:1–120.

Gopalkrishnan, K. S. (1951). Notes on the morphology of the genus *Hemileia*. *Mycologia* **43**:271–283.

Hiratsuka, N. (1958). "Revision of Taxonomy of the Pucciniastreae." Kasai Publ. and Printing Co., Tokyo.

Hiratsuka, N. (1969). Notes on the genus *Miyagia* Miyabe ex Sydow. *Trans. Mycol. Soc. Jap.* **10**:89–90.

Hiratsuka, Y., and G. B. Cummins. (1963). Morphology of the spermagonia of the rust fungi. *Mycologia* **55**:487–507.

Holm, L. (1966). Études urédinologiques 4. *Sv. Bot. Tidskr.* **60**:23–32.

Hylander, N., I. Jørstad, and J. Nannfeldt. (1953). Enumeratio Uredinearum Scandinavicarum. *Opera Bot.* **1**:1–102.

Jackson, H. S. (1931). The rusts of South America. *Mycologia* **23**:96–116.

Kern, F. D. (1964). Lists and keys of the cedar rusts of the world. *Mem. N. Y. Bot. Gard.* **10**: 305–326.

Kuhnholz-Lordat, G. (1941). Les Urédinées corbiculées. *Bull. Acad. Montpellier* **71**:91–92.

Kulkarni, U. K. (1968). On the validity of the genus *Kulkarniella* (Uredinales) Gokhale and Patel. *Mycopathol. Mycol. Appl.* **36**:305–310.

Laundon, G. F. (1963). *Uredopeltis* (Uredinales). *Trans. Brit. Mycol. Soc.* **46**:503–504.

Laundon, G. F. (1964). *Angusia* (Uredinales). *Trans. Brit. Mycol. Soc.* **47**:327–329.

Laundon, G. F. (1965). Rust fungi. III. *Mycol. Pap.* **102**:1–52.

Laundon, G. F. (1967). Terminology in the rust fungi. *Trans. Brit. Mycol. Soc.* **50**:189–194.

Laundon, G. F. (1972). Delimitation of aecial from uredinial states. *Trans. Brit. Mycol. Soc.* **58**:344–346.

Mains, E. B. (1938a). Studies in the Uredinales, the genus *Chaconia*. *Bull. Torrey Bot. Club* **65**:625–629.

Mains, E. B. (1938b). The genus *Blastospora*. *Amer. J. Bot.* **25**:677–679.

Mains, E. B. (1939). Studies in the Uredinales, the genus *Maravalia*. *Bull. Torrey Bot. Club* **66**:173–179.

Monoson, H. L. (1969). The species of *Hamaspora*. *Mycopathol. Mycol. Appl.* **37**:263–272.

Peterson, R. S. (1961). Host alternation of spruce broom rust. *Science* **134**:468–469.

Peterson, R. S. (1967). The *Peridermium* species on pine stems. *Bull. Torrey Bot. Club* **94**: 511–542.

Peterson, R. S., and E. A. Cronin. (1967). Non-rosaceous hosts of *Phragmidium* (Uredinales). *Plant Dis. Rep.* **51**:766–767.

Rajendren, R. B. (1967a). A new type of nuclear life cycle in *Hemileia vastatrix*. *Mycologia* **59**:279–285.

Rajendren, R. B. (1967b). The identity and status of the rust on *Dalbergia paniculata*. *Bull. Torrey Bot. Club.* **94**:84–86.

Rajendren, R. B. (1970). *Kernkampella*: A new genus in the Uredinales. *Mycologia* **62**:837–843.

Saccardo, P. A. (1888). *Sylloge Fungorum* **7**:767.

Savile, D. B. O. (1954). Cellular mechanics, taxonomy and evolution in the Uredinales and Ustilaginales. *Mycologia* **46**:736–761.

Savile, D. B. O. (1965). *Puccinia karelica* and species delimitation in the Uredinales. *Can. J. Bot.* **43**:231–238.

Savile, D. B. O. (1968). The case against "uredium." *Mycologia* **60**:459–464.

Scott, K. J., and D. J. Maclean. (1969). Culturing of rust fungi. *Annu. Rev. Phytopathol.* **7**: 123–146.

Sydow, H. von (1921). Die Verwertung der Verwandtschaftsverhaltnisse und des gegenwärtigen Entwicklungsganges zur Umgrenzung der Gattungen bei den Uredineen. *Ann. Mycol. Berlin* **19**:161–175.

Thirumalachar, M. J. (1961). Critical notes on some plant rusts. III. *Mycologia* **52**:688–693.

Thirumalachar, M. J., and G. B. Cummins. (1949). The taxonomic significance of sporogenous basal cells in the Uredinales. *Mycologia* **41**:523–526.

Wilson, M., and G. R. Bisby. (1954). List of British Uredinales. *Trans. Brit. Mycol. Soc.* **37**: 61–86.

Wilson, M., and D. M. Henderson. (1966). "British Rust Fungi." Cambridge Univ. Press, London and New York.

CHAPTER 16

Ustilaginales

Rubén Durán

Department of Plant Pathology
Washington State University
Pullman, Washington

I. INTRODUCTION

As parasites of cereals, grasses, ornamentals, and thousands of non-economic plants, the smut fungi (Ustilaginales) have attracted much attention and been subjects of research for well over a century. Although by no means ubiquitous, they occur to some extent wherever flowering plants grow and, hence, have been studied by mycologists and plant pathologists in most countries of the world.

The smuts like other fungi have a limited number of morphological characters suitable for taxonomic purposes; their host range, however, includes a vast number of angiosperms in over 75 families, a fact which has figured prominently in their classification (Fischer and Holton, 1957). Since each species has a limited host range, this specificity, in conjunction with morphological characters, has served traditionally to characterize and delimit the 1100 or so species. Indeed, experience has shown that morphology and host-specialization are interdependent and mutually complementary in smut fungi systematics. Fischer (1953) emphasized this point when he cautioned that a purely morphological system of classifying smuts would lead inevitably to unrealistic "lumping" of morphologically similar but obviously different species, e.g., *Urocystis colchici* and *U. agropyri* which cause onion and wheat flag smut, respectively. It seems equally appropriate and necessary to caution that overemphasis of host-specialization also would lead to taxonomic chaos, particularly if such specialization were stressed at the host-species level. For example, *Ustilago striiformis*, and other species with extensive host ranges, conceivably would be split into a multiplicity of "species," none of which would be morphologically distinguishable.

Realizing the importance of host-specialization in the taxonomy of the smuts, Fischer (1953), and later Fischer and Shaw (1953), proposed a species concept based principally on morphology and host-specialization at the host-family level. Intended for smuts in general, this compromise offers a partial solution to the species problem, particularly for taxa with a dearth of distinct morphological characters. In *Entyloma*, for example, morphological similarity among the species makes identification difficult, frustrating, or impossible. Under this proposal, morphologically similar *Entylomas* would be considered different species if each occurred on a different plant family. In support of Fischer's compromise, it is significant to note that all species of smut fungi described thus far are restricted in their host range to members of no more than one flowering plant family (Zundel, 1953).

Morphology and host-specialization must both be stressed in smut taxonomy. In the final analysis, however, no matter how carefully taxonomic criteria are chosen and subsequently applied, taxonomy and human arbitration are difficult to separate. To paraphrase an old cliché, there is an element of truth in the statement that says any system of classification, like a time-piece, is better than none, and that the best may be less than perfect. In this spirit, hoping only to escape reproach, I have attempted to discuss the salient morphological characters of the smut fungi which most ustilaginologists agree are adequate to deal with the species problem in the smuts. Aside from morphology and symptomatology, I have also stressed the importance of host-specialization as have others before me.

II.   FAMILIES OF SMUT FUNGI

Whereas classification of genera and species is based principally on teliospore morphology, symptomatology, as well as host range, family classification up until now has been based strictly on mode of teliospore germination. The two families currently recognized were described almost 125 years ago by the Tulasne brothers (1847) who characterized the Ustilaginaceae to include species which produce transversely septate promycelia with lateral and terminal sporidia; in the Tilletiaceae they included species which produce nonseptate promycelia with only terminal sporidia. Since the Tulasnes described them, the two families have been recognized throughout the world, although Fischer and Holton (1957) expressed the opinion that "... as more and more has been learned about the same and additional species that formed the basis of the Tulasnes' classification, it has become increasingly impractical to continue to do so." Cunningham (1924), in expressing his opposition to the two-family system of classification, stated that if the two methods of germination warrant two families, then, logically a third should be named to include species which produce an infection mycelium without ever producing a promycelium.

Pointing to other inconsistencies in the two-family system, Fischer and Holton (1957) added that the intercalary system of teliospore formation in the Ustilaginaceae versus the acrogenous method in the Tilletiaceae (previously considered distinct for each family), may both occur in the same family and, hence, cannot be considered as grounds for justifying two families. They also considered differences in mode of germination impractical, because many species are difficult or even impossible to germinate as yet and, hence, cannot be classified to family, except by analogy or similarity to other presumably related species. Admittedly, these objections are understandable; however, they do not justify abandoning the current two-family system. For surely, if classification to family must be inferred by analogy, the same must be said as regards generic classification. Moreover, the paramount importance of basidial ontogeny cannot be ignored in the taxonomy of the smuts any more than it can be ignored in the taxonomy of the higher basidiomycetes. As for overlapping similarities in mode of teliospore formation, it is questionable whether this criterion ever served legitimately to justify separate families, simply because it has received very limited study.

Those who would seriously advocate a third family for species which germinate directly must first address themselves to the problem of variation in the heterobasidium attributable to suboptimum conditions, notably excess moisture. For as Rogers (1934) has pointed out, the heterobasidium under these conditions is capable of any of the modifications attributable to ordinary mycelium. Until such time as it can be shown that direct germination in a substantial number of species is an invariable phenomenon, or at least a largely consistent one—even under optimum conditions— establishing a third family of smut fungi seems both premature and undesirable.

It should be emphasized that difficulties in demonstrating spore germination in some species in no way reduce the need to do so. Moreover, it must not be assumed that species in which germination has not been demonstrated are necessarily difficult or impossible to germinate, for such is not the case, as Durán and Safeeulla have shown (1968). To illustrate further, *Ustilago brunkii*, a North American *Andropogon* smut described in 1890, was transferred to *Tolyposporella* in 1902 by G. P. Clinton (Fischer, 1953). I recently collected this smut in Mexico, germinated it easily, and in so doing demonstrated that it has no possible affinities in *Ustilago*, where it was first classified, nor in *Tolyposporella*, since its mode of germination is clearly Tilletiaceous (Durán, 1972). Another case in point is

Tolyposporella nolinae transferred recently to *Melanotaenium* because of its Tilletiaceous mode of germination (Durán and Safeeulla, 1968; Thirumalachar *et al.*, 1967).

These examples clearly show the importance of demonstrating spore germination in classification of the smuts. Moreover, it is very likely that germination will continue to be demonstrated in additional species as fresh collections are made. In this regard, I must add that in my experience many species have proved much more difficult to collect than to germinate.

Hopefully, from my point of view, the present system of family classification based on morphology of the basidium will be retained. For as Rogers has asserted, "The greater part of whatever truth and permanence inherent in the present conceptions of basidiomycetes relationships has been attained through study of the basidium" (Rogers, 1934).

III. MORPHOLOGY

A. Spore Balls

Roughly half of the currently recognized genera of smut fungi produce their spores in clumps or aggregates known as spore balls. These may either consist of fertile spores entirely, or of a combination of fertile spores, accessory sterile cells, and/or sterile pseudoparenchymatous elements. Because there is considerable variation in the way that fertile spores and the accessory elements are arranged in the spore ball, the morphology of spore balls has played an extremely important role in smut systematics, and, in fact, has been the basis for describing a substantial number of genera. For convenience, morphological characteristics of various types of spore balls will be discussed in connection with representative genera.

Sorosporium and *Thecaphora* are examples of genera in which the spore balls consist entirely of fertile spores. Aside from this common characteristic, however, the spore balls are morphologically different in all other respects. In *Sorosporium*, the spore balls are composed of a relatively large number of small spores which tend to retain a polyhedral configuration as a result of internal compression in the spore ball. As a whole, the spore balls are fragile and tend to disintegrate into individual spores. *Thecaphora*, by comparison, has much larger spores, permanent spore balls, and relatively few spores per ball. Moreover, the characteristically wedge-shaped spores tend to be radially arranged in the spore ball. Unlike *Sorosporium*, the outer exposed free surfaces of the spores are very prominently verrucose.

In *Tolyposporium*, the spores in the ball are held together by a curious network of interconnected folds and thickenings which apparently emanate from the exospore walls. For the most part, the balls tend to be permanent or semipermanent. According to Ling (1949), the spore balls of *Dermato-sorus*, by contrast, consist of an outer semipermanent cortex of pseudo-parenchyma made up of small reddish-brown cells 3–6.5 μm in diameter, and an inner mass of fertile loosely united spores chiefly 7–9 μm in diameter.

In *Polysaccopsis*, *Urocystis*, and *Ustacystis* the spore balls contain as few as one or as many as twenty or more fertile spores. In each of these genera, the fertile spores are surrounded by a variable number of bladder-like sterile cells which form an outer cortex. Because the number of sterile cells per spore ball varies, the cortex may be largely continuous or dis-continuous.

One of the most curious and morphologically intricate of spore balls is that described for the genus *Zundelula* (Thirumalachar and Narasimhan, 1952). In this monotypic genus, the very large permanent spore balls consist of several outer layers of large sterile cells which form a cortex around an inner cylinder of smaller fertile spores. In a manner analogous to *Tolypo-sporium*, the inner core of fertile spores are interconnected by a dense system of pseudoparenchymatous elements which hold the spores together. *Zundelula* is the only genus in which the spore balls consist of sterile cells, fertile spores, and mycelial elements arranged as described above.

The generic concept of *Tolyposporella* is largely based on the so-called epispore or saclike structure in which the spore balls are embedded. Due to the opaqueness of the spore balls, the laminated epispore surrounding individual spores or groups of spores is best seen when the spore ball is fractured. Apparently, the epispores are interconnected within the ball in a manner analogous to *Tolyposporium* which tends to hold the spores firmly together.

Unfortunately, the epispore of *Tolyposporella* spp. has been likened to somewhat similar structures in certain smuts which do not have affinities in *Tolyposporella*. For example, Fischer (1953) listed three species of *Tolyposporella* for North America, namely, *T. chrysopogonis* (the type species), *T. nolinae*, and *T. brunkii*. Recent spore germination studies have clearly shown that the latter species are members of the Tilletiaceae, and, hence, do not belong in *Tolyposporella* (Durán and Safeeulla, 1968; Durán, 1972; Thirumalachar *et al.*, 1967).

Morphologically, the most bizarre spore balls are probably those of *Narasimhania* (Thirumalachar and Pavgi, 1952). In the leaves of *Alisma*, the spore balls appear as tiny black points which occupy the entire space between the epidermal layers. Within the balls, the fertile spores are arranged in such a way as to form irregular clathroid structures interconnected by pseudoparenchymatous tissue. The result is that in surface view the clathroid arms tend to give the spore balls a star-shaped appearance. In that they are permanently embedded in leaf tissue, the spore balls of *Narasimhania* have been likened to those of other leaf smuts, e.g., *Doassansia* and *Burrillia*.

In *Doassansia*, spore balls generally are large, conspicuous, very firm, and usually permanently embedded in mesophyll leaf tissue. A sterile outer cortex envelops the spore balls which in their centers contain the masses of fertile spores. The *Burrillia* spore ball has no outer sterile cortex, but instead consists of an outer layer of fertile spores and an inner network of sterile pseudoparenchymatous cells. In *Tracya*, the spore ball consists, for the most part, of a single outer layer of fertile spores and an internal system of plectenchymatous fungus filaments. The *Testicularia* spore ball, on the other hand, consists of several superimposed layers of fertile spores on the outside and an inner mass of sterile pseudoparenchymatous cells. Unlike the previous four genera, the spore balls in *Tuburcinia* are very black and opaque; they tend to be permanent, and are composed entirely of fertile spores.

In addition to the assorted ways that fertile spores, sterile cells, and auxiliary mycelial elements are arranged in spore balls of different genera, spore-ball characters of taxonomic importance also include size, color, sometimes ornamentation, shape, number of fertile spores per ball, whether permanently embedded in host tissue or exposed in the sorus at maturity, and degree of permanence.

B. Teliospores

Although the taxonomy of the smuts is based largely on teliospore morphology, most species cannot be identified or adequately described solely on the basis of teliospore characters. *Ustilago hypodytes* and *U. minima* on *Oryzopsis hymenoides* are difficult to distinguish solely on the basis of teliospore characters, since both species have smooth spores of similar size, shape, and color. The two species can be easily distinguished, however, because of macroscopic differences in soral characters.

Teliospore characters, including size, shape, color, and exospore ornamentation are, nevertheless, criteria which are indispensable to smut systematics. All of the species described thus far have been based on combined study of these characters, along with host range studies, symptomatology, and to some extent, geographic distribution.

For convenience, teliospore morphology will be discussed under Size, Color, Ornamentation, and Shape.

1. Size

Spore size seemingly has greater taxonomic significance among small-spored species of *Ustilago*, *Sphacelotheca*, and *Sorosporium* and other genera than, for example, among the relatively large-spored genera like

Cintractia and *Tilletia*. The spores of *Ustilago minima*, for example, range in diameter from 3.5–5.0 μm; those of *Sorosporium contortun*, 7–11 μm; and those of *Sphacelotheca sorghi* from 5–7 μm (Fischer, 1953). Thus, while the average range in spore size in these examples is limited to a few microns, variations in spore diameters of 12–15 μm are common among individual species of *Tilletia*.

Figures of spore size, no matter how carefully computed, should serve only to supplement other important morphological criteria like teliospore shape, color, and ornamentation, as well as symptomatology, although occasionally, new species have been described principally on the basis of spore size. *Tilletia asperifolioides*, for example, has spores which range in size from 26–33 μm in diameter, while those of *T. asperifolia*, a related species, range from 17–21 μm. Except for marked differences in spore size, the two species are very similar. Both are ovariicolous; both have a common host; and both have reticulate teliospores which, in a sense, makes them more similar than different.

Whether spore size alone in some instances can justify creation of new species must, in the final analysis, be left to the judgment of the individual. In my opinion, spore size probably should be considered of lesser taxonomic importance than teliospore shape, color, or exospore ornamentation.

2. Color

Strictly speaking, color is not a morphological character, but rightfully is included in all descriptions of smut fungi. Although to some extent it varies in every species, spore color usually is sufficiently constant to be of significant taxonomic importance or at least in a diagnostic sense. In *Tilletia pallida*, for example, the specific epithet alludes to the generally pale yellow spores of the *Agrostis* bunt, and in *T. fusca* to the dark brown fuscoid spores of certain *Bromus* and *Festuca* bunts.

In some species, the pigment in the exospore wall is unevenly distributed. The result is that spores may be pigmented in a manner characteristic of certain species or even genera. In various species of *Ustilago*, for example, the spores exhibit equatorial bands and lighter polar areas or caps (Fischer, 1953). This is unusually striking in *Planetella lironis* in which the exospore is darker as well as thicker at the equatorial band, and lighter and much thinner at the polar areas (Savile, 1951). Similarly, in *Tilletia oplismenus-cristati*, the exospore is adorned with unevenly distributed clusters of verruculations, which give the spores an uneven surface coloration pattern (Durán and Fischer, 1961). For unknown reasons, spores of various *Ustilago* species are more heavily pigmented at one pole than at the other.

Regarding the color of the spore mass, it is generally blackish, with various hues of brown, fawn, or pale yellow. It has often been mentioned in descrip-

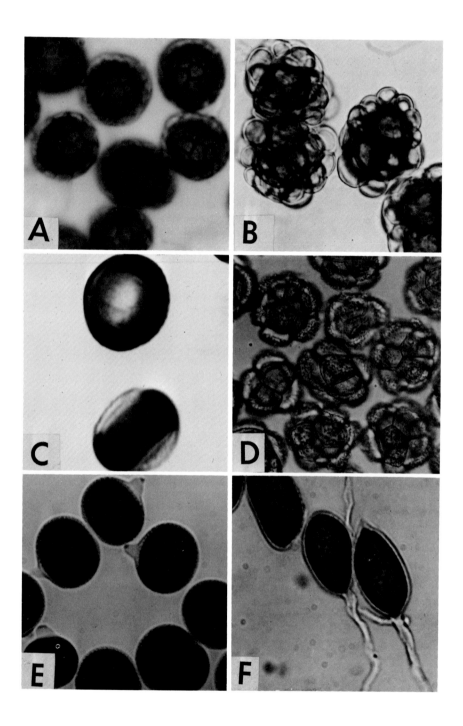

tive works, although the manner in which it has been described has not been standardized. No doubt some of the color differences reported are due not only to a lack of standard lighting techniques, but also to examination of spore masses from sori in different stages of maturity.

The color of individual spores does not necessarily correspond with that of the spore's mass (Durán and Fischer, 1961). Very often spore masses may appear black, whereas single spores may be dark reddish brown, or other shades under transmitted light, especially under high magnification. Although there is no absolute correlation between spore-mass color, and that of individual spores, dark spore masses tend always to contain spores which are correspondingly dark, although they invariably are somewhat lighter. Seldom are the spores from a single sorus the same as regards color intensity. Those which are immature invariably contain less pigment or none at all, although they are always sculptured in a manner characteristic of the species. Possibly, synthesis of melanin pigments fails to occur in some cells of the sporogenous mycelium when reserves are finally exhausted during sporogenesis. Or, possibly, individual sori develop from various sporogenous mycelia with different genetic factors for spore color.

Spore color, admittedly, is particularly subject to quantitative and qualitative interpretation, usually more so, in fact, than other teliospore characters. Nevertheless, because it is one of the few phenotypic characters available to the ustilaginologist, no doubt it will continue to be extremely useful in smut systematics.

Some attempts have been made to identify races of certain species of smut fungi on the basis of slight differences in spore color. I would agree with those who state that slight differences in color are mainly of academic interest and are of little taxonomic value (Fischer and Holton, 1957).

3. Ornamentation

Probably exospore ornamentation has figured more prominently in species delimitation than any other single teliospore character. Exospore

---

FIG. 1. Examples of spore types in the Ustilaginales. A, *Tilletia aegopogonis* R. Durán. A prominent reticulum adorning the exospore wall similar to that of cereal bunts. Magnification: × 950; B, *Urocystis hypoxidis* Thax. Inner fertile teliospores in balls surrounded by translucent bladderlike cells. Magnification: × 850; C, *Planetella lironis* Savile. Spores with flattened poles and equatorial bands. Magnification: × 2000; D, *Thecaphora pulcherrima* R. Durán. Spores in balls or clusters, smooth on contiguous surfaces, prominently verrucose on the free surfaces. Magnification: × 400; E, *Tilletia corona* Scribn. Note the prominent mamillae. Magnification: × 850; F, *Neovossia iowensis* Hume and Hods. Spores invested in a hyline sheath continuous with prominent appendages. Magnification: × 900.

sculpturing consists of several basic patterns, i.e., smooth, reticulate, cerebriform, echinulate, verrucose, tuberculate, etc. For a given species, it has been customary to describe sculpturing of the exospore in terms of size and shape of reticulations, verruculations, tubercles, or other markings. Echinulations and verruculations generally are described in terms of density, since, for the most part, they are so small that they cannot be measured accurately, even with oil immersion lenses.

The basic patterns, moreover, are by no means absolute. Reticulate and cerebriform patterns frequently intergrade; nor can echinulate and verrucose markings always be clearly distinguished. These examples of intergrading sculpture patterns probably represent variable expression of the same basic pattern (Holton *et al.*, 1968). Nevertheless, because it is genetically controlled, exospore morphology is sufficiently constant for each species to permit adequate species delimitation.

For purposes of identification, or before new species are described, the exospore patterns of unidentified specimens should be compared with those of other smuts from the same host-genus. Host-genus indexes, like those of Fischer (1953) and Durán and Fischer (1961), provide relatively easy means for comparing the morphology of unidentified specimens with that of known species. On *Avena sativa*, for example, *Ustilago hordei*, with its perfectly smooth spores, is easily distinguished from the echinulate spores of *U. avenae*. Since these are the only known smuts on *A. sativa*, a specimen with spores other than smooth or echinulate would very likely be considered a new species, regardless of other similarities.

The importance that taxonomists have attributed to the exospore pattern in species delimitation can be further illustrated in the two species of common bunt of wheat, *Tilletia caries* and *T. laevis*. Despite virtual similarity as regards host range, symptomatology, and even life cycle, the two have always been considered distinct species, simply because the spores of *T. caries* are reticulate, while in *T. laevis* they are smooth (Durán and Fischer, 1961).

4. Shape

The spores of many species of smut fungi are commonly globose to subglobose, or occasionally polyhedral. This range in spore shapes applies largely, although not exclusively, to species in which the sporogenous mycelium produces spores singly, as opposed to those produced in balls.

Unfortunately, descriptive terms like "globose to subglobose, or oc-casionally polyhedral," in themselves are of little diagnostic or taxonomic value, simply because such descriptions of spore configuration are common to many species. Much more striking are the configurations assumed by spores of *Neovossia* and *Mykosyrinx* and certain species of other genera. The large (12–18 × 19–30 μm) ellipsoid to elongate spores of *Neovossia iowensis*, for example, are extremely striking because of the attached hyaline appendage which oftentimes is larger than the spore itself (Fischer, 1953). Similarly, the curious shape of the spores of *Mykosyrinx cissi* prompted Fischer (1953) to describe them as "... spores in pairs, dark olive brown, resembling open clamshells, each spore of a pair apparently a complementing half of a sphere."

The spores of certain species of *Tilletia* likewise may have striking configurations, e.g., the prominent hyaline mammilla in *T. corona* (Durán and Fischer, 1961). Judging from descriptions of these species (Fischer, 1953), it is obvious that the authors who published the original descriptions were more impressed by the striking configurations of the spores themselves than by any other criterion.

For species in which the spores are borne in balls, the shape of individual spores is influenced not only by the way they are held together within the ball, but also by the relative number of spores which constitute the ball. In *Sorosporium*, for example, the innermost spores of the ball sustain considerable internal pressure, and, hence, tend to assume a polygonal configuration visible when the spores are freed. In *Ustacystis*, the balls contain a few loosely held spores which tend to be flattened mostly at the contiguous surfaces. In some *Thecaphora* species the spores are radially positioned within the spore balls and tend to be wedge shaped. Notwithstanding the fact that the shape of spores in balls can, to some extent, be correlated with the way they are borne within it, their configuration as a whole is of much less importance than the morphology of the spore ball itself. No doubt this is due to the extreme variability in the shape of spores in such genera as *Burrillia*, *Doassansia*, *Tracya*, *Tuburcinia*, *Urocystis*, *Narasimhania*, and many others.

C. Sterile Cells

To a considerable extent, the characters of the so-called sterile cells, including their size, shape, and color, have been used to characterize certain species of smut fungi, and, at times, even genera. Unfortunately, the term does not refer to structures of common origin. In some *Sphacelotheca* spp., for example, the cellular components of the sorus peridium, which tend to disintegrate into individual cells or groups of cells, are referred to as "sterile cells."

In genera in which the spore balls consist of both fertile and sterile elements, the term "sterile cells" has been applied to the sterile cellular components of the spore ball, as distinguished from the fertile spores (e.g., *Urocystis*). The term of course has no application to the spore balls of genera in which the balls consist entirely of fertile spores (e.g., *Sorosporium*). There is yet another use of the term, and that is in connection with the genus *Tilletia*. In the species of this genus, the fertile spores are almost invariably interspersed with considerable numbers of unsculptured, hyaline or tinted cells of various shapes and sizes which also have come to be known as "sterile cells." In fact, sterile-cell characters have found greater taxonomic application among the species of *Tilletia* than among those of any other genus (Durán and Fischer, 1961).

In some species of *Tilletia* and *Sphacelotheca*, however, the number of sterile cells may be so few, or their morphology so lacking in distinct characters, that describing them serves little or no purpose, either from a diagnostic or taxonomic point of view. In other genera (e.g., *Urocystis, Polysaccopsis, Testicularia*) the sterile cells are of taxonomic importance more because of their arrangement in the balls with respect to the fertile spores than for any unusually striking morphologic characters of their own.

In *Zundelula*, the sterile cells are unusual, not only because of their arrangement in the spore balls, but also because they are darker, much larger, and are adorned with noticeably thicker reticulations than the fertile spores (Thirumalachar and Narasimhan, 1952).

Whether they occur single, as in *Tilletia* and *Spacelotheca*, or as spore-ball components, sterile cells, like teliospores, may be described in terms of size, color, and shape. Moreover, they are often described in terms of wall thickness, whether naked or encased in a hyaline sheath, and, occasionally, in terms of relative numbers.

IV. SYMPTOMATOLOGY

A. Soral Characters

Sori may develop in various parts of the inflorescence, including the ovaries, ovules, anthers, petals, etc., or they may form in leaves, fruits, stems, axillary buds, rhizomes, and even roots. In *Tilletia*, sorus formation occurs mostly in the ovaries (Durán and Fischer, 1961). In *Ustilago* and other genera, however, sori may form in a multiplicity of plant organs.

Because the site of sporulation for a given species or group of species is generally constant, it has served to categorize the smut fungi into inflorescence smuts, leaf smuts, stem smuts, root smuts, etc. However, occasionally stem smuts may sporulate in parts of the inflorescence, or, conversely, inflorescence smuts may sporulate in vegetative organs, e.g., leaves. At times, leaf smuts may even sporulate in rhizomes below ground (Durán, 1968). Adventitious sporulation, however, generally occurs in addition to sporulation at the normal site and seldom, if ever, replaces it (Fischer and Holton, 1957). Despite some variation, the site of soral development very likely will continue to serve as a means of helping delimit species and even genera. In *Entyloma, Burrillia, Doassansia, Tracya,* and *Tuburcinia,* for example, the sori are almost invariably formed in the leaves where they tend to remain embedded, even after maturity. Or, the sori of other leaf smuts when mature may erupt to various degrees as in some species of *Urocystis, Ustilago, Schizonella, Tolyposporella,* and *Tilletia.* Some smuts, however, sporulate on most host organs as does *Mundkurella heptapleuri* which forms sori in fruits, stems, petioles, and leaves (Beer, 1920; Zundel, 1953). *Melanotaenium* spp. similarly sporulate in most above-ground parts of the host, and in roots as well (Zundel, 1953).

Occasionally, sori assume the general configuration of affected plant parts, as in the ovariicolous grass and cereal smuts in which sorus shape more or less resembles that of the caryopsis. In case of flag and stripe smut of grasses, sori assume the parallel pattern of leaf veination. By way of contrast, soral characters are frequently unusually striking due largely to hypertrophy of affected organs. Spurlike bodies are formed, for example, in the ovaries of *Oplismenus compositus* infected by *Tilletia vittata,* and large, hood-shaped galls up to 8 cm long and 2 cm broad from the axillary buds of *Panicum antidotale* infected by *T. tumefaciens* (Durán and Fischer, 1961); or, sori may consist of conspicuous sausage-shaped pustules on the leaves of *Bouteloua gracilis* infected with *Ustilago buchloes* (Fischer, 1953). More common but no less striking are the sori of *U. maydis,* the boil-smut fungus, so named because of the large tumorlike galls formed in ears and meristematic tissues. In yet other examples, sori may consist of fusiform internodal swellings in the stems of *Alopecurus carolinianus* and *Oryzopsis hymenoides* infected by *T. youngii* and *U. minima,* respectively.

Of extreme importance is the morphology of the sorus itself, both as to external and internal characteristics. In *Farysia,* the sorus is traversed by a series of capillitiumlike threads (elaters) which presumably function in spore dispersal. In *Cintractia* and *Anthracoidea,* sori are mostly formed in the ovaries around a central columella of host tissue; unlike *Anthracoidea,* in *Cintractia* a series of sterile stromatic elements radiate through the sori in a manner suggestive of *Farysia* (Kukkonen, 1963). It is interesting to note that many *Anthracoidea* smuts were classified in *Cintractia* until Kukkonen (1963) showed that the sterile stromatic tissue of the *Cintractia* sorus is lacking in *Anthracoidea* and that these genera are, therefore, distinct. Fullerton and Langdon (1969), on the other hand, in demonstrating structural and developmental similarities in sori of the *Ustilago* species attacking *Echinochloa* spp., concluded that a number of *Ustilago* smuts on *Echinochloa* could be referred to one species, namely, *U. trichophora.*

Other soral characters of diagnostic use include the consistency of the sorus itself, i.e., whether it be dusty or agglutinated; whether it is enveloped in a peridium of fungus or host cells; formed around a central columella of host tissue; formed in galls in saccate cavities; embedded permanently in host tissue, or exposed at maturity.

Clearly, developmental studies of the sorus need additional emphasis. Such emphasis should serve not only to better define currently accepted genera and species but those which at present are ill defined as well. It is significant that Kukkonen's (1963) comparative studies of sorus morphology in *Anthracoidea* and *Cintractia* helped resurrect Brefeld's *Anthracoidea* which for years had been laid to rest as a synonym of *Cintractia*. Fullerton and Langdon's (1969) studies of soral characters in some of the *Echinochloa* smuts also serve to indicate the need for further emphasizing soral ontogeny in the classification of the smut fungi.

V.   ORDER USTILAGINALES AND KEY

A.   General Characteristics

This order is made up of parasites of angiosperms; the soma consists of a haplophase and a dikaryophase of varying duration. The former is capable in some species of prolonged yeastlike reproduction on nonliving substrata; the latter, for the most part, is obligately parasitic, sporulating in various host organs and giving rise to resting spores (teliospores) produced in sori of varying complexity, either singly or in balls. The balls consist of fertile spores exclusively, or fertile spores, accessory sterile cells, and/or sterile hyphal elements. Resting-spore germination is mostly indirect, i.e., by means of transversely septate promycelia bearing lateral basidospores (Ustilaginaceae), or else the basidiospores are borne in clusters at the apices of nonseptate promycelia (Tilletiaceae). Occasionally germination is direct in species of either family, i.e., the promycelium gives rise to mycelia without producing basidiospores.

B.   Important Literature

Important floristic and monographic studies of the smuts include: Plowright (1889), Great Britain; Massee (1899), the world; Clinton (1904, 1906), North America; McAlpine (1910), Australia; Liro (1924, 1938), Finland; Cunningham (1924), New Zealand; Ciferri (1938), Italy; Zundel (1938, 1939, 1953), South Africa and the world; Gutner (1941), Russia; Hirschhorn (1942), Argentina; Viennot-Bourgin (1949), France; Ainsworth and Sampson (1950), Great Britain; Mundkur and Thirumalachar (1952), India; Ling (1953), China; Fischer (1953), North America; Săvulescu (1957), Rumania; Durán and Fischer (1961), the world; and Kukkonen (1963), the world. For an extensive guide to the literature dealing with the smuts see (Fischer, 1951). The generic review by Thirumalachar (1966) is also useful.

C. Key to Genera

The fact that it has been possible to demonstrate germination in representative species of almost every genus of smuts serves to emphasize the importance of attempting to obtain such information for all species. Only then can family affinities be determined, and generic affinities more accurately hypothesized. Too often, supposed generic affinities have been disavowed by germination studies (Durán and Safeeula, 1968; Durán, 1972). Nevertheless, the morphological similarity of the promycelia and sporidia among the Ustilaginaceae would serve little to differentiate the genera of this family; the same may be said of the Tilletiaceae. Hence, the key which follows stresses germination characters only when morphological characters are insufficient to distinguish between genera like *Glomosporium*, *Tuburcinia*, and *Thecaphora*.

KEY TO GENERA OF USTILAGINALES

1. Sori containing 1- and 2-celled spores; on *Araliaceae* . . . . **Mundkurella**
     A monotypic genus from India; Thirumalachar (1944).

1′. Sori containing 1-celled spores only 2

    2(1′) Spores mostly single, not arranged in balls 3

    2′(1′) Spores in aggregates of 2, 3, or more spores, or in regular balls containing many spores (*Urocystis* may have 1 fertile spore only; if so, surrounded by a cortex of bladderlike sterile cells) 17

3(2) Sori more or less dusty at maturity 4

3′(2) Sori more or less agglutinated at maturity, or remaining in host cells until freed by physical action 8

    4(3) Spores with a large hyaline appendage (as long or longer than the spores); on *Gramineae* . . . . **Neovossia**
     Three species. Zundel, 1953; Fischer, 1953; Mundkur and Thirumalachar, 1952.

    4′(3) Spores without a large appendage, sometimes with a remnant of the sporogenous mycelium attached to the spores (with a short mamilla in *Tilletia corona*) 5

5(4′) Spores large, mostly 16–54 μm in diameter, exospore reticulate, smooth, verrucose, tuberculate or otherwise sculptured, usually in the ovaries but occasionally in the leaves; probably limited to *Gramineae* . . . . **Tilletia**
     Seventy-six species, fide Durán and Fischer, 1961. Two new species on *Gramineae* from Mexico are described by Durán 1971.

5′(4′) Spores mostly smaller, 4–16 μm in diameter, exospore mostly echinulate, smooth, verrucose, or occasionally reticulate, etc., sori in various parts of the host 6

    6(5′) Spores in the sorus intermixed with elaterlike hyphal strands, borne in chains which early disintegrate; on *Cyperaceae* . . . . **Farysia**
     Syn. *Elateromyces* Bubák. Sixteen species and several varieties.

6'(5') Spores in the sorus not intermixed with elaters, nor borne in chains   7

7(6') Sori enveloped in a peridium of fungus cells, forming around a central columella of host
   tissue, at times with the vestiges of vascular elements remaining . . . . .   **Sphacelotheca**
   137 species, fide Fischer and Holton (1957). Hirschhorn (1939) makes no distinction
   between *Sphacelotheca* and *Ustilago*. Fischer (1953) contends the fungus peridium
   and basipetal formation of spores around a central columella differentiates *Sphacelotheca* from *Ustilago* as well as other smut genera. See also Fullerton and Langdon,
   (1969).

7'(6') Sori almost never enveloped in a fungus peridium, nor formed around columella of host
   tissue . . . . **Ustilago**[1]

   8(3') Sori forming black, cylindrical, agglutinated galls or columns 8–12 mm long and
      0.5–0.75 mm thick, surrounded by a thin layer of hyaline hyphae, arising mostly from
      the undersides of the leaves; spores with prominent germ pores; on *Cyperaceae*
      . . . . **Cintractiella**
      A monotypic genus from New Guinea; Boedijn (1937).

   8'(3') Sori otherwise, spores without germ pores   9

9(8') Sori forming fingerlike galls or swellings in roots; on *Cyperaceae* and *Juncaceae*
   . . . . **Entorrhiza**
      Eight species; root-infecting; see Zundel (1953), in which sori of *E. isoetes* are said to
      form in and between the microspores at the leaf bases of *Isoetes lacustris*.

9'(8') Sori mostly forming in host parts other than roots   10

   10(9') Sori tending to remain embedded in host tissue after maturity, forming more or less
      distinct spots in leaves or other aboveground parts . . . .   **Entyloma**
      Fischer and Holton (1957) list 151 species; Savile (1947); Fischer (1953).

   10'(9') Sori conspicuous, variously erumpent at maturity   11

11(10') Sori invested in a hard, dark peridium; spores with dark equatorial bands and lighter
   polar areas; on *Cyperaceae* . . . . **Planetella**
      Monotypic; Savile (1951).

11'(10') Sori otherwise, spores more or less uniformly pigmented, or at least without equatorial
   bands 12

   12(11') Sori in the axillary buds, causing a witches' broom; on *Convolvulaceae*
      . . . . **Georgefischeria**
      Monotypic; Narasimhan *et al.* (1963).

   12'(11') Sori in host parts other than axillary buds, not causing a witches' broom
      13

13(12') Sori mostly forming in the ovaries around a central columella of host tissue, the spores
   intermixed with sterile stromatic tissue, portions of the latter reaching through the sorus
   . . . . **Cintractia**
      Fischer and Holton (1957) list 58 species.

13'(12') Sori in various parts of the host, lacking sterile stromatic tissue, the sporogenous
   mycelium as a whole converted into spores 14

   14(13') Sori forming in hard, gall-like masses; mostly in the inflorescence

---

[1] In *U. minima* the sori are enveloped in a peridium of fungus mycelium which often endures
long after the spores are shed. On *Oryzopsis hymenoides* it forms an "annulus" around the
culms and sometimes persists in the field from one year to another.

. . . . **Melanopsichium**
Seven species; Fischer (1953) discusses N. and S. American species.

14'(13') Sori forming either in the ovaries or vegetative parts, but not in galls 15

15(14') Sori in the ovaries; on *Cyperaceae* . . . . **Anthracoidea**
Superficially similar to *Cintractia* except lacking sterile stromatic tissue. For a dis-
ˉcussion of the distinctions between *Cintractia* and *Anthracoidea* see Kukkonen
(1963) who lists 23 species.

15'(14') Sori in vegetative parts 16

16(15') Sori forming conspicuous agglutinated tar-spots on leaves; on *Gramineae*
. . . . **Jamesdicksonia**
Monotypic; Thirumalachar *et al.* (1960).

16'(15') Sori at first covered by host epidermis, later rupturing; in stem, leaf, or tissues of
the upper roots . . . . **Melanotaenium**
Eighteen species. A number of smuts whose generic affinities are in doubt have been
ascribed to *Melanotaenium*. For a generic concept see Beer (1920).

17(2') Spores chiefly in pairs 18

17'(2') Spores in regular spore balls, somewhat indefinite in *Ustacystis* 20

18(17) Sori agglutinated, confluent, forming striae on leaves; on *Cyperaceae*
. . . . **Schizonella**
Two species or fide Fischer (1953) monotypic.

18'(17) Sori powdery to granular, not forming striae on leaves 19

19(18') Sori forming in the pedicels, and peduncles, causing hypertrophy of the tissues; on
Vitaceae . . . . **Mycosyrinx**
Apparently monotypic; Fischer (1953).

19'(18') Sori destroying the seeds, forming in the funiculi and placentae; on *Scrophulariaceae*
. . . . **Schroeteria**
Three species; Zundel (1953).

20(17') Sori dusty or granular at maturity 21

20'(17') Sori more or less agglutinated at maturity or embedded in host tissue 29

21(20) Spore balls consisting entirely of fertile spores 22

21'(20) Spore balls consisting of both fertile spores and sterile cells, or the balls consisting of
fertile spores only surrounded by a thin layer of pseudoparenchyma 25

22(21) Spore balls evanescent or tending to fragment into single spores
. . . . **Sorosporium**
Fischer and Holton (1957) list 115 species; Zundel (1953).

22'(21) Spore balls permanent, not tending to fragment 23

23(22') Spores in the balls radially positioned, typically sculptured on exposed surfaces only,
germinating directly, or the transversely septate promycelium bearing a single apical sporid-
ium . . . . **Thecaphora**[2]
Thirty species.

[2]Since the spore balls of *Thecaphora, Glomosporium,* and *Tuburcinia* are very similar, it is
sometimes difficult to separate these genera without germinating the spores. If germination
is unsuccessful, possibly they can be distinguished on the basis of host-family specialization
[see Fischer and Holton (1957) for distribution of the species by host family].

23′(22′) Spores held in the balls as above, or held together by a series of folds or thickenings of the exospore wall; germination not as above 24

24(23′) Spores in the balls radially positioned, germination as in *Tilletiaceae*
. . . . **Glomosporium**[2]
Two species; Hirschhorn (1945).

24′(23′) Spores in the balls held together by interconnected thickenings of the exospore walls, germination as in *Ustilaginaceae* . . . . **Tolyposporium**
Thirty species.

25(21′) Spore balls formed in saccate cavities in galls, the balls consisting of 1 to several fertile spores surrounded by bladderlike sterile cells as in *Urocystis*; on *Solanaceae*
. . . . **Polysaccopsis**
Monotypic; Hennings (1898); Zundel (1953).

25′(21′) Spore balls neither formed in cavities nor surrounded by sterile cells of the *Urocystis* type, except in *Ustacystis* 26

26(25′) Spore balls with an outer cortex of larger sterile cells surrounding an inner core of smaller fertile spores, the latter traversed or interconnected with pseudoparenchymatous tissue; on *Cyperaceae* . . . . **Zundelula**
Monotypic; Thirumalachar and Narasimhan (1952).

26′(21′) Spore balls lacking sterile cells but surrounded by a thin delicate cortex of pseudoparenchymatous tissues; or if surrounded by sterile cells, the latter smaller than the fertile spores and of the *Unocystis* type 27

27(26′) Spore balls surrounded by a thin layer of pseudoparenchyma; on *Cyperaceae*
. . . . **Dermatosorus**
Monotypic; Ling (1949).

27′(26′) Spore balls surrounded by sterile bladderlike cells, lightercolored and mostly smaller than the fertile spores 28

28(27′) Spore balls with 1–20 fertile spores and a cortex of sterile bladderlike cells; spore germination Tilletiaceous . . . . **Urocystis**
Eighty species; Zundel (1953).

28′(27′) Spore balls with relatively few fertile spores (1–3); sterile cells as in *Urocystis* except fewer; spore germination Ustilaginaceous . . . . **Ustacystis**
Monotypic; Fischer (1953).

29(20′) Spore balls variable, sometimes indefinite, the spores with laminated epispores, up to 10 μm thick. . . . . **Tolyposporella**
Six species; Atkinson (1897); Zundel (1953).

29′(20′) Spore balls much more definite, the spores not encased in a thick epispore 30

30(29′) Spore balls in the host forming clathrate masses traversed by yellowish-brown pseudoparenchymatous cells; on *Alismataceae* . . . . **Narasimhania**
Monotypic; Thirumalachar and Pavgi (1952).

30′(29′) Spore balls not forming a clathrate mass in the host 31

31(30′) Spore balls with an outer cortex of sterile cells. . . . . **Doassansia**
Thirty-five species; Setchell (1892); Fischer (1953).

31′(30′) Spore balls without an outer cortex of sterile cells 32

32(31′) Spore balls consisting entirely of dark-colored fertile spores, occasionally enveloped in a thin layer of hyphal fragments . . . . **Tuburcinia**[2]
Six species; Zundel (1953); Fischer (1953).

REFERENCES

Ainsworth, G. C., and K. Sampson. (1950). "The British smut fungi (Ustilaginales)." Commonwealth Mycol. Inst., Kew, Surrey, England.

Atkinson, G. F. (1897). Some fungi from Alabama. *Cornell Univ. Bull.* **3**:16.

Beer, R. (1920). On a new species of *Melanotaenium* with a general account of the genus. *Trans. Brit. Mycol. Soc.* **6**:331–343.

Boedijn, K. B. (1937). A smut causing galls on the leaves of *Hypolytrum. Bull. Jard. Bot. Buitenzorg* [3] **14**:368–372.

Ciferri, R. (1938). Ustilaginales. *Flora Ital. Cryptogama* Part 1, No. 17.

Clinton, G. P. (1904). North American Ustilagineae. *Proc. Bost. Soc. Natur. Hist.* **31**:329–529.

Clinton, G. P. (1906). Ustilaginales. *N. Amer. Flora* **7**:1–82.

Cunningham, G. H. (1924). The Ustilagineae, or "smuts" of New Zealand. *Trans. Proc. N.Z. Inst.* **55**:397–433.

Durán, R. (1968). Subterranean sporulation by two graminicolous smut fungi. *Phytopathology* **58**:390.

Durán, R. (1971). Some host and distribution records of Mexican smut fungi. *Mycologia* **62**:1094–1105.

Durán, R. (1972). Aspects of teliospore germination in North American smut fungi. II. *Can. J. Bot.* **50**:2569–2573.

Durán, R., and G. W. Fischer. (1961). "The genus *Tilletia*." Washington State University, Pullman.

Durán, R., and K. M. Safeeulla. (1968). Aspects of teliospore germination in some North American smut fungi. I. *Mycologia* **60**:231–243.

Fischer, G. W. (1951). "The Smut Fungi. A Guide to the Literature with Bibliography." Ronald Press, New York.

Fischer, G. W. (1953). "Manual of the North American Smut Fungi" Ronald Press, New York.

Fischer, G. W., and C. S. Holton. (1957). "Biology and Control of the Smut Fungi." Ronald Press, New York.

Fischer, G. W., and C. G. Shaw. (1953). A proposed species concept in the smut fungi, with application to North American species. *Phytopathology* **43**:181–188.

Fullerton, R. A., and R. F. N. Langdon. (1969). A study of some smuts of *Echinochloa* spp. *Proc. Linn. Soc. N. S. W.* **93**:281–293.

Gutner, L. S. (1941). "The Smut Fungi of the U.S.S.R." Lenin, State Publ. Dept., Sect. Agr. (in Russian, transl. title).

Hennings, von P. (1898). Die Gattung *Diplotheca* sowie einige interessante und neue, von E. Ule gesammelte Pilze aus Brasilien. *Hedwigia Beibl.* **37**:206.

Hirschhorn, E. (1939). Refundición del género "*Sphacelotheca*" en "*Ustilago.*" *Physis, Buenos Aires* 15:103–111.

Hirschhorn, E. (1942). Revision de las especies de *Tilletia* de la Argentina. *Rev. Mus. La Plata, Bot. Sect.* [N. S.] 5:1–20.

Hirschhorn, E. (1945). Two new species of the *Tilletiaceae* from Argentina. *Mycologia* 37: 278–283.

Holton, C. S., J. A. Hoffmann, and R. Durán. (1968). Variation in the smut fungi. *Annu. Rev. Phytopathol.* 6:213–242.

Kukkonen, I. (1963). Taxonomic studies of the genus *Anthracoidea* (Ustilaginales). *Ann. Bot. Soc. Zool. Bot. Fenn. "Vanamo"* 34:1–122.

Ling, L. (1949). A second contribution to the knowledge of the Ustilaginales of China. *Mycologia* 41:252–269.

Ling, L. (1953). The Ustilaginales of China. *Farlowia* 4:305–351.

Liro, J. I. (1924). Die Ustilagineen Finnlands. I. *Ann. Acad. Sci. Fenn., Ser. A.* 17:1–636.

Liro, J. I. (1938). Die Ustilagineen Finnlands. II. *Ann. Acad. Sci. Fenn., Ser. A* 42:1–720.

McAlpine, D. (1910). "The Smuts of Australia, Their Structure, Life History, Treatment, and Classification." Melbourne.

Massee, G. E. (1899). A revision of the genus *Tilletia. Bull. Misc. Inform. Roy. Bot. Gard.* pp. 141–159.

Mundkur, B. B., and M. J. Thirumalachar. (1952). "Ustilaginales of India." Commonwealth Mycol. Inst., Kew, Surrey, England.

Narasimhan, M. J., M. J. Thirumalachar, M. C. Srinivasan, and H. C. Govindu. (1963). *Georgefischeria,* a new genus of the Ustilaginales. *Mycologia* 55:30–34.

Plowright, C. B. (1889). "A Monograph of the British Uredineae and Ustilagineae." London.

Rogers, D. P. (1934). The basidium. *Stud. Natur. Hist. Iowa Univ.* 16:160–183.

Savile, D. B. O. (1947). A study of the species of *Entyloma* on North American composites. *Can. J. Res., Sect.* C 25:105–120.

Savile, D. B. O. (1951). Two new smuts on *Carex* in Canada. *Can. J. Bot.* 29:326–328.

Săvulescu, T. (1957). "Ustilaginalelle din republica populara romina," Vols. I and II. Bucharest.

Setchell, W. A. (1892). An examination of the species of the genus *Doassansia* Cornu. *Ann. Bot. (London)* 6:1–47.

Sydow, von H., and P. Sydow. (1901). Mycologische Mittheilungen. *Hedwigia Beibl.* 40:3.

Thirumalachar, M. J. (1944). A new genus of smuts. *Mycologia* 36:591–597.

Thirumalachar, M. J. (1966). Conspectus of our knowledge in the genera of Ustilaginales. *Indian Phytopathol.* 19:3–13.

Thirumalachar, M. J., and M. J. Narasimhan. (1952). *Zundelula,* a new genus of smuts. *Sydowia* [2] 6:407–411.

Thirumalachar, M. J., and M. S. Pavgi. (1952). Notes on some Indian Ustilagineae. V. *Sydowia* [2] 6:389–395.

Thirumalachar, M. J., M. S. Pavgi, and M. M. Payak. (1960). *Jamesdicksonia,* a new genus of the Ustilaginales. *Mycologia* 52:475–489.

Thirumalachar, M. J., M. D. Whitehead, and M. J. O'Brien. (1967). Studies in the genus *Tolyposporella. Mycologia* 59:389–396.

Tulasne, L. R., and C. Tulasne. (1847). Mémoire sur les Ustilaginées comparées aux Urédinées. *Ann. Sci. Natur.* [3] 7:12–127.

Viennot-Bourgin, G. (1949). "Les champignons parasites des plantes cultivées," Vol. 2, pp. 795–902. Masson, Paris.

Zundel, G. L. (1938). The Ustilaginales of South Africa. *Bothalia* 3:283–330.

Zundel, G. L. (1939). Additions and corrections to Ustilaginales. *N. Amer. Flora* 7:971–1030.

Zundel, G. L. (1953). "The Ustilaginales of the World." School of Agriculture, Pennsylvania State University, State College, Pennsylvania.

Basidiomycotina
Hymenomycetes
Phragmobasidiomycetidae, Chapter 17
Holobasidiomycetidae, Chapters 18–23

Phragmobasidiomycetidae: Tremellales, Auriculariales, Septobasidiales

R. F. R. McNabb[1]

Microbiology Department
Lincoln College
Canterbury, New Zealand

I. SUBCLASS PHRAGMOBASIDIOMYCETIDAE

The basidiocarp in this subclass is gelatinous, waxy, or arid; phragmo-basidiate metabasidium is completely or incompletely divided by primary septa. The probasidia can either persist as distinct structures or not, and the spores are often repetitive.

This subclass corresponds to Patouillard's (1887, 1900) "Hétérobasidiés" with the exclusion of the rusts and smuts which were included in the Auriculariaceae, and the holobasidiate Caloceraceae (=Dacrymycetaceae) and Tulasnellaceae.

The first major contribution toward an understanding of the group was that of Tulasne (1853; Tulasne and Tulasne, 1871) whose investigations revealed the great variation in basidial morphology within these fungi. Although the tremellaceous, auriculariaceous, and dacrymycetaceous basidial types were recognized, no taxonomic readjustments were proposed. Patouillard (1887, 1900) departed from the Friesian classification by dividing the Basidiomycetes into two subclasses, the "Homobasidiés" and "Hétéro-basidiés," and in so doing laid the foundation of a classification that has remained unchallenged until recent years.

The Heterobasidiomycetes, as defined by Patouillard, was characterized by four criteria: septate basidia, repetitive basidiospores, swollen sterigmata, and more or less gelatinous basidiocarps. However, an outstanding feature

[1]Deceased.

303

of Patouillard's system was its elasticity, and thus the Heterobasidiomycetes contained not only the phragmobasidiate Auriculariaceae and Tremellaceae, but two holobasidiate families, the Dacrymycetaceae and Tulasnellaceae.

Following Talbot's (1965) proposal to abolish Patouillard's subclasses because of the absence of characters by which they could consistently be distinguished, Lowy (1968) restricted the Heterobasidiomycetes to taxa with completely divided basidia, swollen sterigmata, and repetitive basidiospores, and erected a new subclass, the Metabasidiomycetidae, to accommodate taxa possessing characters intermediate between the two existing subclasses.

The validity of the four criteria considered diagnostic of the Heterobasidiomycetes has been reexamined by Lowy (1968, 1969) and Talbot (1968, 1970). Interest has centered mainly on septation of the metabasidium. The value of this character depends largely on whether septation is primary and therefore associated with nuclear division, or adventitious and associated mainly with progressive vacuolation of the basidium at spore formation.

There has been no critical reassessment of the subclass comparable to Donk's (1964) conspectus of the families of the Aphyllophorales, and opinions differ as to which families should be included in the subclass, and their rank.

Orders are distinguished on basidial morphology and, in the case of the Septobasidiales, the parasitic-symbiotic habit.

Lack of uniformity in basidial terminology has created many difficulties (Neuhoff, 1924; Rogers, 1934; Martin, 1957; Talbot, 1954, 1968; Donk, 1954, 1958b, 1964). Publications encompassing the subclass in general include Martin (1945, 1952), Pilát (1957), Neuhoff (1936), Kobayasi and Tubaki (1965), Olive (1958), Donk (1966), and Raitviir (1967).

KEY TO ORDERS OF PHRAGMOBASIDIOMYCETIDAE

Metabasidia globose, pyriform, clavate, or rarely spindle-shaped, longitudinally or obliquely
    cruciate-septate or septate . . . . **Tremellales** p. 304
Metabasidia more or less cylindrical, transversely septate. Saprobic, or parasitic on phanero-
    gams, cryptogams or other fungi . . . . **Auriculariales** p. 310
Symbiotic-parasitic on scale insects on living plants . . . . **Septobasidiales** p. 313

II. TREMELLALES

Tremellales are saprobic and occasionally parasitic on other fungi. The basidiocarp is gymnocarpous, resupinate, pustulate, sessile-pileate, stipitate-pileate, clavate or clavarioid; in some internal parasites it is absent. They are gelatinous, waxy, fleshy, arid, or coriaceous. The hymenium is unilateral or amphigenous, and has dikaryophyses and paraphysoids; thick-walled cystidia and gloeocystidia are present or absent. The probasidia are globose, ovate, pyriform or clavate, or rarely spindle-shaped and catenulate. Metabasidia replace probasidia, incompletely or completely longitudinally or obliquely cruciate-septate into four cells and occasionally into two cells. Sterigmata are elongate or subulate, occasionally deciduous, spiculate or, rarely, nonspiculate. The spores are thin-walled, or occasionally thick-walled, smooth, rarely spinose or verrucose, inamyloid, aseptate, repetitive. Germination occurs by germ tubes, conidia, or blastospores. Hymenial conidia are occasionally present and separate conidial fructifications are rare.

The order is characterized by the typically cruciate-septate metabasidium, although irregularities in the position and number of septa are not infrequent. In *Patouillardina*, septation is occasionally transverse and the basidium approaches the auriculariaceous type in appearance. Considerable interest is centered on the type and position of septa in *Tremellodendropsis* and *Pseudotremellodendron*, and the relationship of these genera to the clavariaceous *Aphelaria* (Corner, 1966a). It is apparent from the literature that further examination of living material is required.

Families are distinguished by basidial characters and genera separated on a variety of macroscopical and microscopical criteria. The scarcity of stable characters makes generic separation of effused forms difficult, and resupinate or effused genera are frequently distinguished by the presence or absence of a single hymenial element, texture, or by characters generally considered of minor taxonomic importance at generic level (Fig. 1). Many effused, poorly characterized species have been assigned to the artificial genus *Sebacina* sensu lato. The transfer of related groups of species at present residing in *Sebacina* to more naturally defined genera is the major task confronting taxonomists interested in Tremellales. In recent years, recognition of the myxarioid sphaeropedunculate basidium as a distinct basidial type has added a valuable taxonomic character (Wells, 1961; Donk, 1966). The distribution of this type of basidium within the order has not been fully determined, but its use in defining such genera as *Myxarium* and *Protodontia* indicates its value in generic separation.

Revisions of smaller genera and regional monographs of some large genera have been undertaken (see Donk, 1966), but there is a lack of information on some critical species, particularly from tropical areas. Until these have been reexamined in the light of characters now considered of taxonomic importance, it will be difficult to arrive at a satisfactory classification.

KEY TO FAMILIES AND GENERA OF TREMELLALES

1. Basidia catenulate, maturing basipetally, protosterigmata sporoid, deciduous
   (Sirobasidiaceae) . . . . **Sirobasidium**
   Kobayasi (1952) recognizes 6 spp.

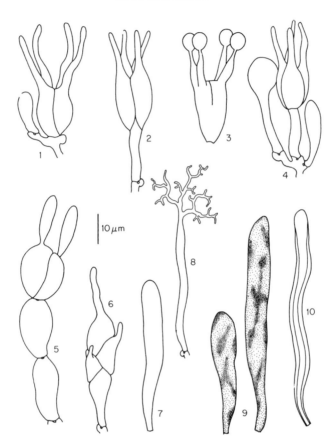

FIG. 1. Tremellales: morphological characters of importance in identification. 1, *Sebacina novae-zelandiae*, globose, sessile, cruciate-septate metabasidium with developing probasidium; 2, *Bourdotia petiolata*, sphaeropedunculate metabasidium with stalk delimited by basal divergence of cruciate septa; 3, *Metabourdotia tahitiensis*, incompletely cruciate-septate metabasidium [After Olive (1957). Reproduced by permission of *American Journal of Botany*]; 4, *Myxarium* sp., myxarioid sphaeropedunculate metabasidium with stalk delimited by a septum, and developing probasidia; 5, *Sirobasidium sanguineum*, catenulate basidia with deciduous protosterigmata. 6, *Patouillardina cinerea*, spindle-shaped metabasidium with oblique first-formed septum and 2 subsequent septa at right angles; 7, *Sebacina umbrina*, paraphysoid; 8, *Sebacina umbrina*, dendroid dikaryophysis; 9, *Ductifera sucina*, gloeocystidia with granular contents; 10, *Heterochaetella dubia*, thick-walled cystidium.

1′. Basidia not catenulate, protosterigmata not deciduous 2

    2(1′) Spores nonapiculate, borne symmetrically on cylindrical, filiform, nonspiculate sterigmata **(Hyaloriaceae)** 3

    2′(1′) Spores apiculate, borne asymmetrically on spiculate sterigmata . . **(Tremellaceae)**
    4

3(2) Basidiocarp stipitate and pileate, tough-gelatinous; basidia myxarioid sphaeropedunculate, overtopped by dikaryophyses; spores developing thick, gelatinous walls
· **Hyaloria**
Monotypic: *H. pilacre* restricted to tropical South America, Wells (1969).

3'(2) Basidiocarp effused, arachnoid; basidia not sphaeropedunculate; spores thin-walled
· **Xenolachne**
Monotypic: *X. flagellifera* parasitic on hymenium of the discomycete *Hyaloscypha atromaria* in North America, Rogers (1947).

4(2') Basidiocarp erumpent, pustulate, sessile-pileate, or stipitate-pileate, sometimes effused with age, occasionally lacking in parasitic spp. but then spores globose, subglobose, or ovate, margins determinate 5

4'(2') Basidiocarp resupinate, effused, adpressed, often anastomosing or coalescing, occasionally lacking in parasitic spp. but then spores elongate, cylindrical, or allantoid, margins indeterminate 16

5(4) Basidiocarp erect, simple or branched, clavarioid, coralloid or thelephoroid in appearance
· 6

5'(4') Basidiocarp sessile-pileate, stipitate-pileate, not as above 9

6(5) Basidiocarp cartilaginous to gelatinous, simple or branched; on wood
· **Holtermannia**
Kobayasi (1937) describes and keys 6 spp.

6'(5) Basidiocarp tough-fleshy to subcoriaceous; on ground 7

7(6') Probasidia subglobose to pyriform; metabasidium cruciate-septate; basidiocarp branched and often basally anastomosing, occasionally simple and clavate . . . . **Tremellodendron**
Bodman (1942) describes and keys 8 spp. restricted to North America.

7'(6') Probasidia elongate-clavate; metabasidium often divided into a stalk cell and a fertile, semi-inflated apical cell by a transverse septum, apical cell when present completely or incompletely cruciate-septate 8

8(7') Apical portion of metabasidium completely cruciate-septate and delimited from stalk prior to spore formation by a transverse septum; basidiocarp much branched; spores repetitive . . . . **Pseudotremellodendron**
Reid (1956) recognizes a single species, *P. pusio* of Corner (1966a).

8'(7') Apical portion of metabasidium incompletely cruciate-septate by invagination of sterigmatic bases and sometimes delimited from stalk subsequent to spore formation by a traverse septum; basidiocarp often palmately branched . . . **Tremellodendropsis**
Monotypic: *T. tuberosa* cf. Corner (1966a). See also p.354.

9(5') Basidiocarp stipitate-pileate or substipitate-pileate 10

9'(5') Basidiocarp sessile-pileate, cerebriform, convoluted, lobate, pustulate or pulvinate
· 11

10(9) Hymenial surface spinose, inferior; basidiocarp stipitate and flabellate, or dimidiate
· **Pseudohydnum**
Monotypic: *P. gelatinosum* ± cosmopolitan.

10'(9) Hymenial surface smooth or faintly venose, inferior; basidiocarp substipitate to stipitate, infundibuliform to spathulate . . . . **Phlogiotis**
Monotypic: *P. helvelloides* of North Temperate distribution. Syn.: *Tremiscus* (Donk, 1966).

11(9') Gelatinous pycnidia preceding or accompanying basidiocarp as separate fructifications

. . . . **Craterocolla**
1 or possibly 2 spp. of predominantly European distribution. Syn.: *Ditangium* (Donk, 1966).

11'(9') Gelatinous pycnidia absent 12

12(11') Gloeocystidia with yellow-brown granular contents present; branched dikaryophyses overtopping basidia, forming a firm superficial layer . . . . **Ductifera**
Wells (1958) keys 4 spp.

12'(11') Gloeocystidia with yellow-brown contents absent 13

13(12') Basidia myxarioid sphaeropedunculate; basidiocarp initially pustulate, becoming
subglobose, pulvinate, or convoluted . . . . **Myxarium**
Segregated from *Exidia* by Donk (1966). Raitviir (1967) recognizes 4 spp.

13'(12') Basidia not myxarioid sphaeropedunculate 14

14(13') Spores globose, subglobose or ovate; branched dikaryophyses usually absent,
if present rarely forming a firm superficial layer; basidiocarp gelatinous, lacking in some
parasitic spp. . . . . **Tremella**
A large, cosmopolitan genus, mainly saprobic but a few spp. parasitic on or
within fructifications of other fungi. See·Donk (1966) for regional monographs, also
Torkelsen (1968). Syn.: *Naematelia*.

14'(13') Spores cylindrical to allantoid 15

15(14') Basidiocarp pustulate, minute, gregarious, often coalescing to form reticulate masses,
gelatinous; dikaryophyses not forming a firm superficial layer . . . . **Pseudostypella**
Monotypic: *P. nothofagi* restricted to New Zealand, McNabb (1969).

15'(14') Basidiocarp pustulate, tuberculate, lobate, foliose or cerebriform, often anastomosing,
firm- to tough-gelatinous; branched dikaryophyses overtopping basidia, interwoven apically
to form a firm superficial layer . . . . **Exidia**
A large, cosmopolitan genus. See Donk (1966) for regional monographs.

16(4') Basidia spindle-shaped, first-formed septum regularly oblique-transverse, the 2
later formed septa at right angles to the first . . . . **Patouillardina**
Monotypic: *P. cinerea* occurs in tropical America and islands of the Pacific, Olive
(1958).

16'(4') Basidia not combining these characters 17

17(16') Metabasidia incompletely cruciate-septate 18

17'(16') Metabasidia completely cruciate-septate or occasionally longitudinally 1-septate
. . . . 19

18(17) Sterigmata inflated, tulasnelloid, basally septate; gloeocystidia absent
. . . . **Pseudotulasnella**
Monotypic: *P. guatemalensis* restricted to Guatemala, Lowy (1964).

18'(17) Sterigmata stout but not tulasnelloid, basally aseptate; gloeocystidia with yellow
contents present . . . . **Metabourdotia**
Monotypic: *M. tahitiensis* restricted to Tahiti, Olive (1957).

19(17') Basidiocarp shallow or deeply poroid, or merulioid 20

19'(17') Basidiocarp not poroid 21

20(19) Basidiocarp coriaceous, deeply poroid . . . . **Aporpium**
Monotypic: *A. caryae* widely distributed but mainly temperate, Teixeira and Rogers
(1955).

20′(19) Basidiocarp soft-fleshy or waxy, merulioid . . . . **Protomerulius**
Some 3 tropical spp. described.

21(19′) Projecting, thick-walled cystidia present, conspicuous, cylindrical; basidia myxarioid sphaeropedunculate . . . . **Heterochaetella**
Luck-Allen (1960) recognizes and keys 3 spp.

21′(19′) Projecting, thick-walled cystidia absent 22

22(21′) Basidiocarp tuberculate, papillate, dentate or spinose, resupinate and bearing spines or pegs 23

22′(21′) Basidiocarp not as above 26

23(22) Hymenium pierced by sterile spines or pegs originating in subhymenial layers and composed of parallel or interwoven hyphae . . . . **Heterochaete**
Bodman (1952) describes and keys 29 spp., mainly extra-European.

23′(22) Hymenium continuous over apices of spines or tubercules, or interrupted at extreme apices where sterile axial elements protrude 24

24(23′) Sterile axial elements absent, hymenium continuous over thick, blunt spines; basidia not myxarioid sphaeropedunculate, stalk delimited by basal divergence of the longitudinal cruciate septa . . . . **Protohydnum**
Monotypic: *P. cartilagineum* occurs in tropical America, Martin (1941).

24′(23′) Sterile axial elements present; basidia myxarioid sphaeropedunculate 25

25(24′) Axial elements consisting of unspecialized or slightly modified hyphae, protruding apically . . . . **Protodontia**
Two or 3 spp., Donk (1966).

25′(24′) Axial elements consisting of large gloeocystidia . . . . **Stypella**
Martin (1934) recognizes 2 or possibly 3 spp. Donk (1966) restricts the genus to *S. papillata.*

26(22′) Basidiocarp ± coriaceous, basal layer well-developed, composed of thick-walled hyphae, margins often reflexed when dry . . . . **Eichleriella**
A small genus of some 6 spp.; congeneric with *Exidiopsis* by Wells (1961). See Donk (1966).

26′(22′) Basidiocarp not combining these characters 27

27(26′) Clamp connections constantly absent; basidiocarp cartilaginous to coriaceous, rarely waxy-gelatinous, firm to tough, resupinate or occasionally lobate, encrusting . . **Sebacina**
In the restricted sense of Ervin (1957) and Wells (1961), *Sebacina* contains 4 or 5 spp. of Donk (1966).

27′(26′) Clamp connections present; basidiocarp arid-waxy, waxy-gelatinous or gelatinous, resupinate, effused 28

28(27′) Gloeocystidia with yellow-brown contents absent; basidiocarp gelatinous to arid-waxy, lacking in some parasitic spp. . . . . **Exidiopsis**
As here circumscribed, this genus is a temporary repository for many spp. described under *Sebacina* s. lat. See McGuire (1941), Wells (1961), Luck-Allen (1963), and Donk (1966).

28′(27′) Gloeocystidia with yellow-brown contents present 29

29(28′) Basidiocarp arid to arid-waxy . . . . **Basidiodendron**
Luck-Allen (1963) recognizes 10 spp. and assigns to this genus the arid spp. included in *Bourdotia* by Wells (1959).

29′(28′) Basidiocarp gelatinous to waxy-gelatinous . . . . **Bourdotia**
Wells (1959) recognizes 11 spp., many of which were transferred to *Basidiodendron* by Luck-Allen (1963).

III. AURICULARIALES

Auriculariales are saprobic or parasitic on phanerogams, cryptogams, or other fungi. The basidiocarp is gymnocarpous, resupinate, pustulate, sessile-pileate, stipitate-pileate, clavate or clavarioid, and is absent in some internal parasites. They are gelatinous, waxy, fleshy, or arid. The hymenium is unilateral or amphigenous and dikaryophyses are present or absent. The probasidia are persistent or nonpersistent, thin- or thick-walled. The metabasidia are more or less cylindrical, straight, curved or occasionally apically coiled, and transversely (0-)1-3-septate. Sterigmata are elongate or reduced to spicula and the spores are thin-walled or occasionally thick-walled, smooth, inamyloid, aseptate or rarely septate, and repetitive. Germination occurs by germ tubes or conidia. Hymenial conidia are occasionally present. Separate conidial fructifications are rare.

The transversely septate metabasidium is characteristic of the order. Very occasionally, as in *Platygloea unispora*, it may remain aseptate. A wide variation in basidial morphology exists within the Auriculariales, ranging from the unspecialized basidium of *Auricularia* to the strongly modified basidium reminiscent of the Uredinales found in *Cystobasidium* (Fig. 2).

One family (Auriculariaceae) is recognized, although *Phleogena* and allied genera with stipitate and capitate basidiocarps are often separated as the Phleogenaceae. Genera are based on a variety of criteria including parasitic habit, basidiocarp shape and texture, septation of the metabasidium, and presence or absence of persistent probasidia. The effused genera *Platygloea*, *Helicobasidium* and *Helicogloea* are artificially delimited and act as repositories for poorly characterized species that cannot be accommodated in other more readily defined genera. Morphological distinctions between the probasidium and metabasidium assume importance in the Auriculariales since the probasidium may persist as a thin- or thick-walled structure after formation of the metabasidium. Although the distribution of the persistent probasidium throughout the order has yet to be fully determined, it is an important taxonomic character often correlated with parasitism.

A number of generic revisions have been undertaken (see Donk, 1966) but the absence of adequate microscopical descriptions of many species makes classification difficult. This applies equally to tropical and temperate species, some of which are the types of generic names.

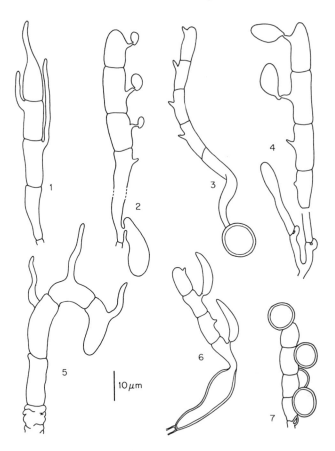

FIG. 2. Auriculariales and Septobasidiales: basidial types. 1, *Auricularia polytricha*, unspecialized metabasidium; 2, *Helicogloea lagerheimii*, metabasidium with thin-walled, lateral probasidium; 3, *Cystobasidium proliferans*, metabasidium with thick-walled, persistent, basal probasidium [After Olive (1952). Reproduced by permission of *Mycologia*]; 4, *Mycogloea macrospora*, deciduous metabasidium detached from stalk cell, and developing probasidium; 5, *Helicobasidium brebissonii*, apically recurved metabasidium with collapsed basal probasidium; 6. *Septobasidium* sp., metabasidium with thick-walled, persistent, basal probasidium; 7, *Phleogena faginea*, metabasidium with thick-walled, sessile basidiospores.

KEY TO GENERA OF AURICULARIALES

1. Parasitic on sporophytes or gametophytes of mosses, or leaves of vascular plants . . . 2

1′. Saprobic, occasionally parasitic on other fungi or on roots and stems of vascular plants
. . . . 6

    2(1) Parasitic on mosses 3

2'(1) Parasitic on vascular plants   4

3(2) Basidiocarp on tips of gametophytic shoots, erect, clavate; probasidia thin-walled
. . . . **Eocronartium**

   Monotypic: *E. muscicola* ± cosmopolitan, Fitzpatrick (1918).

3'(2) Basidiocarp on sporophyte setae and capsules, pustulate, small; probasidia thin-walled
. . . . **Jola**

   Three spp. occurring in the tropics of both hemispheres.

4(2') Parasitic on leaves of Cyperaceae; basidiocarp gelatinous, erumpent, pustulate
. . . . **Xenogloea**

   Monotypic: *X. eriophori* occurs in Europe and North America, Kao (1956). Donk
   (1958a) considers *Kriegeria* Bres. the correct name.

4'(2') Parasitic on leaves of ferns or Dicotyledons; basidiocarp arid to subcartilaginous
. . . .   5

5(4') Basidiocarp small, discrete, readily separable from host; composed mainly of sterile
hyphae; probasidia thick-walled; parasitic on ferns . . . . **Platycarpa**
   Two spp. occurring in Jamaica and South America, Couch (1949).

5'(4') Basidiocarp effused, adnate, composed mainly of basidia; probasidia thin-walled;
parasitic on ferns or Dicotyledons . . . . **Herpobasidium**
   Three spp. of temperate distribution recognized. *H. deformans* has a conidial state,
   Glomopsis lonicerae, Donk (1966).

6(1') Basidiocarp pileate, erect; probasidia thin-walled and nonpersistent except in
Neotyphula   7

6'(1') Basidiocarp resupinate, effused, or occasionally pustulate; probasidia sometimes
persisting as distinct structures 15

7(6) Basidiocarp clavarioid, branched, coriaceous; hymenium absent, basidia parallel to
surface with sterigmata perpendicular . . . . **Paraphelaria**
   Monotypic: *P. amboinensis* occurs in the tropical Pacific area, Corner (1966b).

7'(6) Basidiocarp not combining these characters   8

8(7') Basidiocarp clavate or stipitate and capitate, simple or rarely branched   9

8'(7') Basidiocarp nodulose, lobate, cupulate, auriform, foliose or sessile-pileate . . . 13

9(8) Probasidia persistent, thick-walled, pyriform; basidiocarp clavate, stipitate
. . . . **Neotyphula**

   Monotypic: *N. guianensis* restricted to tropical America, Martin (1948).

9'(8) Probasidia nonpersistent; basidiocarp stipitate and capitate 10

10(9') Metabasidia transversely 1-septate; dikaryophyses projecting beyond basidia,
pseudoperidium absent . . . . **Stilbum** Tode ex Merat sensu Juel
   Monotypic: *S. vulgare* of north temperate distribution, Juel (1898).

10'(9') Metabasidia transversely 3-septate; dikaryophyses projecting beyond basidia
and sometimes forming a pseudoperidium 11

11(10') Basidiocarp gelatinous or waxy-gelatinous . . . . **Hoehnelomyces**
   Two spp. of tropical distribution recognized.

11'(10') Basidiocarp dry or fleshy 12

12(11') Basidiocarp dry; pileus ± globose, dikaryophyses branched and interwoven
distally forming a brittle pseudoperidium . . . . **Phleogena**
   Possibly monotypic: *P. faginea* ± cosmopolitan, Shear and Dodge (1925).

12′(11′) Basidiocarp fleshy; pileus ± discoid, dikaryophyses projecting to form a loose covering . . . . **Pilacrella**
  Monotypic: *P. solani* collected once in Europe, Pilát (1957).

13(8′) Basidiocarp nodulose, firm-gelatinous, radially fibrillose, hard when dry
  . . . . **Mylittopsis**
  Monotypic: *M. marmorata* occurs in southern U.S.A. and Malaya, Rogers and Martin (1955).

13′(8′) Basidiocarp foliose, cupulate, auriform or resupinate with reflexed margins, firm-gelatinous, drying horny 14

  14(13′) Basidiocarp foliose, erect to lobate, firm-gelatinous; hymenium amphigenous
    . . . . **Phyllogloea**
    Two spp. recognized by Lowy (1959).

  14′(13′) Basidiocarp discoid, cupulate, auriform, or resupinate with reflexed margins; hymenium unilateral, abhymenial surface bearing sterile hairs . . . . . . **Auricularia**
    Ten spp. recognized and described by Lowy (1952). Donk (1966) segregates *Hirneola* on growth form.

15(6′) Probasidia lateral, saccate, thin-walled, conspicuous; basidiocarp soft-gelatinous to floccose . . . . **Helicogloea**
  Baker (1946) recognizes 11 spp. Donk (1966) restricts the genus to gelatinous spp. and retains *Saccoblastia* for floccose spp.

15′(6′) Probasidia basal, persistent or not 16

  16(15′) Probasidia persistent, globose to ovate, moderately thick-walled; basidiocarp arachnoid, waxy or indefinite . . . . **Cystobasidium**
    Olive (1952) recognizes 3 spp.

  16′(15′) Probasidia persistent or not, if persistent thin-walled or slightly thick-walled and irregular in shape 17

17(16′) Metabasidia readily detached, cylindrical, borne on subbasidial cells of smaller diameter . . . . **Mycogloea**
  McNabb (1965) recognizes 3 spp.

17′(16′) Metabasidia not readily detached 18

  18(17′) Basidiocarp arid, floccose or fleshy-fibrous, composed of ± loosely interwoven hyphae; metabasidia emergent, typically circinnately coiled apically; parasitic on subterranean parts of plants . . . . **Helicobasidium**
    Donk (1966) restricts the genus to *H. brebissonii* (violet root rot) and a few extra-European spp.

  18′(17′) Basidiocarp waxy to gelatinous, absent in some parasitic spp.; metabasidia straight or curved; probasidia persisting as thin-walled structures of irregular size and shape in some spp.; saprophytic, or parasitic on fructifications of other fungi
    . . . . **Platygloea**
    In a preliminary survey of the genus, Bandoni (1956) recognizes and keys 23 spp. Donk (1966) considers *Achroomyces* an earlier name.

IV. SEPTOBASIDIALES

This order is symbiotic-parasitic on scale insects on living plants. Its basidiocarps are gymnocarpous, resupinate, smooth, nodulose, creviced or occasionally spinose. Septobasidiales are arid, crustaceous or spongy, and often composed of a basal subiculum with an intermediate context of hyphal pillars or ridges, and a hymenium which has a context sometimes compacted throughout. The hyphae are septate, and clamp connections are absent. The hymenium is unilateral and made up of basidia and sterile hyphae. The probasidia are nonpersistent, or persistent and ovate, subglobose, pyriform or subcylindrical, and thin- or thick-walled. The metabasidia are more or less cylindrical, straight, curved or coiled, transversely (0-)1-3-septate. The sterigmata are elongate or reduced to spicula and the spores are hyaline, thin-walled, smooth, inamyloid, septate, and rarely repetitive. Germination occurs by conidia or blastospores and hymenial conidia are occasionally present.

This order shares with the Auriculariales the transversely septate metabasidium, but is distinguished by its unique relationship with scale insects (Fig. 2). As in Auriculariales, basidial morphology varies greatly.

The order contains a single monogeneric family, the Septobasidiaceae. The systematic position of these fungi was established by Patouillard (1892) who erected the genus *Septobasidium*, and the biological relationship between fungus and insect recognized some years later by von Höhnel and Litschauer (1907). It is to Couch that we owe the greater part of our knowledge of the taxonomy and biology of *Septobasidium*. His monograph (Couch, 1938) summarizes his own investigations and those of earlier authors, and remains the basic work on the genus. The species concept is based on relatively stable microscopical characters including arrangement of context hyphae, haustorium type, persistence of probasidia, and septation of metabasidia.

Couch (1938) described and keyed 165 species of *Septobasidium*. The genus is mainly of tropical to warm temperate distribution.

REFERENCES

Baker, G. E. (1946). Addenda to the genera *Helicogloea* and *Physalacria*. *Mycologia* **38**: 630–638.

Bandoni, R. J. (1956). A preliminary survey of the genus *Platygloea*. *Mycologia* **48**:821–840.

Bodman, M. C. (1942). The genus *Tremellodendron*. *Amer. Midl. Natur.* **27**:203–216.

Bodman, M. C. (1952). A taxonomic study of the genus *Heterochaete*. *Lloydia* **15**:193–233.

Corner, E. J. H. (1966a). The clavarioid complex of *Aphelaria* and *Tremellodendropsis*. *Trans. Brit. Mycol. Soc.* **49**:205–211.

Corner, E. J. H. (1966b). *Paraphelaria*, a new genus of Auriculariaceae (Basidiomycetes). *Persoonia* **4**:345–350.

Couch, J. N. (1938). "The Genus *Septobasidium*." Univ. of North Carolina Press, Chapel Hill.

Couch, J. N. (1949). The taxonomy of *Septobasidium polypodii* and *S. album*. *Mycologia* **41**: 427–441.

Donk, M. A. (1954). A note on sterigmata in general. *Bothalia* **6**:301–302.

Donk, M. A. (1958a). The generic names proposed for Hymenomycetes. VIII. Auriculariaceae, Septobasidiaceae, Tremellaceae, Dacrymycetaceae. *Taxon* 7:164–207, 193–207, 236–250.

Donk, M. A. (1958b). Notes on the basidium. *Blumea, Suppl.* 4:96–105.

Donk, M. A. (1964). A conspectus of the families of Aphyllophorales. *Persoonia* 3:199–324.

Donk, M. A. (1966). Check list of European hymenomycetous Heterobasidiae. *Persoonia* 4:145–335.

Ervin, M. D. (1957). The genus *Sebacina*. *Mycologia* 49:118–123.

Fitzpatrick, H. M. (1918). The life history and parasitism of *Eocronartium muscicola*. *Phytopathology* 8:197–218.

Juel, H. O. (1898). *Stilbum vulgare* Tode ein bisher verkannter Basidiomycet. *Kgl. Sv. Vetenskapsakad., Handl.* 24(3):1–15.

Kao, C. J. (1956). The cytology of *Xenogloea eriophori*. *Mycologia* 48:288–301.

Kobayasi, Y. (1937). On the genus *Holtermannia* of Tremellaceae. *Sci. Rep. Tokyo Bunrika Daigaku, Sect. B* 3:75–81.

Kobayasi, Y. (1952). Revision of *Sirobasidium*, with description of a new species found in Japan. *Trans. Mycol. Soc. Jap.* 4:29–34.

Kobayasi, Y. and K. Tubaki. (1965). Studies on cultural characters and asexual reproduction of Heterobasidiomycetes. I. *Trans. Mycol. Soc. Jap.* 6:29–36.

Lowy, B. (1952). The genus *Auricularia*. *Mycologia* 44:656–692.

Lowy, B. (1959). New or noteworthy Tremellales from Bolivia. *Mycologia* 51:840–850.

Lowy, B. (1964). A new genus of the Tulasnellaceae. *Mycologia* 56:696–700.

Lowy, B. (1968). Taxonomic problems in the Heterobasidiomycetes. *Taxon* 17:118–127.

Lowy, B. (1969). Septate holobasidia. *Taxon* 18:632–634.

Luck-Allen, E. R. (1960). The genus *Heterochaetella*. *Can. J. Bot.* 38:559–569.

Luck-Allen, E. R. (1963). The genus *Basidiodendron*. *Can. J. Bot.* 41:1025–1052.

McGuire, J. M. (1941). The species of *Sebacina* (Tremellales) of temperate North America. *Lloydia* 4:1–43.

McNabb, R. F. R. (1965). Some auriculariaceous fungi from the British Isles. *Trans. Brit. Mycol. Soc.* 48:187–192.

McNabb, R. F. R. (1969). New Zealand Tremellales. III. *N.Z. J. Bot.* 7:241–261.

Martin, G. W. (1934). The genus *Stypella*. *Stud. Natur. Hist. Iowa. Univ.* 16(2):143–150.

Martin, G. W. (1941). New or noteworthy tropical fungi. I. *Lloydia* 4:262–269.

Martin, G. W. (1945). The classification of the Tremellales. *Mycologia* 37:527–542.

Martin, G. W. (1948). New or noteworthy tropical fungi. IV. *Lloydia* 11:111–122.

Martin, G. W. (1952). Revision of the North Central Tremellales. *Stud. Natur. Hist. Iowa Univ.* 19(3):1–122.

Martin, G. W. (1957). The tulasnelloid fungi and their bearing on basidial terminology. *Brittonia* 9:25–30.

Neuhoff, W. (1924). Zytologie und systematische Stellung der Auriculariaceen und Tremellaceen. *Bot. Arch.* 8:250–297.

Neuhoff, W. (1936). Die Gallertpilze Schwedens (Tremellaceae, Dacrymycetaceae, Tulasnellaceae, Auriculariaceae). *Ark. for Bot. A* 28(1):1–57.

Olive, L. S. (1952). A new species of *Cystobasidium* from New Jersey. *Mycologia* 44:564–569.

Olive, L. S. (1957). Two new genera of the Ceratobasidiaceae and their phylogenetic significance. *Amer. J. Bot.* 44:429–435.

Olive, L. S. (1958). The lower Basidiomycetes of Tahiti. I. *Bull. Torrey Bot. Club* 85:5–27, 89–110.

Patouillard, N. (1887). "Les Hyménomycètes d'Europe." Klincksieck, Paris.

Patouillard, N. (1892). *Septobasidium*, nouveau genre d'Hyménomycètes hétérobasidiés. *J. Bot. Paris* 6:61–64.

Patouillard, N. (1900). "Essai taxonomique sur les familles et les genres des Hyménomycètes." Duclume, Lons-le-Saunier.

Pilát, A. (1957). Uebersicht der europaischen Auriculariales und Tremellales unter besonderer Berucksichtigung der tschechoslowakischen Arten. *Sbo. Narod. Mus. Praze, Ser. B* No. 3, 535–540.

Raitviir, A. G. (1967). "A Key to the Heterobasidiomycetidae Found in the U.S.S.R." Nauka, Leningrad. (English translation of title.)

Reid, D. A. (1956). New or interesting records of Australasian Basidiomycetes. II. *Kew Bull.* No. 3, 535–540.

Rogers, D. P. (1934). The basidium. *Stud. Natur. Hist. Iowa Univ.* **16**(2):160–182.

Rogers, D. P. (1947). A new gymnocarpous heterobasidiomycete with gasteromycetous basidia. *Mycologia* **39**:556–564.

Rogers, D. P., and G. W. Martin. (1955). The genus *Mylittopsis*. *Mycologia* **47**:891–894.

Shear, C. L., and B. O. Dodge. (1925). The life history of *Pilacre faginea* (Fr.) B. & Br. *J. Agric. Res.* **30**:407–417.

Talbot, P. H. B. (1954). Micromorphology of the lower Hymenomycetes. *Bothalia* **6**:249–299.

Talbot, P. H. B. (1965). Studies of "*Pellicularia*" and associated genera of Hymenomycetes. *Persoonia* **3**:371–406.

Talbot, P. H. B. (1968). Fossilized pre-Patouillardian taxonomy. *Taxon* **17**:620–628.

Talbot, P. H. B. (1970). The controversy over septate holobasidia. *Taxon* **19**:570–572.

Teixeira, A. R., and Rogers, D. P. (1955). *Aporpium*, a polyporoid genus of the Tremellaceae. *Mycologia* **47**:408–415.

Torkelsen, A. E. (1968). The genus *Tremella* in Norway. *Nytt Mag. Bot.* **15**:225–239.

Tulasne, E. L. (1853). Observations sur l'organisation des Trémellinées. *Ann. Sci. Natur. Bot. Biol. Veg.* [3] **19**:193–231.

Tulasne, E. L., and C. Tulasne. (1871). New notes upon the tremellineous fungi and their analogues. *J. Linn. Soc. London, Bot.* **13**:31–42.

von Höhnel, F., and V. Litschauer. (1907). Beiträge zur Kenntnis der Corticieen. *Sitzungsber. Kaiserl. Akad. Wiss. Wien, Math. Naturwiss. Kl., Abt.* **116**:739–852.

Wells, K. (1958). Studies of some Tremellaceae. II. The genus *Ductifera*. *Mycologia* **50**: 407–416.

Wells, K. (1959). Studies of some Tremellaceae. III. The genus *Bourdotia*. *Mycologia* **51**: 541–563.

Wells, K. (1961). Studies of some Tremellaceae. IV. *Exidiopsis*. *Mycologia* **53**:317–370.

Wells, K. (1969). New or noteworthy Tremellales from southern Brazil. *Mycologia* **61**:77–86.

CHAPTER 18

Holobasidiomycetidae:

EXOBASIDIALES, BRACHYBASIDIALES, DACRYMYCETALES

R. F. R. McNabb[1]

Microbiology Department
Lincoln College
Canterbury, New Zealand

TULASNELLALES

P. H. B. Talbot

Waite Agricultural Research Institute
University of Adelaide
Glen Osmond, South Australia

I. HOLOBASIDIOMYCETIDAE

KEY TO ORDERS OF THE HOLOBASIDIOMYCETIDAE

1. Internal leaf parasites of higher plants 2

1'. Not internal parasites of higher plants or if parasitic having a distinct external fruit-body
. . . . 3

    2(1) Basidia emerge to form a hymenium . . . . **Exobasidiales** p. 318

    2'(1) Basidia emerge from the leaf in tufts . . . . **Brachybasidiales** p. 319

3(1') Basidia furcate (sterigmata 2); spores not repetitive . . . . . **Dacrymycetales** p. 319

3'(1') Basidia not furcate (sterigmata usually 4) 4

    4(3') Basidia subspherical to broad with stout fingerlike or inflated sterigmata; spores
    repetitive . . . . **Tulasnellales** p. 322

    4'(3') Sterigmata relatively small, uninflated; spores not repetitive 5

5(4') Fruit-body development gymnocarpous; hymenium unilateral or amphigenous, smooth
or covering dentate processes or lining tubes; if tubular tubes firmly united to the basidiocarp
. . . . **Aphyllophorales** p. 327

5'(4') Fruit-body development hemiangiocarpous or gymnocarpous; hymenium covering
lamellae on the lower surface of the pileus or lining tubes which are easily separable from
the pileus . . . . **Agaricales** p. 423

[1]Deceased.

II. EXOBASIDIALES

The order is either parasitic on stems, leaves, or flower buds of phanerogams, inciting leaf spots or hypertrophy of infected parts, or systemic. The mycelium is inter- and/or intracellular. The basidiocarp is rudimentary or absent. The basidia are holobasidiate, cylindrical to subclavate, two-eight-sterigmate, isolated, occurring in fascicles or forming a continuous hymenium at maturity; they are emergent through stomata, or between epidermal or cortical cells. The spores are thin-walled, smooth, inamyloid, aseptate or septate, and nonrepetitive. Germination takes place by conidia and/or germ tubes.

Exobasidiales are characterized by the unspecialized holobasidium, absence of well-defined basidiocarps, and the strictly parasitic habit. It contains a single family, the Exobasidiaceae.

The simply constructed fructifications provide few characters of value in generic separation. Genera are defined mainly on spore septation, host relationship, whether basidia emerge through stomata or between epidermal cells, and presence or absence of a subepidermal "stroma." Differences in interpretation of what constitutes a stroma reduces the value of the last-named character. As the oldest and largest genus, *Exobasidium* contains a number of insufficiently described species as well as species of doubtful exobasidiaceous affinities. Dismemberment of the genus has occurred to a limited extent. While the generic segregate *Muribasidiospora* has a sound morphological basis, the same cannot be said of *Arcticomyces* and *Exobasidiellum*, which are based on host family rather than morphological characters.

Useful references not cited in the key include Donk (1956a, 1966), Graafland (1953), McNabb (1962), and Sündstrom (1960, 1964).

KEY TO GENERA OF EXOBASIDIALES

1. Basidia in fascicles, emergent through stomata, occasionally between epidermal cells, arising from aggregated hyphae in substomatal air-spaces 2

1'. Basidia emergent between epidermal cells, forming a ± continuous hymenium at maturity, occasionally solitary and emergent through stomata 4

    2(1) Spores muriform; inciting unhypertrophied leaf spots on *Rhus* and *Celtis*; mycelium intercellular . . . . **Muribasidiospora**
        Rajendren (1968) recognises 3 spp.

    2'(1) Spores aseptate or transversely septate 3

3(2') Inciting unhypertrophied leaf spots on Commelinaceae; elongate paraphyses usually present; mycelium intracellular . . . . **Kordyana**
    Some 6 spp.; Gäumann (1922).

3'(2') Inciting systemic infections of Saxifragaceae; paraphyses absent; mycelium inter- and intracellular . . . . **Arcticomyces**
    Monotypic: *A. warmingii* of arctic distribution; Savile (1959b).

4(1') Basidia isolated, not forming a hymenium; on Gramineae . . . . **Exobasidiellum**
  Two spp. tentatively placed in Tulasnellaceae by Donk (1966); cf. Reid (1969).

4'(1') Basidia forming a ± continuous hymenium at maturity; on Ericaceae, Epacridaceae,
  Empetraceae, Symplocaceae, Lauraceae, Theaceae, and possibly some other families
  . . . . **Exobasidium**
    Some 50 spp., Savile (1959a).

III. BRACHYBASIDIALES

Brachybasidiales are parasitic on leaves of phanerogams, inciting leaf spots. The basidiocarp is minutely pustulate to discoid and is emergent through stomata or erumpent through the epidermis, and ± gelatinous. The basidia are holobasidiate, the probasidia are persistent, clavate to subpyriform, and thin- to thick-walled, with the metabasidia arising apically; these are cylindrical, bisterigmate, with short sterigmata. The spores are thin-walled, smooth, inamyloid, aseptate or septate, and nonrepetitive. Germination occurs by conidia and/or germ tubes.

The order is distinguished by the persistent probasidium, bisterigmate metabasidium, and strictly parasitic habit. Its single family, the Brachybasidiaceae, has traditionally included only *Brachybasidium*. *Dicellomyces* is here tentatively referred to the family on the basis of its many morphological similarities to *Brachybasidium*. Cytologically, *Dicellomyces* is reported to possess stichic basidia as opposed to the chiastic basidia of *Brachybasidium*.

KEY TO GENERA OF BRACHYBASIDIALES

1. Basidiocarp pustulate, emergent through stomata, composed of basidia only; on *Pinanga*
  (Palmae) . . . . **Brachybasidium**
    Monotypic: *B. pinangae* restricted to Java, Gäumann (1922).

1'. Basidiocarp pulvinate to discoid, erumpent through epidermis, composed of basidia,
  gloeocystidia, and thick-walled vesicular elements; on *Arundinaria* (Gramineae)
  . . . . **Dicellomyces**
    Monotypic: *D. gloeosporus* restricted to southeastern U.S.A., Olive (1945, 1951).

IV. DACRYMYCETALES

Dacrymycetales are saprobic and lignicolous. The basidiocarp is gymnocarpous, effused, pustulate, pulvinate, discoid, cupulate, attenuate-cylindrical, sessile-pileate, stipitate-pileate, and clavate or clavarioid. They are gelatinous and occasionally waxy or arid. The hymenium is unilateral or amphigenous with dikaryophyses present or absent. The basidia are holobasidiate, cylindrical-subclavate, clavate or occasionally obclavate, bisterigmate, with stout, elongate sterigmata. The spores are thin- or thick-walled, smooth, inamyloid, aseptate or more often septate,

and nonrepetitive. Germination takes place by conidia and/or germ tubes;
hymenial conidia are occasionally present. Separate conidial fructifications
are rare.

The Dacrymycetales is a remarkably homogeneous order characterized
by the furcate basidium. Apart from *Cerinomyces*, which is considered
better placed in the Corticiaceae sensu lato by some authors (Donk, 1966),
the order is sharply delimited from the remainder of the subclass. The single
family (Dacrymycetaceae) has traditionally been included with other jelly
fungi in subclass Phragmobasidiomycetidae, usually in the Tremellales.
Although this disposition of the Dacrymycetaceae is still widely accepted,
there is a growing body of opinion favoring the transfer of the family to
the Holobasidiomycetidae (Donk, 1964a, 1966; Talbot, 1968).

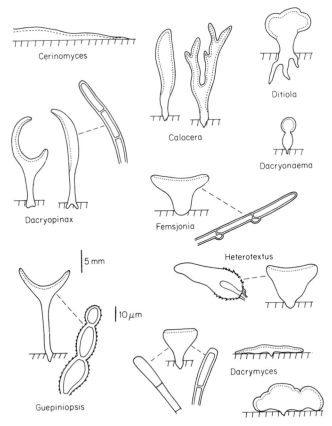

FIG. 1. Dacrymycetales: longitudinal sections of basidiocarps of typical representatives
of the genera. Position of hymenium indicated by dotted lines and diagnostic cortical structures
by broken lines. Diagrammatic.

A critical reevaluation of genera was undertaken by Kennedy (1958a). Because of their simple construction and uniform basidial morphology, there are few criteria of value in generic separation, and genera are defined primarily on basidiocarp shape, presence or absence of a cortex, and hyphal characters (Fig. 1). Despite renewed interest in the family over recent years, generic limits are poorly defined in *Dacrymyces*, *Heterotextus*, and *Guepiniopsis*.

Differences in basidial terminology create problems similar to those encountered in the Phragmobasidiomycetidae. Useful general references not cited in the key include Neuhoff (1936), Kobayasi (1939a,b), Martin (1952), Donk (1964b, 1966), Kennedy (1958a), and Raitviir (1967).

KEY TO GENERA OF DACRYMYCETALES

1. Basidiocarp originating as fertile patches on a loose subiculum, effused, radicating base or definite point of attachment absent; arid, soft-waxy or waxy-gelatinous
   . . . . **Cerinomyces**
   McNabb (1964) describes and keys 6 spp.

1'. Basidiocarp originating as pustules, discs or cylinders, individually attached to the substratum; gelatinous to cartilaginous 2

    2(1') Internal construction heterogenous; basidiocarp stipitate and pileate, stipe of thick-walled hyphae, pileus of thin-walled, branched hyphae . . . . **Ditiola**
   Kennedy (1964) recognizes 1 sp.; McNabb (1966) 2 spp.

    2'(1') Internal construction homogeneous 3

3(2') Basidiocarp attenuate-cylindrical, clavate, narrowly spathulate or petaloid, simple or branched, occasionally stipitate with a rugose pileus; hymenium amphigenous; organized into 3 zones in transverse section . . . . **Calocera**
   McNabb (1965a) describes and keys 11 spp. Syn.: *Dacryomitra*.

3'(2') Basidiocarp not attenuate-cylindrical, occasionally clavate, spathulate or petaloid but then hymenium unilateral or inferior; not obviously zoned in transverse section 4

    4(3') Basidiocarp stipitate-pileate, pileus spathulate, petaloid, cupulate, obliquely cupulate, flabellate or foliose, occasionally morchelloid; hymenium inferior, unilateral or confined to interior of cup, abhymenial surface covered with thick-walled or occasionally thin-walled hairs . . . . **Dacryopinax**
   McNabb (1965b) describes and keys 7 spp.

    4'(3') Basidiocarp pustulate, discoid, pezizoid, cupulate or stipitate-pileate; hymenium amphigenous or restricted to superior surface 5

5(4') Hymenial surface pruinose or farinaceous; stipe cartilaginous, becoming dark brown and varnished . . . . **Dacryonaema**
   Monotypic: *D. rufum* restricted to northern Europe, Nannfeldt (1947).

5'(4') Hymenial surface gelatinous; stipe present or absent, gelatinous or tomentose, never varnished 6

    6(5') Basidiocarp exclusive of cortical hairs composed of thick-walled hyphae
   . . . . **Femsjonia**
   McNabb (1965e) describes and keys 2 spp.

6′(5′) Basidiocarp exclusive of cortical hairs composed of thin-walled hyphae 7

7(6′) Basidiocarp applanate, pustulate, cerebriform, or occasionally stipitate with a spathulate, subglobose or lobed pileus, hymenium amphigenous; occasionally turbinate, discoid or cupulate with hymenium restricted to the superior surface and abhymenium covered with unspecialized, thin- or thick-walled hairs . . . . **Dacrymyces**
Kennedy (1958b) describes and keys 8 spp.

7′(6′) Basidiocarp cupulate or discoid, stipitate or substipitate; hymenium superior, restricted to interior of cup or surface of disc; abhymenium of thick-walled terminal cells or thick-walled, septate hairs 8

8(7′) Abhymenial surface of septate hyphae, the individual cells of which become thick-walled, inflated, and appear catenulate . . . . **Guepiniopsis**
One more or less cosmopolitan sp. recognized by McNabb (1965c).

8′(7′) Abhymenial surface of thick-walled, broadly cylindrical, obclavate, obpyriform or obovate terminal cells forming a palisade . . . . **Heterotextus**
McNabb (1965d) recognizes and keys 4 spp.

V. TULASNELLALES

Tulasnellales are saprobes or facultative parasites, especially on plant parts near soil, or they occur as aerial web-blights. Several species are mycorrhizal in terrestrial orchids. The basidiocarp is effused, mucoid, waxy, gelatinous, arid, pruinose, arachnoid, corticioid to hypochnoid, white, brownish, grey or pink, violet or lilac. The order is euhymenial, and the hymenium is smooth, discontinuous or found as cymose or racemose clusters of basidia which rarely form a distinct palisade. The hyphae are monomitic, with or without clamps, thin-walled and hyaline or with the basal hyphae often colored, thicker-walled, rather wide, and branching at a wide angle, sometimes with laminated walls. Substromatic clusters of wide, branched moniliform hyphae are often present in the culture. Sclerotial and *Rhizoctonia* states are present or absent. Gloeocystidia and/or thin-walled emergent cystidia are rarely present. The basidia are subglobose, obovate, pyriform, to capitate-clavate or broad-cylindrical and barrel-shaped, with aseptate metabasidium, but sometimes they develop adventitious septa in or at the base of the sterigmata. There are from two-four-seven sterigmata which are swollen, sporelike, and often deciduous, or subcylindrical to fusoid and not deciduous, and capable of great variation in length. The spores are repetitive, not amyloid, smooth, thin-walled, hyaline to yellowish or pinkish-lilac, subglobose, ellipsoid, oblong, or cylindrical.

There is considerable disagreement on the basidial morphology and the relationships and systematic position of the tulasnelloid fungi (see Donk, 1956b, 1958, 1964a, 1966; Lowy, 1968, 1969; Martin, 1957; Olive, 1957a,b; Rogers, 1933, 1935, 1943; Talbot, 1965, 1968, 1970). As comprehended here, the definitive characters of the order are: holobasidia (sometimes with adventitiously septate sterigmata), swollen variable sterigmata and repetitive spores, and an effused basidiocarp with a euhymenium.

KEY TO FAMILIES AND GENERA OF TULASNELLALES

1. Protosterigmata strongly swollen and sporelike, subglobose, ellipsoid or obpyriform at first, becoming elongated apically into a fusiform or subulate part bearing the spiculum, cut off from the metabasidium by an adventitious septum at the base and then often deciduous
· **Tulasnellaceae**
 A single genus **Tulasnella**. Syn.: *Gloeotulasnella, Prototremella, Muciporus, Pachysterigma*.

1′. Protosterigmata stout, subcylindrical or fusoid, more or less straight and digitate, not deciduous, rarely developing adventitious septa and then not at the base
· **Ceratobasidiaceae** 2

 2(1′) Metabasidia little wider than their supporting hyphae or pedicels · · · · · · · · 3

 2′(1′) Metabasidia 2–3 times as wide as their supporting hyphae or pedicels · · · · · 4

3(2) Basidia short subcylindrical to barrel-shaped or obovoid; spores ellipsoid with 1 side flattened or rarely obpyriform to obovate, hyaline to fawn-colored; hyphal cells multinucleate; parasitic on plant parts near soil, or saprobic on soil and wood; sclerotial or sterile mycelial (*Rhizoctonia*) states present · · · · · · · · · · · · · · · · · **Thanatephorus**
 T. cucumeris, pathogen of extremely wide range of hosts, perfect state of *Rhizoctonia solani*. Syn.: *Pellicularia* Cooke *sensu* Rogers, *pro parte*.

3′(2) Basidia cylindric-clavate; spores typically biapiculate, fusoid, citriform or broad ellipsoid, hyaline becoming yellowish; saprobic on decaying wood and litter; no known conidial, sclerotial or rhizoctonia states · · · · · · · · · · · · · · · · · · · **Uthatobasidium**

 4(2′) Basidia subglobose to obpyriform, abruptly narrowed at the pedicels; gloeocystidia and cystidia absent; hyphae with binucleate cells; sclerotia present or absent; often forming pruinose web-blights of tropical crops, also saprobic in soil or wood, or mycorrhizal in orchids · **Ceratobasidium**
  Ceratobasidium anceps on bracken; *C. cornigerum* mycorrhizal. Syn.: *Koleroga, Pellicularia* Cooke *sensu* Rogers *pro parte*.

 4′(2′) Basidia obovate or broadly clavate on a tapering narrow pedicel; gloeocystidia and/or cystidia usually present; sclerotia absent; saprobic in soil · · · · · **Oliveonia**
  Syn.: *Heteromyces, Hydrabasidium*.

REFERENCES

Donk, M. A. (1956a). The generic names proposed for Hymenomycetes. VI. Brachybasidiaceae. Cryptobasidiaceae, Exobasidiaceae. *Reinwardtia* **4**:113–118.

Donk, M. A. (1956b). Notes on resupinate Hymenomycetes II. The tulasnelloid fungi. *Reinwardtia* **3**:363–379.

Donk, M. A. (1958). Notes on resupinate Hymenomycetes. V. *Fungus* **28**:16–36.

Donk, M. A. (1964a). A conspectus of the families of Aphyllophorales. *Persoonia* **3**:199–324.

Donk, M. A. (1964b). On some old species of Dacrymycetaceae. *Proc. Kon. Nede. Akad. Wetensch., Ser. C* **67**:1–18.

Donk, M. A. (1966). Check list of European hymenomycetous Heterobasidiae. *Persoonia* **4**:145–335.

Gäumann, E. (1922). Uber die Gattung *Kordyana* Rac. *Ann. Mycol.* **20**:257–271.

Graafland, W. (1953). Four species of *Exobasidium* in pure culture. *Acta Bot. Neer.* **1**:516–522.

Kennedy, L. L. (1958a). The genera of the Dacrymycetaceae. *Mycologia* **50**:874–895.

Kennedy, L. L. (1958b). The genus *Dacrymyces*. *Mycologia* **50**:896–915.

Kennedy, L. L. (1964). The genus *Ditiola*. *Mycologia* **56**:298–308.

Kobayasi, Y. (1939a). On the *Dacrymyces* group. *Sci. Rep. Tokyo Bunrika Daigaku, Sect. B* **4**:105–128.

Kobayasi, Y. (1939b). On the genera *Femsjonia*, *Guepinia* and *Calocera* from Japan. *Sci. Rep. Tokyo Bunrika Daigaku, Sect. B* **4**:215–227.

Lowy, B. (1968). Taxonomic problems in the Heterobasidiomycetes. *Taxon* **17**:118–127.

Lowy, B. (1969). Septate holobasidia. *Taxon* **18**:632–634.

McNabb, R. F. R. (1962). The genus *Exobasidium* in New Zealand. *Trans. Roy. Soc. N.Z. Bot.* **1**:259–68.

McNabb, R. F. R. (1964). Taxonomic studies in the Dacrymycetaceae. I. *Cerinomyces* Martin. *N.Z. J. Bot.* **2**:415–424.

McNabb, R. F. R. (1965a). Taxonomic studies in the Dacrymycetaceae. II. *Calocera* (Fries) Fries. *N.Z. J. Bot.* **3**:31–58.

McNabb, R. F. R. (1965b). Taxonomic studies in the Dacrymycetaceae. III. *Dacryopinax* Martin. *N.Z. J. Bot.* **3**:59–72.

McNabb, R. F. R. (1965c). Taxonomic studies in the Dacrymycetaceae. IV. *Guepiniopsis* Patouillard. *N.Z. J. Bot.* **3**:159–169.

McNabb, R. F. R. (1965d). Taxonomic studies in the Dacrymycetaceae. V. *Heterotextus* Lloyd. *N.Z. J. Bot.* **3**:215–222.

McNabb, R. F. R. (1965e). Taxonomic studies in the Dacrymycetaceae. VI. *Femsjonia* Fries. *N.Z. J. Bot.* **3**:223–228.

McNabb, R. F. R. (1966). Taxonomic studies in the Dacrymycetaceae. VII. *Ditiola* Fries. *N.Z. J. Bot.* **4**:546–558.

Martin, G. W. (1952). Revision of the North Central Tremellales. *Stud. Nat. Hist. Iowa Univ.* **19**: 1–122.

Martin, G. W. (1957). The tulasnelloid fungi and their bearing on basidial terminology. *Brittonia* **9**:25–30.

Nannfeldt, J. A. (1947). *Sphaeronema rufum* Fr., a misunderstood member of Dacrymycetaceae. *Sv. Bot. Tidskr.* **41**:321–338.

Neuhoff, W. (1936). Die Gallertpilze Schwedens (Tremellaceae, Dacrymycetaceae, Tulasnellaceae, Auriculariaceae). *Ark. Bot. A* **28**:1–57.

Olive, L. S. (1945). A new *Dacrymyces*-like parasite of *Arundinaria*. *Mycologia* **37**:543–552.

Olive, L. S. (1951). Taxonomic notes on Louisiana fungi. III. Additions to the Tremellales. *Mycologia* **43**:677–90.

Olive, L. S. (1957a). Two new genera of the Ceratobasidiaceae and their phylogenetic significance. *Amer. J. Bot.* **44**:429–435.

Olive, L. S. (1957b). Tulasnellaceae of Tahiti. A revision of the family. *Mycologia* **49**:663–679.

Raitviir, A. G. (1967). "A Key to the Heterobasidiomycetidae Found in the U.S.S.R." Nauka, Leningrad (in Russian).

Rajendrên, R. B. (1968). *Muribasidiospora*—a new genus of the Exobasidiaceae. *Mycopathol. Mycol. Applic.* **36**:218

Reid, D. A. (1969). New or interesting British plant diseases. *Trans. Brit. Mycol. Soc.* **52**: 19–38.

Rogers, D. P. (1933). A taxonomic review of the Tulasnellaceae. *Ann. Mycol.* **31**:181–203.

Rogers, D. P. (1935). Notes on the lower Basidiomycetes. *Stud. Nat. Hist., Iowa Univ.* **17**: 1–43.

Rogers, D. P. (1943). The genus *Pellicularia* (Thelephoraceae). *Farlowia* **1**:95–118.

Savile, D. B. O. (1959a). Notes on *Exobasidium*. *Can. J. Bot.* **37**:641–656.

Savile, D. B. O. (1959b). The botany of Somerset Island, District of Franklin. *Can. J. Bot.* **37**:959–1002.

Sundström, K-R. (1960). Physiological and morphological studies of some species belonging to the genus *Exobasidium*. *Phytopathol. Ze.* **40**:213–217.

Sundström, K-R. (1964). Studies of the physiology, morphology and serology of *Exobasidium*. *Symb. Bot. Upsal.* **18**:1–89.

Talbot, P. H. B. (1965). Studies of "*Pellicularia*" and associated genera of Hymenomycetes. *Persoonia* **3**:371–406.

Talbot, P. H. B. (1968). Fossilized pre-Patouillardian taxonomy? *Taxon* **17**:620–628.

Talbot, P. H. B. (1970). The controversy over septate holobasidia. *Taxon* **19**:570–572.

CHAPTER 19

Aphyllophorales I: General Characteristics; Thelephoroid and Cupuloid Families

P. H. B. TALBOT

Waite Agricultural Research Institute
University of Adelaide
Glen Osmond, South Australia

I. ORDER APHYLLOPHORALES

A. General Characteristics

Aphyllophorales are mostly saprobes on soil, wood, and litter; if they are internal parasites they then form a distinct external basidiocarp with a hymenophore and supporting tissues. The basidiocarp is basically gymnocarpous, appressed, effused, effuso-reflexed, sessile-pileate, stipitate-pileate, discoid, cupulate, clavate or coralloid with radial or flattened branching. The hymenophore is unilateral or amphigenous, smooth, dentate, plicate (with fertile edges to the folds) or tubular, with discrete tubes, or coalesced and firmly united with a woody, corky or coriaceous-membranous (not fleshy) context. The hymenium is a more-or-less thickening euhymenium or a catahymenium, often with associated sterile hyphidia, cystidia, gloeocystidia, setae or gloeovessels. The basidia are holobasidiate, and rarely have adventitious septa in the metabasidium or sterigmata; they also rarely proliferate in apical linear succession. The sterigmata are typically curved, slender, tapering, rarely digitate, and straight and stout. The basidiospores are forcibly discharged and not repetitive (if they are repetitive and holobasidiate, see Tulasnellales), hyaline or colored, smooth or ornamented, sometimes amyloid, dextrinoid or cyanophilous, very rarely septate, and thin- or thick-walled. The hyphae are combined in monomitic, dimitic, or trimitic systems, and are hyaline or colored, with or without clamp connections, rarely amyloid, dextrinoid or cyanophilous.

This order comprises the Hymenomycetes of Fries (1874) with the exclusion of phragmobasidiate fungi, Dacrymycetales, Tulasnellales, Agaricales, and the biophilous parasitic fungi such as the Exobasidiales and Brachybasidiales. The Friesian classification stressed field characters, particularly the shape of the basidiocarp and the configuration and orientation of the hymenophore. We owe to Patouillard (1900) the major part of our present system based on basidial morphology, microscopic analysis, and an appreciation that genera with diverse hymenial configuration could well be placed in more natural series than the families of Fries. Patouillard's system was adopted by Bourdot and Galzin (1928), whose subgenera based largely on textural features have since formed the nuclei of many newly proposed genera. Corner (1932a,b) introduced hyphal analysis into the taxonomy of the Aphyllophorales and has since repeatedly demonstrated the artificiality of a classification that gives too little attention to the problems posed by homoplasy. Studies of enzymes produced in culture, and of the cytology of mycelia and fructifications, received great impetus from the work of Boidin (1958a). While many authors have contributed generic revisions, and a few have proposed new families cutting across traditional lines, there was not until 1964 (Donk, 1964) any overall conspectus of the Aphyllophorales; this work, and that of Parmasto (1968), together list the most important literature references to genera of Aphyllophorales exclusive of the Polyporaceae. Families are now defined by several macroscopic and microscopic characters in combination, instead of by a few obvious but deceptive external features. The more stable microscopic characters are especially important. What formerly defined a family may now be common to several families and thus lessened in value when considered alone. The wide range of variability and overlap in families precludes the construction of short keys, and a truly dichotomous separation is all but impossible. Disagreement in matters of terminology is a special problem. For terminology of basidia compare Neuhoff (1924), Rogers (1934), Donk (1931, 1954, 1958, 1964), Martin (1938, 1957), and Talbot (1954, 1968). For terminology of cystidia and other sterile structures compare Romagnesi (1944), Lentz (1954), Talbot (1954), Singer (1962), Donk (1964), and Smith (1966).

KEY TO FAMILIES OF APHYLLOPHORALES

1. Spores colored, often apically truncated, with an outer hyaline layer pierced by spines from an inner brown layer; hymenophore tubular; basidiocarp pileate, dimidiate or stipitate; trimitic; with clamp connections . . . . **Ganodermataceae** p. 401

The Thelephoraceae as circumscribed by Donk (1964) are here dispersed. *Thelephora* and *Scytinopogon* (hymenophore smooth), see pp. 354–360; *Hydnellum*, *Hydnodon*, *Sarcodon*, and *Polyozellus* (hymenophore more or less toothed), see pp. 377–379; *Boletopsis* and *Lenzitopsis* (hymenophore more or less poroid), see pp. 400–418. The three genera included by Donk unaccounted for elsewhere are those in which the fruitbody is strictly effused: *Tomentella* (including *Pseudotomentella* and *Tomentellastrum*) and *Kneiffiella* P. Karst. (hymenophore smooth or warty, *K.* being distinguished from *T.* by the protruding hyphaelike cystidia) and *Caldesiella* (hymenophore has teeth with sterile tips).

11′(10) Not combining these characters 12

   12(11′) Spores smooth or minutely sculptured; hymenophore dentate or lamellate (lacerate-dentate); gloeocystidia darkening in sulfoaldehyde; generative hyphae thin-walled, clamped . . . . **Auriscalpiaceae**   p. 286

   12′(11′) Not combining these characters 21

13(10′) Basidiocarp pileate or clavarioid; hymenophore smooth or dentate; spores smooth or minutely asperulate; gloeocystidia not darkening in sulfoaldehyde; generative hyphae thin-walled or thick-walled, with clamps . . . . **Hericiaceae**   p. 352

13′(10′) Without these combined characters 14

   14(13′) Hymenophore dentate; corky or woody; generative hyphae thin-walled to thick-walled and clamped (pseudodimitic with some generative hyphae resembling skeletal hyphae); cystidia present, encrusted or not; gloeocystidia absent; spores smooth, hyaline . . . . **Echinodontiaceae**   p. 376

   14′(13′) Without these combined characters 21

15(9′) Hymenophore dentate; basidiocarp stipitate-pileate 16

15′(9′) Hymenophore not dentate or if somewhat hydnoid then the basidiocarp not stipitate-pileate, rarely with pseudodentate sterile emergent fascicles of hyphae 17

     16(15) Generative hyphae clamped, thin-walled; spores smooth
     . . . . **Hydnaceae**   p. 376

     16′(15) Generative hyphae without clamps, thin-walled or thick-walled; spores echinulate
     . . . . **Bankeraceae**   p. 377

17(15′) Basidiocarp infundibuliform or tubular, fleshy or membranous to coriaceous; generative hyphae thin-walled, inflating; hymenophore smooth, wrinkled or plicate; spores smooth, hyaline . . . . **Cantharellaceae**   p. 365

17′(15′) Not combining these characters 18

   18(17′) Basidia regularly 2-spored; sterigmata generally stout, strongly curved; fruitbody clavarioid; generative hyphae inflating . . . . **Clavulinaceae**   p. 355

   18′(17′) Not combining these characters 19

19(18′) Basidiocarp clavarioid, erect, simple or branched; hymenophore amphigenous, smooth or becoming wrinkled . . . . **Clavariaceae**   p. 352

19′(18′) Not combining these characters 20

   20(19′) Basidiocarp erect, branching into wavy or flattened lobes; hymenophore smooth; hymenium inferior on horizontal parts of lobes or amphigenous on erect surfaces; monomitic, with inflating hyphae and vascular hyphae
   . . . . **Sparassidaceae**   p. 359

   20′(19′) Not with these features 21

21 (12′ and 20′) Hymenophore tubular, irpicoid or rarely radially or concentrically lamellate; edges of dissepiments sterile; fruitbody effused, reflexed, dimidiate to stipitate-pileate, never clavarioid, sometimes perennial; monomitic, dimitic or trimitic; hyphae hyaline to brown, with or without clamps; setae absent; spores hyaline or cream-colored, rarely ornamented, rarely pseudoamyloid or amyloid . . . . **Polyporaceae**   p. 401

21′ (12′ and 20′) Not combining these characters 22

   22(21′) Fruitbody effused, spathulate or clavarioid, dimitic, usually with dextrinoid pale yellow-brown thick-walled dichohyphidial binding hyphae or asterosetae; cystidia or gloeocystidia often present; spores hyaline, smooth, verrucose or echinulate, amyloid or not . . . . **Lachnocladiaceae** p. 342

22′(21′) Without such binding dichohyphidia or asterosetae; monomitic, dimitic or rarely trimitic 23

23(22′) Basidiocarp strictly effused or discoid to patelliform; hymenophore smooth, merulioid (with fertile edges to the folds), tuberculate, dentate (with sterile tips to the teeth) or pseudo-dentate (with sterile emergent fascicles of hyphae, originating deep in the context); monomitic or dimitic with skeletal hyphae; without setae; with or without clamps; various sorts of cystidia, gloeocystidia and hyphidia present or absent; spores usually hyaline or bright-colored, smooth or ornamented, amyloid or not, even in outline.
. . . . **Corticiaceae** pp. 332, 401

23′(22′) Basidiocarp not strictly effused; not combining the above characters 24

24(23′) Basidiocarp spathulate, infundibuliform, pseudoinfundibuliform or merismatoid, distinctly stipitate; hymenophore smooth, tuberculate or radially ribbed; monomitic, dimitic or trimitic, with narrow generative hyphae, rarely inflating; cystidia or gloeo-cystidia often present; spores smooth, hyaline, not amyloid
. . . . **Podoscyphaceae** p. 342

24′(23′) Basidiocarp effuso-reflexed, flabellate, conchate, dimidiate or resupinate; hymeno-phore smooth to rugulose or tuberculate, exceptionally irpicoid; dimitic with skeletal hyphae, or exceptionally trimitic; usually with a dark cortex and often a trichoderm on the abhymenial side, an intermediate medullary layer, and the hymenium; setae absent; with or without clamps; cystidia or gloeocystidia often present; spores hyaline, smooth, thin-walled, amyloid or not . . . . **Stereaceae** p. 345
[From Talbot, "Principles of Fungal Taxonomy" (1971), with permission of Macmillan and Co.]

II. CONIOPHORACEAE

Important Literature

Nannfeldt and Eriksson (1953); Cooke (1957); Lentz (1957); Pouzar (1958); Donk (1964); Parmasto (1968).

KEY TO GENERA OF CONIOPHORACEAE

1. Hymenophore smooth to tuberculate 2

1′. Hymenophore variously convoluted, not smooth when fresh 5

2(1) Without cystidia; corticioid . . . . **Coniophora**
Coniophora puteana causes dry-rot of structural timber.

2′(1) With cystidia; corticioid or hypochnoid 3

3(2′) Cystidia brown, septate, without clamps . . . . **Coniophorella** (Fig. 10)

3′(2′) Cystidia hyaline 4

4(3′) Cystidia aseptate; basidia long, clavate-cylindrical, sinuous, often somewhat utri-form; spores sometimes with intermembranal space between wall layers at one or both ends . . . . **Jaapia**
Syn.: Coniobotrys.

4′(3′) Cystidia with clamped septa; basidia short, barrel-shaped to obovate; spores sub-hyaline without an intermembranal space . . . . **Suillosporium**

5(1′) Spores small, hyaline or subhyaline, doubtfully cyanophilous 6

5′(1′) Spores moderately large to large, rusty, ochraceous or brown, cyanophilous 7

    6(5) Basidiocarp erect, bearing several superimposed pilei at intervals on a common simple or branched stipe; hymenophore radially or irregularly rugose, verrucose or pustulate . . . . **Podoserpula**
        Doubtful as Coniophoraceae.

    6′(5) Basidiocarp resupinate or with revolute margin; hymenophore merulioid; rhizomorphic strands common . . . . **Leucogyrophana**
        Doubtful as Coniophoraceae; ? Corticiaceae.

7(5′) Hymenophore merulioid, alveolate or gyrose-plicate . . . . **Serpula**
    Syn.: *Gyrophana*; *Gyrophora*; *Merulius* pro parte; *Meruliporia*; *Sesia*; *Xylomyzon*; *Xylophagus*. *Serpula lacrimans* causes dry-rot of structural timber.

7′(5′) Hymenophore hydnoid, with elongated or fasciculate spines, or warted
. . . . **Gyrodontium**

III.   CORTICIACEAE

Important Literature

Donk (1964); Parmasto (1968).

KEY TO GENERA OF CORTICIACEAE

1. Marine fungus; basidia narrow clavate; sterigmata developing into long apically branched diaspores . . . . **Digitatispora**

1′. Not marine, nor with such spores 2

    2(1′) Catahymenial; dendrophyphidia, acanthohyphidia, moniliform or gloeocystidial pseudohyphidia, or skeletal paraphysoids (cystidiohyphidia) present alone or in combination 3

    2′(1′) Euhymenial; mostly lacking well-developed hyphidia but sparse dendrohyphidia or simple hyphidia sometimes present; gloeocystidia present or not 13

3(2) Spores amyloid, smooth or asperulate to aculeate 4

3′(2) Spores not amyloid, smooth 8

    4(3) Basidia pleurobasidiate, constricted in the middle, aculeate below; acanthohyphidia present; spores with warts soluble in KOH . . . . **Acanthobasidium**

    4′(3) Basidia not pleurobasidiate, rarely aculeate 5

5(4′) Acanthohyphidia present; spores aculeate or asperate, rarely smooth, usually large, ellipsoid or globose, rarely cylindrical; on living trees . . . . **Acanthophysium**
    Acanthophysellum with smooth, cylindric, allantoid or ellipsoid, generally small spores, occurring on dead wood and bark, may be congeneric.

5′(4′) Without acanthohyphidia 6

    6(5′) Dendrohyphidia nil; apically encrusted skeletal paraphysoids (cystidiohyphidia) present; spores verrucose . . . . **Aleurocystidiellum**

    6′(5′) Dendrohyphidia present 7

7(6′) Dendrohyphidia abundant, encrusted, yellowish; gloeocystidial pseudohyphidia nil;

spores smooth . . . . **Dendrophysellum**

7'(6') Dendrohyphidia often poorly differentiated; pseudohyphidia present; spores usually aculeate or asperate, rarely smooth . . . . **Aleurodiscus**
  Syn.: *Nodularia*; *Gloeosoma*; *Cyphella*.

  8(3') Basidiocarp discoid to patelliform or shallow cupulate, coriaceous to membranous-gelatinous, drying horny 9

  8'(3') Basidiocarp effused or effuso-reflexed 10

9(8) Metuloid cystidia present, fusoid, thick-walled, encrusted; spores broad ellipsoid, ovoid or subglobose . . . . **Aleurocystis**
  Syn. *Aleurocystus*; *Matula nom. anam.*

9'(8) Metuloids absent; spores cylindrical to allantoid; dendrohyphidia leaden-colored in sulfovanillin . . . . **Cytidia**
  Syn.: *Lomatia*; *Lomatinia.*

  10(8') Basidia with a vesicular base, long slender neck of variable length, and an expanded apex, often showing apical linear proliferation; sterigmata (2-)4 . . . . . . **Galzinia**

  10'(8') Basidia not so 11

11(10') Effused or narrowly reflexed, pale but vividly colored (rose, lilac) when fresh, ochraceous or greyish dry; dendrohyphidia nodulose, thin-walled; cystidia and gloeocystidia absent; basidia dimerous, often ventricose at base . . . . **Laeticorticium**
  Syn.: *Lyomyces*; *Minnsia nom. anam.*; *Hyphelia nom. anam.*

11'(10') Effused, white or pale colored (not vivid) 12

  12(11') Hymenium with sterile dentate papillae composed of dendrohyphidia; fruitbody crustose to woody . . . . **Dendrothele**
  Syn.: *Aleurocorticium.*

  12'(11') Hymenium without such papillae; dendrohyphidia abundantly encrusted; fruitbody ceraceous-corneous to subsuberose; basidia usually clavate-utriform, small to very large . . . . **Vuilleminia**

13(2') With sterile fascicles of agglutinated hyphae usually originating deep in the context or from basal hyphae, usually traversing the context and emerging through the hymenium to form discrete spines or papillae or linear ridges or subporoid networks, seldom not emergent 14

13'(2') Without such hyphal fascicles 18

  14(13) Hyphal fascicles emergent as small discrete spines or papillae . . . . . . **Epithele**

  14'(13) Hyphal fascicles united linearly to form fine lines or shallow subporoid networks above the hymenium 15

15(14') Context containing long cylindrical gloeocystidia arising from basal hyphae and becoming secondarily septate . . . . **Gloiothele**

15'(14') Gloeocystidia nil 16

  16(15') Hyphal fascicles surrounded by large crystals, immersed or emergent as ridges or tubercles, not subporoid . . . . **Grammothele**

  16'(15') Hyphal fascicles not encased in crystals, forming linear ridges then becoming subporoid 17

17(16') Pseudopores minute, isodiametric . . . . **Porogramme**

17'(16') Pseudopores elongated . . . . **Hymenogramme**

18(13') Bi- or multiradicate lyocystidia present, usually dissolving or distorting in KOH, originating laterally from basal hyphae, sometimes encrusted with minerals or sheathed with climbing hyphae; if such cystidia are absent then the basidia are pleurobasidiate with 2–4 spores (very exceptionally 5) 19

18'(13') Such cystidia always absent; basidia regularly with 4–8 spores if pleurobasidiate 24

19(18) Lyocystidia absent; basidia pleurobasidiate with 2–4(-5) spores 20

19'(18) Lyocystidia present; basidia pleurobasidiate, podobasidiate or clavate 21

20(19) Basidia bisporous; spores angular-pyramidical, not amyloid . . . . **Xenosperma**

20'(19) Basidia not bisporous; spores not pyramidical, smooth or ornamented, amyloid or not . . . . **Xenasmatella** (Fig. 3)

21(19') Lyocystidia conical, with a very narrow lumen not or hardly widening at the apex 22

21'(19') Lyocystidia cylindrical, thin-walled with a wide lumen or thick-walled with an abruptly expanded lumen at the apex, very rarely conical and then thin-walled 23

22(21) Basidia podobasidiate to clavate; cystidia amyloid or pseudoamyloid . . . . **Tubulixenasma** (Fig. 5)
   Syn.: *Tubulicium.*

22'(21) Basidia pleurobasidiate; cystidia not amyloid or pseudoamyloid . . . . **Litschauerella**

23(21') Basidia pleurobasidiate with 4 or more sterigmata; spores warted, the warts soluble in KOH; capitate cystidioles sometimes present . . . . **Xenasma**

23'(21') Basidia clavate with 4 or fewer sterigmata; spores smooth; cystidial wall often more or less amyloid . . . . **Tubulicrinis**

24(18') Spores repetitive; sterigmata large, digitate, mostly straight, often subfusoidly swollen about the middle, variable in length . . (see Tulasnellales, Ceratobasidiaceae)

24'(18') Spores not repetitive; sterigmata curved or straight, not digitate nor inflated, not especially variable in length except in *Cejpomyces* 25

25(24') Basidia not constricted about the middle, short and wide, subglobose, obovoid, obpyriform, obconical (rarely oblong or barrel-shaped and then pleurobasidiate); fruitbody thin, hardly visible 26

25'(24') Basidia either constricted (utriform or urniform) or if not then barrel-shaped (but not pleurobasidiate) or short to long cylindrical, or clavate 29

26(25) Basidia formed in apical linear succession within empty wall of preceding basidium 27

26'(25) Basidia not formed thus 28

27(26) Cystidia small, thin-walled, septate, usually with clamps at the septa; sterigmata 4 . . . . **Repetobasidium** (Fig. 1)

27'(26) Cystidia long, subulate, not septate, firm-walled, with discrete, regularly-arranged, flattened, elongated crystals; sterigmata 2–4 . . . . **Subulicystidium**

28(26') Sterigmata more than 4; basidia obovate, obconical, obpyriform (rarely barrel-shaped and pleurobasidiate); cystidia nil . . . . **Paullicorticium**

28'(26') Sterigmata 4; basidia subglobose; small thin-walled leptocystidia present or absent . . . . **Sphaerobasidium**

29(25') Basidiocarp pruinose, tufted, hypochnoid to submembranous, white or yellowish to isabelline; basal hyphae (often scanty) wide, often reaching 10 μm or more, often thick-walled and colored; ascending hyphae short-celled, not ampoullaceous, branched at a wide angle oppositely or subcymosely, staining strongly in aniline blue; basidia in terminal discontinuous subcymose clusters, rarely continuous 30

29'(25') Not so 33

     30(29) Spores hyaline, becoming yellow-brown, with short to long conical or cylindrical obtuse spines; basidia not constricted; sterigmata (2-)4; no sclerotial or conidial states known . . . . **Botryohypochnus**

     30'(29) Spores smooth (rarely minutely asperulate ?), hyaline 31

31(30') Basidia clavate, not constricted about the middle; sterigmata 4, stout and digitate at first, later elongating to a variable length; spores navicular, developing 1–3 transverse septa when mature . . . . **Cejpomyces**

31'(30') Basidia mostly constricted, suburniform; sterigmata 4 or more, not especially stout nor elongating to a variable length 32

     32(31') Subbasidial branches frequently involute or circinate; sterigmata 4; spores oblong to broad ellipsoid and flattened (rarely 1–2 septate ?); cystidia absent; hyphae not clamped; no known conidial states but orange, rosy or brownish sclerotia present in culture . . . . **Waitea**

     32'(31') Subbasidial branches typically cymose, not involute or circinate; sterigmata (4–)6(–8); spores amygdaliform, fusoid or subnavicular; cystidia present or absent; hyphae clamped or not; sclerotia absent; conidial state, when present, *Oidium*
     . . . . **Botryobasidium**
       Syn.: *Oidium* Fr. *em.* Linder, *nom. anam.*

33(29') Basidia either urniform (swollen probasidium emits a short subcylindrical tube flattening and expanding at the apex into a corona of 6–8 sterigmata; stichic), or if not urniform then with more than 4 sterigmata 34

33'(29') Basidia either utriform (similar to urniform but the corona often narrower than the probasidium and with 4 or fewer sterigmata; chiastic), or clavate (probasidium narrow-clavate, elongating as a whole without a contrasting proliferation) with 4 or fewer sterigmata
. . . . 36

     34(33) Basidia distinctly urniform 35

     34'(33) Basidia not urniform, without basal swelling, narrowly clavate to evenly tubular and apically truncate; sterigmata 6–8; hyphae clamped, not ampoullaceous
     . . . . **Sistotremastrum**

35(34) Spores smooth; hyphae clamped, often ampoullaceous; basidiocarp corticioid, grandinioid, hydnoid or poroid; gloeocystidia present or absent . . . . . **Sistotrema** (Fig. 2)
     Syn.: *Heptasporium.*

35'(34) Spores strongly echinulate, globose, hyaline; hyphae with rare clamps, not ampoullaceous; basidiocarp resupinate; hymenophore of parallel sinuous plates; cystidial structures absent . . . . **Echinotrema**

     36(33') With smooth or asperulate, amyloid to pseudoamyloid spores and gloeocystidia
     37

     36'(33') Without this combination of characters 38

37(36) Strictly resupinate; hymenophore smooth, tuberculate or rarely finely dentate; spores

asperulate, rarely smooth; cystidia sometimes present, thick-walled, encrusted; hyphae not especially loosely woven; gloeocystidia reacting or not with sulfoaldehyde
. . . . **Gloeocystidiellum**

37'(36) Effused, effuso-reflexed to dimidiate; abhymenial surface brown, sulcate, tomentose; hymenophore smooth; spores asperulate in Melzer reagent; cystidia nil; hyphae very loosely intertexed; gloeocystidia not reacting with sulfoaldehyde; chlamydospores present in mycelium. . . . . **Laxitextum**
Hericiaceae rather than Corticiaceae?

38(36') Whole basidiocarp or at least the hymenophore mucoid, waxy or subgelatinous in texture, becoming hard and brittle (usually vernicose or horny, rarely cartilaginous) on drying; hymenophore smooth, radially rugose, plicate or merulioid-poroid, or dentate; hyphae rather indistinct, often nodulose, often with gelatinized walls; basidia very thin-walled, densely arranged, narrowly clavate to cylindrical, long; spores never brightly colored in a print 39

38'(36') Texture byssoid, pellicular, soft- or corky-membranous, coriaceous or crustose, not becoming vernicose, corneous or cartilaginous on drying; hymenophore smooth, odontioid, raduloid, if plicate to merulioid then also pellicular and often separable as a layer from the context; hyphae not gelatinized nor nodulose, distinct or not; basidia short or long-clavate, frequently more or less utriform; spores sometimes brightly colored in a print 51

39(38) Basidia with a slender base and an inflated cylindrical apex, stichic, with (1–)2(–3) sterigmata; fruitbody corticioid, waxy, adherent, with smooth hymenophore; hyphae thin-walled, clamped; spores spherical, multiguttulate, not amyloid . . . . **Clavulicium**
Possibly Clavulinaceae?

39'(38) Not so 40

40(39') Basidiocarp not stratose, or only 2-layered in section with the basal hyphae parallel and agglutinated; resupinate 41

40'(39') Basidiocarp stratose, 3-layered, usually with a thick velutinate or cottony ab-hymenial layer, a gelatinized middle layer and the hymenium; resupinate-reflexed or sessile-pileate. 45

41(40) With thick- or thin-walled cylindrical cystidia arising terminally from basal hyphae, not radicate, emergent from hymenium, distorted by KOH, not amyloid; hymenophore smooth to odontioid-hydnoid, sometimes exuding an opalescent droplet from apices of denticles, becoming areolate on drying . . . . **Dacryobolus**

41'(40) Not so 42

42(41') Cystidia thin-walled, subcylindrical or fusoid, constricted near the capitate resin-covered apex; hymenophore granulose or with short denticles, whitish to grayish-yellow . . . . **Resinicium**

42'(41') Not so 43

43(42') Hymenophore smooth, radiately rugose or plicate or reticulate, rarely tuberculate, whitish or often notably colored (orange, red, vinous, purple, leaden); cystidia absent or if present emergent and thin-walled at first, or becoming embedded and thick-walled
. . . . **Phlebia**
Syn.: *Ricnophora.*

43'(42') Hymenophore nodulose, tuberculate or strongly denticulate, typically yellow to pale rosy-cream; cystidia few or absent 44

44(43') Denticles well developed, entire, commonly yellow to orange . . . . **Mycoacia**

44′(43′) Nodulose-tuberculate with pale sulfur- or rose-colored fasciculate denticles; mycelium sulfur-colored . . . . **Sarcodontia**

45(40′) Basidiocarp patelliform to cupulate   46

45′(40′) Basidiocarp resupinate, effuso-reflexed to sessile-pileate (conchate, applanate, sub-flabellate) 47

46(45) Patelliform, centrally attached, adherent with revolute tomentose margin, later coalescent; hymenium smooth to slightly dentate, fawn, yellow-brown or gray-brown; context hyphae not or only slightly gelatinized . . . . **Cytidiella**

46′(45) Cupulate then often flattening; exterior tomentose; hymenium flesh-colored to brown, smooth or somewhat radiately veined; context hyphae rather strongly gela-tinized . . . . **Auriculariopsis**

47(45′) With vesicular gloeocystidia in context, elongated near the hymenium; fruitbody effuso-reflexed, stereoid; tomentum of clamped thick-walled hyphae; hymenium more or less smooth, violet, purplish or reddish-brown . . . . **Chondrostereum**
    Chondrostereum purpureum, cause of silverleaf disease of fruit trees.

47′(45′) Without vesicular gloeocystidia 48

48(47′) Resupinate or (exceptionally) narrowly reflexed; hymenophore poroid with sterile dissepiments, then becoming somewhat plicate; hyphae without or with rare clamps . . . . **Meruliopsis**
    Syn.: *Merulioporia*; ? *Caloporus*; ? *Caloporia*.

48′(47′) If resupinate then also widely reflexed, otherwise sessile-pileate 49

49(48′) Hymenophore smooth becoming somewhat plicate, rugose or subtuberculate; thin-walled, smooth emergent leptocystidia present; fruitbody sessile, orbicular then conchate to subflabellate, sometimes imbricate, gelatinous . . . . **Gloeostereum**

49′(48′) Hymenophore radiately plicate, reticulate-poroid, or poroid; leptocystidia absent or not well-developed 50

50(49′) Hymenophore with small rounded pores, usually gray, vinous or yellowish; pilei sessile, solitary or imbricate, conchate, applanate or effuso-reflexed
. . . . **Gloeoporus**

50′(49′) Hymenophore radiately plicate or reticulate-poroid; fruitbody effuso-reflexed or pileate . . . . **Merulius**

51(38′) Spores amyloid or pseudoamyloid 52

51′(38′) Spores not amyloid or pseudoamyloid 54

52(51) Resupinate then effuso-reflexed to sessile-dimidiate or laterally substipitate; hymenophore partly smooth, partly aculeate or with irpicoid flattened sinuous plates; cystidia absent; texture soft ceraceous (*Irpicodon*) or coriaceous . . . . . **Plicatura**

52′(51) Remaining resupinate or with only a revolute margin; cystidia present or absent; hymenophore smooth or merulioid 53

53(52′) Effused; pellicular to submembranous and separable; hymenophore smooth; cystidia sometimes present; spores ovoid or cylindrical . . . . **Amylocorticium**

53′(52′) Effused, sometimes with revolute velutinate margin; hymenophore merulioid, waxy; cystidia absent; spores subglobose . . . . **Leucogyrophana**
    ? Coniophoraceae.

54(51′) Hyphae with strongly ampoullaceous swellings at clamps or septa 55

54′(51′) Hyphae not ampoullaceous 58

65(64) Hymenium tufted or granulose; spores yellowish-olive viewed in water, deep violet in alkali, smooth, rather thick-walled . . . . **Hypochnopsis**

65'(64) Hymenium with subincised spines to 2 mm long; (spores controversial, said to be deep violet when fresh, smooth; also said to be aculeolate.) (Possibly not different from *Hypochnopsis*) . . . . **Amaurodon**

66(64') Spores deep violet, smooth, elliptical; hymenium violaceous or lilac becoming gray-brown or chocolate when dry; fruitbody very thin, hypochnoid to soft membranous; cystidioles short, obtuse, often with scanty crystals; hyphae rather wide (5–9 μm), without clamps (at least in basal hyphae); subhymenial hyphae often slightly encrusted with small rounded crystals . . . . **Hypochnella**

66'(64') Spores not violet but pink, salmon or lilac in a mass; fruitbody fleshy or coriaceous to densely agglutinated in texture, not very thin; cystidia and/or gloeocystidia often present 67

67(66') Pseudosetae abundantly formed in palisade, cylindrical, dark brown, cyanophilous, with paler often rugose apex, rarely branching near apex; large ventricose or fusoid gloeocystidia present or not; hyphal system monomitic or dimitic with skeletals, with or without clamps, most hyphae brown; hymenium ochraceous, fulvous to chestnut . . . **Duportella**

67'(66') Pseudosetae absent; metuloid cystidia (lamprocystidia), gloeocystidia, and dendrohyphidia commonly present in various combinations; hyphae usually densely agglutinated, hyaline to brown, clamps present or absent; hymenium rosy, orange to reddish, lilac, slate-gray, cinereous, or purple-brown; spores ellipsoid to cylindrical . . . . **Peniophora**
Syn.: *Cryptochaete. Sterellum*, gelatinous, with (?) white spores, a possible synonym.

68(61') Spores pyramidical, triangular in outline, angular to nodulose; cystidia, gloeocystidia and hyphidia absent; hyphae clamped; basidia clavate, 2–4-spored; fruitbody thin, whitish, felted, with fibrillose margin and smooth hymenium . . . . **Tylospora**
Syn.: *Tylosperma.*

68'(61') Spores not so 69

69(68') Cystidia long (to 120 μm), narrow (to 7 μm), thin- but firm-walled, subulate, aseptate, with discrete, regularly-arranged elongated flattened tuberculate crystals; base of cystidium the same width as the hyphae or somewhat ventricose, arising laterally or terminally; hyphae hyaline, regularly clamped; basidia sometimes showing apical linear proliferation; fruitbody white, arachnoid to cottony-membranous . . . . **Subulicystidium**

69'(68') Without such cystidia; basidia not proliferating apically 70

70(69') Spores thick-walled, smooth or often punctate, or finely verrucose or asperulate, subglobose, broad ellipsoid or obovate; without cystidia or with thin-walled fusoid leptocystidia or gloeocystidia; basidia subutriform; hyphae distinct, clamped . . . . **Hypochnicium**

70'(69') Spores thin-walled but sometimes with a large central guttule, smooth . . . 71

71(70') Basidiocarp small, pulvinate, thick (to 2 mm), stratose, pallid buff; hymenium smooth; hyphae usually thick-walled, clamped; cystidia subcylindrical, aseptate, large (80–300 × 6–12 μm), not encrusted, with thickened walls; spores cylindrical to fusoid, somewhat curved, large (10–21 μm long) . . . . **Chaetoderma**

71'(70') Without these combined characters 72

72(71') Cystidia (metuloids, lamprocystidia) clavate or fusoid-conical, thick-walled with a narrow lumen expanding towards the apex, encrusted; fruitbody usually perennial, white to brownish; hymenium smooth to short-aculeate or verrucose; hyphae usually branching from clamps; spores ellipsoid, rarely subcylindrical, small (3–6 μm long)

. . . . **Metulodontia**
    Syn.: *Scopuloides*; ? *Grandiniella*.

72'(71') Cystidia absent or, if present, either thin-walled or firm-walled with a wide lumen,
    smooth or encrusted 73

73(72') Context hyphae loosely intertexed, distinct, not conglutinate; fruitbody usually loosely
    adherent, arachnoid, byssoid, floccose or pellicular, rarely membranous and then easily
    separable 74

73'(72') Context hyphae conglutinate, or distinct but densely intertexed and compact; fruitbody
    more or less firmly adherent, membranous, coriaceous, ceraceous, fleshy or crustose . . 81

    74(73) Hymenophore becoming tuberculate to dentate (hydnoid); hymenium yellow
        75

    74'(73) Hymenophore remaining smooth or becoming merulioid-porose; hymenium
        of various colors 76

75(74) Septate hyphocystidia present, not clamped; margin fibrillose or indistinct; hyphae
    branched dichotomously or verticillately, without clamps; basidia clavate; spores cylindrical,
    curved . . . . **Odonticium**

75'(74) Cystidia absent; margin with yellow rhizomorphs; hyphae with rare clamps, granulose
    encrusted; basidia clavate; spores broad ellipsoid . . . . **Hydnophlebia**

    76(74') Hymenophore becoming merulioid-porose and persisting thus on drying, cream,
        yellow or reddish; usually without clamps; cystidia absent or thin-walled; spores
        ellipsoid to cylindrical; basidiocarp quite thick (to 2 mm), effused or effuso-reflexed
        . . . . **Byssomerulius**

    76'(74') Hymenophore smooth, or merulioid-porose only when fresh and moist . . . 77

77(76') Basidiocarp pellicular 78

77'(76') Basidiocarp byssoid to submembranous   80

    78(77) Spores thin-walled, hyaline; hyphae usually with clamps, rarely without; basidia
        with (2–)4 sterigmata 79

    78'(77) Spores with thickened, often yellowish, walls; hyphae always without clamps;
        basidia with (2–)4 or rarely 6 sterigmata . . . . **Piloderma**

79(78) Hymenophore often merulioid when fresh; spores ovate, pip-shaped, ellipsoid or sub-
    globose; hyphae with wide-angled branching; clamps rarely absent; septate, clamped
    hyphocystidia sometimes present; sclerotia sometimes present . . . . . **Athelia** (Fig. 9)
        Sclerotium rolfsii is an important plant pathogenic sclerotial state of an *Athelia*
        species.

79'(78) Hymenophore smooth; spores cylindrical; hyphae clamped; leptocystidia obclavate,
    obtuse, often subventricose at the base; septocystidia present in some spp.
        . . . . **Atheloderma**

    80(77') Clamps absent; cystidia absent or septate but lacking clamps; hyphae wide, with
        wide-angled branching; fruitbody submembranous . . . . **Athelidium**

    80'(77') Clamps abundant; clamped septocystidia present; hyphae narrow, sometimes
        crystal-encrusted; fruitbody byssoid to floccose-membranous, with sterile hyphae
        forming thin rhizomorphs . . . . **Amphinema**

81(73') Clamps absent or confined to basal hyphae only (but usually present in culture, often

paired or whorled); hyphae branched subarticulately and at a wide angle; basal hyphae usually thick-walled; hymenophore smooth, dentate or merulioid; basidiocarp membranous to ceraceous; cystidia present in most species . . . . **Phanerochaete**
  Syn.: *Membranicium*; *Xerocarpus*.

81'(73') Clamps regularly present and not confined to basal hyphae 82

  82(81') Basidiocarp stratose or at least 3-layered 83

  82'(81') Basidiocarp neither stratose nor clearly layered 85

83(82) Context 3-layered; spores hyaline 84

83'(82) Context stratose; spores becoming stained yellowish, short-cylindrical; basidiocarp crustose, indurated; hymenophore smooth, yellowish to ochraceous-reddish; cystidia cylindrical to conical or clavate, thick-walled at base, sometimes 1–3-septate, not encrusted; basidioles numerous . . . . **Crustoderma**

  84(83) Basidiocarp arising subcortically in wood, fleshy to ceraceous-membranous; hymenophore smooth or hydnoid; leptocystidia sometimes present; basal hyphae horizontal; intermediate hyphae erect, loosely arranged; spores cylindrical, oblong or obovate . . . . **Basidioradulum**

  84'(83) Basidiocarp fleshy then strongly ceraceous; hymenophore smooth, apricot color; cystidia absent; basal hyphae erect, loosely arranged, indistinct, intermediate hyphae erect, conglutinate; spores long ellipsoid to cylindrical . . . . **Cerocorticium**

85(82') Cystidia and gloeocystidia absent; hyphidia little branched, flexuous, often rather nodulose, sometimes slightly fusoid or capitate; hyphae narrow (1–4 µm); spores ellipsoid to globose; hymenophore smooth to raduloid or hydnoid, ceraceous-membranous
  . . . . **Radulomyces**

85'(82') Cystidia and/or gloeocystidia present in most spp.; hyphidia absent 86

  86(85') Hyphae narrow (3–4 µm), fibrous, often branched from clamps in three equal branches; basidia small (10–15 µm long), subutriform-clavate; cystidia usually present, mostly small, smooth or encrusted especially at the apex, subulate, capitate, lageniform, rarely with thickened walls and hyphalike or moniliform, occasionally septate; hymenophore smooth, odontioid or hydnoid, soft-membranous to ceraceous . . **Hyphodontia**

  86'(85') Hyphae not especially narrow or fibrous; basidia large, subutriform-clavate; cystidia and/or gloeocystidia usually present; cystidia firm-walled, sometimes encrusted, sometimes septate; gloeocystidia clavate-cylindric or in form of stephanocysts; hymenophore smooth, dentate or raduloid in a few spp., compact-fleshy, not ceraceous
  . . . . **Hyphoderma**

Residual Corticiaceae

1. Hymenophore granular, tuberculate or somewhat dentate
  . . . . "**Odontia**" (incl. "**Grandinia**")

1'. Hymenophore smooth 2

  2(1') Neither cystidia nor gloeocystidia present . . . . "**Corticium**"

  2'(1') Either cystidia or gloeocystidia, or both, present 3

  3(2') Without cystidia but with gloeocystidia . . . . "**Gloeocystidium**"

  3'(2') With cystidia; gloeocystidia sometimes also present . . . . "**Peniophora**"

IV. LACHNOCLADIACEAE

Important Literature

Reid (1965); Corner (1966, 1968).

KEY TO GENERA OF LACHNOCLADIACEAE

1. Basidiocarp corticioid, effused or effuso-reflexed 2

1′. Basidiocarp stipitate, clavarioid or spathulate to flabelliform 5

    2(1) With asterosetae occupying most of the context . . . . **Asterostroma**
    Syn.: *Stellatostroma*.

    2′(1) Without asterosetae; context composed of generative hyphae and abundant narrow
    dichohyphidia or narrow somewhat dendroid hyphidia, sometimes dextrinoid . . . 3

3(2′) Basidiocarp thick (1 mm), membranous; hyphae dimitic with a basal layer of aseptate,
thick-walled, nondextrinoid, branched, tapering binding hyphae; dendrohyphidia and long
thin- to thick-walled septate cystidia present; spores large, thick-walled, subglobose, smooth,
not amyloid . . . . **Licrostroma**
    Syn.: *Michenera nom. anam.*?; *Artocreas nom. anam.* ?

3′(2′) Not so 4

    4(3′) Context duplex, composed of a horizontal basal layer of very narrow (3 μm) thick-
    walled dextrinoid hyphae and a stratose layer of dichohyphidia or dendrohyphidia;
    gloeocystidia present or absent; spores amyloid or not; cystidia absent
    . . . . **Scytinostroma**

    4′(3′) Context not duplex, composed mainly of strongly or weakly dextrinoid dendrohy-
    phidia or thick-walled dichohyphidia and crystals; cystidia and gloeocystidia present or
    absent; spores amyloid or not; some spp. with oedocephaloid conidial states in culture
    . . . . **Vararia** (Fig. 6)
    Syn.: *Asterostromella*; *Dichostereum*; *Langloisula p. parte. nom. confusum*?

5(1′) Basidiocarp clavarioid, with unilateral hymenophore on underside of flattened branches;
dichohyphidia abundant in mycelium and basidiocarp but not in the hymenium; gloeocystidia present; spores hyaline, often drying yellowish . . . . **Lachnocladium**

5′(1′) Basidiocarp spathulate or laterally stipitate and flabelliform 6

    6(5′) Hymenophore with cantharelloid folds; context fleshy with inflating hyphae;
    dichohyphidia limited to stipe and base of pileus, not present in hymenium; hymenium
    composed of compact basidia and gloeocystidia in palisade; spores not amyloid
    . . . . **Dichantharellus**

    6′(5′) Hymenophore smooth; context coriaceous with uninflated hyphae; dichohyphidia
    present in hymenium together with scattered basidia and gloeocystidia; spores amyloid
    . . . . **Dichopleuropus**

V. PODOSCYPHACEAE

Important Literature

Reid (1965); Whelden (1966).

KEY TO GENERA OF PODOSCYPHACEAE

1. Monomitic, or rarely dimitic with skeletal hyphae near the base of the fruitbody only . . .   2

1'. Dimitic in all parts of the fruitbody, or trimitic   5

2(1) Thin-walled, cylindrical, emergent cystidia present in hymenium; hymenophore smooth, minutely setulose under lens; monomitic; without clamps . . . . .   **Cotylidia**
Syn.: *Bresadolina*; *Craterella*.

2'(1) Cystidia absent   3

3(2') Hyphae encrusted with purple-brown pigment insoluble in KOH; monomitic; context blackish; gloeocystidia absent; cuticular layer of erect, densely arranged agglutinated coralloid hyphae . . . .   **Coralloderma**

3'(2') Hyphae not thus encrusted; context not black; cuticular layer absent or not of coralloid hyphae   4

4(3') Dimitic at base of basidiocarp, monomitic elsewhere; generative hyphae clamped, with thickening walls and with parts becoming strongly inflated (to 28 μm); skeletal hyphae long, with almost obliterated lumen; gloeocystidia absent; hymenophore smooth
. . . .   **Inflatostereum**

4'(3') Monomitic, without inflating generative hyphae; clamps present or absent; gloeocystidia present or absent; hymenophore smooth to radiately rugose . . . . .   **Stereopsis**
Possibly Gomphaceae.

5(1') Gloeocystidia present; metuloid cystidia present or absent; dimitic or trimitic . . .   6

5'(1') Gloeocystidia and metuloid cystidia absent; trimitic; basidiocarp infundibuliform, its upper surface finely tomentose and bearing conspicuous flattened branched antlerlike processes especially near the base . . . .   **Aquascypha**

6(5) Basidiocarp usually large and thick; surface bearing sharp bladelike crests often covered with a thick tomentum of clamped hyphae; hymenophore with smooth to tuberculate branching ribs or undulating folds; metuloid cystidia in some spp.; caulocystidia and pilocystidia absent . . . .   **Cymatoderma**
Syn.: *Actinostroma*; *Beccariella*; *Cladoderris*.

6'(6) Basidiocarp fairly small, thin and often translucent; surface glabrous, pruinose or with a tomentum of hyphae lacking clamps, without crests but sometimes with antlerlike processes; hymenophore smooth; metuloids, caulocystidia and pilocystidia present in some species . . . .   **Podoscypha**   (Fig. 4)
Syn. *Stereogloeocystidium*.

VI.   PUNCTULARIACEAE

A single genus, *Punctularia*, with characters of the family. Syn.: *Phaeophlebia*; ? *Auricula*. Literature: Talbot (1958); Donk (1964).

VII.   SCHIZOPHYLLACEAE
AND
RESIDUAL "CYPHELLACEAE"

Important Literature

Donk (1959, 1962); Cooke (1961); Singer (1962); Reid (1963).

KEY TO GENERA OF SCHIZOPHYLLACEAE AND RESIDUAL "CYPHELLACEAE"

Schizophyllaceae

1. Basidiocarp originally cupulate, attached by a reduced base, sometimes becoming appress·d or flabellate, proliferating from marginal clefts to form longitudinally split radiating lobes resembling agaric gills, undergoing hygroscopic movements; spores smooth; hyaline, smooth hymenial cystidia present in some spp. . . . . **Schizophyllum**

1'. Not proliferating to form split gill-like structures 2

    2(1') Hymenophore becoming plicate with radiating, branching folds; basidiocarp cupulate, campanulate, reniform or flabellate, attached by a short stipe on the dorsal side . . . . **Plicaturopsis**

    2'(1') Hymenophore smooth; basidiocarps cupulate to tubular, sessile, often on a subiculum or stroma 3

3(2') Stroma membranous-fibrous, well-developed, weakly dimitic, bearing densely gregarious basidiocarps at first globose with a small apical pore, later tubular, cupulate or flattened; hyphae monomitic in the fruitbodies . . . . **Stromatoscypha**
    Syn.: *Porotheleum.*

3'(2') Stroma, if present, an arachnoid thin subiculum; basidiocarps scattered or densely gregarious, tubular or barrel-shaped, not globose at first; surface with dichohyphidial hairs at least near the margins; monomitic . . . . **Henningsomyces**
    Syn. *Solenia.*

Some residual genera of "Cyphellaceae"

1. Spores yellow or brown, smooth or asperulate 2

1'. Spores hyaline, smooth 3

    2(1) Cupulate basidiocarps embedded in a thick (to 2 mm) stroma; surface of brown encrusted hyphae; spores ellipsoid, smooth or asperulate . . . . . **Phaeoporotheleum**

    2'(1) Cupules not in a stroma; surface hairs elongated and simple or floccose, branched, anastomosed; spores globose to ovate, asperulate . . . . **Asterocyphella**

3(1') Basidiocarp obconical, attached by reduced base; hymenium convex; cystidia abundant, emergent from all parts of the basidiocarp, not encrusted, with thickened echinulate walls except at base and apex, fusiform to ventricose, slightly amyloid; spores not amyloid
    . . . . **Wiesnerina**

3'(1') Hymenium not convex; cystidia absent 4

    4(3') Basidiocarps discrete, pale brown, glabrous, pendulous, bell-shaped, with a very short stipe; hyphae monomitic, without clamps, thin-walled and brown in the context; spores ellipsoid . . . . **Phaeodepas**
    Compare *Glabrocyphella*, insufficiently described.

    4'(3') Basidiocarps gregarious to crowded 5

5(4') Basidiocarps on a common stroma; surface hairs branched, coralloid, brown-walled at least at the base, encrusted with brown granules . . . . **Stromatocyphella**

5'(4') Basidiocarps lacking a stroma or subiculum; surface covered with yellow hairs
    . . . . **Woldmaria**

VIII. STEREACEAE

Important Literature

Boidin (1958b, 1959a,b); Pouzar (1959); Lentz (1960); Donk (1964); Parmasto (1968).

KEY TO GENERA OF STEREACEAE

1. Hymenium with emergent hyphal pegs (sterile fascicles of agglutinated brown or purplish hyphae); spores large (16–24 μm long), not amyloid, ellipsoid to cylindrical 2

1'. Without hyphal pegs 3

   2(1) Pileus sessile or substipitate, conchate, tough gelatinous to membranous, thin; surface tan color, not notably velutinate; hyphal pegs purple, conical then fimbriate to granulose . . . . **Mycobonia**
      Syn.: *Hirneola* Fr. 1848; *Grandinioides.*

   2'(1) Pileus sessile, dimidiate or conchate, coriaceous-corneous, hard, brittle, not especially thin; surface brown, velutinate or strigose, sulcate-zonate; hyphal pegs dark brown
. . . . **Veluticeps**

3(1') Spores amyloid 4

3'(1') Spores not amyloid 7

   4(3) Spores minutely asperulate, somewhat thick-walled, subglobose to broad ellipsoid; metuloid cystidia in context and hymenium, hyaline to pale yellowish, encrusted, thick-walled, fusiform; dimitic or trimitic; oedocephaloid conidiophores or chlamydospores in culture . . . . **Laurilia** (Fig. 7)
      Possibly Echinodontiaceae. *Laurilia sulcata* and *L. taxodii* cause white rots of conifers.

   4'(3) Spores smooth; hyphal system never trimitic 5

5(4') Without hyphidia; cystidia brown, thick-walled, with encrusted slightly rugose apices, arising as long apical modifications of skeletal hyphae throughout the thickening fruitbody but progressively shorter and more fusoid nearer the hymenium, often bearing small brown lateral hemispherical nodules; clamps present in the fruitbody; arthrospores formed in culture by one species (*A. areolatum*) . . . . **Amylostereum**
      Syn.: *Trichocarpus* P. Karst. 1889; *Lloydellopsis. Amylostereum areolatum* and *A. chailletii* carried in mycangia by *Sirex* woodwasps.

5'(4') Acanthohyphidia abundant, scarce, or absent; cystidia formed only from ends of skeletal hyphae in or near the hymenium, subcylindrical, with smooth, hyaline to subhyaline walls and yellowish to brownish contents that sometimes "bleed" red on contact with air; clamps rare or absent in the fruitbody 6

   6(5') Acanthohyphidia very abundant, colored, thick-walled; clamps rare or absent in the basidiocarp, present but never opposed or whorled in culture; negative oxidase reaction; context brown . . . . **Xylobolus** (Fig. 8)

   6'(5') Acanthohyphidia scanty or absent, thin-walled with few aculeate processes, subhyaline; clamps absent in basidiocarp, simple, opposed or whorled in culture; positive oxidase reaction: context pallid . . . . **Stereum** (Fig. 11)
      Syn.: *Haematostereum*; *Stereoasterostromella.*

7(3') Gloeocystidia abundant, occupying most of the stratose context, thin-walled, vesicular,

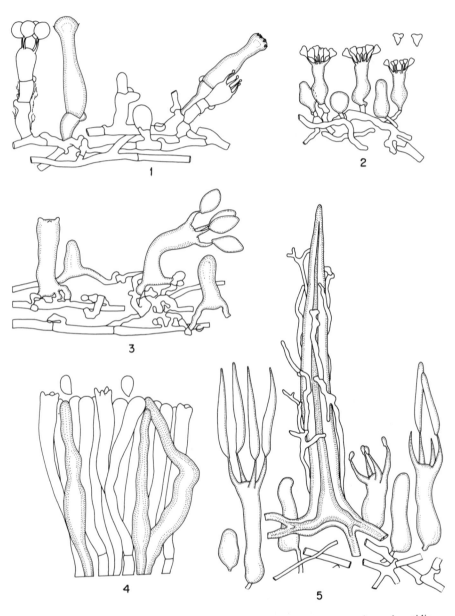

FIGS. 1–5,   Aphyllophorales. Fig. 1, *Repetobasidium mirificum*, septate, clamped cystidia, and a basidium with percurrent proliferation; Fig. 2, *Sistotrema subtrigonosperum*, dimerous urniform basidia; Fig. 3, *Xenasmatella filicina*, pleurobasidia; Fig. 4, *Podoscypha elegans*, hymenium with basidia and gloeocystidia; Fig. 5, *Tubilixenasma vermiferum*, podobasidia and 1 radicate lyocystidium with climbing hyphae.

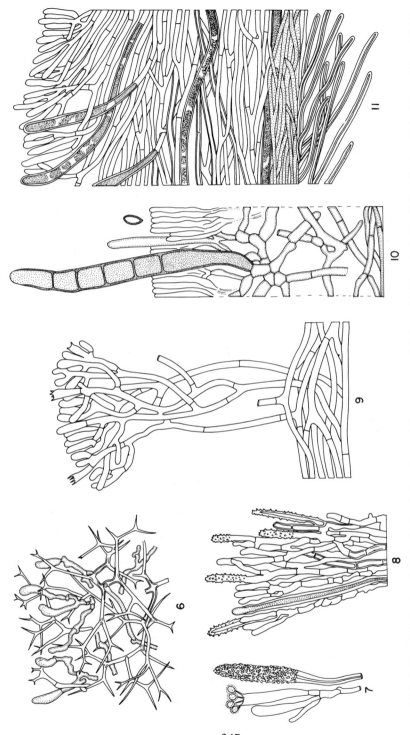

FIGS. 6–11. Aphyllophorales. Fig. 6, *Vararia investiens*, catahymenium, and dichohyphidial context; Fig. 7, *Laurilia taxodii*, basidia, and 1 metuloid cystidium; Fig. 8, *Xylobolus frustulatus*, acanthohyphidial hymenium; Fig. 9, *Athelia epiphylla*, showing byssoid to pellicular context; Fig. 10, *Coniophorella olivacea*, section showing septate cystidium; Fig. 11, *Stereum sanguinolentum*, showing steroid tissue distribution (trichoderm, cortex, medullary layer, and hymenium) and skeletocystidia.

ovoid to ellipsoid or somewhat elongated, with homogeneous contents or often "empty"; dimitic with clamped generative and sparse, pale-colored skeletal hyphae; fruitbody effused or narrowly reflexed with woody indurated texture . . . . **Cystostereum**

7'(3') Without gloeocystidia 8

    8(7') Without cystidia; thin-walled encrusted cystidioles abundant in sterile but rare in fertile hymenia; hyphae often encrusted with brown material that greens in KOH; clamps absent . . . . **Boreostereum**

    8'(7') With cystidia; no hyphae greening in KOH 9

9(8') Cystidia smooth or roughened but not encrusted, thick-walled but with lumen of appreciable width, long, cylindrical to subclavate, obtuse, deep brown, emergent to 80–90 µm; context and hymenium brown, cortex almost black; hymenium smooth to pruinose; basidia very long (to 100 µm) . . . . **Columnocystis**
    Syn.: *Chaetocarpus*.

9'(8') Cystidia usually strongly encrusted with crystals, thick-walled with a narrow lumen, often subfusoid near apex, not strongly emergent, subhyaline to brownish; hymenium smooth or with anastomosing crests or shallow pores; basidia long but not exceptionally so
. . . . **Lopharia**
    Syn.: *Licentia*; *Lloydella*; *Porostereum*; *Thwaitesiella*.

REFERENCES

Boidin, J. (1958a). Essai biotaxonomique sur les Hydnés résupinés et les Corticiés. Thesis No. 202, University of Lyon.

Boidin, J. (1958b). Hétérobasidiomycètes saprophytes et Homobasidiomycètes résupinés. V. Essai sur le genre *Stereum* Pers ex S. F. Gray. *Rev. Mycol.* **23**:318–346.

Boidin, J. (1959a). Hétérobasidiomycètes saprophytes et Homobasidiomycètes résupinés. VI. Essai sur le genre *Stereum* sensu lato. *Rev. Mycol.* **24**:197–225.

Boidin, J. (1959b). Hétérobasidiomycètes saprophytes et Homobasidiomycètes résupinés. VII. Essai sur le genre "*Stereum* sensu lato". *Bull. Mens. Soc. Linn. Lyon* **28**:205–222.

Bourdot, H., and A. Galzin. (1928). "Hyménomycetes de France." Sceaux.

Cooke, W. B. (1957). The genera *Serpula* and *Meruliporia*. *Mycologia* **49**:197–225.

Cooke, W. B. (1961). The cyphellaceous fungi. A study in the Porotheleaceae. *Sydowia, Beih.* **4**:1–144.

Corner, E. J. H. (1932a). The fruit-body of *Polystictus xanthopus* Fr. *Ann. Bot.* (*London*) **46**: 71–111.

Corner, E. J. H. (1932b). A *Fomes* with two systems of hyphae. *Trans. Brit. Mycol. Soc.* **17**: 51–81.

Corner, E. J. H. (1966). "A Monograph of Cantharelloid Fungi," Ann. Bot. Mem. No. 2. Oxford Univ. Press, London and New York.

Corner, E. J. H. (1968). Mycology in the tropics—apologia pro monographia sua secunda. *New Phytol.* **67**:219–228.

Donk, M. A. (1931). Revisie van de Nederlandse Heterobasidiomycetae en Homobasidiomycetae-Aphyllophoraceae. Deel 1. *Meded. Ned. Mycol. Ver.* **18–20**.

Donk, M. A. (1954). A note on sterigmata in general. *Bothalia* **6**:301–302.

Donk, M. A. (1958). Notes on the basidium. *Blumea, Suppl.* **4**: (H. J. Lam Jubilee Vol.), 96–105.

Donk, M. A. (1959). Notes on "Cyphellaceae." I. *Persoonia* **1**:25–110.

Donk, M. A. (1962). Notes on "Cyphellaceae." II. *Persoonia* **2**:331–348.

Donk, M. A. (1964). A conspectus of the families of Aphyllophorales. *Persoonia* **3**:199–324.

Fries, E. M. (1874). "Hymenomycetes Europaei." Uppsala.

Lentz, P. L. (1954). Modified hyphae of Hymenomycetes. *Bot. Rev.* **20**:135–199.

Lentz, P. L. (1957). Studies in *Coniophora*. I. The basidium. *Mycologia* **49**: 534–544.

Lentz, P. L. (1960). Taxonomy of *Stereum* and allied genera. *Sydowia* **14**:116–135.

Martin, G. W. (1938). The morphology of the basidium. *Amer. J. Bot.* **25**:682–685.

Martin, G. W. (1957). The tulasnelloid fungi and their bearing on basidial terminology. *Brittonia* **9**:25–30.

Nannfeldt, J. A., and J. Eriksson. (1953). On the Hymenomycetous genus *Jaapia* Bres. and its taxonomical position. *Sv. Bot. Tidskr.* **47**:177–189.

Neuhoff, W. (1924). Zytologie und systematische Stellung der Auriculariaceen und Tremellaceen. *Bot. Arch.* **8**:250–297.

Parmasto, E. (1968). "Conspectus systematis Corticiacearum." Tartu.

Patouillard, N. (1900). Essai taxonomique sur les familles et les genres des Hyménomycètes. Thesis, Lons-le-Saunier, Paris.

Pouzar, Z. (1958). Nova genera macromycetum. II. *Ceska Mykol.* **12**:31–36.

Pouzar, Z. (1959). New genera of higher fungi. III. *Ceska Mykol.* **13**:10–19.

Reid, D. A. (1963). Notes on some fungi of Michigan. I. "Cyphellaceae." *Persoonia* **3**:97–154.

Reid, D. A. (1965). A monograph of the stipitate stereoid fungi. *Nova Hedwigia, Beih.* **18**.

Rogers, D. P. (1934). The basidium. *Stud. Natur. Hist. Iowa Univ.* **16**:160–182.

Romagnesi, R. (1944). La cystide chez les Agaricacées. *Rev. Mycol.* **9**: Suppl., 4–21.

Singer, R. (1962). "The 'Agaricales' (Mushrooms) in Modern Taxonomy," 2nd ed. Cramer, Weinheim.

Smith, A. H. (1966). The hyphal structure of the basidiocarp. *In* "The Fungi" (G. C. Ainsworth and A. S. Sussman, eds.), Vol. 2, pp. 151–177. Academic Press, New York.

Talbot, P. H. B. (1954). Micromorphology of the lower Hymenomycetes. *Bothalia* **6**:249–299.

Talbot, P. H. B. (1958). Studies of some South African resupinate Hymenomycetes. Part II. *Bothalia* **7**:131–187.

Talbot, P. H. B. (1968). Fossilized pre-Patouillardian taxonomy? *Taxon* **17**:620–628.

Whelden, A. L. (1966). *Stereum radicans, Clavariadelphus* and the Gomphaceae. *Brittonia* **18**:127–131.

CHAPTER 20

Aphyllophorales II: The Clavarioid and Cantharelloid Basidiomycetes

RONALD H. PETERSEN

Department of Botany
University of Tennessee
Knoxville, Tennessee

I. INTRODUCTION

The study of cantharelloid and clavarioid basidiomycetes has a long and illustrious history. Both groups have roots into pre-Friesian literature and, as such, were long caught in a historical trap which equated hymenial shape and surface features with natural relationships. The early workers (Tournefort; Vaillant, 1727; Micheli, 1729; Linnaeus, etc.) accepted the clavarioid fungi as consisting of two groups, those whose basidiocarps were branched or coralloid, and those whose basidiocarps were unbranched, simple clubs, occuring either singly or in fascicles. None was pileate, and little confusion existed, therefore, between these and other groups of fleshy fungi. Only later was the true heterogeneity of the group discovered, when microscopy became an important tool and when asci were associated with *Geoglossum* and its relatives, and the peculiar basidia and spores of *Dacrymyces* uncovered. Although less misunderstood through the years, *Cantharellus* had a prior name, *Merulius*, which was variously interpreted as a genus of resupinate basidiomycetes, pileate basidiomycetes, or erect ascomycetes (when used as including species now placed in *Morchella*).

Through the years, additional genera were described and laid aside, and the generic complexes placed in various groupings, but only in the last two decades have both groups come into their own, and, in both cases, the results have been significant to the understanding of the taxonomy and phylogeny of the basidiomycetous fungi and illustrative of the inadequacy of the presumed natural groups based on gross hymenial characters.

For background knowledge of the anatomy of basidiocarps in this and other basidiomycetous fungi, the reader is referred to the papers of Talbot (1954) and Lentz (1954, 1971). The suprageneric scheme I advocate for these fungi falls somewhere between that of Donk (1964b) and Corner (1970), and the history of generic names is accepted as that written by Donk (1954, 1964b). Because the family placement of many genera differs with the authors involved, no family scheme is followed below, but the various dispositions are often indicated under each genus.

II. CLAVARIOID HOLOBASIDIOMYCETES

The familial lines for this group have not yet been drawn clearly enough to be stated authoritatively. From the traditional rendering of a single family (Clavariaceae) to accommodate all species having upright fruit bodies with amphigenous hymenium, various authors segregated generic complexes, until Donk (1964b) distributed the clavarioid species throughout several families. Corner (1950) had included a number of hydnoid taxa together with clavarioid based on hyphal analysis of basidiocarp construction, but the definition of the clavarioid fungi includes the concept of erect or upright basidiocarp, and a few of the hydnoid genera have been excluded from later treatments (Corner, 1970). Petersen (1971a) has suggested that if basidiocarp habit is important in some groups, then it should be important in others as well, so that genera such as *Myxomycidium, Hormomitaria, Deflexula, Allantula, Delentaria*, and the like, and the family Hericiaceae, should not be considered clavarioid by definition (and herein are not).

The alliances between genera whose species produce large, highly complex basidiocarps are more easily perceived than between taxa with much reduced, generally characterless fruit bodies. Thus, *Ramaria, Clavariadelphus, Clavaria, Pterula*, and such others are easily conceived as the nuclei for family groupings. But the placement, for example, of *Pistillaria, Typhula, Pistillina*, and *Ceratellopsis* is considerably more difficult, largely due to the reduction in basidiocarp stature and construction.

The number of significant papers providing keys and descriptions to this group for limited areas is increasing rapidly. The following partial list may be used as a guide.

North America: Coker (1923, 1939, 1947), Burt (1922), Doty (1944–Pacific northwestern United States), Leathers (1955, Michigan), Marr (1968–*Ramaria* for western Washington), Henry (1967–Pennsylvania), Olexia (1968–*Clavaria*), Dodd (1970–*Clavicorona*).

Europe: Britain–Cotton and Wakefield (1919), Holland–Donk (1933); France–Bourdot and Galzin (1927), Perreau (1969); Denmark–Christiansen (1967); Czechoslovakia–Pilát (1958); Finland–Karsten (1881, 1889, etc.).

USSR: Parmasto (1958, 1970), Shvarzman (1964), Phylimonova (1965).

Asia: Japan–Imai (1929, 1930, 1931, 1934); India–Thind (1961), and others; and see under Corner; Solomon Islands–Corner (1967); Java–van Overeem (1923); Philippines–Dogma(1966a,b, 1967); Himalaya Mountains– Anonymous (1970); Corner in Balfour-Browne (1955); Bonin Islands–Ito and Imai (1937).

Africa: South Africa–van der Byl (1932); Congo–Corner and Heinemann (1967).

Australia: Cleland (1931, 1935); Fawcett (1939b, and previous).

South America: Argentina–Corner (1957).

World: Corner (1950, 1970).

A. Use of the Key

The taxonomy of the clavarioid Basidiomycetes is currently in transition. Old genera are being split, while some of the more recent segregates are under question. All this makes construction of an adequate key to the genera a difficult task. All too often key choices must be based on characters quite familiar to those engaged in research on the group, while quite ambiguous to nonclavariologists. Some of these uncertainties may be remedied before presentation of the key.

Because the clavarias are so totally heterogeneous at the suprageneric level, little uniformity may be expected in basidiocarp construction. Most species produce basidiocarps composed only of generative (thin- to very slightly thick-walled, usually hyaline, normally branched) hyphae, and these are termed *monomitic.* A variety of secondary hyphal types may also be encountered, however, and when a single second hyphal type is present the fruit body is considered *dimitic.* Corner (1950) conceives of the term as indicating only the presence of generative and skeletal (thick-walled, non-septate, elongated hyphal tips of indefinite length) hyphae, but other workers (see Petersen, 1971a) have protested that other hyphal types (i.e., gloeoplerous hyphae) may occur with generatives without the presence of skeletals, and that the use of the term dimitic (or trimitic for that matter) should always be followed by an explanatory phrase revealing the specific hyphal types to be found. Likewise, a trimitic fruit body is one in which three hyphal types are found, tetramitic four, etc. For example, in *Ramaria* can be found monomitic basidiocarps, dimitic with skeletals, dimitic with gloeoplerous hyphae, trimitic with skeletals and binding hyphae, trimitic with skeletals and gloeoplerous hyphae, and tetramitic with skeletals, binding hyphae, and gloeoplerous hyphae.

Skeletal hyphae may be relatively undifferentiated (as in *Ramaria*) or modified into setal hymenial elements (*Clavariachaete*), dichohyphidia (*Lachnocladium*), or other forms. Gloeoplerous hyphae may stain in sulfobenzaldehyde or sulfovanillin (*Clavicorona*) or in cotton blue (*Ramaria*). Spore ornamentation (when present) may be cyanophilous (staining in cotton blue as in *Ramaria*), amyloid (staining in Melzer's reagent as in *Amylaria*), or inert to stains (*Clavulinopsis* and *Ramariopsis*). Basidiocarps may turn green when aqueous iron salts are applied (*Clavariadelphus*, etc.) or may be inert to the chemical (*Clavaria*, etc.). For application of several additional macrochemicals in the taxonomy of *Ramaria*, see Marr (1968), and for formulae for macro- and microchemicals, see Singer (1962).

For investigation of modern taxonomy in the group, the two major publications of Corner (1950, 1970) are indispensible. Although the second is cited where pertinent, for it is more restricted in scope than the first, the former is cited after each genus accepted as clavarioid. A citation "EC+" indicates that Corner (specifically in 1950, but not necessarily in later writings) accepted the genus in the same general terms as used in the key. Conversely, "EC−" shows that Corner did not include the genus in his major publication. Three other works are similarly cited, namely Coker (1923) by WC+ or WC−, Bourdot and Galzin (1928) by BG+ or BG−, and Patouillard (1900) by P+ or P−. Where these authors have treated the species now placed in the cited genus, but under another generic name this is indicated (i.e., under *Ramaria*−"WC as *Clavaria*").

A number of authors, from Fries to Corner, have included genera with the clavarioid fungi which are either not Holobasidiomycetes, or not clavarioid by definition. I have tried to include these in the key for completeness, but those genera which I do not accept as fulfilling the definition of a clavarioid holobasidiomycete I have enclosed in parentheses, even though in some cases some small explanation is included as well.

KEY TO THE GENERA OF CLAVARIOID BASIDIOMYCETES

1. Basidia septate, either transversely or cruciately (for incompletely or secondarily-septate basidial forms, see below 2

    Here are included members of the Auriculariales and Tremellales with basidia primarily septate. *Clavulina*, with postpartal septation, may be found below.

1'. Basidia undivided, secondarily or partially divided 5

    2(1) Basidium transversely septate 3

    2'(1) Basidium cruciately septate 4

3(2) Basidiocarp clavarioid, free-standing . . . . **(Paraphelaria** p. 312)
    Tropical, with a single species, the genus obviously belongs in the Auriculariales. Ref.: Corner, 1966c.

3'(2) Basidiocarp crustose, resupinate, or sometimes forming clavarioid outgrowths
    . . . . **(Septobasidium** p. 314)
    Hardly clavarioid, this genus must be sought under Septobasidiales p. 315

    4(2') Basidial initial segregating into a stalk portion and apical basidium which becomes cruciately septate . . . . **(Tremellodendropsis** p. 307 subg. *Transeptia*)

Described as a subgenus, this taxon probably deserves generic rank. Ref.: Crawford, 1954.

4′(2′) Basidia typically tremellaceous, globose to subglobose, cruciately septate
. . . . **(Tremellodendron** p. 307)
Although several important sources of literature include the clavarioid fungi, the genus exhibits typical tremelloid basidia, and must be sought under Tremellales, p. 304.

5(1′) Basidia of the "tuning fork" type . . . . **(Calocera** p. 320)
Although microscopically distinct, especially *C. viscosa* is easily mistaken for a *Clavulinopsis* in the field. Classical taxonomy may be found elsewhere, but antibiotics have also been used as characters. Ref.: Schwalb, 1966.

5′(1′) Basidia ovoid, clavate, or cylindrical, but not of the "tuning fork" type 6

6(5′) Basidia clavate, often becoming partially apically longitudinally septate, usually with many intermediate forms almost to typically tremellaceous basidia
. . . . **(Tremellodendropsis** subg. *Tremellodendropsis* p. 307)

6′(5′) Basidia variously shaped, but not longitudinally divided 7

7(6′) Meiotic nuclear spindles stichic; basidia usually secondarily transversely septate, with 2 cornute sterigmata 8

7′(6′) Meiotic nuclear spindles chiastic or in an unknown position; basidia very rarely transversely septate; sterigmata number usually four 9

8(7) Basidiocarp resupinate; basidia not secondarily septate . . . . **(Clavulicium)**

8′(7) Basidiocarp clavarioid; basidia often secondarily septate . . . . **Clavulina**
One of the earliest splits from *Clavaria*; the sole clavarioid representative of the Clavulinaceae; until recently based on (a) 2 sterigmata per basidium, and (b) usual postpartal basidial septation. *Clavulina amazonensis* has 4 sterigmata per basidium, and several species exhibit nonseptate basidia. Only stichic nuclear divisions now seem to define the genus. The *C. rugosa-C. cristata* complex seems cosmopolitan. Ref.: van Overeem, 1923; Donk, 1933, 1964b; Juel, 1916; Bauch, 1927; Corner, 1970; Dogma, 1967: EC+/WC as *Clavaria*/BG as *Clavaria*/P as *Clavaria*.

9(′7′) Spores amyloid, roughened; branches positively or negatively geotropic 10

9′(7′) Spores not amyloid; smooth or roughened 15

10(9) Spore sculpturing of warts, lines or crests, resembling *Bondarzewia* spores; branches negatively geotropic . . . . **Amylaria**
Apparently limited to the Himalaya Mts., Corner linked this with *Hericium*, but it may well be a clavarioid member of the Bondarzewiaceae. Ref: Corner, in Balfour-Browne, 1955; Smith and Leathers, 1967: EC−/WC−/BG−/P−.

10′(9) Spore sculpturing of small spines or spores smooth; branches positively or negatively geotropic 11

11(10′) Branches positively geotropic 12

11′(10′) Branches of fruit bodies negatively geotropic, often pyxidate or truncate, or fruit-body simple . . . . **Clavicorona**
Now fully known for North America. Formerly thought to possess species with amyloid, nonamyloid, and dextrinoid spores, all are now known to be amyloid. The most cosmopolitan species appear to be *C. pyxidata* (Northern Hemisphere) and *C. turgida* (Asia, South America, southeastern U.S. Ref.: Doty, 1947; Dodd, 1970; Smith and Leathers, 1967: EC+/WC as *Clavaria*/BG as *Clavaria*/P as *Clavaria*.

12(11) Basidiocarps composed of branches producing downward-directed spines . . 13

12'(11) Basidiocarps individual, small, downward-directed spines with little or no basal subiculum 14

13(12) Context dimitic with gloeoplerous hyphae: basidiocarp flesh amyloid; major branches separate, immersed in a tomentum, or confluent . . . . (**Hericium** p. 378)

13'(12) Context dimitic with skeletal hyphae; basidiocarp dark-colored; major branches embedded in a heavy tomentum . . . . (**Gloiodon** p. 378)

14(12') Gloeoplerous hyphal system present . . . . (**Dentipratulum** p. 378)

14'(12') Gloeoplerous hyphal system absent . . . . (**Mucronella** p. 378)
The whole of the Hericiaceae (p. 330) should be sought elsewhere. Donk has included *Clavicorona* as a clavarioid member of the family, which now seems to include a whole series of degenerate taxa from *Hericium* to *Mucronella*. Ref.: Donk, 1964b.

15(9') Basidiocarp flesh turning black in KOH; skeletal hyphae present, undifferentiated, or as dichohyphidia or asterosetae 16

15'(9') Basidiocarp flesh not turning black in KOH: skeletal hyphae sometimes present, but not differentiated 17

16(15) Skeletal hyphae appearing as dichohyphidia . . . . **Lachnocladium**
Traditionally used to include clavarioid fungi with leathery consistency, Corner clearly defined the genus. Reid made it the type of the Lachnocladiaceae, while Donk included it in the Hymenochaetaceae. Mostly tropical. Ref.: Corner, 1950; Reid, 1965; Donk, 1964b; Parmasto, 1970: EC+/WC+p.p. /BG−/P+p.p.

16'(15) Skeletal hyphae not differentiated; hymenial setae present . . . . **Clavariachaete**
With 2 species, this is the clavarioid representative of the Hymenochaetaceae. Ref.: Corner, 1950: EC+/WC−/BG−/P−.

17(15') Spores complex-warted or tuberculate; fruit body hyphae not inflated; branches flattened 18

17'(15') Spores smooth or rough, not complex-warted; fruit body hyphae various 19

18(17) Spores white or pale; fruit-bodies white . . . . **Scytinopogon**
Well-defined originally, the inclusion of additional species has made the generic parameters less distinct. There seem to be similarities to *Ramariopsis*. Ref.: Singer, 1945; Corner, 1970: EC+/WC as *Clavaria*/BG−/P−.

18'(17) Spores brown or fuscous; fruit-body flesh often turning blackish green in KOH . . . . (**Thelephora**)

19(17') Basidiocarp hyphae dimitic, of noninflated, thick-walled skeletals and thin-walled generatives; spores smooth 20

19'(17') Basidiocarp hyphae monomitic (and then spores rough or smooth), di- or trimitic (and then spores rough) 25

20(19) Fruit bodies sausage-shaped structures, decumbent, connected by slender, ropelike rhizomorphic strands . . . . (**Allantula**)
Although included in the clavarioid fungi by Corner, the genus must be excluded by definition. Ref.: Corner, 1952a.

20'(19) Fruit bodies branched, hydnoid or clavarioid 21

21(20') Basidiocarps clavarioid; cystidia of 2 different types . . . . **Dimorphocystis**
Corner has included this in the Pterulaceae. Tropical, presently with 3 species. Ref.: Boedijn, 1959, as *Actiniceps*; Corner, 1961b: EC +/WC −/BG −/P −.

21′(20′) Basidiocarps clavarioid or hydnoid; cystidia present or absent, but when present then of a single type 22

22(21′) Basidiocarp branches bending downward . . . . **(Deflexula)**
Tropical or subtropical. Corner has placed this in the Pterulaceae, where it is a hydnoid representative. Ref.: Corner (1952a).

22′(21′) Basidiocarp erect 23

23(22′) Fertile corticioid mat present at basidiocarp base, or present but separate from fruit body 24

23′(22′) Fertile corticioid mat absent; basidiocarps usually branched . . . . **Pterula**
Although one of the oldest splits from *Clavaria*, the genus remained obscure until the work by Lloyd. Apparently tropical or subtropical, and most common in the New World. One species produces fruit bodies from sclerotia. Ref.: Lloyd, 1919; Corner, 1952b; Berthier, 1967: EC+/WC+/BG+/P+.

24(23) Gloeocystidia present . . . . **Parapterulicium**
Corner has pointed out that the phylogenetic placement of the genus in the pteruloid fungi is by default, although it bears certain similarities to the "Xanthochroic series." Ref.: Corner, 1952a, 1970: EC−/WC−/BG−/P−.

24′(23) Gloeocystidia absent . . . . **Pterulicium**
Basidiocarps of *Lentaria* also produce a fertile mat, but are not dimitic. Ref.: Corner, 1952b: EC+/WC−/BG−/P−.

25(19′) Basidiocarps turning slate green to deep green in $FeSO_4$ 26

25′(19′) Basidiocarps not turning green in $FeSO_4$ 34
This macrochemical reaction has been questioned by Corner, but thus far it has been used with good results. Note the splitting of *Clavulinopsis* on this basis, however. Ref.: Corner, 1970.

26(25) Spores ornamented; ornamentation cyanophilous 27
Under this character are found representatives of the Gomphaceae. Ref.: Donk 1961, 1964b.

26′(25) Spores smooth or rough, but not cyanophilous 30

27(26) Basidiocarp resupinate . . . . **(Ramaricium)**

27′(26) Basidiocarp hydnoid, clavarioid, cantharelloid, or merismatoid 28

28(27′) Basidiocarp hydnoid with copious subiculum . . . . **(Kavinia)**
Petersen has split *Hydnocristella* from *Kavinia* on spore and hyphal characters, but neither genus is clavarioid, although both may be related to such genera. Ref.: Petersen, 1971d.

28′(27′) Basidiocarp clavarioid, cantharelloid or merismatoid 29

29(28′) Basidiocarp clavarioid, coralloid; spore ornamentation composed of spines, low warts, or longitudinal striae . . . . **Ramaria** p.p.
Here are found the most common species of the genus, although some species with smooth spores will be found under key choice 33. Typification of the genus has been under question, but now *R. botrytis*, with striate spores has been accepted formally. Ref.: Donk, 1941, 1949, 1964b; Doty, 1948b; Petersen, 1968b. The cyanophilous reaction of the spore ornamentation has been used as a major character in species separation. Ref.: Petersen, 1967b; Corner, 1970. The genus has been used as the type for the Ramariaceae, but more conservatively has been placed in the Gomphaceae. Ref.: Donk, 1961, 1964b; Corner, 1970. Fruit bodies may be mono-, di-, or trimitic,

and Petersen has suggested that this condition, together with spore differences, might be cause for segregation of several separate genera. Ref.: Corner, 1961a, 1970; Petersen, 1967b: EC+/WC as *Clavaria*/BG as *Clavaria*/P as *Clavaria*.

29′(28′) Basidiocarp cantharelloid or merismatoid . . . . **(Gomphus)**
    Although not clavarioid, *Gomphus* is important to the phylogenetic placement of many clavarioid taxa, such as *Ramaria*.

    30(26′) Basidiocarp branched; spores elongated, thin- or thick-walled 32

    30′(26′) Basidiocarp simple to fasciculate, rarely subcantharelloid 31

31(30′) Ampulliform swellings present on hyphae in juxtaposition to clamp connections; basidiocarp trama usually spongy or loosely fibrous; spores smooth, thin-walled
. . . . **Clavariadelphus**
    Variously placed in the Clavariaceae, Gomphaceae, or Clavariadelphaceae, the genus has been held to be a degenerate offshoot from *Gomphus*. Ref.: Corner, 1950; Donk, 1964b; Petersen, 1971b. Recent works have added knowledge of a few species, and Petersen has segregated the *C. fistulosa* complex into a separate genus, not yet published at this writing. Ref.: Corner, 1970; Smith, 1971; Wells and Kempton, 1968; Petersen, 1972: EC+/WC as *Clavaria*/BG as *Clavaria*/P as *Clavaria*.

31′(30′) Ampulliform swellings absent, although tramal hyphae usually inflated; basidiocarp trama solid or hollow, but rarely spongy; spores rough or smooth, somewhat thick-walled
. . . . **Clavulinopsis** p.p.
    This disposition of *Clavulinopsis* is not natural, since it is based on basidiocarp gross morphology. Since Corner's major resurrection of the genus, much additional knowledge of its taxonomy has been forthcoming, including the use of carotenoid pigmentation and macrochemical reaction in FeSO$_4$. Ref.: Corner, 1950, 1970; Petersen, 1968a, 1971c; Fiasson *et al.*, 1970. The distinctions between *Clavulinopsis* and *Clavaria* have been questioned, but *Clavulinopsis'* place in the Clavariaceae has not. Ref.: Petersen and Olexia, 1969; Corner, 1970: EC+/WC as *Clavaria*/BG as *Clavaria*/P as *Clavaria*.

    32(30) Basidiocarp decurved . . . . **Delentaria** see p.378 and 37 below
        Although described as a clavarioid genus, the basidiocarps are hydnoid. Ref.: Corner, 1970.

    32′(30) Basidiocarps erect 33

33(32′) Basidiocarp fleshy, bulky; spores with measurably thick walls, usually guttulate, cream, or ochraceous; hyphae often with ampulliform swelling in juxtaposition to septa
. . . . **Ramaria** (p.p.)
    Although most species of the genus are found above in key choice 29, some species (i.e., *R. obtussisima*, *R. pinicola*) with smooth spores must be keyed out here. Ref.: Petersen, 1967c.

33′(32′) Basidiocarp tough or fibrous, small; spores thin-walled, usually aguttulate, white to cream; hyphae without ampulliform swellings . . . . **Lentaria**
    Basidiocarps also are often accompanied by a fertile basal mat. One species, *L. surculus*, seems almost cosmopolitan, with the North Temperate extension usually separated as *L. byssiseda*. Corner has included the genus in the Ramariaceae. Petersen segregated some species into *Multiclavula*. Ref.: Petersen, 1967a: EC+/WC as *Clavaria*/BG as *Clavaria*/P as *Clavaria*.

    34(25′) Basidiocarps arising from sclerotia . . . . **Typhula**

Restricted by this character, the taxonomy of the genus remains in confusion. Berthier has described a *Pterula* with sclerotia, and has questioned the validity of the amyloid spore reaction in *Typhula*. One of the few clavarioid genera of economic significance, these have been called "snow molds," and a separate literature in plant pathology may be sought. Ref.: Remsberg, 1940; Berthier, 1966, 1967: EC+/WC+/BG+/P+.

34'(25') Basidiocarps not arising from sclerotia 35

35(34') Basidiocarps composed of 1 or more petalloid processes with inferior hymenium
. . . . **(Sparassis)**
The sole representative of the Sparassidaceae, the inferior hymenium excludes the genus from the clavarioid fungi. Ref.: Donk, 1964b: EC−/WC−/BG+/P+.

35'(34') Basidiocarps with amphigenous hymenium 36

36(35') Basidiocarps tough to tough-gelatinous; tramal hyphae somewhat thick-walled, uninflated, agglutinated 37

36'(35') Basidiocarps fibrous to brittle, but not though-gelatinous, and if so, then tramal hyphae thin-walled 39

37(36) Basidiocarps decurved . . . . **(Delentaria)**
(see above under key choice 32)

37'(36) Basidiocarps erect 38

38(37') Tramal hyphae hyaline, with or without clamps . . . . **Aphelaria**
Crawford segregated *Tremellodendropsis*. Ref.: Corner, 1970: EC+/WC as *Lachnocladium*/BG as *Thelephora*/P as *Thelephora*.

38'(37') Tramal hyphae brownish, clamped; subhymenial hyphae without clamps
. . . . **Phaeoaphelaria**
The family Aphelariaceae needs further definition, but within the family, only this genus exhibits dark hyphae. Ref.: Corner, 1970: EC−/WC−/BG−/P−.

39(36') Basidiocarps downward-directed toothlike processes 40

39'(36') Basidiocarps erect 41

40(39) Oleocystidia present; consistency of basidiocarp waxy to fleshy, but not watery-gelatinous . . . . **(Hormomitaria)**
Although initially included in the clavarioid fungi, the genus is hydnoid. Ref.: Corner, 1950.

40'(39) Oleocystidia absent; consistency of basidiocarp watery-gelatinous, often with a waterdrop exuded apically . . . . **(Myxomycidium)**
Basidiocarps are pendulous, and so the genus must be considered hydnoid. Linder investigated basidial development, Ref.: Linder, 1934; Rogers, 1971.

41(39') Apical portion of basidiocarp inflated, conical, obconical to spherical 42

41'(39') Apical portion of basidiocarp cylindrical to narrowly fusiform 44

42(41) Oleocystidia or thick-walled cystidia present in hymenium 43

42'(41) Cystidia absent . . . . **(Caripia)**

43(42) Inflated club portion hollow, capitate to bullate . . . . **(Physalacria)**
These species have been considered degenerate agarics by Singer and others. Ref.: Singer, 1962; Baker, 1941

43'(42) Inflated club portion solid, conical to cylindrical . . . . **Pseudotyphula**
Originally described to accomodate 2 species, *P. ochracea* and *P. tenuipes*, the latter

was transferred to *Hormomitaria*. The fruit-body orientation of the former is unknown. Ref.: Corner, 1953; Berthet and Berthier, 1966: EC−/WC−/BG−/P−.

44(41′) Sterile tissue percurrent through hymenium . . . . **Ceratellopsis**
A substitute name for Patouillard's *Ceratella*, a homonym. Although placed in the Clavariadelphaceae, its affinities are still obscure. Ref.: Corner, 1970: EC+/WC−/ BG as *Ceratella*/P as *Ceratella*.

44′(41′) Hymenium extended to basidiocarp apex 45

45(44′) Basidiocarps small, usually under 1.5 cm high, usually simple 46

45′(44′) Basidiocarps larger, usually over 2 cm high 49

46(45) Basidia broadly ovoid when immature; mycelium usually phycophilous or bryo-philous . . . . **Multiclavula**
Petersen segregated several species from *Clavaria*, *Clavulinopsis*, and *Lentaria* into *Multiclavula*. Oberwinkler has accepted the genus; Corner has questioned it. Meiotic nuclear spindles may be stichic, and if so, *Stichoclavaria* may be a prior name. Ref.: Petersen, 1967a; Corner, 1970; Oberwinkler, 1970: EC as *Clavaria*, *Clavulinopsis*, etc./WC as *Clavaria*/BG as *Clavaria*/P as *Clavaria*.

46′(45) Basidia short-cylindrical to clavate 47

47(46′) Thick-walled cystidia present in hymenium . . . . **Chaetotyphula**
Similar in gross morphology to *Pistillaria* and *Typhula*, the genus has been placed in the Clavariadelphaceae. Ref.: EC+/WC−/BG−/P−.

47′(46′) Thick-walled cystidia absent from hymenium 48

48(47′) Club portion discoid, with superior hymenium . . . . **Pistillina**
Doubtfully clavarioid. The tramal hyphae agglutinate. The genus has been placed in the Clavariadelphaceae, although this is open to question. Ref.: Corner, 1970: EC+/WC−/BG+/P+.

48′(47′) Club portion cylindrical, clavate or globose, with amphigenous hymenium . . . . **Pistillaria**
A heterogeneous assemblage of reduced forms, separated often from *Typhula* on the absence of sclerotia. Ref.: Berthier, 1966: EC+/WC−/BG+/P+.

49(45′) Tramal hyphae without clamp connections, often secondarily septate, often agglutinated . . . . **Clavaria**
Much reduced in size from its classical concept, the genus still serves as the type of the Clavariaceae. Two subgenera are included, and separation from *Clavulinopsis* has been questioned. Ref.: Olexia, 1968; Petersen and Olexia, 1969: EC+/WC+/ BG+/P+.

49′(45′) Tramal hyphae clamped (rare septa without clamps in some species), not secondarily septate; agglutination present or absent 50

50(49′) Basidia less than 25 µm long; spores usually less than 4.5 µm long, smooth to asperulate; tramal hyphae often agglutinated . . . . **Ramariopsis**
Most cosmopolitan species is *R. kunzei*. The genus has been expanded significantly recently, and a key to North American species is available. Ref.: Petersen, 1966, 1969c; Corner, 1970; Donk, 1933: EC+/WC as *Clavaria*/BG as *Clavaria*/P as *Clavaria*.

50′(49′) Basidia over 40 µm long; spores usually over 5 µm long, smooth or (rarely) echinulate-warty . . . . **Clavulinopsis**
See under key choice 31′.

REFERENCES

Anonymous. (1970). "Collection, Description and Identification of the Specified Fungal Flora in the Western and Central Himalayas up to Kathmandu in Nepal. Clavariaceae," Final Tech. Rep. P.L. 480 (mimeo.). Chandigarh.

Baker, G. E. (1941). Studies in the genus *Physalacria. Bull. Torrey Bot. Club* **68**:265–288.

Balfour-Browne, F. L. (1955). Some Himalayan fungi. *Bull. Brit. Mus. (Natur. Hist.)* **1**:197–218

Banerjee, S. N. (1947). Fungus flora of Calcutta and suburbs. I. *Bull. Bot. Soc. Bengal* **1**:37–54.

Bataille, F. (1948). "Les reactions macrochimiques chez les champignons." Le Chevalier, Paris.

Bauch, R. (1927). Untersuchungen über zweisesporige Hymenomyceten. II. Kerndegeneration bei einigen *Clavaria*-arten. *Arch. Protistenk.* **58**:285–299.

Boedijn, K. B. (1959). The genus *Actiniceps* Berk. & Br. *Persoonia* **1**:11–14.

Berthier, J. (1966). Amyloïdie, pseudoamyloïdie et cyanophilie chez *Pistillaria* Fr. et les *Typhula* Fr. (Hyménomycètes). *C. R. Acad. Sci.* **262**:2022–2024.

Berthier, N. (1967). Une nouvelle clavariacée a sclérote: *Pterula scleroticola* nov. sp. *Bull. Soc. Mycol. Fr.* **83**:731–737.

Berthet, P., and J. Berthier (1966). Une curieuse Clavariacee africaine: *Hormomitaria eburnea* nov. sp. *Rev. Mycol.* **31**:160–166.

Bourdot, H., and A. Galzin. "1927" (1928). "Hyménomycetes de France." Sceaux.

Britzelmayr, M. (1886). "Hymenomyceten aus Südbayern," Vol. VIII, pp. 273–306. Berlin.

Burt, E. A. (1922). The North American species of *Clavaria* with illustrations of the type specimens. *Ann. Mo. Bot. Gard.* **9**:1–78.

Christiansen, M. P. (1967). Clavariaceae Danicae. *Friesia* **8**:117–160.

Cleland, J. B. (1931). Australian fungi: Notes and descriptions. No. 8. *Trans. Roy. Soc. S. Aust.* **55**:152–160.

Cleland, J. B. (1935). "Mushrooms, Toadstools and Other Larger Fungi of South Australia." Adelaide.

Cleland, J. B., and E. Cheel. (1916). Records of Australian fungi. *Proc. Linn. Soc. N.S.W.* **41**:853–870.

Coker, W. C. (1923). "The Clavarias of the United States and Canada." Chapel Hill, North Carolina.

Coker, W. C. (1927). New or noteworthy Basidiomycetes. *J. Elisha Mitchell Sci. Soc.* **42**:251–257.

Coker, W. C. (1939). New or noteworthy Basidiomycetes. *J. Elisha Mitchell Sci. Soc.* **55**:373–387.

Coker, W. C. (1947). Further notes on clavarias, with several new species. *J. Elisha Mitchell Sci. Soc.* **63**:43–67.

Corner, E. J. H. (1950). A monograph of *Clavaria* and allied genera. *Ann. Bot. Mem.* **1**:740 p.

Corner, E. J. H. (1952a). Addenda Clavariacea. I. Two new Pteruloid genera and *Deflexula. Ann. Bot. (London)* [N. S.] **16**:269–291.

Corner, E. J. H. (1952b). Addenda Clavariacea. II. *Pterula* and *Pterulicium. Ann. Bot. (London)* [N. S.] **16**:531–569.

Corner, E. J. H. (1953). Addenda Clavariacea. III. *Ann. Bot. (London)* [N. S.] **17**:347–368.

Corner, E. J. H. (1957). Some clavarias from Argentina. *Darwiniana* **11**:193–206.

Corner, E. J. H. (1961a). Dimitic species of *Ramaria* (Clavariaceae). *Trans. Brit. Mycol. Soc.* **44**:233–238.

Corner, E. J. H. (1961b) A note on *Wiesnerina* (Cyphellaceae). *Trans. Brit. Mycol. Soc.* **44**:230–232.

Corner, E. J. H. (1966a). Species of *Ramaria* (Clavariaceae) without clamps. *Trans. Brit. Mycol. Soc.* **49**:101–113.

Corner, E. J. H. (1966b). The clavarioid complex of *Aphelaria* and *Tremellodendropsis*. *Trans. Brit. Mycol. Soc.* **49**:205–211.

Corner, E. J. H. (1966c). *Paraphelaria*, a new genus of Auriculariaceae (Basidiomycetes). *Persoonia* **4**:345–350.

Corner, E. J. H. (1967). Clavarioid fungi of the Solomon Islands. *Proc. Linn. Soc. London* **178**:91–106.

Corner, E. J. H. (1970). Supplement to "A monograph of *Clavaria* and allied genera." *Beih. Nova Hedwigia* **33**:1–299.

Corner, E. J. H., and P. Heinemann. (1967). Clavaires et *Thelephora*. "Flore iconographique des Champignons du Congo," Fasc. 16. Brussels.

Corner, E. J. H., K. S. Thind, and G. P. S. Anand. (1956). The Clavariaceae of the Mussoorie Hills (India). II. *Trans. Brit. Mycol. Soc.* **39**:475–484.

Corner, E. J. H., K. S. Thind, and S. Dev. (1957). The Clavariaceae of the Mussoorie Hills (India). VII. *Trans. Brit. Mycol. Soc.* **40**:472–476.

Corner, E. J. H., K. S. Thind, and S. Dev. (1958). The Clavariaceae of the Mussoorie Hills (India). IX. *Trans. Brit. Mycol. Soc.* **41**:203–206.

Cotton, A. D., and E. M. Wakefield. (1919). Revision of the British Clavariae. *Trans. Brit. Mycol. Soc.* **6**:164–198.

Crawford, D. A. (1954). Studies on New Zealand Clavariaceae. I. *Trans. Roy. Soc. N.Z.* **82**:617–631.

Dodd, J. L. (1970). The genus *Clavicorona*, with emphasis on North American species. Ph. D. Dissertation (ined.), University of Tennessee, Knoxville.

Dogma, I. J. (1966a). Philippine Clavariaceae. The pteruloid series. *Philipp. Agr.* **49**:844–861.

Dogma, I. J. (1966b). Philippine Clavariaceae. II. The thelephoroid and xanthochroic series and *Clavicorona*. *Philipp. Agr.* **50**:147–164.

Dogma, I. J. (1967). Additions to the genus *Clavulina*, Clavariaceae. *Philipp. Agr.* **50**:771–778.

Domanski, S. (1965). Wood inhabiting fungi in Bialowieza virgin forest in Poland. II. The mucronelloid fungus of *Hericium* group: *Dentipratulum bialoviesense* gen. et. sp. nov. *Acta Mycol.* **1**:5–11.

Donk, M. A. (1933). Revision der Niederländischen Homobasidiomycetes-Aphyllophorales. II. *Meded. Bot. Mus. Rijksuniv. Utrecht* **9**:1–278.

Donk, M. A. (1941). Nomina generica conservanda and confusa for Basidiomycetes (Fungi). *Bull. Jard. Bot. Buitenzorg* [3] **17**:155–197.

Donk, M. A. (1949). New and revised nomina generica conservanda proposed for Basidiomycetes (Fungi). *Bull. Jard. Bot. Buitenzorg* [3] **18**:83–168.

Donk, M. A. (1954). The generic names proposed for Hymenomycetes. III. "Clavariaceae." *Reinwardtia* **3**:441–493.

Donk, M. A. (1961). Four new families of Hymenomycetes. *Persoonia* **1**:405–407.

Donk, M. A. (1964a). Nomina conservanda proposita 1964. (130) *Ramaria* (Fr.) Bon. *Regn. Veg.* **34**:38–40.

Donk, M. A. (1964b). A conspectus of the families of the Aphyllophorales. *Persoonia* **3**:199–324.

Doty, M. S. (1944). "*Clavaria*, the Species Known from Oregon and the Pacific Northwest." Oregon State Coll. Press, Corvallis.

Doty, M. S. (1947). *Clavicorona*, a new genus among the clavarioid fungi. *Lloydia* **10**:38–44.

Doty, M. S. (1948a). A preliminary key to the genera of clavarioid fungi. *Bull. Chicago Acad. Sci.* **8**:173–178.

Doty, M. S. (1948b). Proposal and notes on some genera of clavarioid fungi and their types. *Lloydia* **11**:123–128.

Fawcett, S. G. M. (1938). Studies on the Australian Clavariaceae. Part I. *Proc. Roy. Soc. Victoria*, [N. S.] **51**:1–10.

Fawcett, S. G. M. (1939a). Studies on the Australian Clavariaceae. Part II. *Proc. Roy. Soc. Victoria,* [N. S.] **51**:265–280.

Fawcett, S. G. M. (1939b). Studies on the Australian Clavariaceae. Part III. *Proc. Roy. Soc. Victoria,* [N. S.] **53**:153–163.

Fiasson, J. L., R. H. Petersen, M. P. Bouchez, and N. Arpin. (1970). Contribution biochimique a la connaissance taxinomique de certains champignons cantharelloïd et clavarioïd. *Rev. Mycol.* **34**:357–364.

Fries, E. M. (1821). "Systema mycologicum," Vol. 1. Lund.

Gray, S. F. (1821). "A Natural Arrangement of British Plants," Vol. 1. London.

Henry, L. K. (1967). A review of the Clavariaceae (coral fungi) of western Pennsylvania. *Ann. Carnegie Mus.* **39**:125–142.

Holmskjold, T. (1970). "Beata ruris otia fungorum Danicis," Vol. I. Copenhagen.

Imai, S. (1929). On the Clavariaceae of Japan. I. *Trans. Sapporo. Natur. Hist. Soc.* **9**:38–45.

Imai, S. (1930). On the Clavariaceae of Japan. II. *Trans. Sapporo Natur. Hist. Soc.* **9**:70–76.

Imai, S. (1931). On the Clavariaceae of Japan. III. *Trans. Sapporo Natur. Hist. Soc.* **12**:9–21.

Imai, S. (1934). On the Clavariaceae of Japan IV. *Trans. Sapporo Natur. Hist. Soc.* **13**:385–387.

Ito, S., and S. Imai. (1937). Fungi of the Bonin Islands. II. *Trans. Sapporo Natur. Hist. Soc.* **15**:52–59.

Juel, H. O. (1916). Cytologische pilzstudien. I. Die basidien der gattungen *Cantharellus, Craterellus* und *Clavaria. Nova Acta Regiae Soc. Sci. Upsal.* [4] **4**:1–34.

Karsten, P. A. (1881). Enumeratio Thelephorearum Fr. et Clavariarum Fr. Fennicarum, systemate novo dispositarum. *Rev. Mycol.* **3**:21–23.

Karsten, P. A. (1889). Kritisk öfversight af Finlands Basidsivampar. *Birdr. Kann. Finlands Natur. Folk* **48**:1–470.

Leathers, C. R. (1955). The genus *Clavaria* Fries in Michigan. Dissertation (ined.), University of Michigan, Ann Arbor.

Lentz, P. L. (1954). Modified hyphae of Hymenomycetes. *Bot. Rev.* **20**:135–199.

Lentz. P. L. (1971). Analysis of modified hyphae as a tool in taxonomic research in the higher Basidiomycetes. *In* "Evolution in the Higher Basidiomycetes" (R. H. Petersen, ed.) University of Tennessee Press, Knoxville, Tennessee.

Linder, D. H. (1934). The genus *Myxomycidium. Mycologia* **26**:332–343.

Lloyd, C. G. (1919). The genus *Pterula. Mycol. Notes* **60**:863–870.

Maire, R. (1914). La Flore mycologiques des forêts de Cedres de l'Atlas *Bull. Soc. Mycol. Fr.* **30**:199–220.

Marr, C. D. (1968). *Ramaria* of western Washington. Dissertation (ined.), University of Washington, Seattle.

Martin, G. W. (1940). Some tropical American clavarias. *Lilloa* **5**:191–197.

Micheli, P. A. (1729). "Nova plantarum genera." Florence.

Oberwinkler, F. (1970). Die Gattungen der Basidiolichenen. *Deut. Bot. Ges.* [N. S.] **4**:139–169.

Olexia, P. D. (1968). The genus *Clavaria* in North America. Dissertation (ined.), University of Tennessee, Knoxville.

Parmasto, E. C. (1958). "Key to the Clavariaceae of the U.S.S.R." "Nauka," Moskva, Leningrad (Russian title).

Parmasto, E. C. "1970." (1971). "The Lachnocladiaceae of the Soviet Union." Akad. Nauk Estonian S.S.R. (Russian title).

Patouillard, N. (1900). "Essai taxonomique sur les familles et les genres des Hyménomycètes." Lucien, Lons-le-Saumier.

Perreau, Jacqueline. (1969). Les clavaires. *Rev. Mycol.* **33**:396–415.

Petch, T. (1925). Notes on Ceylon Clavariae. *Ann. Roy. Bot. Gard. Peradeniya* **9**:329–338.

Petersen, R. H. (1964). Notes on clavarioid fungi. II. Corrections in the genera *Ramariopsis* and *Clavaria. Bull. Torrey Bot. Club* **91**:274–280.

Petersen, R. H. (1966). Notes on clavarioid fungi. V. Emendation and additions to *Ramariopsis. Mycologia* **58**:201–207.

Petersen, R. H. (1967a). Notes on clavarioid fungi. VII. Redefinition of the *Clavaria vernalis—C. mucida* complex. *Amer Midl. Natur.* **77**:205–221.

Petersen, R. H. (1967b). Evidence on the interrelationships of the families of clavarioid fungi. *Trans. Brit. Mycol. Soc.* **50**:641–648.

Petersen, R. H. (1967c). Notes on clavarioid fungi. VIII. The *Ramaria pinicola* complex. *Bull. Torrey Bot. Club* **94**:417–422.

Petersen, R. H. (1967d). Notes on clavarioid fungi. VI. Two new species and notes on the origin of *Clavulina. Mycologia* **59**:39–46.

Petersen, R. H. (1968a). The genus *Clavulinopsis* in North America. *Mycologia Mem.* **2**:1–39.

Petersen, R. H. (1968b). *Ramaria* (Holmskjold) S. F. Gray versus *Ramaria* (Fries) Bonordon. *Taxon* **17**:279–280.

Petersen, R. H. "1967" (1969a). Type studies in the Clavariaceae. *Sydowia* **21**:105–122.

Petersen, R. H. (1969b). Type studies in clavarioid fungi. III. The taxa described by J. B. Cleland. *Bull. Torrey Bot. Club* **96**:457–466.

Petersen, R. H. (1969c). Notes on clavarioid fungi. X. New species and type studies in *Ramariopsis*, with a key to species in North America. *Mycologia* **61**:549–559.

Petersen, R. H. ed. (1971a). "Evolution in the Higher Basidiomycetes." University of Tennessee Press, Knoxville, Tennessee.

Petersen, R. H. (1971b). Familial interrelationships in the clavarioid and cantharelloid fungi. *In* (R. H. Petersen. ed.). "Evolution in the Higher Basidiomycetes". University of Tennessee Press, Knoxville, Tennessee.

Petersen, R. H. (1971c). Notes on clavarioid fungi. IX. Addendum to *Clavulinopsis* in North America. *Persoonia* **6**:219–229.

Petersen, R. H. (1971d) A new genus segregated from *Kavinia* Pilát. *Ceska Mykol.* **25**:129–134.

Petersen, R. H. (1972). Notes on clavarioid fungi. XII. Miscellaneous notes on *Clavariadelphus*, with a new segregate genus. *Mycologia* **64**:137–152.

Petersen, R. H., and P. D. Olexia. (1967). Type studies in the clavarioid fungi. I. The taxa described by Charles Horton Peck. *Mycologia* **59**: 767–802.

Petersen, R. H., and P. D. Olexia. (1969). Notes on clavarioid fungi. XI. Miscellaneous notes on *Clavaria. Can. J. Bot.* **47**:1133–1142.

Phylimonova, N. M. (1965). Geographicae elementa Clavariaceae fungorum e Kasachstania. *Not. Syst. Herb. Inst. Bot. Acad. Sci. Kasachstanicae* **3**:74–84.

Pilát, A. (1948). Velenovskyi species novae Basidiomycetum. *Opera Bot.* **6**:1–317.

Pilát, A. (1958). Ubersicht der Europäischen Clavariaceen unter besonderer Berücksichtigung der tschecoslowakischen Arten. (Czech title). *Sb. Narod. Mus. Praze* **14**:129–255.

Reid, D. (1965). A monograph of the stipitate stereoid fungi. *Beih. Nova Hedwigia* **18**:1–382.

Remsberg, R. E. (1940). Studies in the genus *Typhula, Mycologia* **32**:52–96.

Rogers, D. P. (1971). Patterns of evolution to the homobasidium. *In* "Evolution in the Higher Basidiomycetes" (R. H. Petersen, ed.), pp. 241–257. University of Tennessee Press, Knoxville, Tennessee.

Schwalb, M. (1966). Antibiotic production in the Dacrymycetaceae. *Nature (London)* **211**:311–312.

Shvarzman, C. P. (1964). "Flora Sporovych Rastenij Kazachstana." Acad. Sci., Kazach SSR, Alma Ata. (Russian title).

Singer, R. (1945). New genera of fungi. II. *Lloydia* **8**:139–144.

Singer, R. (1962). "The Agaricales in modern taxonomy," 2nd ed. Cramer. Weinheim.

Smith, A. H. (1971). Some observations on selected species of *Clavariadelphus. Mycologia* **63**:1073–1076.

Smith, A. H., and C. R. Leathers. (1967). Two new species of clavarioid fungi. *Mycologia* **59**:456–462.

Talbot, P. H. B. (1954). Micromorphology of the lower Hymenomycetes. *Bothalia* **6**:249–299.

Teodoro, N. G. (1937). An enumeration of Philippine fungi. *Comm. Philipp. Dep. Agr. Manila, Tech. Bull.* **4**:1–585.

Thind, K. S. (1961). "The Clavariaceae of India." India Dept. Agric. New Delhi.

Thind, K. S., and G. P. S. Anand. (1956a). The Clavariaceae of the Mussoorie Hills. I. *J. Indian Bot. Soc.* **35**:92–102.

Thind, K. S. and G. P. S. Anand. (1956b). The Clavariaceae of the Mussoorie Hills. III. *J. Indian Bot. Soc.* **35**:171–180.

Thind, K. S., and G. P. S. Anand. (1956c). The Clavariaceae of the Mussoorie Hills. IV. *J. Indian Bot. Soc.* **35**:323–332.

Thind, K. S., and S. Dev. (1956). The Clavariaceae of the Mussoorie Hills. V. *J. Indian Bot. Soc.* **35**:512–521.

Thind, K. S., and S. Dev. (1957a). The Clavariaceae of the Mussoorie Hills. VI. *J. Indian Bot. Soc.* **36**:92–103.

Thind, K. S., and S. Dev. (1957b). The Clavariaceae of the Mussoorie Hills. VIII. *J. Indian Bot. Soc.* **36**:475–485.

Thind, K. S., and G. S. Raswan. (1958). The Clavariaceae of the Mussoorie Hills. X. *J. Indian Bot. Soc.* **37**:455–469.

Thind, K. S., and S. S. Rattan. (1967). The Clavariaceae of India. XI. *Proc. Indian Acad. Sci.* **66**:143–156.

Ulbrich, E. (1928). Die Höheren Pilze Basidiomycetes. *Kryptogamenflora.*

Vaillant, S. (1797). "Botanicon Parisiense." Leyde.

van der Byl, P. A. (1932). Oor enige Suid-Afrikaanse *Clavaria*-soorte of Knotsswamme. *S. Afr. J. Sci.* **29**:317–323.

van Overeem, C. (1923). Beiträge zur Pilzflora von Niederländischen Indien. *Bull. Jard. Bot. Buitenzorg* [3] **5**:247–293.

Velenovsky, J. (1947). Novitates mycologicae novissimae. *Opera Bot.* **4**:1–167.

Wehmeyer, L. E. (1935). Contributions to a study of the fungus flora of Nova Scotia I. *Pap. Mich. Acad. Sci., Arts Lett.* **20**:233–266.

Wells, V., and P. E. Kempton. (1968). A preliminary study of *Clavariadelphus* in North America. *Mich. Bot.* **7**:35–57.

III. CANTHARELLOID HOLOBASIDIOMYCETES

Surely the most comprehensive treatise on these fungi to date is that by Corner (1966). Generally I agree with the broad usage of "cantharelloid" to indicate those agaricoid basidiocarps with hymenial folds, ridges or merulioid surface, but many of the genera taken up under this adjective by Corner are so blatantly related to agaric groups that to include them here (even in the key) is to do an injustice to the truly cantharelloid members of the group. Thus the following genera included by Corner are not to be found in this chapter, but should be sought under the Agaricaceae or Thelephoraceae: *Caripia, Geopetalum, Hygrophoropsis, Leptoglossum, Mycenella, Nyctalis, Pterygellus, Rimbachia* and *Trogia.*

Virtually no monographs of this group are available for restricted areas. Under various taxonomic arrangements, however, the most common taxa are included in several major works on agaricoid fungi. Notable exceptions are the works of Corner (1966), Smith (1968), Smith and Morse (1947), and Petersen (1971a), in which various genera are taken up for subcontinental areas.

KEY TO THE GENERA OF CANTHARELLOID HOLOBASIDIOMYCETES

1. Dichophyses abundant in tramal tissues . . . . **Dicantharellus**
   This seems to be the only cantharelloid representative of the Lachnocladiaceae. Ref.: Corner, 1966.

1′. Dichophyses not present in tramal tissues 2

   2(1′) Spores roughened; cream, ochraceous or rusty in prints 3

   2′(1′) Spores smooth; white, cream, yellow, salmon or pink 4

3(2) Gloeoplerous hyphal system present, terminating as cystidial elements
   . . . . **Gloeocantharellus**

3′(2) Gloeoplerous hyphal system rudimentary, or if extensive, not terminating in cystidial elements . . . . **Gomphus**
   The type genus of the Gomphaceae, and related to *Ramaria* of the clavarioid group, basidiocarps here often are multipileate, and pleuro- to mesopodal. Spores are ornamented, the ornamentation cyanophilous. Although world-wide, species diversity seems most intense in far western North America. Ref.: Corner, 1966, 1969; Smith and Morse, 1947; Smith, 1968; Petersen, 1969, 1971a,b; Donk, 1933: Synonyms include *Neurophyllum*.

   4(2′) Consistency of basidiocarp extremely watery . . . . **Goossensia**
   Although apparently distinct to the original author, the watery consistency has been a puzzle to other authors. Its separation from *Cantharellus* appears questionable. Ref.: Heinemann, 1958; Corner, 1966.

   4′(2′) Consistency of basidiocarp normal, fleshy 5

5(4′) Tramal hyphal septa clamped . . . . **Cantharellus**
   Species without clamps have been included under this genus, but I am at a loss to separate *Cantharellus* from *Craterellus* and *Pseudocraterellus* on any other constant character. Although several species are gastronomically prized, the taxonomy of the genus is still confused, but several excellent treatises are now available. Chemotaxonomy has also been applied with success. Ref.: Corner, 1957, 1966, 1969; Donk, 1958, 1969; Petersen, 1971b; Petersen and Ryvarden, 1972; Arpin, 1966; Arpin and Fiasson, 1971; Fiasson and Arpin, 1967; Fiasson and Bouchez, 1968; Fiasson *et al.*, 1970. Phylogenetic relationships are with *Hydnum s. str.*, through microanatomical characters and stichic meiotic nuclear spindle orientation. Ref.: Donk, 1964. Synonyms include *Merulius* and *Alectorolophoides*.

5′(4′) Tramal hyphal septa without clamps 6

   6(5′) Tramal hyphae commonly secondarily septate . . . . **Pseudocraterellus**
   This genus is typified by *P. sinuosus*, but the diagnostic character varies in that species, and is also present in some species of *Craterellus*, raising doubts as to the genus's taxonomic validity. Ref.: Corner, 1957, 1966; Petersen, 1968.

6'(5') Tramal hyphae rarely or never secondarily septate . . . . **Craterellus**
Most species produce dark-colored, pervious basidiocarps. The genus limitations
are indistinct, with *Cantharellus* and *Pseudocraterellus* the nearest relatives. Ref.:
Corner, 1957, 1966; Petersen, 1968; Smith, 1968.

REFERENCES

Most surveys of agaricoid basidiomycetes include species of cantharelloid fungi. The best
summary of these may be found in Corner (1966).

Arpin, N. (1966). Recherches chimiotaxinomiques sur les champignons. Sur la présence de
carotènes chez *Clitocybe venustissima* (Fries) Sacc. *C. R. Acad. Sci.* **262**:347–349.

Arpin, N., and J. L. Fiasson. (1971). The pigments of Basidiomycetes: Their chemotaxonomic
interest. *In* "Evolution in the Higher Basidiomycetes" (R. H. Petersen, ed.).
University of Tennessee Press, Knoxville, Tennessee.

Coker, W. C. (1919). *Craterellus, Cantharellus* and related genera in North Carolina. *J. Elisha
Mitchell Sci. Soc.* **35**:24–48.

Corner, E. J. H. (1957). *Craterellus* Pers., *Cantharellus* Fr. and *Pseudocraterellus* gen. nov.
Sydowia, Beih. **1**:266–276.

Corner, E. J. H. (1966). A monograph of cantharelloid fungi. *Ann. Bot. Mem.* **2**:1–255.

Corner, E. J. H. (1969). Notes on cantharelloid fungi. *Nova Hedwigia* **18**:783–818.

Donk, M. A. 1933. Revision der Niederländischen Homobasidiomycetes-Aphyllophorales.
Meded. Bot. Mus. Rijksuniv. Utrecht **9**:1–278.

Donk, M. A. (1957). The generic names proposed for Hymenomycetes. VII. "Thelephoraceae."
Taxon **6**:17–28.

Donk, M. A. (1958). The generic names proposed for Hymenomycetes. IX. "Meruliaceae"
and *Cantharellus* s. str. *Fungus, Wageningen* **28**:7–15.

Donk, M. A. (1964). A conspectus of the families of Aphyllophorales. *Persoonia* **3**:199–324.

Donk, M. A. (1969). Notes on *Cantharellus* sect. *Leptocantharellus. Persoonia* **5**:265–284.

Fiasson, J. L., and N. Arpin. (1967). Recherches chimiotaxinomiques sur les champignons.
V. Sur les carotenoïdes mineurs de "*Cantharellus tubaeformis*" Fr. *Bull. Soc. Chim. Biol.*
49:537–542.

Fiasson, J. L., and M. P. Bouchez. (1968). Recherches chimiotaxinomiques sur les champignons.
Les carotènes de *Omphalia chrysophylla* Fr. *C. R. Acad. Sci.* **266**:1379–1381.

Fiasson, J. L., R. H. Petersen, M–P Bouchez, and N. Arpin. (1970). Contribution biochimique
a la connaissance taxinomique de certains champignons cantharelloïdes et clavarioïdes.
Rev. Mycol. [N. S.] **34**:357–364.

Heim, R. (1954). A propos de trois chanterelles americaines. *Rev. Mycol.* **19**:46–56.

Heinemann, P. (1958). Champignons recoltes au Congo Belge par Madame M. Goossens-
Fontana. III. Cantharellineae. *Bull. Jard. Bot. Brux.* **28**:385–438.

Heinemann, P. (1959). "Flore Iconographique de Champignons du Congo," Fasc. 8. Brussels.

Peck, C. H. (1892a). New York species of *Cantharellus. N. Y. State Mus. Bull.* **1**:34–43.

Peck, C. H. (1892b). New York species of *Craterellus. N. Y. State Mus. Bull.* **1**:44–48.

Petersen, R. H. (1968). Notes on cantharelloid fungi. I. *Gomphus* S. F. Gray, and some clues
to the origin of ramarioid fungi. *J. Elisha Mitchell Sci. Soc.* **84**:373–381.

Petersen, R. H. (1969). Notes on cantharelloid fungi. II. Some new taxa, and notes on *Pseudocraterellus. Persoonia* **5**:211–223.

Petersen, R. H. (1971a). The genera *Gomphus* and *Gloeocantharellus* in North America. *Nova
Hedwigia* **22**:1–112.

Petersen, R. H. (1971b). Interfamilial relationships in the clavarioid and cantharelloid fungi.
In "Evolution in the Higher Basidiomycetes" (R. H. Petersen, ed.). University of Tennessee
Press, Knoxville, Tennessee.

Petersen, R. H., and L. Ryvarden. "1971" (1972). Notes on cantharelloid fungi. IV. Two new species of *Cantharellus*. *Sve. Bot. Tidskr.* **65**:399–405.

Singer, R. (1945). New genera of fungi. II. *Lloydia* **8**:139–144.

Singer, R. (1947). Coscinoids and coscinocystidia in *Linderomyces lateritius*. *Farlowia* **3**: 155–157.

Singer, R. (1963). Two new genera of fungi for South America. *Vellozia* **1**:14–19.

Smith, A. H. (1968). The Cantharellaceae of Michigan. *Mich. Bot.* **7**:143–183.

Smith, A. H., and E. E. Morse. (1947). The genus *Cantharellus* in the western United States. *Mycologia* **39**:497–534.

CHAPTER 21

Aphyllophorales III: Hydnaceae and Echinodontiaceae

KENNETH A. HARRISON

University of Michigan Herbarium
Ann Arbor, Michigan

I. INTRODUCTION

The character used to recognize the Hydnaceae originally was the presence of positively geotropic spines. These ranged from small granular warts, individual spines, and clusters of spines, to spines supported by various types of pileate basidiocarps. Since 1821 the family has been subdivided and rearranged many times, often when new genera are being proposed. Almost all hydnums are rare, and many only produce fruit bodies at intervals of several years.

It has been found that endemic species occur in every region in North America, and there is good reason to estimate that less than half the species present in Canada and the United States have been found and described. In addition, there are vast regions in the tropics and Asia that are completely unknown mycologically. Even in Europe old species not seen for years are being rediscovered and redescribed, and occasionally new species are found that make it necessary to revise classifications. For this reason the proposed systematic arrangement in this chapter is conservative, and the grouping of species into genera and subfamilies is based largely on easily recognized macroscopic characters.

The complexity of the family, the incompleteness of even local floras, and the difficulty of recognizing many species makes it uncertain what early mycologists were describing. This is particularly true of prestarting point records, so a strong plea is being made to designate modern, adequately described neotypes that will stabilize the nomenclature in this poorly understood family.

The classification to species and genera is based on differences in appearance, growth habit, habitat, color, context, odor, and taste, and microscopic differences in the spores, hyphal characters, clamps, cystidia, gloeocystidia, and reactions to chemical reagents.

A.   Development of the Basidiocarp

The growth patterns exhibited by the basidiocarps in the Hydnaceae vary. In the centrally stipitate genera *Hydnellum* and *Phellodon*, the margins of the pilei grow in an indeterminate manner. The pileus continues to expand as long as growing conditions are favorable. In *Hydnum* and *Bankera*, the growth is of a determinate pattern with the pileus margin having a limited capacity to expand. Associated with these growth patterns are differences in context, as the flesh of the latter is fleshy, rather brittle, and more or less homogenous, while in *Hydnellum* and *Phellodon* the flesh is tough, sessile, duplex, and zonate.

In the four genera mentioned above the basidiocarp develops from a rounded pad of mycelium. At a certain stage the hyphae grow upward, forming a compact column that emerges as a young stipe that looks like a young *Clavaria*. In *Hydnellum* and *Phellodon* the young stipes thicken and lengthen to a height slightly above the level of the surrounding duff, or when among lichens, grass, etc., may be only partially above. At this critical height the marginal hyphae bend outward and grow horizontally. The upper surface thickens, with the hyphae of the upper part expanding laterally, while those of the lower section turn downward to form a thin parenchymatous layer that becomes the base of the spines and supports the hymenium between. The positively geotropic spines appear directly beneath the overhang of the margin—often so close that the first spines do not have space to project downward free of the stipe. The margin of the pileus, as it expands laterally, surrounds small projecting objects such as twigs, grass, lichens or pine needles. Any contact with a solid obstacle results in an eccentric basidiocarp, and when it encounters other pilei there are fusions which result in irregular concrescent sporophores. In *Bankera*, *Hydnum*, and *Dentinum* the young basidiocarps are pileate earlier than in the indeterminate group. The expansion of the pileus is much more rapid due to the inflation of the individual cells of the tramal tissues. Another difference is that the margin does not grow around objects and debris, and the context is free of foreign material. The young margins are thick or incurved, and do not fuse with other pilei even when they are in contact. Often in a cespitose group of pilei only one or two become dominant, while the others collapse without developing.

The context of the hydnums with a lignicolous habitat varies from woody, through fibrous, coriaceous, waxy to membranaceous, and may be conchate, imbricate, resupinate, or reflexed resupinate, or a few may be eccentrically stipitate and irregularly pileate. The spines are of many types, suggesting that they have originated from divergent lines. The color of the hydnums varies widely, with browns rather common. Odor and taste are extremely difficult to describe. Some hydnums have unpleasant, mealy, bitter, or medicinal tastes, while odors may be unpleasant, fragrant, mealy, or of fenugreek. Species of *Bankera* and *Phellodon* have a distinctive odor of slippery elm (fenugreek? coumarine?) after drying that has been used as a generic character. These are all helpful in recognizing species once they are familiar to the mycologist, but are confusing when used in a key with terms that cannot be standardized for general use.

B.   Habitats

The terricolous Hydnaceae are associated with well-drained coniferous or deciduous woods especially where oaks are common. Some are abundant under *Pinus* spp. on sand plains, while others are more frequently found under *Picea* and *Abies* spp. on heavier soils. They grow best in moderately cool weather when there are frequent showers. In the northern parts of the United States and Canada some of the hydnellums will continue to grow for periods of two or three months, expanding when moist, and surviving moderate periods of dry weather until killed by the first sharp frost in the fall. In the mountains of North Carolina in 1971, hydnums were growing nicely in mid-August, but soon became infested with insect larvae and started to rot so that most fruit bodies had collapsed by mid-September. Hydnums in North America range from Florida in the south to the cool northern conifer forests of Alaska, and may be strangely absent from large areas while abundant in a few restricted spots. Recent collecting indicates that there is also a southern flora concentrated for a short time in certain regions. Lignicolous species are widely distributed throughout the world, and are usually associated with certain types of woody substrates.

C.   Microscopic Characters

The importance of hyphal characters for the study of fungi was established by Corner (1932) and has proved useful in the Polyporaceae. Maas Geesteranus (1962, 1963), in a series of studies, has described details of the structure of many type species in a number of genera. *Bankera, Phellodon, Hydnum, Hydnellum,* and *Dentinum* are monomitic with some binding hyphae. Hyphal walls thicken somewhat and oleiferous hyphae can be recognized following an irregular pathway through the tissues.

Clamps are present in some species in all genera except *Bankera*. They have recently been demonstrated in a new species of *Phellodon* (Harrison, 1972). In some, it is not always easy to recognize functional clamps because binding hyphae develop as buds in much the same position and resemble them from many angles. It is also difficult to be certain that clamps are not present. Several cases are known where the clamp at the base of a basidium gives rise in succession to a new basidia, forming fascicles without apparent clamps. They are most easily seen in a section of an immature hymenium when the first and second basidia are forming. Many patterns of clamps are found in the various species. They range from large medallion types with large loops, to small ones easily overlooked in the angles at the septa of large inflated hyphae.

Gloeocystidia and cystidia have not been recognized in terricolous species, but are common in lignicolous species and genera, where their presence or absence may be significant at the generic level. In one spined genus cystidiallike hyphae form the axis of the spines. These characters are variable. Maas Geesteranus (1964) has documented a very interesting geocline in *Steccherinum rawakense*, showing the variation occurring in different parts of the world.

D.   Spore Characters

The spores of hydnums vary widely between the different genera, but are often so similar in size that they are of little use, except to the specialist, for identification at the species level. In *Phellodon, Hericium*, and related genera, the uniformity of size is associated with very fine roughenings or echinulations. Recent illustrations using the scanning microscope indicate that it may be possible to recognize species and develop generic concepts from features that are not resolved by ordinary light microscopes. *Phellodon* was included in the Bankeraceae Donk, but Grand and Moore (1970, Figs. 14–17) show that its spores have the basic structure of *Hydnum* and *Hydnellum*, and this is strong support for retaining the four genera in one family.

E.   Chemical Reactions

Melzer's reagent may give various color reactions apparent either macroscopically or microscopically: A dextrinoid reaction is reddish brown, shades of bluish black to grayish black are amyloid, and when grayish it is "apparent amyloid" (Harrison, 1964). The latter is a slow reaction with the granules faintly colored at first, intensifying, and increasing in size on standing. These changes may be present in the hyphae or spores or both. KOH solutions vary in strength but 3% is commonly used. The usual color reactions are the intensification of the brown color of hyphal walls (xanthochroic), or greenish blue, dull, or dark bluish to black on the hyphae or tramal tissues. Reactions, especially with 3% potash, are most useful at the species or stirps level, but with the reservation that the color changes can be produced by a large number of chemicals. The "apparent amyloid" reaction is partially linked to a faint blue-green color reaction in the same hyphae with KOH.

F.   Problems in Classification

The problems facing mycologists at the present time are that fungi have evolved from ancestral lines which probably had their origins, at the latest, in the early carboniferous age. Since then, fungi have been under the environmental pressures produced by fluctuations of climate; until now fungi are adapted to the multitudinous seasonal variations found throughout the world. Fungi appear to adapt readily, so that it is not uncommon to find apparent endemic species in regions that have emerged from the most recent ice age.

It has not been possible to separate the Hydnaceae clearly from thelephoroid or polyporoid families. The positively geotropic spines have microscopic characters found in negatively geotropic clavarias, and similar chemical reactions occur in several genera of hydnums, polypores, and clavarias. These, and spore features, are all useful in separating species and genera, but it is uncertain which should be considered the dominant character for the purposes of classification. As a result, systems are changed at will to indicate the preferences of individual mycologists. *Clavaria* specialists include borderline genera in their groups, while enthusiasts for the thelephoras claim many resupinate forms. Splinter genera are being proposed, with conflicting claims for the validity of characters being used in delimiting a particular group.

The dividing line between a hymenium that is smooth (corticioid) or of granular warts (grandinioid) is an arbitrary one. Variations in the amounts of roughness range from almost smooth through rounded knobs and warts (raduloid), and on to thick or thin sharp-pointed spines (hydnoid). When subulate spines are examined, the tips are sterile, and in *Hydnum* and *Hydnellum* these continue to lengthen throughout the lifetime of the fruit body, and are often the last parts to die. In the genus *Odontia* s. l., which is also a good corticioid fungus, the spine tips appear fimbriate—an effect produced by the protruding tips of skeletal cystidia forming the axis.

The origin of the spines varies in the different genera. Usually the youngest surface on the margin of a pileus is sterile and smooth, then tiny knobs form in a random pattern and grow into the shape typical of the species. Sometimes reticulations or outlines of pores appear (poroid), and these increase unevenly in height, with spines finally developing from the elevations. In a few cases the pore walls also split, forming flattened teeth, or the spines are flattened and irregular as they grow (irpicoid). In *Hericium* the spines are the ends of branches (hericioid), although in *H. ramosum* they are a special type growing along the lower surface of all main branches. As the various types of hymenophores are not exclusively limited to one genus, there is a trend to consider microscopic resemblances as more significant than gross morphological similarities. In this treatment, gross morphology is the dominant character because of its long use. Many recent proposals are interesting, but new finds and studies are resulting in a continuous succession of changes which need to be evaluated over a period of time.

G.   *Theories on Pathways of Development*

It is logical to assume that the first fungus colony was a flattened pad of mycelium. Any physical or physiological variation permitting survival under the fierce competition that occurs in nature would have an advantage. One of the factors that has profoundly influenced the form of fungi has been that much of their life is spent in habitats where water conservation is not essential for survival. The fungi as a whole have failed to develop any outstanding method for preventing water loss, and the adaptions that do exist are almost exclusively confined to fruit-body development. All need high humidity, rainy days, or damp locations for the production and release of vast quantities of spores necessary for the continuation of the species.

It is postulated that a primitive hymenophore is one with a palisade of basidia forming a smooth covering to a flat surface, and there will always be a built-in tendency to revert to such a form. The modifications giving an advantage are those that increased the area and the longevity of the fruiting surface. Fruiting areas can be increased by the production of elevations of any shape, and spines probably would be one of the earlier types of support for basidia. Elevations can evolve into stalks or clubs or branches (*Clavaria*), clubs can expand into vases (*Gomphus*), or stalks can grow and then marginal growth develop laterally at the apices into pilei with the hymenophore positively geotropic. The most sophisticated development in the Hydnaceae is the evolution of an elevated fruit body from the apex of a stipe. In *Hydnellum* this involves genetic control for the regulation of mycelial growth that is at first negatively geotropic from an ageotropic subiculum, then changing to diageotropic for the formation of the pileus, and finally positively geotropic for the spines. It is all but impossible to establish that any special structure is derived, reduced, or an improved modification of a similar structure in another species or, in other words, to decide what is primitive or advanced in structures or characters.

It would appear that a number of widely different basidiocarps (clavarioid, boletoid, agaricoid) are more or less equal in survival values. However, because of the rarity of spinulose forms and their wide variability, it is considered that they are relatively inefficient, and that spined fungi have a lower chance of survival. A few species have been moderately successful because of having gained an advantage by establishing themselves in specialized niches (*Auriscalpium vulgare*). Spined forms are not confined to one family or even one order. One of the most striking is *Pseudohydnum gelatinosum* in the Tremellales.

H.   History of Nomenclature

The name *Hydna* was introduced by Linnaeus in 1735, changed to *Hydnum* (1737a) to replace *Erinaceus*, first used for the spined fungi by Dillenius (1719) and later by Micheli (1729) for a group of eight stipitate hydnums. Linnaeus (1753) confirmed the use for stipitate hydnums in "Species Plantarum," when he published *Hydnum* with four species: *H. imbricatum* L., *H. repandum*, *H. tomentosum*, and *H. auriscalpium* in this sequence. *Hydnum* L. was used by Persoon (1801) for a group of ten species, and was further expanded by Fries (1821) when he included 88 species under this name, and grouped them in tribes: Mesopus, Pleuropus, Merisma, Apus, and Resupinatus. Persoon and Fries also continued the use of *H. imbricatum* as the first species in the genus, and this was the only species included under the generic name *Hydnum* by S. F. Gray (1821) when he subdivided the hydnums into six genera. By including *H. imbricatum*, Gray maintained an original concept of the genus. The same group was later typified by the same species under the name *Sarcodon* by Cooke and Quélet (1878) and Karsten (1881).

The need to subdivide the original group of hydnums was recognized by Fries, and he proposed five genera in 1825. By 1874 he had divided and enlarged it to a family of eleven genera: *Hydnum, Hericium, Tremellodon, Sistotrema, Irpex, Radulum, Phlebia, Grandinia, Odontia, Kneiffia*, and *Mucronella*. Saccardo (1888) accepted the Friesian genera and added *Caldesiella, Lopharia*, and *Grammothele*. Of the fourteen genera *Sistotrema, Tremellodon, Irpex*, and *Phlebia*, plus the three added by Saccardo, are now usually omitted, while *Kneiffia* has disappeared into other genera.

The establishment of the Code of Nomenclature, and careful research on the priority of various names and their typification, has resulted in many name changes. Also, the taxonomy is constantly changing as a result of new discoveries in morphology and species. The latest complete attempt to classify the Hydnaceae was by Nikolajeva (1961), who kept them in one family of two subfamilies of seventeen genera. Since then, a number of new genera have been proposed, and the following is suggested as a possible arrangement of the family. This treatment differs from that of Nikolajeva because, by adhering to the Code, the type of *Hydnum* L. ex Fr. is *H. imbricatum*, not *H. repandum*, and permits the retention of several genera derived from the original *Hydnum* Fr. in the Tribe Hydneae Fr.

I. Classification

HYDNACEAE

I. Subfamily: Hydnoideae.
   1. Tribe: Hydneae.
   Genera: *Hydnellum, Hydnum, Phellodon, Bankera.*
Related genera with hydnoid characters: *Boletopsis, Polyozellus.*
   2. Tribe: Dentineae.
   Genus: *Dentinum.*
   3. Tribe: Hydnodoneae.
   Genus: *Hydnodon.*
II. Subfamily: Hericioideae.
   1. Tribe: Hericieae.
   Genera: *Hericium, Creolophus, Dentipratulum.*
   2. Tribe: Mucronelleae.
   Genera: *Mucronella, Dentipellis, Delentaria, Hormomitaria.*
A related genus: *Deflexula.*
   3. Tribe: Gloiodoneae.
   Genera: *Gloiodon, Auriscalpium.*
Related genera: *Beenakia, Hydnopolyporus.*
   4. Tribe: Climacodoneae.
   Genera: *Climacodon, Mycoleptodonoides, Cautinia.*
Related genera: *Gloeodontia, Kavinia.*
   5. Tribe: Odontieae.
   Genera: *Odontia, Sarcodontia, Steccherinum, Stecchericium, Mycoacia, Radulum, Basidioradulum, Grandinia.*
Related genera: *Irpex, Hydnochaete, Sistotrema, Gyrodontium.*

ECHINODONTIACEAE

Genus: *Echinodontium.*

II. KEY

Many of the proposed newer genera are included in the following key, and can be located by reference to chemical reactions or microscopic characters used to identify them. *Boletopsis* and *Polyozellus* are included but are out of place in a spined family key. The first is a fleshy polypore with brown tuberculate spores, the other is a cantharelloid species with hyaline tuberculate spores. They are included because the shape of the spores is hydnoid. *Polyporus tomentosus* was not included although it was found in some North American herbaria as an unknown *Hydnellum.*

This may seem strange, but many hydnums have been found with the spines fused so that the hymenophore appears more or less poroid.

KEY TO HYDNACEAE AND SOME RELATED GENERA

Basidiocarp variable, supporting positively geotropic spines of various types. Hymenophore hydnoid, hericioid, odontoid, irpicioid, grandinioid, or nearly corticioid.

1. Basidiocarp almost exclusively terrestrial, probably associated with roots of trees (mycorrhizal) 2

1'. Basidiocarp almost exclusively lignicolous, with a very few on forest duff. (Probably exclusively saprobes in that they break down woody tissues) 11

2(1) Basidiocarp stipitate, context tough or fleshy, spores white or colored 3

2'(1) Basidiocarp variable, hymenophore poroid, smooth or obscurely spined . . . 7

3(2) Spores brown, tuberculate 4

3'(2) Spores hyaline, smooth, or echinulate 5

4(3) Basidiocarp with context tough, scissile, growth indeterminate
. . . . **Hydnellum** p. 381

4'(3) Basidiocarp with context fleshy, brittle, growth determinate . . **Hydnum** p. 380

5(3') Basidiocarp fleshy, spores smooth, thin-walled . . . . **Dentinum** p. 382

5'(3') Basidiocarp fleshy or tough, spores echinulate 6

6(5') Basidiocarp fleshy, context brittle . . . . **Bankera** p. 381

6'(5') Basidiocarp tough, context scissile . . . . **Phellodon** p. 381

7(2') Basidiocarp with irregular spines having evidence of pores on margin, or poroid (Polyporaceae) 8

7'(2') Basidiocarp with hymenium smooth or plicate, with thelephoroid characteristics . . 9

8(7) Basidiocarp poroid, spores brown, hyaline under microscope, tuberculate as in *Hydnellum* . . . . **Boletopsis** p. 383

8'(7) Basidiocarp floccose, spines irregular on ridges, spores hyaline, smooth, basidia urniform . . . . **Sistotrema** p. 392

9(7') Basidiocarp cuneately segmented with spines pectinately arranged
. . . . **Hydnopolyporus** p. 386

9'(7') Basidiocarp irregular or more or less cantharelloid 10

10(9') Context light-colored, slightly gelatinous, spines subulate, spores pink
. . . . **Hydnodon** p. 383

10'(9') Context dark-colored, brittle; hymenium smooth to plicate, spores hyaline
. . . . **Polyozellus** p. 383

11(1') Basidiocarp range from resupinate corticioid forms with smooth hymenium, to spines, including pileate and branched forms 12

11'(1') Basidiocarp of single spines, aggregates, or clusters from a common base, subiculum not evident 15

12(11) Basidiocarp a tubercle or system of branches supporting spines which are positively geotropic, but may project horizontally at times 13

12'(11) Basidiocarp not as above 14

13(12) Context monomitic, context and spores strongly amyloid . . . . . **Hericium** p. 384

13'(12) Context dimitic, spores and context inamyloid. Relationship with *Pterula* suggested
. . . . **Deflexula**   p. 386

   14(12') Basidiocarp perennial, resupinate to ungulate, hymenophore smooth, coarse
   spines or tubercles; spores hyaline, amyloid
. . . . **(Echinodontiaceae) Echinodontium**   p. 392

   14'(12') Basidiocarp not as above 18

15(11') Basidiocarp of individual spines with a short sterile stipe. Spores inamyloid. Tropical
. . . . **Hormomitaria**   p. 385

15'(11') Basidiocarp of individual or rarely branched spines. Spores amyloid or not, no sterile
stipe 16

   16(15') Individual spines without gloeocystidia, spores faintly amyloid
. . . . **Mucronella**   p. 385

   16'(15') Spines clustered, from a common base 17

17(16') Tramal tissue of spines with gloeocystidia, spores strongly amyloid
. . . . **Dentipratulum**   p. 385

17'(16') Tramal tissue of spines with gloeocystidia, spores inamyloid . . **Delentaria**   p. 385

   18(14') Context darkening in KOH (xanthochroic) 19

   18'(14') Context not darkening in KOH 20

19(18) Asterosetae abundant in subiculum, hymenium with setae . . . . . **Asterodon**   p. 414

19'(18) Asterosetae absent, hymenium with setae . . . . **Hydnochaete**   p. 392

   20(18') Basidiocarp resupinate, hymenophore hydnoid or corticioid (also check 32–34).
. . . . 21

   20'(18') Basidiocarp reflexed, imbricate, conchate, dimidiate, or stipitate, hymenophore
   hydnoid 25

21(20) Hymenium hydnoid, reflexed resupinate; spores thick-walled, amyloid
. . . . **Dentipellis**   p. 385

21'(20) Hymenium corticioid; spores thin-walled 22

   22(21') Spores ochraceous, inamyloid (cyanophilus) . . . . **Kavinia**   p. 389

   22'(21') Spores hyaline, amyloid 23

23(22') Context dimitic . . . . **Gloeodontia**   p. 389

23'(22') Context monomitic 24

   24(23') Basidiocarp resupinate . . . . **Gloeocystidellum**   p. 389

   24'(23') Basidiocarp reflexed resupinate . . . . **Laxitextum**   p. 389

25(20') Spores amyloid 26

25'(20') Spores inamyloid 29

   26(25) Basidiocarp eccentric to centrally stipitate, usually on cones of *Pinus* spp.
. . . . **Auriscalpium**   p. 387

   26'(25) Basidiocarp usually reflexed resupinate, imbricate, or dimidiate 27

27(26') Basidiocarp of branches within a mat of tomentum, spines from lower side of branches
. . . . **Gloiodon**   p. 386

27'(26') Basidiocarp reflexed pileate or imbricate 28

   28(27') Gloeocystidia as in *Hericium* (European) . . . . **Creolophus**   p. 384

III.   FAMILIES, SUBFAMILIES, TRIBES, AND GENERA

A.   Hydnaceae

The sporophore of Hydnaceae is variable, resupinate, branched, stipitate, fleshy, coriaceus, ceraceous, or floccose. The hymenium is found on positively geotropic spines, granules, warts, or elevations. Spores are white or colored, smooth, echinulate, or tuberculate. The family is terricolous or lignicolous.

This is a family of extremely diverse elements that have obviously evolved over a long period of time from very ancient forms. It is highly probable that the genera are the result of parallel evolution, and that spines have arisen independently a number of times. Hydnums have been classified in many ways depending on the weight given to the different characters. Here the family is divided into two subfamilies, Hydnoideae and Hericioideae.

1. Subfamily. Hydnoideae

The sporophore is pileate, centrally stipitate, and fleshy to fibrous. The hymenium is found on subulate spines. Spores are white or brownish, smooth or echinulate, or coarsely tuberculate. Hydnoideae are terricolous and the type is *Hydnum* (L. ex Fr.) S. F. Gray.

This subfamily contains two of the larger genera and three with limited numbers of species. There has been a trend to stress the connection of the Hydneae with the Thelephoraceae because of tuberculate brown spores in genera of both families. Corner (1968, p. 25) has stated, "The thelephoroid spore has become the criterion of affinity.... this spore brings in many misgivings because of its apparent modifications; that is, this kind of spore has evidently not been immutable." Also, on page 26, he says, "There is in fact no evidence that any fruit body of *Thelephora* can become hydnoid, polyporoid or lamellate."

Fries' original treatment of *Hydnum* pointed out the interesting parallelism between hydnums and polypores, and he mentioned six pairs that were similar (Fries 1821, p. 398): *H. cyathiforme* (=*H. zonatum*) with *Polyporus perennis*; *H. subsquamosum* with *P. subsquamosum*; *H. repandum* with *P. ovinus*; *H. melaleucum* with *P. carbonarium*; *H. velutinum* with *P. tomentosum*, and *H. ochraceum* with *P. abietinum*. Harrison (1971a) mentions six additional pairs based on American species of *Hydnums*.

a.   Tribe. Hydneae.   The sporophore characters are as in the subfamily. Spores are white echinulate or brown tuberculate. The tribe includes four genera: *Bankera* and *Phellodon* with white echinulate spores, and *Hydnum* and *Hydnellum* with brown tuberculate spores. This is closer to the original concept of hydnaceous fungi that goes back to Linnaeus (1753) and Micheli (1729).

i.   Hydnum.   This is a large genus and the limits have been subjected to numerous changes over the years. The Friesian concept included all fungi with positively geotropic spines and warts, and thus contained species now placed in other orders and families. The type is *Hydnum imbricatum*.

S. F. Gray (1821) was the first to subdivide the genus *Hydnum*, validated by Fries, by making six genera to include the British species. He included only *H. imbricatum* in his genus *Hydnum* L. and by doing so made a very strong point for it to be the type of genus. In doing this he followed Linnaeus, Persoon and Fries who used it first in their treatment of *Hydnum* L. In 1878 Quélet (in Cooke and Quélet), used the same type for the genus *Sarcodon* (validated later by Karsten), and thus *Sarcodon* becomes a synonym of *Hydnum* (L. ex Fr.) S. F. Gray. Species with white echinulate spores were included until recently, but these have now been transferred to *Bankera*. At the present time the species included are quite uniform in appearance and microscopic details. A few of the taxa may be confused with *Hydnellum* because of occasional traces of zonate flesh following severe ecological stress, or because the margin may continue to expand and become thin under exceptionally favorable growing conditions. However, microscopically the cells of the trama of the pileus are always inflated, and the tramal tissues are brittle rather than fibrous as in *Hydnellum*.

ii. Hydnellum. This a large genus with a great diversity of forms and colors. Reactions are obtained when a weak solution of potash is applied to the teeth and context of some of the species. Growth takes place under conditions of high humidity and moderate temperature, and variations in these factors cause quite striking changes in appearance. Growth is by radial expansion and is indeterminate. It can continue indefinitely, surrounding objects as the pileus expands, or fusing with adjacent fruit bodies until unfavorable conditions terminate all growth. Apparently all species, when growing very rapidly and when the humidity is high, can exude drops of colored liquid. These are usually reddish, but species are known that produce yellowish or brownish droplets. The related genus *Phellodon* has been observed with blackish droplets. Miller and Boyle (1943) used the name *Calodon*, and included *Phellodon* in the same genus. A request to conserve *Calodon* against *Hydnellum* was refused by the committee on nomenclature.

iii. Bankera. This genus originally contained two North American species, but Maas Geesteranus has since added several European species including *B. mollis* and *B. violascens*, while considering *B. carnosa* a synonym of the former. We still maintain that *B. carnosa* is a distinct North American taxon that should not be included in either a white or a violet-colored European species. Once again the problem is to obtain sufficient material of these rare fungi in all stages of growth, so that the puzzling variations that exist can be recognized.

iv. Phellodon. Over the years, the white-spored stipitate hydnums have been placed in several genera. Fries treated them under *Hydnum*. Quélet in 1878 (in Cooke and Quélet) included them in *Calodon*, and Karsten (1881), when he validated *Calodon* Quél., selected three to form the genus *Phellodon*. Banker added a new fleshy, white-spored hydnum, *P. carnosus*, in 1913, and Miller and Boyle (1943) considered that they should be under *Calodon* as the differences were not sufficient to warrant separate genera. Coker and Beers (1951) selected a fleshy species to form the genus *Bankera*. Since then *Phellodon* has been restricted to species with tough, fibrous flesh.

Recently Donk (1961) proposed the family Bankeraceae for *Phellodon* and *Bankera*, based on white spores that were echinulate, suspected lack of thelephoric acid, odor, and lack of clamps. This year two species have been found with clamp connections. Grand and Moore (1970) have published photographs of spores of *Phellodon niger* var. *alboniger* taken with the scanning electron microscope, showing that what have previously been considered echinulations on the surface are in reality fine tubercles, differing only in size from those on spores of *Hydnellum*. Also, Brady has stated that thelephoric acid was found in species of *Phellodon* as well as *Hydnellum*, collected in the Pacific Northwest (Harrison, 1971a, p. 392). The remaining characters of white spore color and an odor are not considered adequate for distinguishing a family, and *Phellodon* and *Bankera* are being retained in *Hydneae*.

b. Tribe. Dentineae trib. nov. Basionym: Dentinaceae Kotl. et Pouz. See *Ceska Mycologie* **26**:217, 1972.

Dentinum. The genus includes a group of some ten species, and this group is so distinctive that there is no indication of its connection with other families or genera of hydnums.

Erinaceus Dillenius (1719) was the earliest name used for spined fungi and it was probably associated with *Hydnum repandum*, though the plate might represent one of several. The name was next used for a group of stipitate hydnums by Micheli (1729), but Linnaeus changed it to *Hydna* (1735) and *Hydnum* (1737a). In 1753, the starting point for naming the higher plants, Linnaeus used *Hydnum* describing four species binomially, with *H. imbricatum* the first and *H. repandum* the second. The name was accepted by Persoon (1801) and by Fries (1821). Although each in turn greatly expanded the genus, both kept *H. imbricatum* as the first species. In 1821 S. F. Gray selected two of the Persoonian species (Fries also included them), *H. repandum* and *H. rufescens*, to be his new genus *Dentinum*, while he kept *H. imbricatum* as the only species in *Hydnum*. By these selections he established the limits of the genera *Hydnum* L. ex Fries and *Dentinum*.

It has been argued by Donk (1956) that the generic name *Hydnum* has to follow *H. repandum*. It is not felt that this is justified by any of the sequence of events listed above. The starting point for the naming of fungi has been decreed to be January 1, 1821, and the first revisions should be considered authoritative in this case as well as in all other taxonomic procedures, unless the right of conservation is invoked and agreed to (Harrison, 1971b). Also, by following Donk's interpretation, we eliminate Hydnaceae as the family for most of the hydnums, and thus lose a concept that has been associated with this group of fungi during the lifetime of Fries up to the time of Donk.

c. Tribe. Hydnodoneae trib. nov.   Pileus carnosus, repandus vel sinuso-lobatus; aculei brevei, stipes irregularis, excentricus; caro subgelatinosus; sporae corallinae, subglobosae, tuberculatae. Ad terram in silvus. Typus *Hydnodon* Banker.

Hydnodon.   This is a peculiar genus without known affinities. When based on the spines and tuberculate spores it is a hydnum, but the pink spore color suggests Thelephoraceae.

d. Genera with Hydnoid Characters from Other Families.   Four genera previously included in other families because of macroscopic characters have similar spores, and it has been suggested that they should be in the same series with *Hydnellum* and *Hydnum*. *Boletopsis* is a poroid genus that has spores with brown tubercles. *Polyozellus* is a cantharelloid element. It also has stereoid affinities. Two other genera with rough brown spores, belonging to the Thelephoraceae (*Caldesiella* and *Tomentella*), are not discussed.

i. Boletopsis.   Fayod originally suggested a relationship with the boletes. This is a polyporoid genus with tuberculate spores so different from *Polyporus* that Donk (1933) proposed it as a tribe, while Bondarzew and Singer have raised it to the rank of a monogeneric family Boletopsidaceae. Singer (1962) points out the similarity of the spores to those of *Hydnum, Hydnellum,* and *Polyozellus.*

Peck named *B. grisea*, a light-colored taxon that is widely distributed in eastern North America, while the Pacific Coast area has a very dark form which gives a blue-black reaction with weak potash solutions. It resembles the European species *B. leucomelaena*. Two typical collections of *B. grisea* were made in the Marquette area of Michigan in the late fall of 1970, and gave heavy spore deposits that were buffy brown in color. Under the microscope they were hyaline.

ii. Polyozellus.   This genus is distinguished by having tuberculate spores similar to those of *Hydnum* and *Hydnellum*. The type species is a dark bluish-black color, and KOH solution gives a strong blue-green reaction which leaches out, giving the liquid an olive black shade. A similar reaction occurs in some *Hydnellums* and *Hydnums*. It has been suggested that, because of the tuberculate spores, this genus should be included in the same family as *Hydnum, Hydnellum,* with *Boletopsis* a polyporoid element. However, until chemical and electron microscopic studies have been carried out, it seems best to consider this another case of convergent evolution.

2. *Subfamily. Hericioideae subfam. nov.*

Basionym: Hericiaceae Donk, *Persoonia* 3:269, 1964.

This subfamily has an extremely variable sporophore, which is resupinate, reflexed, dimidiate, irregularly stipitate, or branching from a basal tubercle. There are tramal tissues of interwoven strands that are free branches in *Hericium.* The context is monomitic or dimitic, and amyloid (one genus). Gloeocystidia or cystidia are present or absent. Spores are smooth, echinulate, amyloid (one tribe), or inamyloid.

This subfamily includes species with basidiocarps that range from resupinate forms to large, complexly branched forms. The diversity in structure and appearance is the basis for maintaining five tribes.

a. *Tribe. Hericieae.* The sporophore is variable, has single spines, and is dimidiate or arises from a tubercle as branches. Gloeocystidia are present and the context amyloid in one genus. Spores are amyloid in all genera and the type is *Hericium* (Pers.) S. F. Gray.

This tribe contains three taxa with spines on the ends of positively geotropic branches originating in the tramal context, subiculum, or the substrate. Occasionally some spines on ascending branches may be diageotropic.

i. *Hericium.* Besides the type, three species are recognized in North America. *Hericium erinaceus* causes a serious heart rot of oaks; *H. abietis* is a heart rot of conifers in western mountains; *H. coralloides* is found on rotting deciduous wood in eastern North America, and *H. ramosum* on deciduous wood throughout the region. It was only recently that four species were recognized in America, and distribution may be different when correct identifications are available from all regions. The most extensive studies to date have been made in Russia by Nikolajeva (1961). She uses names not recognized in North America. The type is *Hericium coralloides* (Pers.) S. F. Gray.

ii. *Creolophus.* The genus was revived by Donk (1956) with a question mark, because of the uncertainty of recognizing the type species *H. corrugatum.* He suggested that *H. corrugatum* might have been a form of *H. cirrhatum.* As a result of this suggestion, Maas Geesteranus (1962) has used *H. cirrhatum* as the species to illustrate the hyphal structure, and this is the basis for the generic description. The genus, as introduced, corresponded with *Hydnum, tribus* Apus Fr. "Pileo carnoso molli" Fries (1874; see Donk, 1956). Karsten included three species: *H. cirrhatum, H. diversidens,* and *H. fulgens* in addition to the type.

Banker (1913) considered *Climacodon* as a synonym.

iii. Dentipratulum. This is a very interesting genus which has the morphology of *Hericium*, the appearance of *Mucronella*, and is an intermediate between the two.

b. Tribe. Mucronelleae. The sporophore is solitary, gregarious, or cespitose, and the spines have an arachnoid or floccose subiculum. Gloeocystidia are present or lacking. The spores are amyloid or inamyloid. The type of the tribe is *Mucronella.*

This tribe includes four genera of which one, *Hormomitaria*, with three tropical species, is somewhat different, and Corner (1970) suggests is related to *Physalacria.*

i. Mucronella. This genus, of worldwide distribution, has very few of the characters of the genus *Hericium*, other than the positively geotropic spines and the slight amyloid reaction. The discovery of *Dentipratulum*, an intermediate genus, indicates that *Mucronella* should be considered a reduced form that has developed on rotten wood. A number of other species have been included in the genus but are inconspicuous and rarely collected. *Mucronella ulmi* has been found to possess dimitic flesh with gloeocystidia. It has been transferred to *Deflexula* by Corner (1950).

ii. Dentipellis. One species, *D. separans*, was included in addition to the type, and it was mentioned that two more species remained to be described.

There are three genera which have one or more characters in common with *Hericium*, while varying in growth form. *Dentipellis* is resupinate with a definite subiculum, but has the gloeocystidia in the spines and the amyloid reaction in the spores. *Mucronella* has only the faintest traces of hyphae as a subiculum and a slight amyloid reaction to the spores, while *Dentipratulum* has similar spines without an obvious subiculum, but does have gloeocystidia and a strong amyloid reaction to the spores. When included in the same subfamily as *Hericium*, one thinks of them as having evolved from a common ancestor by loss of characters.

iii. Delentaria. This genus was proposed for one species of doubtful position. It has positively geotropic branching spines, and narrow basidia. It would be easier to locate if associated with genera having the positively geotropic character, so it is placed here. Corner (1970) placed it in Ramariaceae, a negatively geotropic clavarioid family.

iv. Hormomitaria. This is a genus with positively geotropic spines that outwardly resembles *Mucronella*. Corner (1970) considers it to be related to *Physalacria*, and states that it would be a *Pistillaria* if negatively geotropic. *Hormomitaria tenuipes* (synonym, *Mucronella tenuipes*) does not have the moniliform rows of secondary septate cells in the context typical of the genus, and may not be cogeneric.

c. Genera with Hydnoid Characters from Other Families. Two related genera are introduced at this point because of a possible relationship to taxa in this tribe. *Deflexula* is dimitic, but *D. ulmi* is so close that it has been known as a *Mucronella* for some years. The other, *Hydnopolyporus*, is a very peculiar monospecific genus of obscure affinities.

i.   Deflexula.   This genus has the dimitic structure of *Pterula*, but the spines are positively geotropic. Some species were originally in *Hydnum* and *Mucronella*. The species are largely tropical. *Deflexula ulmi* has been recorded from North America and Russia, and is outwardly a *Mucronella*, but does have skeletals and hyphoids.

ii.   Hydnopolyporus.   Reid (1962) doubtfully refers this genus of two species to the Polyporaceae, but the hymenium has spines as many Hydnaceae.

d.   Tribe.   Gloiodoneae trib. nov.   Basionym: Gloiodonoideae Donk, Meded. Ned. Mycol. Ver. **18–20**: *190, 1931.* The sporophore is pileate, stipitate or dimidiate, and grows from a subiculum. The context is dimitic, dark, fused or consisting of separate strands or of separate stipe and pileus. Spores are echinulate and amyloid. The genus *Gloiodon* is the type.

This tribe consists of two genera that are not too closely related, but with characteristics in common. *Gloiodon* is the more primitive, and the structure is basically that of a *Hericium*. In the other, *Auriscalpium*, the strandlike development has been modified by generic controls that regulate negative, diageotropic, and positive geotropic growth in an orderly sequence.

i.   Gloiodon.   The genus was originally based on one species that is rather rare but apparently widespread. It has an amazing variability in form, and one Michigan collection is stipitate, as in *Phellodon*, while another collection from Oregon has the ends of the ramifying branches on the margin protruding free. The original notes (A. H. Smith) stated that it is "probably a Hericium." The commonest form is resupinate to laterally sessile. Its general structure, of branches growing out of matted tomentum, indicates the closeness to *Hericium*, while its spores indicate that we should consider very carefully its relationship with the group containing *Phellodon*. Maas Geesteranus (1963) has placed it in the family Auriscalpiaceae which includes *Lentinellus*, but while it is easy to see its connection with *Auriscalpium*, the morphological differences are so distinct that it seems best to consider the relationship to be with Hericieae. Both *Gloiodon* and *Auriscalpium* are examples of primitive fungi, most probably relics of early steps in evolution from a common ancestor, that were able to survive because of their ability to exist on specialized substrates: cones of *Pinus* species for *Auriscalpium*, and water-saturated wood for *Gloiodon*. Basically, all are derived from a basal pad of mycelium that protects the primordia of developing branches. In *Auriscalpium* there is a single negatively geotropic branch that grows out of the pad; in *Gloiodon*, numerous branches within a pad; in *Hericium*, the origin is a tubercle that may be a fused clump of branches or separate branches, or merely a root. Finally in *Dentipratulum*, only a weft of mycelium is present supporting the spines.

ii. Auriscalpium. Four species are known. *A. vulgare,* known since the time of Linnaeus, is almost exclusively found on *Pinus* cones. Recently the lignicolous *A. villipes* and *A. fimbriato-incisum* have been described from South America and China, respectively. *Auriscalpium umbella* has been described from New Zealand growing in moss under *Nothofagus* spp. While Maas Geesteranus' (1963) arguments for creating a family Auriscalpiaceae are very interesting, I still have reservations about uniting spined and lamellate fungi.

The genetic codes necessary for the control of the intricate and precise development of the outward forms of spined and lamellate structures of fruit bodies would appear to be a more sophisticated development than that involved with the strengthening of cell walls to stand environmental stresses, or that producing specialized structures for the elimination of waste products. Macroscopic features are therefore considered to be as important as microscopic, and they are infinitely easier for most people to recognize and handle when working with fungi at our present level of knowledge.

iii. A related genus. A similar development of form has taken place in *Beenakia* Reid, (1955) but this is a rather distant connection as the structure and spores appear to be very different.

Beenakia is a primitive genus. Its peculiar habitat suggests that this fungus' association with tree ferns may be very ancient. *Beenakia* resembles *Auriscalpium* in arising from a matted subiculum. Hyphal morphology and spore characters are different. Maas Geesteranus (1971) has placed it in the Gomphaceae because of the cyanophylic reaction of the spores, and makes a suggestion of a relationship with *Ramaria*. It is considered that *Beenakia* fits best within a group that has positively geotropic development of the spines. What its real connections are remains to be discovered.

e. Tribe. Climacodoneae trib. nov. Pileus dimidiatus vel suborbicularis, irregularis stipitatus, carnosus vel fibrosus, interdum succosus ubi juventibus, caro intertextus filiformis, monomiticus; gloeocystidia et cystidia praesentia; sporae albae parvae leves, haud amyloideae. Lignatiles. Typus *Climacodon* Karst.

This is a group of three genera that differ from *Hericieae* in not having any amyloid reaction. It was included in *Steccherinum* by Miller and Boyle (1943). *Cautinia* is placed here with some hesitation because of the scanty information available.

i. Climacodon. Originally the genus contained only the type species. Banker (1906) reduced it to *Steccherinum* S. F. Gray sensu lato, but later (1913) accepted Karsten's earlier name *Creolophus*, typified by *H. corrugatum,* as the proper genus for *H. septentrionale*. This was because Banker could not find a collection of *C. corrugatum* named by Fries, and all the later collections so named at Uppsala proved to be *H. septentrionale*. Donk (1956) and others since have felt that *H. corrugatum* is probably a form of *H. cirrhatum*, so now it is considered that the two genera are not typified by the same taxon. If this concept of *H. corrugatum* can be maintained, *Climacodon* is the correct name for *H. septentrionale* when dividing the genus *Steccherinum*. The characters for separating *Donkia* from *Climacodon* are based mostly on multiple clamps on the hyphae, and the presence of cystidia in one and gloeocystidia in the other. The two genera are very much alike, and we are following Nikolajeva (1961) in reducing *Donkia* to synonymy.

ii. Donkia. *Donkia* is a monotypic genus. The type species was placed in *Steccherinum* by Banker (1906), and in *Creolophus* in 1913, and more recently in *Climacodon* by Nikolajeva (1961). Maas Geesteranus (1962) at first considered the two species as belonging to "widely different genera," but in 1971 he accepted Nikolajeva's treatment.

iii. Mycoleptodonoides. This genus is based on a Russian species with a most peculiar structure of variably inflated and branched hyphae that are interwoven in a very intricate manner. [Maas Geesteranus (1962, Figs. 32–34, p. 393).] *M. aitchisonii* has been placed in this genus because of similarities in structure, though it does not have as tough a consistency. Nikolajeva (1961) also included *Hydnum adustum* (synonym, *Steccherinum adustum*), but its addition has been questioned by Maas Geesteranus (1962), who has proposed *Mycorrhaphium* for this species and the closely related *S. pusillum*. However, the differences are in degree not in fundamental characters, and, for the present, it seems best to follow Nikolajeva.

iv. Mycorrhaphium. This genus was described for two species common in North America that are quite distinctive in appearance, but, except for the variable presence of skeletals in the spines, can hardly be considered as possessing distinctive generic characters. The presence of a substance in the tips of the spines of *S. adustum* that blackens on bruising, is an interesting character that does not occur in *S. pusillum*. The normal appearance of the tramal tissues in this genus, in contrast to the weirdly contorted hyphae in *Mycoleptodonoides*, is the only distinctive character separating these two genera. Even *M. aitchisonii* is an intermediate connecting the two; so we are following Nikolajeva and leaving *S. adustum* in *Mycoleptodonoides*. *Mycoleptodonoides pusillum* (Fr.) K. Harrison comb. nov. (basionym *Hydnum pusillum* Brot. ex Fr., *Syst. Mycol.* 1:407, 1821) is a related species that resembles *M. adustum*.

v. Cautinia. This genus urgently needs reexamination from fresh or properly dried material. Bresadola (1925) considered the type as similar to *H.*

septentrionale. However, Maas Geesteranus (1967b) considered that it had the polyporoid characters of *Spongipellis*.

f. Genera with Hericioid Characters from Other Families. The genera possessing amyloid spores and included in Corticiaceae, are *Gloeodontia*, *Laxitextum*, and *Gloeocystidellum*. Their similarity to hydnaceous fungi has been noted, but only *Gloeodontia* is discussed, while the other two are in the key. Another genus, *Kavinia* Pilat, is described, but the spores are colored and rough. This has been placed by Donk in Gomphaceae, together with *Beenakia* Reid.

i. Gloeodontia. This genus was proposed for *Irpex discolor*, but there are other species that appear cogeneric according to Gilbertson (1971). It has been pointed out that the microscopic characters are similar to those of the Auriscalpiaceae, and that this is a corticioid element of such a family.

ii. Kavinia. Donk has suggested a relationship with *Ramaria* (clavarioid) and *Beenakia* (hydnoid), both of which he considers as in the family Gomphaceae. *Kavinia himantia* (synonym, *Mycoacia himantia*) included by Erikssen is different in having smooth white spores.

g. Tribe. Odontieae. The sporophores are resupinate to reflexed and the spines are subulate to granulose. Gloeocystidia and cystidia are present or not. Spores are smooth to rough and amyloid in only one species. It is typified by *Odontia* Fr.

This is a tribe of diverse elements which are gradually being described as new monospecific genera because of microscopic characters. Some of the genera have encrusted cystidia, dimitic context, or gloeocystidia, while others are monomitic, and one, *Stecchericium*, has amyloid spores. It connects *Steccherinum* to *Hericium*, and possibly does not rate generic status. Three of the genera are monospecific. *Dacrybolus* was discarded by Fries (1874) when he transferred *D. sudans* to *Grandinia*. It may not be possible to maintain a tribe based on *Odontia*, as published by Nikolajeva (1961). The presence of *Sarcodontia* in the same tribe would permit the use of tribe Sarcodontieae Nikol., and if it, in turn, were removed, Steccherineae Parmasto (1968) would become the correct name.

i. Odontia. The genus is a large one of worldwide distribution, with innumerable variations within each region. It is separated arbitrarily from *Peniophora*, has been likened to *Steccherinum*, and the synonymy of many species indicates the uncertainty of mycologists—for species have been placed at various times in *Hydnum*, *Acia*, *Kneiffia*, *Mycoacia*, or *Sarcodontia*.

Odontia has been used for various groups of resupinate spined fungi. The name was introduced by Persoon for two species, and later reduced to a section of *Hydnum* (1801) containing ten species. Fries included the Persoonian species in his section *Resupinatus* in 1821. S. F. Gray validated *Odontia* Pers. as a generic name in 1821 with two British species. Fries

(1838) decided to create two new genera from his 1821 section *Resupinatus*, but used the name *Odontia* Pers. and typified it with *Hydnum fimbriatum*. This has been used by mycologists, although the name cannot stand with the Friesian limits unless conserved against *Odontia* S. F. Gray. Rea (1922), Cunningham (1959), and Miller and Boyle (1943) have published on British, New Zealand and Iowa species respectively with *Odontia fimbriata* as the type for the genus. *Odontia* Pers. ex S. F. Gray, when typified by H. *ferruginea* Fr., reduces *Caldesiella* Sacc. to synonymy. Strictly speaking, the name cannot be used when the Code is followed.

    ii. Steccherinum. This is the original name for many resupinate hydnaceous fungi, but has been gradually reduced by the transfer of taxa to new genera or other groups.

    Maas Geesteranus (1962) has confirmed that *S. fimbriatum*, *S. laeticolor*, *S. litschaueri*, *S. murashkinskyi*, *S. rhois*, and *S. setulosum* Miller are co-generic with the type species.

    Parmasto (1968) has recently transferred to *Steccherinum* several new species of *Mycoleptodon*, which were described by Pilat from Czechoslovakia.

    iii. Stecchericium. This is a monospecific genus where the species has been removed from *Steccherinum* by Reid (1963) because of the amyloid reaction and the abundant oleiferous hyphae. Outwardly it strongly resembles *S. ochraceum*, and until relatives have been found, is better treated in its original genus. It is considered that the monomitic structure and amyloid reaction, though important at the species level, hardly warrants generic significance for only one species.

    iv. Sarcodontia. The resupinate genera of hydnaceous fungi have a very long and complicated history. *Sarcodontia* was proposed in 1866, but was overlooked by mycologists until mentioned by Donk (1952). In the meantime, Karsten (1881) proposed *Acia* with *Hydnum fuscoatrum* as the type for a closely related group. This name cannot be used as it was applied earlier to a genus in the Rosaceae. Donk (1931) changed the name to *Mycoacia*, and used the same type species referring to *Acia* Karst. em Pat., and included cystidiate forms. Later, Miller (1933), unaware of Donk's change, proposed *Oxydontia* type *Hydnum setosum* for those species without cystidia.

    Sarcodontia originally contained one species *S. mali*, which has since been placed in synonymy with *Hydnum croceum* (synonym *H. setosum*). When *H. setosum* is included in the circumscription, *Mycoacia* becomes a synonym of *Sarcodontia*.

    v. Mycoacia. Donk (1931) and Miller (1933) independently introduced new names to replace *Acia* Karst., which was preoccupied by an earlier name for a genus of Rosaceae. *Oxydontia*, although since treated as a synonym of *Mycoacia*, is not identical because they are based on different type species. Miller and Boyle (1943), when accepting *Mycoacia* for *Oxydontia*, rejected *M. fuscoatrum* as the type, and retained *H. setosum* instead, thus excluding cystidiate forms. This is not permissable under the Code. As *H. setosum* (=*S. crocea*) is the type of *Sarcodontia*, *Oxydontia* has to be reduced to synonymy with the latter. Recently Gilbertson (1971) has transferred *H. fuscoatrum* to *Steccherinum*, thus eliminating any use of *Mycoacia*, unless *H. fuscoatrum* can be shown as not cogeneric with *Steccherinum*.

vi. Radulum. This is a genus that appeared to be distinct to mycologists until the original circumscription was critically examined by Donk (1956). Taxonomically, it was based on resupinate forms that were not supposed to have gloeocystidia or cystidia in the tissues. The genus was not clearly delimited from *Corticium*. In an attempt to typify it, Banker (1902) selected *H. pendulum*, which since has been identified with *Corticium subcostatum*. A third species, *H. radula* (=*R. orbiculare*), was selected as the type by Clements and Shear (1931) and Miller and Boyle (1943), but Fries (1828) noted that it was not typical of the genus, and if this opinion is followed, it cannot be selected as the type. Donk has argued that the name cannot be preserved by any of the species originally included by Fries in 1825, or again in 1828.

Nobles (1967) proposed *Basidioradulum* for *R. orbiculare*, as *Radulum* can only be used for a genus of ascomycetes.

vii. Basidioradulum. Nobles, while making exhaustive studies of wood-rotting fungi, found that three resupinate species, named because of differences in the hymenophore, were identical in microscopic morphology and cultural characters and were interfertile. The oldest name available was *Hydnum radula*. As the generic name had been typified by an ascomycete, Nobles proposed *Basidioradulum* as the new name. She included only one species, but mentions that a number of taxa previously included in *Radulum* were not cogeneric. The monilioid hyphal ends present on parts of the subiculum on growing margins, or in culture, are unusual characters that are diagnostic for the species.

viii. Grandinia. This is one of the older genera based on a granulose hymenial surface, which is a character that grades into several other genera. It is also not possible to typify it by one of the original Friesian species and maintain a modern concept of the genus.

Nikolajeva (1961) typified it with *G. helvetica* and included three of the species listed by Miller and Boyle, and two Fresian species of 1874. Donk's (1956) opinion of the name was that it was a nomen dubium when typified by *G. granulosa* because of uncertainty as to the identity of the type species. The material in Fries' herbarium has the dichophyses of *Asterostromella*. The name has been widely used but many of the species have been transferred to other genera. Probably it cannot be used unless *G. polycocca* is redis-discovered and redescribed in modern terms.

   h.   Genera with Hydnoid Characters from Other Families. Polyporaceae: Irpex, Hydnochaete and *Sistotrema. Coniophoraceae: Gyrodontium.*

   i.   Irpex.   This genus contains species with spines that are flattened to subulate, and are considered to belong to the Polyporaceae because of the poroid pattern of the margin. It is most interesting that Maas Gee-steranus (1963) has reported that the microscopic characters of *I. lacteus* are identical with *Steccherinum.* This is another case of a conflict in the choice of the family, when using macroscopic versus microscopic characters in classifying the fungi, and determining their natural relationships.

   ii.   Sistotrema.   This is one of the genera recognized by Fries and in-cluded in Hydnei. Since then, the peculiar basidia have been recognized as distinctive. The tendency of the young hymenium to develop ridges and pores before forming the irregularly shaped spines has led many mycologists to associate this genus with the Polyporaceae. It has also been placed in the family Corticiaceae Donk (1964). *Sistotrema confluens* is a very primitive fungus, and invariably, when first found by a young mycologist, is looked for among the spined Hydnaceae.

   iii.   Gyrodontium.   This genus was established for two peculiar spined species growing on wood in the tropics. Most collections had very little information about their appearance when fresh, and had few distinctive microscopic characters. Maas Geesteranus (1964) has reduced a number of later names to synonymy with the type species, even when the material con-sisted of little more than chips of wood with attached fragments of pileus and spines. This is another "*Hydnum,*" except for the spores, and because of them it has been placed in the Coniophoraceae.

B.   Echinodontiaceae

   Donk em Gross. *Mycopathol. Mycol. Appl.* **24**:4 (1964). A description of this family is similar to one of *Echinodontium,* except that the family description has been expanded to include two species previously in *Stereum,* and one in *Radulum.*

   Echinodontium.   The type species was first considered to be a *Fomes,* then a *Hydnum,* and finally was given the name *Echinodontium.* One month later the name *Hydnofomes* P. Henn. was used for another species found in Japan. Since then, similarities with other fungi have been noted, but it has been maintained in the Hydnaceae until Donk (1961) gave it family rank as Echinodontiaceae. Gross (1964) emended both the genus and family and added two species of *Stereum* and one *Radulum,* making a total of six taxa. These differ widely in hymenial configuration, but are connected by being perennial and having echinulate amyloid spores. Five are associated with rots in conifers, and the sixth with oaks.

REFERENCES

Banker, H. J. (1902). A historical review of the proposed genera of the Hydnaceae. *Bull. Torrey Bot. Club* **29**:436–448.

Banker, H. J. (1906). A contribution to a revision of the North American Hydnaceae. *Mem. Torrey Bot. Club* **12**:99–194.

Banker, H. J. (1912). Type studies in the Hydnaceae. I. The genus *Manina*. II. The genus *Steccherinum*. *Mycologia* **4**:271–278, 309–318.

Banker, H. J. (1913). III. The genus *Sarcodon*. IV. The genus *Phellodon*. V. The genus *Hydnellum*. VI. The genera *Creolophus*, *Echinodontium*, *Gloiodon* and *Hydnodon*. *Mycologia* **5**:12–17, 62–66, 194–205, 293–298.

Banker, H. J. (1914). VII. The genera *Asterdon* and *Hydnochaete*. *Mycologia* **6**:231–234.

Bourdot, H., and A. Galzin. (1928). "Hyménomycètes de France." Le Chevalier, Paris.

Bresadola, G. (1925). New species of fungi. *Mycologia* **17**:68–77.

Bresadola, G. (1932). "Iconographia mycologica," Vol,. 21 and 22. Milan.

Chevallier, F. F. (1826). "Flore générale des Environs de Paris, selon la méthode Naturelle," Vol. I. Paris.

Christiansen, M. P. (1960). Danish resupinate Fungi. II. Homobasidiomycetes *Dan. Bot. Ark.* **19**:61–388.

Clements, F. E., and C. L. Shear. (1931). "The Genera of Fungi." Wilson Co. New York.

Coker, W. C., and A. H. Beers. (1951). "The Stipitate Hydnums of the Eastern United States." Univ. of North Carolina Press, Chapel Hill.

Cooke, M. C., and L. Quélet. (1878). "Clavis synoptica Hymenomycetum Europaeorum." Hardwicke and Bogue. London.

Corner, E. J. H. (1932). The fruit body of *Polystictus xanthopus* Fr. *Ann. Bot. London* **46**:71–111.

Corner, E. J. H. (1950). A monograph of *Clavaria* and Allied Genera (*Ann. Bot. Mem.* **1**:1–740).

Corner, E. J. H. (1968). A monograph of *Thelephora* (Basidiomycetes). *Beih., Nova Hedwigia* **27**:1–110.

Corner, E. J. H. (1970). Supplement to "A Monograph of *Clavaria* and Allied Genera." *Beih., Nova Hedwigia* **33**:1–299.

Cunningham, G. H. (1958). Hydnaceae of New Zealand. I. The pileate genera: *Beenakia*, *Dentinum*, *Hericium*, *Hydnum*, *Phellodon* and *Steccherinum*. *Trans. Roy. Soc. N. Z.* **85**: 585–601.

Cunningham, G. H. (1959). Hydnaceae of New Zealand II. The genus *Odontia*. *Trans. Roy. Soc. N. Z.* **86**:65–103.

Dillenius, J. J. (1719). "Catalogus plantarum sponte circa Giessam nascentium." Maximilianum á Sande, Frankfurt am Main.

Donk, M. A. (1931). Revisie van de Nederlandse Heterobasidiomycetae en Homobasidiomycetae-Aphyllophoraceae. *Meded. Ned. Mycol. Ver.* **18–20**:67–200.

Donk, M. A. (1933). Revision der neiderländischen Homobasidiomycetae-Aphyllophoraceae. II. *Meded. Ned. Mycol. Ver.* **22**:1–278.

Donk, M. A. (1952). The status of the generic name *Oxydontia* L. W. Miller ("Hydnaceae"). *Mycologia* **44**:262–263.

Donk, M. A. (1956). The generic names proposed for Hymenomycetes. V. Hydnaceae. *Taxon* **5**:69–80, 95–115.

Donk, M. A. (1961). Four new families of Hymenomycetes. *Persoonia* **1**:405–407.

Donk, M. A. (1964). A conspectus of the families of Aphyllophorales. *Persoonia* **3**:199–324.

Fries, E. M. (1821). "Systema mycologicum," Vol. 1. Lund.

Fries, E. M. (1825). "Systema Orbis Vegetabilis," Vol. I. Lund.

Fries, E. M. (1828). "Elenchus fungorum," Vol. I. Griefswald.

Fries, E. M. (1838). "Epicrisis systematis mycologici." Bering. Uppsala.

Fries, E. M. (1874). "Hymenomycetes Europaei." Bering. Uppsala.

Gilbertson, R. L. (1971). Phylogenetic Relationship of Hymenomycetes with Resupinate Hydnaceous Basidiocarps. *In* "Evolution in the Higher Basidiomycetes," (R. H. Petersen, ed.) pp. 275–307 Univ. of Tennessee Press, Knoxville.

Grand, L. F., and R. T. Moore. (1970). Ultracytotaxonomy of Basidiomycetes. I. Scanning electron microscopy of spores. *J. Elisha Mitchell Sci. Soc.* **86**:106–117.

Gray, S. F. (1821). "A Natural Arrangement of British Plants," Vol. 1. Baldwin, Cradock, and Joy. London.

Gross, H. L. (1964). The Echinodontiaceae. *Mycopathol. Mycol. Appl.* **24**:1–26.

Harrison, K. A. (1961). The Stipitate Hydnums of Nova Scotia. *Can. Dep. Agr., Publ.* **1099**: 1–60.

Harrison, K. A. (1964). New or little known Stipitate Hydnums. *Can. J. Bot.* **42**:1205–1233.

Harrison, K. A. (1968). Studies on the Hydnums of Michigan. I. Genera *Phellodon, Bankera, Hydnellum. Mich. Bot.* **7**:212–264.

Harrison, K. A. (1971a). The evolutionary lines in the fungi with spines supporting the Hymenium. *In* "Evolution of the Higher Basidiomycetes," (R. H. Petersen, ed.) pp. 375–392. Univ. of Tennessee Press, Knoxville.

Harrison, K. A. (1971b). *Dentinum* S. F. Gray or *Hydnum* L. ex Fr. *Mycologia* **63**:1067–1072.

Harrison, K. A. (1972). A new species of *Phellodon* possessing clamp connections. *Can. J. Bot.* **50**:1219–1221.

Karsten, P. A. (1879). Symbolae ad mycologiam fennicam. *Medd. Soc. Fauna, Flora fenn.* **5**:15–46.

Karsten, P. A. (1881). Enumeratio hydnearum Fr. fennicarum. systemate novo dispositarum. *Rev. Mycol.* **3**:19–21.

Linnaeus, C. (1735). "Systema naturae." Leiden.

Linnaeus, C. (1737a). "Flora lapponica," pp. 1–372. Amsterdam.

Linnaeus, C. (1737b). "Genera Plantarum." 384 pp. Leiden.

Linnaeus, C. (1753). "Species Plantarum." 2. Stockholm.

Maas Geesteranus, R. A. (1956). The Stipitate Hydnums of the Netherlands I. *Sarcodon* P. Karst. *Fungus* **26**:44–60.

Maas Geesteranus, R. A. (1957). The Stipitate Hydnums of the Netherlands. II. *Hydnellum* P. Karst. *Fungus* **27**:50–71.

Maas, Geesteranus, R. A. (1958). The Stipitate Hydnums of the Netherlands. III. *Phellodon* P. Karst. and *Bankera* Coker and Beers ex Pouzar. *Fungus* **28**:48–61.

Maas Geesteranus, R. A. (1959). The Stipitate Hydnums of the Netherlands. IV. *Auriscalpium* S. F. Gray, *Hericium* Pers. ex S. F. Gray, *Hydnum* L. ex Fr. and *Sistotrema* Fr. em Donk. *Persoonia* **1**:115–147.

Maas Geesteranus, R. A. (1960). Notes on *Hydnums* I. *Persoonia* **1**:341–384.

Maas Geesteranus, R. A. (1961). A Hydnum from Kashmir. *Persoonia* **1**:409–413.

Maas Geesteranus, R. A. (1962). Hyphal structures in Hydnums. *Persoonia* **2**:377–405.

Maas Geesteranus, R. A. (1963). Hyphal structures in Hydnums. II, III, and IV. *Proc. Kon. Ned. Akad. Wetensch. Ser.* **C66**:426–457.

Maas Geesteranus, R. A. (1964). Notes on Hydnums. II. *Persoonia* **3**:155–192.

Maas Geesteranus, R. A. (1966). Notes on Hydnums. III. *Proc. Kon. Ned. Akad. Wetensch., Ser.* **C69**: 24–36.

Maas Geesteranus, R. A. (1967a). Notes on Hydnums. V and VI. *Proc. Kon. Ned. Akad. Wetensch., Ser.* **C70**:50–72.

Maas Geesteranus, R. A. (1967b). Notes on Hydnums. VII. *Persoonia* **5**:1–13.

Maas Geesteranus, R. A. (1971). Hydnaceous fungi of the eastern old world. *Verh. Kon. Ned. Akad. Wetensch., Ser. 2* **60**:1–176.

Micheli, P. A. (1729). "Nova Plantarum Genera." Florence.

Miller, L. W. (1933). The Genera of Hydnaceae. *Mycologia* **25**:286–302.

Miller, L. W. (1933). The Hydnaceae of Iowa. I. The Genera *Grandinia* and *Oxydontia*. *Mycologia* **25**:356–368.

Miller, L. W. (1934). II. The Genus *Odontia*. III. The Genera *Radulum, Mucronella, Caldesiella* and *Gloiodon*. *Mycologia* **26**: 13–32, 212–219.

Miller, L. W. (1935). IV. The Genera *Steccherinum, Auriscalpium, Hericium, Dentinum, Calodon*. *Mycologia* **27**:357–372.

Miller, L. W., and J. S. Boyle (1943). The Hydnaceae of Iowa. *Stud. Natur. Hist. Iowa Univ.* **18**:1–92.

Nikolajeva, T. L. (1961). Familia Hydnaceae. *Flora Plant. Cryptogamarum URSS*. Vol. **6**:1–432.

Nobles, M. K. (1967). Conspecificity of *Basidioradulum (Radulum) radula* and *Carticum hydrans*. *Mycologia* **59**:192–209.

Parmasto, E. (1968). "Conspectus Systematis Corticiacearum." Tartu, Estonia.

Patouillard, N. (1900). "Essai taxonomique sur les familles et les genres des Hyménomycètes." Lucien Declume. Lons-le-Saunier.

Persoon, C. H. (1801). "Synopsis methodica fungorum." H. Dieterich. Gottingen.

Persoon, C. H. (1825). "Mycologia Europaea," Vol. 2. J. J. Palm. Erlangen.

Rea, C. (1922). "British Basidiomycetae." Cambridge. Univ. Press, London and New York.

Reid, D. A. (1962). Notes on the fungi that have been referred to *Thelephoraceae* sensu lato. *Persoonia* **2**:109–170.

Reid, Derek A. (1956). New or interesting records of Australasian Basidiomycetes. *Kew Bull*. **4**, 1955: 631–648.

Reid, Derek A. (1963). New or interesting records of Australasian Basidiomycetes. V. *Kew Bull*. **17**:267–308.

Saccardo, P. A. (1888). "Sylloge Fungorum," Vol. 6. Padua.

Singer, R. (1962). "The Agaricales in Modern Taxonomy." Cramer, Weinheim.

CHAPTER 22

Aphyllophorales IV: Poroid Families

Polyporaceae, Hymenochaetaceae, Ganodermataceae, Fistulinaceae
Bondarzewiaceae

D. N. Pegler

The Herbarium
Royal Botanic Gardens
Kew, Surrey, England

I. INTRODUCTION

The term "polypore" has been extensively used as a convenient collective term to accommodate all members of the Aphyllophorales producing a poroid or tubular hymenophore. Such a hymenophore configuration has developed in a number of otherwise diverse lineages within the Basidiomycotina. Polyporoid fungi are also extremely polymorphic in their gross morphology, which has resulted in considerable difficulty in arriving at a natural system of classification. The tendency has always been to fall back upon form-genera because these allow for easy recognition in the field. Form-genera are still widely used today and they may be broadly defined as follows: *Polyporus*—annual, fleshy; *Polystictus*—annual, coriaceous; *Fomes*—perennial, corky to woody; *Daedalea*—annual, with labyrinthiform pores; *Trametes*—annual, with tubes of unequal length; *Hexagonia*—annual, with large hexagonal pores; *Lenzites*—annual, with a lamellate hymenophore; and *Poria*—annual or perennial, and resupinate.

Modern concepts of polypore genera require precise studies on the comparative anatomy, cultural features, and ontogeny of each species. This includes an analysis of hyphal systems, pigmententation, clamp-connections, all hymenial structures, and spores. The term "hyphal-system" was first introduced by Corner (1932) and now forms the basis to all generic and specific descriptions. The flesh or context of a polypore may consist of morphologically distinct hyphae, which may be classed into three basic

397

FIG. 1. 1–3, Hyphal-types. 1, Skeletal hypha; 2, Binding hyphae; 3, Generative hyphae; 4–10, Basidiocarp types. 4, Resupinate; 5, Effuso-reflexed; 6, Ungulate; 7, Triquetrous; 8, Dimidiate; 9, Laterally stipitate; 10, Centrally stipitate. 11–16, Spore types. 11, *Polyporus squamosus*, cylindric; 12, *Amyloporia xantha*, allantoid; 13, *Rigidoporus ulmarius*, globose; 14, *Laetiporus sulphureus*, ellipsoid; 15, *Ganoderma lucidum*; 16, *Bondarzewia montana*. 17–19, Cystidioid structures. 17, *Schizopora paradoxa*, cystidioles; 18, *Oxyporus populinus*, incrusted cystidium; 19, *Inonotus cuticularis*, setigerous elements of the pileal surface.

groups, namely generative, skeletal, and binding hyphae. Generative (Fig. 1.3) are the main hyphal type which ultimately give rise to the basidia, and directly or indirectly to all other structures; they are therefore present in any basidiocarp. They are produced directly from the secondary mycelium and so frequently possess clamp-connections. In many species, however, the hyphae have only simple septa presumably because of a difference in their genetical constitution. It is a characteristic of the family Hymeno-chaetaceae that clamp-connections are never produced. Generative hyphae are of unlimited growth, possessing abundant protoplasmic contents which stain readily; their wall is initially thin but may ultimately become thickened or gelatinized. In dried specimens, the thin-walled hyphae soon collapse and may be difficult to find. Specialized branches may be produced by the generative hyphae; these are the skeletal and binding hyphae. Typical skeletal hyphae (Fig. 1.1) may be easily recognized by their thickened wall and correspondingly narrow lumen, a complete absence of primary septation, lack of branching, and absence of clamp-connections. They function as strengthening structures lending rigidity to the basidiocarp, and often develop in considerable numbers to become the dominant hyphal-type, e.g., *Phellinus*. A modification of the simple skeletal-hyphae is the *arboriform* skeletal hypha found in the Ganodermataceae and *Dichomitus*, which differs in possessing an apex with repeated dichotomous branching. Binding hyphae (or ligative hyphae) (Fig. 1.2) may be produced as an alternative to skeletal hyphae, e.g., *Laetiporus*, but more often accompany skeletal hyphae, e.g., *Coriolus*. They are of limited growth, often highly and irregularly branched, and function to weave the other hyphae together. When only generative hyphae are present in the basidiocarp, the hyphal-system is called monomitic; (Fig. 2.1–2) when a second hyphal-type is present then it is called dimitic (Fig. 2.3–4), and with three hyphal-types are present it is called trimitic (Fig. 2.5–6).

The gross morphology is of only limited use at the generic level, as in many cases species ranging from resupinate, effuso-reflexed to dimidiate may occur in the same genus. However, it is clear that a number of totally resupinate genera do occur, as well as a number of stipitate genera. There is a similar variation in the hymenophore configuration which may occur even at the specific level.

II. IMPORTANT LITERATURE

Bourdot and Galzin, "Hyménomycètes de France," 1928; Overholts, "The Polyporaceae of the United States, Alaska and Canada," 1953; Murrill, Polyporaceae, *N. Amer. Fl.* **9**:1–131, 1907–08; Bondartsev, *Trutovye griby evropeiskei chasti SSSR i Kavkaza* [Polyporaceae of European Russia and the Caucasus]," 1953; Bondartsev and Singer, Zur systematik der Polyporaceen, *Ann. Myc., Berlin* **39**:43–65, 1941; Pilát, "Atlas des Champignons de l'Europe III. Polyporaceae," 1936–42; Domanski, Grizby (Polyporaceae 1, Mucronoporaceae 1), 1965; Domanski, Orlos and Skirgello, Grizby III (Polyporaceae pileatae, Mucronoporaceae pileatae, Ganodermataceae, Bondarzewiaceae, Boletopsidaceae, Fistulinaceae), 1967; Cunningham, Polyporaceae of New Zealand, *N.Z. Dep. Sci. Ind. Res.*

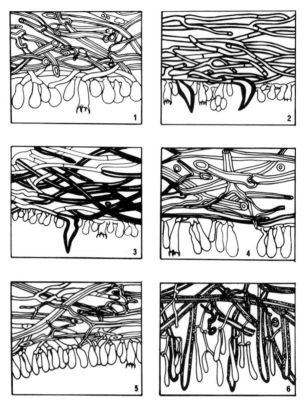

Fig. 2. Vertical sections through the hymenium and context. 1, *Bjerkandera adusta*, monomitic hyphal system with thick-walled generative-hyphae and clamp-connections; 2, *Inonotus radiatus*, monomitic hyphal system with xanthochroic generative hyphae lacking clamp-connections, and setae present in the hymenium; 3, *Phellinus igniarius*, dimitic hyphal system with thin-walled generative hyphae and thick-walled skeletal hyphae; 4, *Polyporus squamosus*, dimitic hyphal system with generative hyphae bearing clamp-connections, and branched binding hyphae; 5, *Coriolus versicolor*, trimitic hyphal system with thin-walled generative hyphae, thick-walled skeletal and binding hyphae; 6, *Daedalea quercina*, trimitic hyphal system, and a catahymenial development with thick-walled skeletocystidia.

Bull. **164**:1–304, 1965; Ito, Polyporales, *Mycol. Flora Jap.* **2**:210–450, 1955; Imazeki, Genera of Polyporaceae of Nippon, *Bull. Tokyo Sci. Mus.* **6**:1–111, 1943.

III. KEY TO FAMILIES

KEY TO POROID FAMILIES OF THE APHYLLOPHORALES

1. Hypobasidia cruciately divided, with subulate or tubular epibasidia—Phragmobasidiomycetidae: Tremellales (polyporoid genus). Subiculum very thin; basidiospores hyaline,

smooth, cylindric allantoid . . . . **Aporpium**

1′. Hypobasidia entire, with apical sterigmata—Holobasidiomycetidae 2

    2(1′) Hymenophore composed of densely crowded free tubes; basidiocarp pileate, laterally stipitate, annual; context monomitic, often with vesicular hyphae, hymenial cystidia absent; spores globose to short ovoid, lignicolous . . . . . . **Fistulinaceae** p. 418

    2′. Hymenophore not composed of free tubes 3

3(2′) Basidiocarp agaricoid, soon decaying; stipitate or sessile; hymenophore poroid or with strongly anastomosing lamellae; context gelatinous or soft fleshy, spores amyloid or not, angiocarpous or hemiangiocarpous . . . . **Agaricales** (see Chap. 23)

3′. Basidiocarp different, gymnocarpous 4

    4(3′) Edge of dissepiments fertile, hymenium continuous, merulioid, reticulate, finally tubular; spore wall smooth 5

    4′. Edge of dissepiments sterile, hymenium discontinuous 6

5(4) Spore wall thick, double, smooth, yellowish brown or darker, with the endosporium strongly cyanophilous . . . . **Coniophoraceae** (Polyporoid genera)

5′. Spores hyaline, thin-walled . . . . **Corticiaceae** (Polyporoid genera)

    6(4′) Spores ornamented; basidiocarps pileate; context never xanthochroic 7

    6′. Spores different 9

7(6) Spores hyaline, globose or nearly so, with conspicuous amyloid ridges or crests (Fig. 1:16); basidiocarps annual, lignicolous . . . . **Bondarzewiaceae** p. 417

7′. Spores inamyloid 8

    8(7′) Spores small, globose, pale brown, bluntly and coarsely tuberculate; basidiocarp fleshy or corky, stipitate or sessile, annual; context firm but not leathery; terricolous . . . . **Thelephoraceae** (Polyporoid genera)

    8 . Spores ellipsoid to globose, with a brown pigmented exosporium beset with a spinose ornamentation piercing into an outer, hyaline, smooth perisporium (Fig. 1:15); annual or perennial; pileal-surface often forming a laccate crust; context pale to purplish brown, corky; hyphal-system trimitic; lignicolous . . . **Ganodermataceae** p. 416

9(6′) Setae present somewhere in the basidiocarp, often in the hymenium; context xanthochroic, rust to ferruginous, permanently darkening in alkaline solution; hyphal-system mono- or dimitic with skeletal-hyphae, clamp-connections never present; spores hyaline or brown . . . . **Hymenochaetaceae** (Polyporoid genera) p. 413

9′. Setae absent; context never xanthochroic; clamp-connections present on generative-hyphae in forms with brown context 10

    10(9′) Basidiocarp resupinate to infundibuliform, white to golden, annual; context reduced, hymenophore at first smooth, finally reticulately poroid; spores small, hyaline, or bluish green, with wall smooth, verrucose or echinulate; basidia 4–8 spored; hyphal-system monomitic . . . . **Corticiaceae** (Polyporoid genera)

    10′. Basidiocarp resupinate to pileate, sessile to stipitate; hymenophore poroid from the first, annual or perennial; context often well-developed, of variable texture from watery fleshy to corky, leathery or ligneous; spores mostly smooth, exceptionally ornamented; basidia 2–4 spored; hyphal-system mono-, di- or trimitic . . . **Polyporaceae** p. 402

IV. POLYPORACEAE

This family is annual or perennial. The basidiocarp is very variable, ranging from totally effused to stipitate, uni- or multipileate. The pileus is dorsiventral and the stipe, when present, is central excentric or lateral. Usually, the hymenophore is tubulate, with the pores being small to large, sometimes trametoid, daedalioid, irpicoid, and rarely lamellate. The tubes are not free, but are sometimes stratified, with the edge of the dissepiments sterile. The context is white to light brown, cinnamon, umbrinous, purplish brown, or rarely red or pink, and the texture is watery-fleshy to coriaceous, corky or woody. The hyphal system is mono-, di-, or trimitic, with clamp-connections present or absent, and a catahymenium or euhymenium present, often thickening. The basidia are two- or four-spored, clavate or suburniform. Hymenial cystidia, skeletohyphidia, or cystidioles are sometimes present, gloeocystidia very rarely present, and setae are absent. Spores are hyaline or pale brown, globose to ellipsoid or cylindric, exceptionally ornamented, and rarely amyloid, dextrinoid or cyanophilous; Polyporaceae are lignicolous, terrestrial, humicolous, and sometimes parasitic.

KEY TO THE POLYPORACEAE

1. Stipitate species; stipe either central, excentric or lateral 2

1'. Resupinate, effuso-reflexed, dimidiate, or ungulate species, lacking stipe but occasionally with an attenuated base 25

2(1) Context brown, yellowish brown, rust, cinnamon, livid or umbrinous 3

2'. Context white- to wood-colored, although may become brownish in old specimens 6

3(2) Hymenophore poroid; spores hyaline 4

3'. Hymenophore radially lamellate; pileus orbicular, infundibuliform; context thin, cinnamon to umbrinous; hyphal system di- or trimitic with clamp-connections; spores hyaline, large, cylindric . . . . **Xerotinus**
        Type: *X. afer.* 2 spp. Africa, S. America.

4(3) Clamp-connections absent; stipe lateral or rarely excentric, incrusted; pileus reniform to spathulate with glabrous crust of palisadic separate hairs; context woody, fulvo-ochraceous with narrow, black hard tissues; hyphae thin-walled, yellow; spores globose; 5–7.5 μm diam . . . . **Pyrroderma**
        Type: *P. sendaiense.* 2 spp. Japan, India. Imazeki (1966).

4'. Clamp-connections present 5

5(4') Stipe excentric, thick, pileus reniform with reddish-brown to black zonations, margin white; context dimitic; tuber short; spores ellipsoid, less than 7μm long. . . . **Podofomes**
        Type: *P. corrugis* Synonym: *Pelloporus.* 1 sp. Europe.

5'. Stipe lateral as an attenuated base to the pileus; pileus flabelliform, thin; spores large (10–15 μm), cylindric; pluricellular hairs present on the edge of tubes . . See 8'. **Elmerina**

6(2') Spores oblong, ellipsoid to cylindric, smooth; clamp-connection present 7

6′. Spores globose to ellipsoid 15

7(6) Prominent sterile structures present in the hymenium; basidiocarp laterally stipitate 8

7′. Sterile hymenial structures absent 9

    8(7) Hymenophore regularly poroid; pileal surface with cystidioid structures; stipe short, tomentose; context white; hyphal-system dimitic; brown stellate cystidia present; spores 7–10 μm long. . . . . **Echinochaete**
        Type: *E. megalopora*. Syn.: *Asterochaete, Dendrochaete*. Four spp. Pantropical. Reid (1963).

    8′. Hymenophore irregular, poroid-lamellate to daedalioid; pileus thin, coriaceous, without a crust; context white becoming brown; hyphal system monomitic; abundant pluricellular hairs at edge of tubes . . . . See 5′. **Elmerina**
        Type: *E. cladophora*. Syn.: *Elmeria*. Three spp. Philippines, Ceylon, E. Africa. Humphrey (1938).

9(7′) Pores showing a basic radial arrangement 10

9′. Pores not distinctly radial; pileus thin; context firm 13

    10(9) Pores distinctly hexagonal, although often radially compressed; edge of dissepiments fimbriate; lateral stipe well developed . . . . **Favolus**
        Type: *F. brasiliensis*. Syn.: *Hexagonia*. About 50 spp. Worldwide. Lloyd (1909).

    10′. Pores not hexagonal 11

11(10′) Basidiocarp, terrestrial, with an excentric stipe arising from an underground sclerotium; pileus reniform, lobate, with surface scales; spores 7–10 μm . . . . See 22′. **Albatrellus**
    Type: *A. albidus*. Syn.: *Scutiger, Caloporus*. Twelve spp. N. Temperate. Pouzar (1966).

11′. Basidiocarp lignicolous, devoid of a sclerotium; stipe with an outer covering layer, often black 12

    12(11′) Context with binding-hyphae with limited branching (Fig. 2/4) . . . . **Polyporus**
        Type *P. tuberaster*. Syn.: *Cerioporus, Melanopus*. About 50 spp. Worldwide. *Polyporus squamosus* causes a heart-rot in many broad-leaved trees, particularly elm (*Ulmus*).

    12′. Context with highly branched binding hyphae of the "Bovista" type . . **Polyporellus**
        Type: *P. brumalis*. Syn.: *Lentus, Leucoporus*. Five spp. Worldwide.

13(9′) Pores distinctly visible, isodiametric; stipe reduced to a small lateral disc; abundant hyphal-pegs present . . . . **Pseudofavolus**
    Type: *P. miguelli*. One sp. Pantropical.

13′. Pores scarcely visible to the naked eye; stipe central or lateral 14

    14(13′) Stipe with a covering layer, attached to the substrate by a basal disc, context firm . . . . **Microporus**
        Type: *M. perulus*. Synonym: *Lignosus*. Fifteen spp. Tropical. Lloyd (1910b). *Microporus xanthopus* is one of the most common pantropical polypores.

    14′. Stipe white or pallid; context tough, fleshy drying bone hard; hyphae thin-walled then inflating to become thick-walled; spores oblong, 4–6 μm . . . . **Osteina**
        Type: *O. obducta*. One sp. Europe, N. America.

15(6′) Spores brownish, subglobose, strongly tuberculate; basidiocarp centrally stipitate; context thick, fleshy, turning violaceous-pink on exposure
         . . . . **Boletopsis** (Thelephoraceae)

15′. Spores hyaline 16

    16(15′) Context duplex with an upper soft layer and a lower rigid layer, monomitic; hymenophore dentate-lacerate to daedalioid; gloeocystidia present but not always constant; asexual spores as well as basidiospores present 17

    16′. Context homogeneous 18

17(16) Basidiospores ellipsoid with a prominent oil-guttule; chlamydospores globose, hyaline . . . . **Heteroporus**
    Type: *H. biennis*. Syn.: *Abortiporus*. Two spp. N. Temperate.

17′. Basidiospores echinulate, without a prominent oil-guttule; conidia small, brown, tuberculate . . . . **Diacanthodes**
    Type: *D. novo-guineensis*. Syn.: *Bornetina*. One sp. Pantropical. O. Fidalgo (1962).

    18(16′) Small thin basidiocarps 19

    18′. Basidiocarps with large flabelliform pilei; basidiocarp simple or compound . . . 21

19(18) Clamp-connections present; basidiocarp up to 4 cm in diameter, infundibuliform, to flabelliform, with a solid, basal stipe; hymenophore alveolar, poroid to subirpicoid; spores subglobose, small; basidia with 4–8 (usually 6) sterigmata . . . . **Sistotrema** (Corticiaceae)

19′. Clamp-connections absent 20

    20(19′) Basidiocarp small, 1–5 mm in diameter, pendant by a short curved stipe attached to the vertex, erumpent from lenticels; context white, fibrous; hyphae gelatinized; spores small, allantoid . . . . **Porodisculus**
    Type: *P. pendulus*. Syn.: *Enslinia*. One sp. N. America.

    20′. Pilei often confluent, forming rosettelike clusters; stipe formed as an attenuated base to the pileus; pileus papillate or irpicoid; context very thin; spores globose to broadly ellipsoid . . . . **Hydnopolyporus**
    Type: *H. palmatus*. Two spp. Tropical and S. America.

21(18′) Context firm; hyphal-system dimitic with skeletal-hyphae . . . . . . **Microporellus**
    Type: *M. dealbatus*. Four spp. N. and S. America.

21′. Context soft, watery-fleshy, rotting easily; hyphal-system monomitic 22

    22(21′) Basidiocarp multipileate 23

    22′. Basidiocarp unipileate; pileus thick, fleshy; stipe central or excentric; tubes very short; hyphae devoid of clamp-connections; spores small, subglobose . . . . See 11. **Albatrellus**

23(22) Hyphae inflating, clamp-connections present; individual pilei small, greyish brown . . . . **Grifola**
    Type: *G. frondosa*. Syn.: *Polypilus, Cladodendron, Cladomeris, Merisma*. One sp. N. Temperate.

23′. Hyphae devoid of clamp-connections 24

    24(23′) Pilei small (1–3 cm in diameter), conchate, superimposed upon a stipe which arises from a submerged pseudosclerotium . . . . **Flabellophora**
    Type: *F. superposita*. One sp. Australasia.

24′. Individual pilei large, flabelliform with brown concentric zoning, forming large rosettelike clusters; lignicolous . . . . **Meripilus**
    Type: *M. giganteus*. Syn.: *Flabellopilus*. Two spp. N. and S. Temperate.

25(1′) Tropical species, at first producing a smooth hymenophore which later becomes disrupted by sterile ridges forming a poroid arrangement, with the basidia restricted to the base of the tubes 26

25′. Hymenium lining the walls of the tubes 27

26(25) Tubes bearing incrusted hyphal fascicles; cystidia absent; spores hyaline, ellipsoid, 4–6 μm long . . . . **Grammothele** (Corticiaceae)

26′. No hyphal fascicles; cystidia present or absent; spores hyaline, ovoid to cylindric, 2.5–17 μm long . . . . **Porogramme** (Corticiaceae)

27(25′) Hymenophore restricted to individual cups borne on a common stroma; hyphal-system dimitic with clamp-connections; spores hyaline, small, narrowly ellipsoid . . . . **Stromatoscypha** (Schizophyllaceae)

27′. Hymenophore reticulate to tubular, tubes nor arising independently except perhaps in *Ceriporia* 28

28(27′) Resupinate species (Fig. 1/4) 29

28′. Basidiocarp effuso-reflexed, dimidiate, flabelliform, reniform or ungulate; single or imbricate 65

29(28) Hymenium continuous over the edge of the dissepiments 30

29′. Hymenium discontinuous 33

30(29) Spores colored; hymenophore reticulate, often dark-colored; clamp-connections present . . . . **Serpula** (Coniophoraceae)

30′. Spores hyaline 31

31(30′) Clamp-connections present; hymenophore reticulate, poroid or irregularly lamellate, pale to orange; spores small, cylindric . . . . See 93. **Merulius** (Corticiaceae)

31′. Clamp-connections absent; hymenophore tubular 32

32(31′) Encrusted conical cystidia present; basidiocarp coriaceous; hymenium finally becoming discontinuous; spores small, ellipsoid, cylindric . . . . **Cystidiophorus** (Corticiaceae)

32′. Cystidia lacking; basidiocarp with a white pubescent margin; context discoloring purplish black at maturity; spores small, arcuate-cylindric . . . . See 93′. **Meruliopsis** (Corticiaceae)

33(29′) Spores colored, clay or bluish-green; clamp-connections absent 34

33′. Spores hyaline; clamp-connections present or absent 35

34(33) Spores small, subglobose, bluish-green; hymenophore granular to reticulate, dark-colored . . . . **Byssocorticium** (Corticiaceae)

34′. Spores pale clay-colored, 7–9 μm long, ovoid with hyaline reticulations and spines; hymenophore with shallow irregular pores, golden-yellow to purplish . . . **Lindtneria** Type: *L. trachyspora*. One sp.

35(33′) Spore wall ornamented 36

35′. Spores smooth, hyaline; clamp-connections present or absent 39

36(35) Spores with amyloid verrucae or ridges, short ellipsoid; hyphal-system dimitic with clamp-connections, skeletal-hyphae dextrinoid, cyanophilous . . . **Wrightoporia** Type: *W. lenta*. Two spp. N. and C. America.

36′. Spores inamyloid 37

37(36′) Spores globose, small (3–6 μm), echinulate 38

37'. Spores large, cylindric with a hyaline perisporium and immersed papillae, cyanophilous; basidiocarp firm to corky, light brown; clamp-connections present . . . . **Pachykytospora**
Type: *P. tuberculosa*. One sp. Europe, Asia, N. America.

38(37) Hymenophore labyrinthiform; basidia urniform with 3–6 sterigmata; clamp-connections absent . . . . **Echinotrema** (Corticiaceae)

38'. Hymenophore regularly poroid; basidia clavate with 4 sterigmata; clamp-connections present . . . . **Cristella** (Corticiaceae)

39(35') Hymenium devoid of cystidia or cystidioles 40

39'. Cystidia, cystidioles, or gloeocystidia present 57

40(39) Tubes stratified 41

40'. Tubes not stratified 44

41(40) Context deep purplish brown; clamp-connections present; spores small, ovoid
. . . . **Melanoporia**
Type: *M. nigra*. One sp. N. America.

41'. Context whitish or pale 42

42(41') Context white becoming yellowish, hard, monomitic; clamp-connections absent
. . . . 43

42'. Context white, yellowish or pink, corky, dimitic; clamp-connections present; basidiocarp normally ungulate with a crust . . . . See 89. *Fomitopsis*

43(42) Spores ovoid, truncate . . . . **Poria**
Type: *P. medulla-panis*. Syn.: *Perenniporia*. About 5 spp. N. Temperate, Australia.

43'. Spores ovoid to subcylindric . . . . **Riopa**
Type: *R. davidii*. One sp. Corsica.

44(40') Spores elongate ellipsoid to cylindric 45

44'. Spores globose, ovoid, or ellipsoid 52

45(44) Context essentially monomitic, although binding-hyphae sometimes found to a limited extent 46

45'. Context dimitic or trimitic; clamp-connections present 49

46(45) Consistency soft-waxy-rigid, or watery-fleshy 47

46'. Consistency hard, horny; basidiocarp often effuso-reflexed . . See 116. **Cartilosoma**

47(46) Context watery-fleshy, white, usually well-developed; clamp-connections present; basidiocarp typically pileate . . . . See 98'. **Tyromyces**

47'. Context reduced to a thin subiculum, waxy; spores small, arcuate cylindric 48

48(47') Clamp-connections absent; subiculum thin, adherent to the substrate
. . . . **Ceriporia**
Type: *C. viridans*. Seven spp. N. and S. Temperate.

48'. Clamp-connections present . . . . **Ceriporiopsis**
Type: *C. gilvescens*. Six spp. N. Temperate.

49(45') Context trimitic, corky, white; spores cylindric, 7–12 μm long . . . See 115. **Trametes**

49'. Context dimitic with thick-walled skeletal-hyphae 50

50(49') Context whitish, clamp-connections present 51

50'. Context pale brown to purplish brown; clamp-connections absent . . **Melanoporella**
Type: *M. carbonacea*. One sp. Cuba.

51(50) Skeletal-hyphae simple, unbranched . . . . See 116′. **Antrodia**

51′. Skeletal-hyphae with much-branched, arboriform apices . . . . **Dichomitus**
Type: *D. squalens.* Two spp. N. Temperate.

52(44′) Context duplex, with a thin dark, gelatinous zone above the tubes; hyphal-system monomitic with clamp-connections; spores cyanophilous, dextrinoid
. . . . **Parmastomyces**
Type: *P. krartzevianus.* One sp. N. Temperate.

52′. Context homogeneous, very thin, readily separable from the substrate 53

53(52′) Clamp-connections absent; subiculum and pores white but discoloring at maturity, hard . . . . See 86′. **Rigidoporous**

53′. Clamp-connections present 54

54(53′) Rhizoids absent; spores cyanophilous, slightly dextrinoid, slightly thick-walled; subiculum soft; hyphal system monomitic; basidia suburniform . . . . **Strangulidium**
Type: *S. sericeomolle.* Two spp. N. Temperate.

54′. Rhizoids present; spores acyanophilous; basidia not suburniform 55

55(54′) Spores amyloid; subiculum soft coriaceous . . . . **Anomoporia**
Type: *A. bombycina.* Three spp. N. Temperate.

55′. Spores inamyloid; margin with byssoid threads or mycelial cords 56

56(55′) Hyphal-system monomitic; subiculum soft, white . . . . **Fibuloporia**
Type: *F. mollusca.* Four spp. Worldwide.

56. Hyphal-system dimitic with skeletal-hyphae; subiculum soft coriaceous, whitish
. . . . **Fibroporia**
Type: *F. vaillantii.* Five spp. N. Temperate.

57(39′) Cystidioles or gloeocystidia only present; clamp-connections present 58

57′. Well-developed cystidia present 61

58(57) Hyphal-system dimitic with conspicuous skeletal-hyphae; cystidioles fusoid; spores minute, allantoid . . . . **Incrustoporia**
Type: *I. stellae.* Four spp. N. Temperate.

58′. Hyphal-system monomitic, white or yellow 59

59(58′) Context thick, with hyphae amyloid or at least dextrinoid in mature basidiocarps; cystidioles apically incrusted; spores small, cylindric (Fig. 1.12) . . . . . . **Amyloporia**
Type: *A. lenis.* Four spp. Worldwide. Eriksson (1949).

59′. Hyphae neither amyloid nor dextrinoid 60

60(59′) Consistency tough; subiculum thin; tubes very irregular, lacerate; cystidioles with capitate apex (Fig. 1.17); short ellipsoid . . . . **Schizopora**
Type: *S. paradoxa.* Synonym: *Xylodon.* Six spp. Worldwide. Domanski (1969).

60′. Consistency watery fleshy; context well-developed; pores regular; cystidioles fusoid when present; gloeocystidia known in one species . . . . . . See 47, 98′. **Tyromyces**

61(57′) Cystidia thin-walled, oblong fusoid, nonencrusted; tubes short; ovoid; basidiocarps typically dimidiate . . . . See 108. **Climacocystis**

61′. Cystidia coated with calcium oxalate crystals (Fig. 1.18) 62

62(61′) Hymenophore always irpicoid; spores small, ellipsoid; context dimitic with skeletal-hyphae; basidiocarp usually effuso-reflexed . . . . See 120′. **Irpex**

62′. Hymenophore poroid; subiculum thin, soft to brittle; spores small, ellipsoid to cylindric 63

63(62′) Clamp-connections absent . . . . See 86. **Oxyporus**

63′. Clamp-connections present 64

   64(63′) Hyphal-system monomitic; cystidia arising in the subhymenium
   . . . . **Chaetoporellus**
       Type: *C. latitans.* Six spp. N. Temperate.

   64′. Hyphal-system dimitic; skeletal-hyphae; cystidia tramal in origin . . . **Chaetoporus**
       Type: *C. euporus.* Seventeen spp. Worldwide.

65(28′) Context changing color with alkali; clamp-connections present 66

65′. Context not noticeably darkening with alkali 69

   66(65) Context turning lilac or crimson with alkali; hyphal-system monomitic; spores
   hyaline, ellipsoid, 4–6 μm long 67

   66′. Context blackish with alkali but discoloration not permanent; hyphal-system dimitic
   68

67(66) Hyphae with incrusting pigment; context soft, fleshy . . . . **Hapalopilus**
       Type: *H. nidulans.* Eight spp. Worldwide.

67′. Hyphae not incrusted; context tough, fibrous; tubes orange becoming resinous when dry
   . . . . **Aurantioporus**
       Type: *A. pilotae.* Two spp. N. Temperate.

   68(66′) Context orange turning purplish black with alkali; yellow skeletal-hyphae present;
   spores truncate, cyanophilous . . . . **Pyrofomes**
       Type: *P. demidoffii.* N. Temperate.

   68′. Context dark fawn; brown skeletal-hyphae present; pores 7–9 per mm; purplish
   brown; spores small (3.5 μm), cylindric . . . . **Nigroporus**
       Type: *N. vinosus.* Three spp. Subtropical-tropical.

69(65′) Context dark-colored from the first, some shade of brown or red 70

69′. Context white, cream or pale wood color, rarely pink, sometimes discoloring at maturity
   84

   70(69) Spores yellowish brown, lavender in print, 11–12 × 6–10 μm, truncated by a
   germ-pore; hymenophore trametoid; context purplish brown to chestnut; hyphal-
   system trimitic; chlamydosporic stage also produced . . . . **Phaeotrametes**
       Type: *P. decipiens.* One sp. Australia, S. America. Wright (1966).

   70′. Spores hyaline or lacking an apical germ-pore 71

71(70′) Spores ellipsoid to cylindric, hyaline or very brown at maturity 72

71′. Spores subglobose to ovoid-ellipsoid, or truncate, hyaline or colored 82

   72(71) Spores over 7 μm long 73

   72′. Spores not exceeding 7 μm in length 79

73(72) Basidiocarp compound, consisting of closely overlapping small pilei arising from a
central solid core; context cinnamon; hyphal-system dimitic with skeletal-hyphae, clamp-
connections absent; spores light brown at maturity, 9–12 μm long . . . . . . **Globifomes**
       Type: *G. graveolens.* One sp. N. America.

73′. Basidiocarp simple, at times imbricate but never with a common central core; clamp-
connections present 74

74(72′) Thick-walled, fusoid hyphal endings present in the hymenium (Fig. 2.6), or spores large 75

74′. Hymenium devoid of skeletohyphidia, although cystidioles may occur 78

75(74) Hymenophore lamellate at maturity, sometimes labyrinthiform; basidiocarp thin, dark brown; spores 13 μm long. . . . . **Gloeophyllum**
     Type: *G. sepiarium*. Syn.: *Lenzitina*. Twelve spp. David (1968).

75′. Hymenophore poroid to labyrinthiform; hyphal-system trimitic 76

76(75′) Hymenophore labyrinthiform with very thick dissepiments (Fig. 1.7); basidiocarp thick, wood-colored to rust-brown; spores 6–9 μm long. . . . . **Daedalea**
     Type: *D. quercina*. Syn.: *Phaeodaedalea*. Two spp. N. Temperate, Brazil. *Daedalea sprucei* produces large chlamydospores in addition to the basidiospores.

76′. Hymenophore regularly hexagonal, pores often very broad, usually with a white pruinose covering to give a waxy appearance 77

77(76′) Pileal-surface brown to black, trichodermial, appearing strigose to velutinate; tropical . . . . see 112′. **Scenidium**
     Type: *S. apiarium*. Syn.: *Pogonomyces*. About 50 spp. This is *Hexagonia* in the broadly accepted sense. Lloyd (1910a), M. K. Fidalgo (1968).

77′. Pileal-surface smooth, glabrous; N. temperate . . . . **Apoxona**
     Type: *A. nitida*. One sp. Mediterranean.

78(74′) Hyphal-system trimitic; basidiocarp effuse-reflexed to dimidiate; cystidioles frequently present . . . . **Coriolopsis**
     Type: *C. occidentalis*. Syn.: *Fomitella*. Seventeen spp. A very common pantropical genus.

78′. Hyphal-system dimitic with skeletal-hyphae; basidiocarp effuso-reflexed, often imbricate; hymenophore poroid or lamellate . . . . **Phaeocoriolellus**
     Type: *P. trabea*. One sp. N. Temperate.

79(72′) Incrusted cystidia in the hymenium 80

79′. Cystidia absent 81

80(79) Cystidia short fusoid, hyaline, apically incrusted; basidiocarp deep umbrinous; pileal-surface densely strigose with stiff, brown-black hairs; context much reduced . . . . **Trichaptum**
     Type: *T. trichomallum*. One sp. N. and S. America.

80′. Cystidia cylindric, brown, tramal in origin, coated with crystals except at the apex; basidiocarp small, effuse-reflexed or conchate; pileal-surface strigose; hyphal-system trimitic; spores ellipsoid . . . . **Metuloidea**
     Type: *M. tawa*. One sp. New Zealand, Australia.

81(79′) Entire basidiocarp some shade of red, thin, coriaceous; pileal-surface smooth, glabrous; hyphal-system trimitic with clamp-connections . . . . **Pycnoporus**
     Type: *P. cinnabarinus*. Three spp. Worldwide. Nobles and Frew (1962).

81′. Basidiocarp reddish brown with a dark, resinous, wrinkled crust; hyphal-system monomitic, lacking clamp-connections . . . . see 97′. **Ischnoderma**
     Type: *I. resinosum*. Two spp. N. Temperate, Australasia.

82(71′) Basidiocarp perennial, often very large, provided with a crust and usually with a stratified hymenophore, hyphal-system trimitic with clamp-connections; context light chestnut brown; spores ellipsoid, 15–25 μm long . . . . **Fomes**

Type: *F. fomentarius.* Syn.: *Placodes, Elfvingiella.* About 20 spp. Worldwide. Lloyd (1915: Sixth General Division).

82′. Basidiocarp smaller, lacking a crust 83

83(82′) Context rust-brown, well-developed, corky, with a strong aniseed odor when fresh; cystidia absent . . . . **Osmoporus**
    Type: *O. odoratus.* Syn.: *Anisomyces, Ceratophora.* three spp. N. Temperate. M. K. Fidalgo (1962).

83′. Context yellowish brown, very thin; tubes bright yellow, often stratified, pores very small; cystidia incrusted; spores minute, 2 × 1.5 μm . . . . **Flaviporus**
    Type: *F. rufoflavus.* Syn.: *Baeostratoporus.* Three spp. Worldwide.

84(69′) Perennial species with stratified hymenophore 85

84′. Annual, or if perennial, devoid of a stratified hymenophore 90

85(84) Hyphal-system monomitic, clamp-connections absent; spores subglobose 86

85′. Hyphal-system dimitic or trimitic; clamp-connections present or absent 87

86(85) Context stratified; apically encrusted cystidia originating in the subhymenial layer (Fig. 1/18) . . . . see 63. **Oxyporus**
    Type: *O. populinus.* Syn.: *Boudiera.* Seven spp. N. Temperate. Cooke (1949).

86′. Context not stratified; cystidia when present, encrusted or not, arising from below the subhymenial layer; pileus often with a crust . . . . see 53. **Rigidoporus**
    Type: *R. zonalis.* Syn.: *Leucofomes, Mensularia, Podoporia.* Fourteen spp. Worldwide. Pouzar (1966). *Rigidoporus lignosus* causes a white root rot common on rubber (*Hevea*) and other tropical perennial crops. *Rigidoporus ulmarius* causes elm (*Ulmus*) butt rot.

87(85′) Spores large, 12–14 μm long, truncate; basidiocarp small, ungulate with an indistinct crust . . . . **Truncospora**
    Type: *T. ochroleuca.* Syn.: *Ungulina.* Two spp. Worldwide.

87′. Spores smaller; basidiocarp generally larger 88

88(87′) Hyphal-system trimitic with clamp-connections; context white, impregnated with crystals; basidiocarp very large, ungulate . . . . **Laricifomes**
    Type: *L. officinalis.* One sp. N. Temperate. Teixeira (1958).

88′. Hyphal-system dimitic with skeletal-hyphae; pileus developing a crust 89

89(88′) Clamp-connections present; basidiocarp applanate to ungulate, at times resupinate; context white to wood-color, sometimes pink; hyphal-system occasionally trimitic; spores small . . . . see 42′. **Fomitopsis**
    Type: *F. pinicola.* Eighteen spp. Worldwide.

89′. Clamp-connections absent; basidiocarp resupinate to dimidiate with an acute margin; spores finely asperulate. . . . . **Heterobasidion**
    Type: *H. annosum.* One sp. Worldwide in Temperate regions. The cause of decay to roots and heart wood of living trees, particularly conifers. Pegler and Waterston (1968, p. 192).

90(84′) Context spongy, watery-fleshy or soft corky 91

90′. Context firm, coriaceous, woody, or cartilaginous 102

91(90) Hymenium continuous over the edge of the dissepiments at least in the early stages; hymenophore reticulate to tubular; spores small, cylindric 92

91′. Hymenium discontinuous, edge of dissepiments sterile 94

92(91) Hymenophore tubular, well-developed; basidiocarp effuso-reflexed to dimidiate, often laterally confluent; context duplex with a lower gelatinous layer, becoming horny when dry . . . . **Gloeoporus** (Corticiaceae)

92′. Hymenophore reticulate or with shallow irregular pores, basidiocarp typically resupinate although may become dimidiate 93

93(82′) Clamp-connections present . . . . See 31. **Merulius** (Corticiaceae)

93′. Clamp-connections absent . . . . See 32′. **Meruliopsis** (Corticiaceae)

94(91′) Pores sulphurine or coral red; basidiocarp applanate, often imbricate; context "cheeselike"; hyphal-system dimitic with binding-hyphae, devoid of clamp-connections . . . . **Laetiporus**
  Type: *L. speciosus.* Syn.: *Cladoporus.* Four spp. Worldwide.

94′. Pores and hyphal-system different 95

95(94′) Amyloid clavate-cylindric cystidia present; pileal-surface strigose; context hyphae amyloid; spores elongate ellipsoid . . . . **Amylocystis**
  Type: *A. lapponicus.* One sp. N. Temperate.

95′. Cystidia absent, although cystidioles or encrusted hyphae may be present 96

96(95′) Hyphal-system essentially monomitic, although occasionally skeletal and binding-hyphae may be found 97

96′. Hyphal-system dimitic or trimitic; basidiocarp dimidiate, reniform or ungulate, corky, ... 99

97(96) Clamp-connections present; context soft, fibrillose fleshy 98

97′. Clamp-connections absent; pileus with a resinous, wrinkled crust; context at first white, finally rust-colored . . . . See 81′. **Ischnoderma**

98(97) Context duplex; hymenophore poroid or irpicoid; spores short ellipsoid with thickened walls . . . . **Spongipellis**
  Type: *S. spumeus.* Syn.: *Irpiciporus.* Ten spp. N. Temperate. Kotlaba and Pouzar (1965), David (1969).

98′. Context never duplex; hymenophore always poroid; cystidioles or gloeocystidia found in some species; spores small, allantoid to ovoid . . . . See 47, 60′. **Tyromyces**
  Type: *T. chioneus.* Syn.: *Postia, Leptoporus.* About 50 spp. Worldwide, common.

99(96′) Margin of basidiocarp developing to form a veillike covering to the hymenophore leaving an opening near the base; pileal-surface with a laccate crust . . . . **Cryptoporus**
  Type: *C. volvatus.* One sp. N. America, Japan.

99′. Hymenophore fully exposed from the first 100

100(99′) Context very thin; basidiocarp small, coriaceous; hyphal-system trimitic, hyphae metachromatic in cresyl blue; spores minute, allantoid . . . . **Leptotrimitus**
  Type: *L. semipileatus.* One sp. N. Temperate.

100′. Context thick, basidiocarp much larger, surface at first covered by a tomentum which becomes weathered to produce a paperlike pellicle 101

101(100′) Hyphal-system of hymenophore dimitic; context with binding-hyphae . . . . **Piptoporus**
  Type: *P. betulinus.* Syn.: *Ungularia.* Six spp.

101′. Hyphal-system of hymenophore monomitic; context with skeletal-hyphae bearing short lateral processes . . . . **Buglossoporus**
  Type: *B. quercinus.* One sp. Europe.

102(90′) Clamp-connections present 103

102′. Clamp-connections absent 117

103(102) Sterile organs present in the hymenium as cystidia or cystidioles 104

103′. Cystidia and cystidioles absent 109

104(103) Hymenophore lamellate; basidiocarp dimidiate 105

104′. Hymenophore poroid, daedalioid, or trametoid; context often duplex 106

105(104) Spores brown, subglobose, verrucose; context monomitic; filiform cystidioles present

. . . . **Lenzitopsis** (Thelephoraceae)

105′. Spores hyaline, cylindric, smooth; context trimitic; thick-walled fusoid cystidia present

. . . . **Lenzites**

        Type: *L. betulina*. Syn.: *Cellularia, Artolenzites, Leucolenzites*. About 60 spp. World-
        wide. Kotlaba (1965).

106(104′) Basidiocarp imbricate, thin; pileal-surface pubescent to strigose 107

106′. Basidiocarp with different characters; spores always less than 8 μm long. . . . 108

107(106) Pileal-surface with a greyish-brown pubescent cuticle; context well-developed or
not; spores arcuate cylindric . . . . **Datronia**
        Type: *D. mollis*. Three spp. N. Temperate. This genus has previously been referred
        to under *Antrodia*.

107′. No distinct cuticular layer; context very thin; hymenophore poroid or irpicoid; spores
ellipsoid cylindric, 9 × 3.4 μm or larger . . . . **Hirschioporus**
        Type: *H. abietinus*. Ten spp. Worldwide. Imazeki (1944).

108(106′) Dimidiate or reniform; context not duplex; cystidia thin-walled, fusoid, spores
ovoid . . . . See 61. **Climacocystis**
        Type: *C. borealis*. One sp. N. Temperate.

108′. Resupinate with numerous small, effuso-reflexed pilei; context duplex with an upper
layer containing skeletal-hyphae; cystidioles only present; spores cylindric

. . . . **Skeletocutis**
        Type: *S. amorphus*. One sp. Europe, Asia.

109(103′) Context duplex with a gelatinized zone (thin black line in section), basidiocarp thin,
imbricate; spores elongate ellipsoid, 4–7 μm long 110

109′. Context not duplex; spores cylindric 111

110(109) Hymenophore poroid, dark grey; hyphal system monomitic (Fig. 2.1)

. . . . **Bjerkandera**
        Type: *B. adusta*. Syn.: *Myriadoporus*. Two spp. Worldwide.

110′. Hymenophore daedalioid to irpicoid; hyphal-system trimitic . . . . . . **Cerrena**
        Type: *C. unicolor*. Syn.: *Phyllodontia, Bulliardia*. Two spp. Van der Westhuizen
        (1963).

111(109′) Spores small, less than 8 μm long; hyphal-system trimitic. 112

111′. Spores more than 8 μm long 114

112(111) Hymenophore never with hexagonal pores 113

112′. Pores small to very large, regularly hexagonal; context usually well-developed
. . . . See 77. **Scenidium**

113(112) Hymenophore pores small, irregular; dissepiments thin; basidiocarp effuso-reflexed
to dimidiate (Fig. 1/8) . . . . **Coriolus**
        Type: *C. versicolor*. Syn.: *Hansenia*. About 20 spp. Worldwide, common.

113′. Hymenophore with radially elongated pores, trametoid; dissepiments thick; context thick . . . . **Pseudotrametes**
> Type: *P. gibbosa*. One sp. Europe, Asia.

> 114(111′) Hyphal-system trimitic 115

> 114′. Hyphal-system monomitic or dimitic 116

115(114) Hymenophore poroid or irregular; context well-developed; pileal-surface glabrescent
. . . . See 48′. **Trametes**
> Type: *T. suaveolens*. Syn.: *Earliella*. About 20 spp. in the restricted sense. Worldwide.

115′. Hymenophore with large, subirpicoid pores; context thin; pileal-surface densely strigose
. . . . **Funalia**
> Type: *F. funalis*. Syn.: *Trametella*. Five spp. Worldwide. Lloyd (1910b).

> 116(114′) Basidiocarp resupinate with a reflexed margin; context soft at first then bone hard; hyphal-system monomitic . . . . See 46′. **Cartilosoma**
> Type: *C. subsinuosum*. One sp. N. Temperate. Kotlaba (1955).

> 116′. Basidiocarp effuse-reflexed, thin, corky; hyphal-system dimitic with ekeletal-hyphae
. . . . See 51. **Antrodia**
> Type: *A. albida*. Syn.: *Coriolellus*. Eleven spp. Worldwide. Sarker (1959).

117(102′) Spores ovoid to ellipsoid; hymenophore trametoid to irpicoid 118

117′. Spores cylindric 121

> 118(117) Spores large, 6–10 μm long, with a double wall . . . .**Leucophellinus**
> Type: *L. irpicioides*. One sp. N. Europe, Asia.

> 118′. Spores smaller, 4–6 μm long, with a thin wall 119

119(118′) Basidiocarp dimidiate to flabelliform; hymenophore trametoid; hyphal·system trimitic; cystidia absent . . . . **Haploporus**
> Type: *H. ljubarskyi*. One sp.

119′. Basidiocarp resupinate with reflexed pilei; hymenophore irpicoid; cystidia present
. 120

> 120(119′) Spores amyloid; hyphal-system monomitic . . . . **Irpicodon**
> Type: *I. pendulus*. One sp. Europe. Possibly better placed in Corticiaceae.

> 120′. Spores inamyloid; hyphal-system dimitic . . . . See 62. **Irpex**
> Type: *I. lacteus*. About 20 spp. Worldwide.

121(117′) Spores 7–10 μm long; basidiocarp large, often imbricate; pileus zoned; hymenophore trametoid to lamellate . . . . **Daedaleopsis**
> Type: *D. confragosa*. Six spp. Worldwide.

121′. Spores 3–7 μm long; pileus with a small sterile cupulate excrescence on upper surface near the base; context thin, often zoned; hymenophore irregularly poroid
. . . . **Poronidulus**
> Type: *P. conchifer*. Two spp. N. America.

V. HYMENOCHAETACEAE

This family is annual or perennial. The basidiocarp is effused to stipitate, corticioid, polyporoid, or clavarioid. When present, the pileus is dorsiventral. The hymenophore is smooth, granuliferous, irpicoid, or tubulate, and is rarely concentrically lamellate but never radially lamellate. When tubulate and perennial, Hymenochaetaceae are then often stratified; the edge of the dissepiments are sterile. The context is yellowish cream to orange, rust brown, or ferruginous, permanently darkening in alkaline solution to give a xanthochroic reaction, and the texture is soft and loose, fibrous, coriaceous, or woody. The hyphal-system is mono- or dimitic with skeletal hyphae, and clamp-connections are always absent. A cata-hymenium or euhymenium is present. The basidia are two- or four-spored and clavate. Hymenial setae are often present and dichohyphidia are sometimes present; cystidia and gloeocystidia occur in a few species. Setae are present as asterosetae, macrosetae, or haplosetae. The spores are hyaline or brown, thin to thick-walled, smooth, rarely ornamented, and occasionally amyloid or cyanophilous. Hymenochaetaceae are lignicolous, sometimes parasitic, and rarely terrestrial.

KEY TO HYMENOCHAETACEAE

1. Dichohyphidia or asterosetae present; hymenial setae absent; gloeocystidia often present
. . . . 2

1′. Dichohyphidia absent; hymenial setae and setal-hyphae often present; gloeocystidia absent; spores inamyloid . . . . **(Hymenochaetoideae)**   4

2(1) Dichohyphidia covering all surfaces, which wholly or partially develop into a cata-hymenium; basidiocarp cream, yellowish to dark-brown . . . . . . **(Varioideae)**   3

2′. Asterosetae present, sometimes resembling dichohyphidia; euhymenium; basidiocarp effused; hymenophore smooth; spores smooth or echinulate, mostly amyloid
. . . . **(Asterostromatoideae) Asterostroma**
Type: *A. apalum*. Seven spp. Tropical and subtropical.

3(2) Basidiocarp clavarioid, highly branched; hymenophore smooth; to granuliferous
. . . . **Vararia**
Type: *V. investiens*. Syn.: *Asterostromella*. About 15 spp. of which Welden (1965) keys out 8 West Indian spp. with notes on extralimital spp. Worldwide.

3′. Basidiocarp clavarioid, highly branched; hymenophore smooth; tropical
. . . . **Lachnocladium**
Type: *L. brasiliense*. Syn.: *Ericladus, Stelligera*. About 25 spp. Tropical and Austra-lasia. Corner (1950).

4(1′) Asterosetae in context; hymenial setae present; basidiocarp effused; hymenophore hydnoid . . . . **Asterodon**
Type: *A. ferruginosus*. Syn.: *Hydnochaete* Peck non Bres., *Hydnochaetella*. N. America, Europe, Japan. Corner (1948).

4′. Setae and/or setal-hyphae present or absent; true asterosetae absent 5

5(4′) Basidiocarp clavarioid or stereoid, stipitate . . . . **Clavariachaete**

5′. Basidiocarp resupinate to pileate 6

6(5′) Hymenophore smooth; setae present; basidiocarp resupinate to stipitate
. . . . **Hymenochaete**
Type: *H. tabacina*. Syn.: *Hymenochaetella*. About 80 spp. Worldwide, but mostly

tropical and subtropical. For N. American species see Burt (1918); Europe, Bourdot and Galzin (1928); Australasia, Cunningham (1963), W. Indies, Reeves and Welden (1967); Japan, Imazeki (1940).

6′. Hymenophore granular to tubular 7

7(6′) Hymenophore raduloid to irpicoid . . . . **Hydnochaete**
    Type: *H. badia*. Three spp. N. and S. America.

7′. Hymenophore tubular, and concentrically lamellate 8

8(7′) Spores ellipsoid with verrucose ornamentation, brown or pale yellow; basidiocarp resupinate or pendant, attached by the vertex; setae absent . . . . **Coltriciella**
    Type: *C. dependans*. 3 spp. Tropical and Australasian.

8′. Spores smooth, hyaline or colored 9

9(8′) Context yellowish-orange to orange; pores very large, irregular; basidiocarp resupinate to dimidiate; setae absent but hyaline cylindric cystidia present; hyphae hyaline, encrusted; spores hyaline . . . . **Pycnoporellus**
    Type: *P. fibrillosus*. Syn.: *Aurantioporellus*. Two spp. N. America, rare in Asia, E. Europe. Kotlaba and Pouzar (1963).

9′. Context yellowish-brown, rust-brown to golden-brown 10

10(9′) Basidiocarp stipitate; hyphal-system monomitic 11

10′. Basidiocarp resupinate, effuse-reflexed, dimidiate to ungulate; context rust-brown to ferruginous; setal structures present or absent 15

11(10) Hymenophore tubular 12

11′. Hymenophore poroid or with irregular concentric lamellae; pileus rust-brown; spores brown, large, ellipsoid . . . . **Cycloporus**
    Type: *C. greenii*. One sp. N. America. Overholts (1953; p. 116)

12(11) Context golden-brown; pileus large flabelliform, with a well-developed cuticle; stipe short sometimes absent; spores small, globose or broadly ellipsoid, with a thickened dull-brown wall . . . . **Aurificaria**
    Type: *A. indica*. Two spp. India, Australia, Philippines.

12′. Context yellowish rust to cinnamon; spores thin-walled 13

13(12′) Haplosetae present; basidiocarp solitary; context spongy fleshy, rust-brown; spores hyaline or pale yellow . . . . **Mucronoporus**
    Type: *M. tomentosus*. Syn.: *Polystictus, Onnia*. Three spp. N. Temperate, India. Lloyd (1908).

13′. Setal-structures absent 14

14(13′) Basidiocarp large, imbricate with a thick short, central or excentric stipe; context fleshy, brittle when dry; hyphae broadly inflated; dark-colored cystidioles present, spores hyaline . . . . **Phaeolus**
    Type: *P. schweinitzii*. Syn.: *Choriphyllum, Romellia, Spongiosus*. Two spp. N. Temperate, Australia.

14′. Basidiocarp smaller, stipe well-developed, central or excentric; context firm, thin; spores hyaline or colored . . . . **Coltricia**
    Type: *C. perennis*. Syn.: *Xanthochrous, Phaeolopsis*. Thirteen spp. Worldwide. Coker (1946), Imazeki and Kobayashi (1966).

15(108) Hyphal-system monomitic; consistency watery fleshy to firm, basidiocarp annual
. . . . 16

15'. Hyphal-system dimitic; consistency hard, leathery or woody; basidiocarp annual or perennial; spores hyaline or colored 17

16(15) Spores minute, hyaline; hymenophore ranging from concentrically lamellate to minutely poroid, not distinct from the context; pileus thin, coriaceous; setae present
. . . . **Cyclomyces**
Type: *C. fuscus.* Syn.: *Cyclomycetella, Cycloporellus.* Twelve spp. N. America, Tropical, Australian. Patouillard (1896), Lloyd (1910b).

16'. Spores larger, hyaline or brown; hymenophore poroid, distinct from the context; setae present or absent (Fig. 2.2) . . . . **Inonotus**
Type: *I. hispidus.* Synonyms: *Inoderma, Inodermus. Cerenella, Xanthoporia, Phaeoporus, Polystictoides.* Forty-two spp. Worldwide. Pegler (1964). *Inonotus hispidus* causes heart-rot of *Fraxinus.*

17(15') Context homogeneous, without an agglutinated layer; pileal-surface with or without a crust; tubes often stratified (Fig. 2.3) . . . . **Phellinus**
Type: *P. torulosus.* Syn.: *Fulvifomes, Ochroporus, Pyropolyporus.* About 80 spp. Worldwide. This is a common genus, previously referred to under *Fomes* because of the production of perennial basidiocarps. *Phellinus pachyphloeus* produces one of the largest known basidiocarps. Several species, such as *P. noxius, P. pomaceus,* are plant pathogens.

17'. Context duplex with an upper spongy layer and a lower form layer separated by a dark agglutinated zone . . . . **Cryptoderma**
Type: *C. ribis.* Syn.: *Daedalioides, Porodaedalea.* Fourteen spp. Worldwide.

VI.   GANODERMATACEAE

Ganodermataceae are annual or perennial; their basidiocarp is sessile or stipitate, but always dorsiventrally pileate. The pileal-surface frequently forms a crust and the hymenophore is always tubulate, with the tubes being narrow and often stratified, the pores minute, and the edge of the dissepiments sterile. Their context is pallid to dark or purplish brown and their texture is corky to woody. The hyphal system is dimitic with skeletal hyphae or trimitic, and the skeletal hyphae often have arboriform apices, clamp-connections, or generative hyphae. A euhymenium is present. The basidia are four-spored; setae and cystidia are absent. The spores are globose to ellipsoid, and brown, but sometimes only faintly so. The wall structure is complex and has a form; the pigmented exoepisporium bears an ornamentation covered by a hyaline perisporium and is inamyloid. The family is lignicolous.

Important Literature

Steyaert, Considérations générales sur le genre Ganoderma et plus specialement sur les espèces Europennes, *Bull. Soc. Roy. Bot. Belg.* **100**: 189–211 (1967); Furtado, Structure of the spore of the Ganodermoideae Donk, *Rickia* **1**:227–241 (1962); Furtado, Relation of microstructures of the taxonomy of the Ganodermoideae with special reference to the cover of the pileal surface, *Mycologia* **57**:588–609, 1965; Heim, L'organisation architecturale des spores de Ganodermes, *Rev. Mycol.* **27**:199–212, 1962; Hansen, On the anatomy of the Danish species of Ganoderma, *Bot. Tidsskrift* **54**:333–352, 1958; Haddow, Studies in Ganoderma, *J. Arn. Arboretum* **12**:25–46, 1931.

KEY TO GENERA OF GANODERMATACEAE

1. Basidiocarp stipitate; stipe central, excentric or lateral 2
1'. Basidiocarp sessile, applanate to ungulate 3
    2(1) Spores broadly ellipsoid with an apical thickening which collapses at maturity to appear truncate, brown (Fig. 1.15) . . . . See 3'. **Ganoderma**
        Type: *G. lucidum.* Syn.: *Dendrophagus, Tomophagus, Trachyderma.* About 60 spp. Worldwide. Patouillard (1889), Steyaert (1961, 1962, 1967), Furtado (1967a).
    2'. Spores globose to broadly ellipsoid, not apically differentiated, pale to deep brown . . . . **Amauroderma**
        Type: *A. schomburgkii.* Syn.: *Whitfordia.* About 45 spp. Tropical and subtropical. Lloyd (1912), Furtado (1967b).
3(1') Hyphal-system dimitic; cuticular hyphae dark-colored . . . . **Elfvingia**
    Type: *E. applanata.* Syn.: *Friesia.* Six spp. Worldwide.
3'. Hyphal-system trimitic; cuticular hyphae light-colored . . . . See 2. **Ganoderma**

VII. BONDARZEWIACEAE

This family is an annual with a stipitate, polyporoid, or clavarioid and branched basidiocarp. The hymenophore is tubulate or smooth and the context is pallid, and firm. The hyphal-system is monomitic and sometimes appears falsely dimitic with elongate, thick-walled hyphal ends; vascular hyphae may be present and clamp-connections are present. There is a euhymenium and the cystidia and gloeocystidia are absent. The spores are globose to short ellipsoid, rather small, hyaline, and ornamented with amyloid spines and crests (Fig. 1.16). Bondarzewiaceae are lignicolous or terricolous.

KEY TO GENERA OF BONDARZEWIACEAE

1. Basidiocarp clavarioid with flabelliform branches; hymenophore smooth, amphigenous . . . . **Amylaria**
    Type: *A. himalayensis.* One sp. Himalayas.
1'. Basidiocarp compound with large rosettelike clusters (up to 50 cm in diameter) of 1–15 lateral pilei; hymenophore tubular . . . . **Bondarzewia**
    Type: *B. montana.* Two spp. N. Temperate, Australasia.

VIII. FISTULINACEAE

Fistulinaceae are annual with a pileate basidiocarp which is laterally stipitate although the stipe may be reduced. The hymenophore consists of densely crowded free tubes which are restricted at their bases. The context is white to cream or reddish, and fleshy and juicy to fibrous or coriaceous. The hyphal-system is monomitic with generative hyphae only or intermixed with vascular hyphae containing a reddish-orange sap. Clamp-connections are present or absent. The basidia are four-spored and cystidia, gloeocystidia, and setae are absent. The spores are hyaline, subglobose to short ovoid, smooth, inamyloid. The family is lignicolous.

A. Important Literature

Lohwag and Follner, Die Hymenophore von *Fistulina hepatica*, *Ann. Mycol.* **34**:456–464, 1936; Seynes "Recherches Vegetaux inferieurs I. Des Fistulines," 1874.

KEY TO GENERA OF FISTULINACEAE

1. Clamp-connections present; basidiocarp with a short, thick lateral stipe . . . . **Fistulina**
    Type: *F. hepatica*. Syn.: *Buglossus*, *Hypodrys*. Worldwide. Three spp. Wright (1961).

1'. Clamp-connections absent; basidiocarp with a well-developed stipe arising from roots; acanthohyphidia present on the pileal-surface. . . . . **Pseudofistulina** 2
    Type: *P. brasiliensis*. One sp. S. America. O. Fidalgo and Fidalgo (1962).

REFERENCES

Bourdot, H. and A. Galzin (1927). "Hyménomyétes de France," 761 pp. Lechevalier, Paris.

Burt, E. A. (1918). The Thelephoraceae of North America X. *Hymenochaete*. *Ann. Mo. Bot. Gard.* **5**:301–372.

Coker, W. C. (1946). The United States species of *Coltricia*. *J. Elisha Mitchell Sci. Soc.* **62**: 95–107.

Cooke, W. B. (1949). *Oxyporus nobilissimus* and the genus *Oxyporus* in North America. *Mycologia* **41**:442–455.

Corner, E. J. H. (1932). The fruit body of *Polystictus xanthopus* Fr. *Ann. Bot. (London)* **46**: 71–111.

Corner, E. J. H. (1948). *Asterodon*, a clue to the morphology of fungus fruit-bodies and with notes on *Asterostroma* and *Asterostromella*. *Trans. Brit. Mycol. Soc.* **31**:234–245.

Corner, E. J. H. (1950). A monograph of *Clavaria* and allied genera. *Ann. Bot. Mem.* **1**:416–437.

Cunningham, G. H. (1963). The Thelephoraceae of Australia and New Zealand. *N.Z. Dep. Sci. Ind. Res., Bull.* **145**:248–292.

David, A. (1968). Caractères culturaux et comportement nucleaire dans le genre *Gloeophyllum* Karst. (Polyporaceae). *Bull. Soc. Mycol. Fr.* **84**:119–126.

David, A. (1969). Caractères culturaux et cytologique d'espèces du genre *Spongipellis* Pat. et affinies. *Bull. Soc. Linn. Lyon* **38**:191–201.

Domanski, S. (1969). Wood inhabiting fungi in Bialowieza virgin forests in Poland VII. *Schizopora paradoxa* (Schrad. ex Fr.) Donk and its diagnose. *Acta Soc. Bot. Pol.* **38**:69–81.

Eriksson, J. (1949). The Swedish species of the "Poria vulgaris group." *Sv. Bot. Tidskr.* **43**: 1–25.

Fidalgo, M. K. (1962). The genus *Osmoporus*. *Rickia* **1**:95–138.

Fidalgo, M. K. (1968). The genus *Hexagona*. *Mem. N.Y. Bot. Gard.* **17**:35–108.

Fidalgo, O. (1962). Type studies and revision of the genus *Diacanthodes* Sing. *Rickia* **1**:145–180.

Fidalgo, O., and M. K. Fidalgo (1962). A new genus based on *Fistulina brasiliensis*. *Mycologia* **54**:342–352.

Furtado, J. S. (1967a). Some tropical species of *Ganoderma* (Polyporaceae) with pale context. *Persoonia* **4**:379–389.

Furtado, J. S. (1967b). Species of *Amauroderma* Murr. with the laccate appearance of *Ganoderma* Karst. *Bull. Jard. Bot. Belg.* **37**:309–317.

Humphrey, C. J. (1938). Notes on some Basidiomycetes from the Orient. *Mycologia* **30**: 327–335.

Imazeki, R. (1940). Studies on the genus *Hymenochaete* of Japan. *Bull. Tokyo Sci. Mus.* **2**: 1–22.

Imazeki, R. (1944). The genus *Hirschioporus*—Polyporaceae of Eastern Asia. II *J. Jap. Bot.* **20**:276–290.

Imazeki, R. (1966). The genus *Pyrroderma* Imazeki. *Trans. Mycol. Soc. Jap.* **7**:3–12.

Imazeki, R., and Y. Kobayashi. (1966). Notes on the genus *Coltricia* S. F. Gray. *Trans. Mycol. Soc. Jap.* **7**:42–44.

Kotlaba, F. (1955). A new species of the mycoflora of Czechoslovakia. *Trametes subsinuosa* Bres. *Ceska Mykol.* **9**:83–90.

Kotlaba, F. (1965). *Lenzites betulina* (L. ex Fr.) Fr. *Ceska Mykol.* **19**:79–82.

Kotlaba, F., and Z. Pouzar (1963). Three noteworthy polypores of the Slovakian Carpathians. *Ceska Mykol.* **17**:174–183.

Kotlaba, F., and Z. Pouzar (1965). Two rare polypores in Czechoslovakia. *Ceska Mykol.* **19**:69–78.

Lloyd, C. G. (1908). *Polystictus* (section *Pelloporus*). *Mycol. Writ.* **3**, Polyporoid Issue **1**:1–5.

Lloyd, C. G. (1909). The genus *Favolus*. *Mycol. Writ.* **3**, Polyporoid Issue **2**:17–21.

Lloyd, C. G. (1910a). Synopsis of the genus *Hexagona*. *Mycol. Writ.* **3**:1–46.

Lloyd, C. G. (1910b). Synopsis of the sections *Microporus, Tabacinus* and *Funales* of the genus *Polystictus*. *Mycol. Writ.* **3**:58–62.

Lloyd, C. G. (1912). Synopsis of the stipitate polyporoids. Section *Amaurodermus*. *Mycol. Writ.* **3**:110–121.

Lloyd, C. G. (1915). Synopsis of the genus *Fomes*. *Mycol. Writ.* **4**:209–288.

Nobles, M. K. and B. P. Frew (1962). Studies in wood inhabiting Hymenomycetes. V. The genus *Pycnoporus*. *Can. J. Bot.* **40**:987–1016.

Overholts, L. O. (1953). The Polyporaceae of the United States, Alaska and Canada. 466pp. Univ. of Michigan Press. Ann Arbor, Michigan.

Patouillard, N. (1889). Le genre *Ganoderma*. *Bull. Soc. Mycol. Fr.* **5**:65–82.

Patouillard, N. (1896). Le genre *Cyclomyces*. *Bull. Soc. Mycol. Fr.* **12**:45–51.

Pegler, D. N. (1964). A survey of the genus *Inonotus* (Polyporaceae). *Trans. Brit. Mycol. Soc.* **47**:175–195.

Pegler, D. N., and J. M. Waterston (1968). Commonwealth Mycological Institute Descriptions of Pathogenic Fungi and Bacteria, Nos. 191–200.

Pouzar, Z. (1966). Studies in the taxonomy of the Polypores. II. *Folia Geobot. & Phyto-Taxon.* **1**:356–375

Reeves, F., and A. L. Welden (1967). West Indian species of *Hymenochaete*. *Mycologia* **59**. 1034–1049.

Reid, D. A. (1963). New or noteworthy Australasian Basidiomycetes. V. *Kew Bull.* **17**:278–280.

Sarker, A. (1959). Studies in wood-inhabiting Hymenomycetes. IV. The genus *Coriolellus* Murr. *Can. J. Bot.* **37**:1251–1270.

Steyaert, R. L. (1961). Genus *Ganoderma* (Polyporaceae). Taxa nova. I. *Bull. Jard. Bot. Brux.* **31**:69–108.

Steyaert, R. L. (1962). Genus *Ganoderma* (Polyporaceae). Taxa nova. II. *Bull. Jard. Bot. Brux.* **32**:89–104.

Steyaert, R. L. (1967). Les *Ganoderma* palmicoles. *Bull. Jard. Bot. Belg.* **37**:469–492.

Teixeira, A. R. (1958). Studies on microstructure of *Laricifomes officinalis*. *Mycologia* **50**: 671–676.

Van der Westhuizen, G. C. A. (1963). The cultural characters, structure of the fruit-body, and type of interfertility of *Cerrena unicolor* (Bull. ex Fr.) Murr. *Can. J. Bot.* **41**:1487–1499.

Welden, A. L. (1965). West Indian species of *Vararia* with notes or extra-limital species. *Mycologia* **57**:502–520.

Wright, J. E. (1961). Del genero *Fistulina* en el hemispherio occidental. *Bol. Soc. Argent. Bot.* **9**:217–228.

Wright, J. E. (1966). The genus *Phaeotrametes*. *Mycologia* **58**:529–540.

CHAPTER 23

Agaricales and Related Secotioid Gasteromycetes

ALEXANDER H. SMITH

Department of Botany, and University Herbarium
University of Michigan
Ann Arbor, Michigan

I. INTRODUCTION

In recent times it has become recognized that most of the species in the family Secotiaceae correspond remarkably well in fundamental characters with the families of the Agaricales as these are set forth in Singer's system (Singer, 1962). To focus sharply on the problem, one must appreciate the manner in which the "mushroom" develops. It consists of pileus, hymenophore, and stipe. In the thousands of species possessing these structures many different patterns of development have been arrived at—most of them concerned in effecting some manner of protection for the developing hymenophore in whatever shape it will eventually take when mature. Protection may involve veils, such as in *Amanita* or a short stipe as in *Russula*. But whatever the differences in the type of basidiocarp development in the Agaricales, the end result at maturity is a more or less expanded structure placing the hymenophore in proper position so that the discharged spores (most of them) fall free from it and are carried away to whatever fate will befall them.

It is true that if the basidiocarp of a mushroom encounters unfavorable conditions of moisture or temperature it will remain in the unexpanded stage (button stage) and either dry out, be eaten by larvae, or if it is retarded simply by cold, the synchronization of the developmental processes will be interrupted to the extent that the hymenium (which is still protected from drying out) will continue to mature even though the stipe and pileus do not expand. This, in a nutshell, is the beginning of a secotioid Gasteromycete. Since the pileus does not expand, it is difficult to ascertain at exactly what stage (or how long it will take in terms of evolution) for a race to develop which will have lost all ability to discharge spores from its basidia. We know it has happened because we have many species in which this capacity no longer is present. The "primitive" secotioid gasteromycete, following this line of reasoning, will thus resemble in gross morphology the unexpanded button of a member of the Agaricales, but will produce mature spores. The presence or absence of a veil appears to be immaterial to the problem of becoming secotioid. At least the evidence indicates that gymnocarpic species connect to secotioid species which are also gymnocarpic in the sense that the margin of the "pileus," though it practically touches the stipe at first, is not intergrown with it; see the Bolbitiaceae and Strophariaceae.

From the standpoint of evolution the picture is the usual one, with each family group (agaricoid plus secotioid members) evolving along a set of relatively fixed characters to produce populations with various additional features. The relatively fixed characters in the present consideration are those on which the families of the Agaricales are based. This is why, in the key to the secotioid genera, they have been keyed out largely on the spore features of the comparable agaric families. This produces a much more "natural" classification than was previously in use.

No detailed account of how to study fleshy fungi is included here. For this the reader is referred to McIlvainea, Vol. 1, No. 1 (The Watling papers), and to Smith (1972). In the following account, references to Singer invariably refer to his "Agaricales in Modern Taxonomy" (Singer, 1962) unless a citation to the contrary is given. The terms used in the keys are those for the most part in current usage.

II. AGARICALES

The Agaricales have basidiocarps which are fleshy to subfleshy or rarely, almost leathery (but if the latter then the hymenophore is not poroid). In small basidiocarps the texture maybe membranous, pliant, or fragile. The hymenophore is smooth (rarely) to lamellate (typically) or poroid (less frequently). The basidia are two- to four- to eight-spored, and one-celled at maturity. The spores are forcibly discharged from the sterigmata at maturity.

This order intergrades with the Cantharellales which feature a hymenophore smooth (typically) or radially wrinkled to lamellate (Cantharellaceae). In the Cantharellales the basidia are often greatly elongated and narrow and bear more than four spores on a basidium in many species. The basidiocarps vary from simple to branched and the stipe lacks an expanded apex or the apex is expanded to a pileus, as in Cantharellus. For agariclike basidiocarps not discharging spores from the sterigmata, see the secotioid Gasteromycetes.

KEY TO FAMILIES OF AGARICALES

1. Trama of pileus and stipe heteromerous; spores with amyloid ornamentation
   . . . . **Russulaceae** p. 440

1'. Trama of pileus and stipe homiomerous 2

    2(1') Hymenophore tubulose to semilamellate (boletinoid) or lamellate (*Phylloporus* and *Phylloboletellus*) but if the latter then the pilear cuticle a bolete-type trichodermium or cutis, and the spores boletoid in shape . . . . **Boletaceae** p. 424

    2'(1') Hymenophore lamellate (or secondarily poroid from cross veins and pilear cuticle not a well-developed trichodermium) 3

3(2') Hymenophore in the form of lamellae with blunt edges; basidia exceptionally long and narrow (50–80 μm long) . . . . **Cantharellaceae**

3'(2') Hymenophore in the form of lamellae with thin sharp edges or in very small species sometimes veined to nearly smooth 4

    4(3') Spores angular or longitudinally striate and pinkish, vinaceous, or reddish cinnamon in deposits . . . . **Entolomataceae** p. 432

    4'(3') Spores not angular or if so white or rusty brown in deposits 5

5(4') Lamellae typically thick and waxy but edges sharp; basidia usually elongate in relation to width at apex; spore deposit white to whitish . . . . **Hygrophoraceae** p. 425

5'(4') Not as above (if lamellae appear waxy and the spores are echinulate see *Laccaria* in the Tricholomataceae) 6

    6(5') Lamellae free or nearly so from stipe, lamellar trama divergent; an outer veil typically present; an inner veil often present and leaving an annulus on the stipe; spores hyaline under microscope . . . . **Amanitaceae** p. 433

    6'(5') Not as above 7

7(6') Hymenophoral trama convergent; spore deposit pink to vinaceous or reddish cinnamon; lamellae free or practically so . . . . **Pluteaceae** p. 433

7'(6') Not with all 3 of the above features 8

    8(7') Spore deposit white, pale yellow, pinkish, or buff (print taken on white paper)
   15

    8'(7') Spore deposit ochraceous fulvous, clay-color to orange-tawny, or to some shade of chocolate brown to black, rarely olive-green or reddish 9

9(8') Spore deposit fuscous to olive-fuscous; spores "boletoid" in shape, dark-brown under microscope; lamellae thick and usually distant and decurrent; species forming mycorrhiza with conifers . . . . **Gomphidiaceae** p. 439

9'(8') If spore deposit blackish to dark-brown the lamellae thin 10

    10(9') Spore deposit some shade of cocoa-brown or chocolate-brown to dull cinnamon-brown, rarely vinaceous to brick-red; spores typically with an apical germ pore; cuticle of pileus in tangential sections appearing cellular to hymeniform
   . . . . **Coprinaceae** p. 436

    10'(9') Not as above (if pileus cuticle is cellular or hymeniform and spores have a germ pore, the color of the spore deposit is ochraceous to rusty-brown or clay-color)
   11

11(10') Lamellae free or practically so; an annulus typically present on the stipe; spore deposit blackish, fuscous, olive-fuscous, or some shade of chocolate-brown; spores (under high dry

objective and unbleached) lacking an apical germ pore; stipe and pileus cleanly separable
. . . . **Agaricaceae** p. 434

11′(10′) Not as above 12

12(11′) Spores in water mounts violaceous fuscous or violet brown but in KOH soon dark yellow-brown; spores with an apical pore and often appearing truncate; if spore deposit dull yellow-brown when fresh then the pilear cutis is composed of appressed filamentous hyphae often gelatinous . . . . **Strophariaceae** p. 435

12′(11′) Not as above 13

13(12′) Cuticle of pileus cellular to hymeniform; spore deposit rusty-brown to clay color; spores with an apical pore . . . . **Bolbitiaceae** p. 434

13′(12′) Cuticle of pileus of filamentose elements; spores lacking a pore 14

14(13′) Hymenophore easily separable from pileus; veil absent . . **Paxillaceae** p. 439

14′(13′) Hymenophore not readily separable from pileus; spore deposit rusty-brown to clay-color; spores frequently rugulose-warty and lacking an apical pore; pilear cuticle of appressed hyphae or a lax turf or its cells inflated somewhat and disarticulating
. . . . **Cortinariaceae** p. 437

15(8) Stipe and pileus readily separable; lamellae free; hymenophoral trama regular; an annulus typically present; spores with an apical pore in some of a mount, hyaline in water mounts but in Melzer's dextrinoid; spore deposit olive-green, off-white or white (see *Melanophyllum* also)
. . . . **Lepiotaceae** p. 433

15′(8) Not with above combination of characters.
Stipe and pileus confluent; hymenophoral trama of various hyphal arrangements; the hyphae thin-walled for the most part; lamellae adnexed, adnate or decurrent but very rarely free (or lamellae lacking in a few minute species); spore deposit white, yellowish, buff, pink, or vinaceous brown but hyaline or nearly so under the microscope; typically soft and fleshy; stipe excentric to lateral or pileus sessile. . . . . **Tricholomataceae** p. 426

A.   Boletaceae

The Boletaceae are fleshy stipitate, pore fungi (rarely lamellate) and they typically have at least a somewhat bilateral arrangement of the hyphae of the hymenophore. The context of the pileus is soft and readily decays. The species generally form mycorrhiza with woody plants. (See also *Filoboletus* in the Tricholomataceae.) The type genus is *Boletus*.

KEY TO GENERA OF BOLETACEAE

1. Pileus typically covered with coarse dry scales gray to blackish at times and becoming matted down to felt (in some specimens); spore deposit umber-brown to blackish-brown and spores reticulate to verrucose and usually globose to subglobose . . . . **Strobilomyces**
In spore characters this genus appears to intergrade with *Boletellus*. There are only a few species in the northern hemisphere but more are found in the tropics (Singer, 1945).

1′. If pileus is squamulose then spores and spore deposit not as above 2

2(1′) Spores longitudinally striate or with longitudinal wings 3

2′(1′) Spores not ornamented as in above choice 4

3(2) Hymenophore lamellate . . . . **Phylloboletellus**

3'(2) Hymenophore tubulose . . . . **Boletellus**

    4(2') With any 2 or more of the following features: (a) stipe glandular dotted; (b) pileus viscid; (c) veil leaving an annulus on the stipe; (d) hymenophore boletinoid; (e) at least some pleurocystidia in bundles and with brown incrustation as revived in KOH . . 5

    4'(2') Not as above 6

5(4) Spore deposit olive, olive-brown, cinnamon, or clay-color . . . . **Suillus**
    As delimited here *Paragyrodon* Singer, *Psiloboletinus* Singer, *Boletinus* Kalchbrenner, and *Gyrodon* Opat. are included in it.

5'(4) Spore deposit avellaneous, purplish-brown, vinaceous brown or lilac-drab
    . . . . **Fuscoboletinus**

    6(4') Hymenophore lamellate to strongly boletinoid 7

    6'(4') Hymenophore distinctly poroid 8

7(6) Hymenophore distinctly lamellate; lamellae thick . . . . **Phylloporus**

7'(6) Hymenophore boletinoid; stipe often excentric to lateral . . . . **Boletinellus**
    Forming mycorrhiza with *Fraxinus* and abundant in eastern North America within the range of that genus.

    8(6') Stipe soon hollow; spores   ellipsoid; spore deposit yellow . . . . . **Gyroporus**

    8'(6') Not with all of the above features 9

9(8') Veil dry and floccose to almost powdery, typically leaving an annular zone on the stipe
    . . . . **Pulveroboletus**

9'(8') Veil if present not as above 10

    10(9') Pileus viscid; hymenophore brilliant yellow; many pleurocystidia with bright-yellow content; spore print ochraceous buff more or less . . . . **Aureoboletus**

    10'(9') Not with above combination of features 11

11(10') Spore deposit gray-brown, reddish-brown, vinaceous, vinaceous brown or purple-brown (if stipe is ornamented with brown to blackish points or squamules, see *Leccinum*)
    . . . . **Tylopilus**
    Porphyrellus is here included in *Tylopilus*.

11'(10') Spore deposit yellow-brown, rusty-brown, rusty-yellow, olive, olive-brown or amber-brown 12

    12(11') Stipe ornamentation as points or squamules or lines which may be blackish at first and remain so or are pallid to brown at first and slowly darken to brown or black over the lower half (see *Gastroboletus* also) . . . . **Leccinum**

    12'(11') Stipe naked to pruinose, furfuraceous, squamulose or reticulate but if ornamented the ornamentation not darkening as in above choice (it may darken after staining blue in some species in which the context and hymenophore also stains blue quickly when injured) . . . . **Boletus**
    Note that *Xerocomus* and *Xanthoconium* are here included in *Boletus*.

B.   *Hygrophoraceae*

The hymenophore is waxy and the basidia are typically elongate, reminding one somewhat of those of the Cantharellaceae from which it is distinguished by the sharp-edged lamellae. Because of possible confusion with genera of the Tricholomataceae, Singer very appropriately gave an "elimination key"

to help sort out the groups that might be confused with it. The type is
Hygrophorus.

KEY TO GENERA OF HYGROPHORACEAE

1. Spores stellate-echinate . . . . **Hygroaster**
         Singer gives the lamellar trama as weakly bilateral. It is a small genus known from
         Trinidad and Europe (1 species in each area).

1'. Spores typically smooth 2

    2(1') Spores amyloid; lamellae vinaceous gray; pileus viscid . . . . . . **Neohygrophorus**
          One subalpine species, Washington State, U.S.A., known.

    2'(1') Not with all 3 of the above features 3

3(2') Lamellar trama divergent from a central strand . . . . **Hygrophorus**
         The genus is most abundant in the conifer areas of the northern hemisphere but a
         number also occur under hardwoods. All appear to be mycorrhiza formers. Some
         are popular esculents (Hesler and Smith, 1963).

3'(2') Lamellar trama parallel to interwoven 4

    4(3') Cheilocystidia present as pseudocystidia; a watery latex usually seen when pileus
    is cut with a sharp razor . . . . **Bertrandia**
          Known from tropical Africa and the Philippines.

    4'(3') Lacking both the above features; lamellar trama parallel to interwoven, spores
    rarely amyloid . . . . **Hygrocybe**
          Humidicutis Singer, *Camarophyllus* Kummer sensu Singer, and *Aeruginospora*
          Höhnel are here treated as synonyms.

C.   Tricholomataceae

The characters of this family are very generalized: the spore deposit is
white to yellow to pink, or rarely, pale tan to vinaceous brown; the spores
generally are thin-walled but may be variously ornamented or smooth, they
lack a germ pore, and may be amyloid, dextrinoid, or inamyloid. The
stipe and pileus are confluent (if a stipe is present). Lamellae are present
(rarely absent) and variously attached.

This key does not include all the generic segregates which have been
proposed for this family. Generic concepts vary widely between investigators,
from those who are committed to describing each distinctive species with its
immediate satellites (a stirps) as a genus, to the ultraconservatives like
myself who believe that the endless multiplication of generic names does more
to confuse our concepts of natural relationships in the family than it does to
clarify them. In view of this situation, the key presented here is designed to
contrast as clearly as possible the majority of the generic segregates, and to
focus on the characters used for their recognition in order to give the user
an introduction as to how the "hyphal approach" to the systematics of the
lamellate fungi is working out. For a detailed account of the genera of the
lamellate fungi, see Singer (1962), and particularly his forthcoming revision,
since the 1962 edition is out of print. It was thought undesirable to try to

write a second "Singer" for the current project. One point which I regard as significant in the following key and in the Singer system is the very large number of monotypic genera and genera with less than five known species. The type genus is *Tricholoma*.

KEY TO SELECTED GENERA OF THE TRICHOLOMATACEAE

1. Habit pleurotoid (stipe typically excentric to lacking) 2

1′. Habit typically tricholomoid or collybioid (stipe typically central) 26

    2(1) Spores amyloid 3

    2′(1) Spores not amyloid 5

3(2) Lamellae serrulate at maturity . . . . **Lentinellus**

3′(2) Lamellae even to somewhat crenulate on the edges 4

    4(3′) With a veil covering the hymenium in young basidiocarps . . . . **Tectella**

    4′(3′) Lacking a veil . . . . **Panellus**

5(2′) Spore deposit pink; pileus ± tomentose . . . . **Phyllotopsis**

5′(2′) Not with both the above features 6

    6(5′) Lamellae soon splitting from the edge to the pileus along their entire length
    . . . . **Schizophyllum**

    6′(5′) Lamellae not splitting as in above choice 7

7(6′) Lamellae serrate at maturity . . . . **Lentinus**

7′(6′) Lamellae with even to crenulate edges 8

    8(7′) Stipe annulate; basidiocarp strongly luminescent . . . . **Lampteromyces**

    8′(7′) Not with both of the above features 9

9(8′) Pileus setose or with pileocystidia; hymenophore smooth to veined or lamellate; habit marasmioid . . . . ′ . . . . **Gliocephala**

9′(8′) Not with above combination of characters 10

    10(9′) Pileus surface formed by a layer of dichophyses (see *Campanella* also)
    . . . . **Asterotus**

    10′(9′) Pileus cuticle not formed by dichophyses 11

11(10′) Hymenophore ± cantharelloid (of blunt-edged narrow lamellae usually connected by veins) 12

11′(10′) Hymenophore truly lamellate (lamellae with sharp edges) 17

    12(11) Dextrinoid lamprocystidia present in the hymenium . . . . **Geopetalum**

    12′(11) Lamprocystidia not present 13

13(12′) Hymenophore ± poroid from anastomoses . . . . **Campanella**

13′(12′) Hymenophore not poroid in appearance 14

    14(13′) Basidiocarp lacking pigment and the hyphae of pilear cuticle showing diverticulae or nodulose outgrowths (see *Marasmiellus* and *Micromphale* also) . . . **Mniopetalum**

    14′(13′) Basidiocarps containing pigment-encrusted hyphae 15

15(14′) Basidiocarps pezizoid when young . . . . **Arrhenia**

15'(14') Basidiocarps omphaloid to pleurotoid when young (rarely pezizoid) 16

16(15') Stipe lacking (or excentric to lateral), pilear cuticle not well differentiated
. . . . **Leptoglossum**

16'(15') Stipe present; pilear cuticle a trichodermium . . . . **Trogia**

17(11') Basidiocarp hyphae with particles which become blue-green in KOH; gills dark-colored (even when fresh) . . . . **Anthracophyllum**

17'(11') Not as above (but see *Marasmiellus* and *Micromphale* with diverticulae on hyphae of pilear cuticle). 18

18(17') A distinct gelatinous layer in the pileus or as its cuticle 19

18'(17') No gelatinous layers present in or on the basidiocarp 20

19(18) Lamprocystidia absent . . . . **Resupinatus**

19'(18) Lamprocystidia present in hymenium . . . . **Hohenbuehelia**

20(18') Basidiocarps fleshy (not tough and readily reviving) context hyphae mostly thin-walled before maturity 21

20'(18') Basidiocarps tough and readily reviving; thick-walled hyphae usually numerous in lamellar trama 24

21(20) Basidiocarps strongly stipitate, stipe 1.5–3 cm thick, excentric to central . . **Hypsizygus**

21'(20) Stipe not as in above choice 22

22(21') Lamellae crowded, adnexed and narrow; subhymenium cellular
. . . . **Pleurocollybia**

22'(21') Not with the above combination of characters 23

23(22') Basidiocarps small and Marasmiuslike; hyphae of pileus cuticle with typically diverticulate hyphae . . . . **Marasmiellus**

23'(22') Basidiocarps mostly larger and fleshier than in above choice . . . . **Pleurotus**

24(20') Stipe absent; pileus with dextrinoid hairs . . . . **Chaetocalathus**

24'(20') Lacking dextrinoid hairs on the pileus 25

25(24') Stipe institious; basidiocarps small . . . . **Nothopanus**

25'(24') Stipe lacking or well-developed but basidiocarps medium to large. . . . . . **Panus**

26(1') Basidiocarps developing on those of other fungi 27

26'(1') Not with parasitic habit as indicated above 28

27(26) Context of basidiocarps breaking up into chlamydospores . . . . . . **Asterophora**

27'(26) Not as above . . . . (see *Volvariella* and *Collybia*)

28(26') Spores amyloid 29

28'(26') Spores nonamyloid 45

29(28) Hymenophore distinctly poroid . . . . **Filoboletus**

29'(28) Hymenophore lamellate 30

30(29') Stipe slender (1–3 mm), pileus when cut exuding watery droplets sometimes staining basidiocarp blackish . . . . **Hydropus**

30'(29') Not with above combination of characters 31

31(30') Spores with a smooth amyloid outer layer (sloughing off at times), and a thicker inner punctate wall . . . . **Fayodia**

31'(30') Spores not as above 32

    32(31') Aspect of basidiocarps usually marasmioid; stipe at base with fulvous, strigose hairs . . . . **Xeromphalina**

    32'(31') Not with above combination of characters 33

33(32') Stipe (3) 4–10 mm or more thick 34

33'(32') Stipe 0.3–3 (5) mm thick 43

    34(33) Lamellae repeatedly dichotomously forked . . . . **Cantharellula**

    34'(33) Lamellae not or only occasionally forked 35

35(34') Spores with amyloid ornamentation (see *Lentinellus* also) 36

35'(34') Spores smooth 37

    36(35) Pileus moist and hygrophanous to subhygrophanous; cystidia of the lamellae often "harpoonlike" from an apical incrustation; stipe strict and lacking conspicuous white mycelium in and around the base . . . . **Melanoleuca**

    36'(35) Pileus dry and fibrillose to scurfy; cystidia not as in above choice; much white mycelium around the stipe and in the duff . . . . **Leucopaxillus**

37(35') Stipe massive; veil thick and membranous often double . . . . **Catathelasma**

37'(35') Stipe and veil not both as in above choice 38

    38(37') Lamellae broadly adnate to decurrent . . . . **Clitocybe**

    38'(37') Lamellae sinuate to adnexed 39

39(38') Veil present leaving scales or an annulus on the stipe . . . . **Floccularia**

39'(38') Veil if present ± granulose but often absent 40

    40(39') Veil composed mostly of sphaerocysts (see *Squamanita* also)

    . . . . **Cystoderma**

    40'(39') Veil absent for all practical purposes (fibrillose in some) 41

41(40') Cuticle of pileus cellular to hymeniform . . . . **Dermoloma**

41'(40') Pileus cuticle not as above 42

    42(41') Pleurocystidia and dermatocystidia present . . . . **Dennisiomyces**

    42'(41') Pleurocystidia absent . . . . **Porpoloma**

43(33') Lignicolous and mostly cespitose; pileus lacking a hypodermium (see *Baespora* also)

. . . . **Clitocybula**

43'(33') Pileus showing some degree of hypodermium development 44

    44(43') Spores small and subglobose (–6 μm); often growing on decaying cones of conifers

    . . . . **Baespora**

    44'(43') Spores larger; habitat various . . . . **Mycena**

45(28') Basidiocarp tough; lamellae with serrate margins . . . . **Lentinus**

45'(28') Not with both of the above features 46

    46(45') Hymenophore of obtuse to foldlike venose lamellae 47

    46'(45') Hymenophore of lamellae with ± acute edges 48

47(46) Lamprocystidia present . . . . **Geopetalum**

47'(46) Lamprocystidia absent; veil of thick-walled cells (see Cantharellaceae also)

. . . . **Delicatula**

48(46') Pileus cutis hymeniform; lamellar trama bilateral . . . . **Cyptotrama**

48'(46') Not with both of above features 49

49(48') Some evidence of a veil present in at least young basidiocarps 50

49'(48') Veil absent for all practical purposes (see *Macrocystidia* also) 60

    50(49) Veil granulose to subgranulose; some veil remains typically showing on the stipe

    . . . . **Cystoderma**

    50'(49) Not as above 51

51(50') Stipe annulate to conspicuously scaly 52

51'(50') Stipe not as above, veil remnants often obliterated by maturity 53

    52(51) Clamps absent; stipe 5 mm or more thick (see *Tricholoma* also)

    . . . . **Floccularia**

    52'(51) Clamps present; stipe scaly and usually swollen at base (if stipe annulate but not

    scaly, see *Oudemansiella* and *Armillaria*) . . . . **Squamanita**

53(51') Spore deposit $\pm$ clay-color; spores subglobose and echinate . . . . **Ripartites**

53'(51') Not as above 54

    54(53') Growing on soil or humus 55

    54'(53') Growing from wood or on soil rich in lignicolous debris; often growing from buried

    wood 58

55(54) Stipe radicating, spores pink in deposit . . . . **Termitomyces**

55'(54) Not as above 56

    56(55') Stipe $\pm$ fleshy; lamellae broadly adnate to decurrent . . . . **Clitocybe**

    56'(55') Stipe fleshy or cartilaginous; gills adnate to sinuate 57

57(56') Stipe fleshy; lamellae sinuate to adnexed . . . . **Tricholoma**

57'(56') Stipe fragile and thin, or cartilaginous . . . . (see *Omphalina, Collybia* and *Mycena*)

    58(54') Cheilocystidia typically large and distinctive; yellow pigment present in most

    species (*T. platyphylla* is an exception); lamellae adnate to decurrent; pileus cuticle not

    hymeniform. . . . . **Tricholomopsis**

    58'(54') Not with above combination of features 59

59(58') Pigments in bright-colored tones (yellow, orange, etc.); pileus cuticle subhymeniform

(the layer not continuous) . . . . **Xerulina**

59'(58') Pigments smoky, blackish, brownish or lacking . . . . **Oudemansiella**

    60(49') Basidiocarps staining gray to black when injured (often changing to ochraceous or

    bluish first); basidia with carminophilous granules; basidiocarp pigmentation typically

    whitish to gray . . . . **Lyophyllum**

    60'(49') Not as above 61

61(60') Basidiocarps large and in large clusters, hymenophore yellow to orange or olivaceous;

luminescent when fresh . . . . **Omphalotus**

61'(60') Not as above 62

    62(61') Lamellae thickish and flesh-colored to vinaceous or violaceous; spores echinulate

    (nearly smooth in 1 species with large elliptic spores) . . . . **Laccaria**

    62'(61') Not with above combination of features 63

63(62') Pileus cuticle hymeniform or cellular or of "broom cells" arranged in a sort of palisade

. . . . 64

63'(62') Pileus cuticle of appressed hyphae gelatinous or otherwise or a trichodermium the elements of which may be $\pm$ tangled in their arrangement and somewhat relaxed . . . 70

64(63) Stipe typically radicating; pleurocystidia gigantic; lamellae broad and subdistant . . . . **Oudemansiella**

64'(63) Not with above combination of features 65

65(64') Growing on humus or soil; pleurocystidia absent; hyphae with clamps; basidiocarps with aspect of a small *Tricholoma* . . . . **Dermoloma**

65'(64') On wood or other plant debris, aspect marasmioid or collybioid 66

66(65') "Broom cells" present in cuticle of pileus; basidiocarps reviving when moistened . . . . **Marasmius**

66'(65') Broom cells absent in pilear cuticle 67

67(66') Clamp connections absent . . . . **Pseudohiatula**

67'(66') Clamp connections present 68

68(67') Basidiocarps tough (reviving somewhat), pileus yellow to orange (see *Marasmius* also) . . . . **Xerulina**

68'(67') Basidiocarps readily withering and not reviving well 69

69(68') Margin of pileus inrolled at first . . . . **Collybia**

69'(68') Margin of pileus straight when young . . . . **Mycena**

70(63') Basidiocarps (when pileus disc is cut with a sharp razor) exuding a watery latex; aspect of basidiocarps mycenoid . . . . **Hydropus**

70'(63') Not with all of the above characters 71

71(70') Basidiocarps collybioid, latex present or absent but numerous conspicuous gloeoplerous hyphae and pseudocystidia present; basidiocarps colored (often) bright red, orange, etc. . . . . **Lactocollybia**

71'(70') Not as above 72

72(71') Pileus reddish to orange, viscid, with pileocystidia; stipe velvety; cespitose on elm and aspen . . . . **Flammulina**

72'(71') Not as above 73

73(72') Spores ornamented (usually warty); aspect of *Mycena* . . . . **Mycenella**

73'(72') If aspect of basidiocarp mycenoid, then the spores smooth 74

74(73') Stipe fleshy and typically over 5 mm thick (often 1–3 cm) 75

74'(73') Stipe thin [0.5–4(6)] mm thick 80

75(74) Terrestrial; lamellae sinuate to adnexed; clamp connections typically absent . . . . **Tricholoma**

75'(74) Not as above 76

76(75') Terrestrial; lamellae adnate to (usually) distinctly decurrent; clamps typically present; spores smooth or ornamented and color of deposit white, yellowish, pink, or vinaceous brown (but spores appearing hyaline under the microscope) . . . **Clitocybe**

76'(75') Not as above 77

77(76') Typically cespitose on soil rich in lignin (around sawdust piles, etc.) spores globose to subglobose; carminophilous granules present in the basidia . . . . **Lyophyllum**

D. Entolomataceae

The stipe and pileus are confluent; a veil typically is absent. The spore deposit is red, to reddish or reddish cinnamon, to brownish vinaceous. Spores are nonamyloid, typically angular or grooved, and lack a pore. The lamellar trama is more or less regular. The pilear cuticle is chiefly of appressed hyphae or hyphae with cystidiiform end-cells more or less ascendent or rarely in a loose trichodermium. The type of the family is *Entoloma*.

KEY TO GENERA OF ENTOLOMATACEAE

1. Spores angular in face view and in profile . . . . **Entoloma**

> The problem of dividing the genus into smaller groups and deciding which names to apply is still under investigation. However, the name *Acurtis* can be discarded now that it has been shown by Mazzer and Smith (1967) that not all white carpophoroid-like bodies are referable to "*Acurtis abortivus.*"

1'. Spores longitudinally striate to grooved or obscurely roughened 2

    2(1') Spores longitudinally striate to grooved . . . . **Clitopilus**

    2'(1') Spores more or less roughened to obscurely angular . . . . **Rhodocybe**

E. *Amanitaceae*

Spore deposit white to only slightly tinted, spores smooth. Lamellae with bilateral trama and free from stipe. Stipe separating cleanly from the pileus. Typically an outer and inner veil are both present.

KEY TO GENERA OF AMANITACEAE

1. Volva typically absent (an outer slime veil may be present) . . . . **Limacella**

1'. Volva typically present, if it is apparently absent check for particles of tissue on young pilei and/or around the base of the stipe . . . . **Amanita**

F. *Pluteaceae*

The spore deposit is pink to vinaceous cinnamon and the spores are typically smooth. The lamellae are free from the stipe and the stipe and pileus are readily separable. The hymenophoral hyphae are convergent (of large elongate cells extending from near subhymenium inward and downward) to the center of the trama as viewed in cross sections of the gills. The type is *Pluteus*.

KEY TO GENERA OF PLUTEACEAE

1. Volva present, annulus absent . . . . **Volvariella**

1'. Volva absent 2

    2(1') Annulus present . . . . **Chamaeota**

       A rare group of a small number of species not well known.

    2'(1') Annulus absent . . . . **Pluteus**

       This is the common genus of the family and is cosmopolitan in distribution.

G. *Lepiotaceae*

The pileus and stipe are cleanly separable and the lamellae are free from the stipe. The lamellar trama is regular to interwoven and the spore deposit is greenish, to olive, to off-white, to white. The spores are typically dextrinoid and an annulus typically present on the stipe. *Lepiota* is the type.

KEY TO GENERA OF LEPIOTACEAE

1. Spore print olive to green 2

1'. Spore print some other color (white to buff or merely off-white) 3

   2(1) Volva present and distinct . . . . **Clarkienda**

   2'(1) Volva lacking . . . . **Chlorophyllum**

3(1') Possessing a well-formed cuplike volva . . . . **Volvolepiota**

3'(1') Lacking a distinct volva 4

   4(3') Spores regularly showing a distinct apical pore . . . . **Leucoprinus**
   Macrolepiota Singer and *Leucoagaricus* (Locquin) Singer are included here under
   Leucoprinus. I do not put much emphasis on the sulcate-pectinate margin of the
   pileus since we often find *L. cepastipes* without it.

   4'(3') Spores lacking a germ pore or a pore sometimes present on a few spores in a mount
   . . . . **Lepiota**
   This is a large genus most abundant in warmer climates such as area bordering the
   Gulf of Mexico. Singer indicates that it is abundant in South America.

H. *Agaricaceae*

The spore deposit is blackish, dark chocolate brown, vinaceous brown, or (rarely) olivaceous brown when fresh. The pileus cuticle is fibrillose to squamulose and rarely composed of sphaerocysts; the stipe and pileus separate cleanly. An annulus is typically present and the lamellae are pallid to pink or rose when immature. *Agaricus* is the type.

KEY TO GENERA OF AGARICACEAE

1. Spore print lacking olive tones when fresh 2

1'. Spore print olivaceous toned when fresh and moist; cuticle of pileus composed of sphaerocysts . . . . **Melanophyllum**
   Singer lists only 2 species but it appears to be cosmopolitan. The color of the spore
   deposit and the changes it undergoes are confusing, as is indicated by the number of
   genera in which the type species has been placed.

   2(1) Spores ovoid to ellipsoid and smooth . . . . **Agaricus**
   This is the largest genus in the family, and is easily recognized at sight, but, as in
   Russula, the species are very difficult. It is cosmopolitan with many species limited
   to grasslands and many others to forest situations. It is a popular genus to mycopha-
   gists, but poisonous species are known.

   2'(1) Spores subangular to nodose in outline . . . . **Cystoagaricus**
   According to Singer, only 1 species is known, but its features are odd to say the least.
   Florida to northern Argentina is the area for it given by Singer.

I. *Bolbitiaceae*

The stipe and pileus are somewhat separable to confluent. The pilei have a hymeniform cuticle or a cellular layer. The spore deposit is bright to dull rusty brown (sometimes earth brown), and the spores have a distinct apical pore or, in a few, the pore is very minute. The stipe is central. Bolbitiaceae grow on humus and organic debris. The type is *Bolbitius*.

KEY TO GENERA OF BOLBITIACEAE

1. Spore print more or less dull date-brown to dull rusty-brown . . . . **Agrocybe**
   The genus is widely distributed over the grasslands of the world as well as in conifer or hardwood forests. It fruits during the spring, summer, and fall. Some species are common in the North Temperate areas.
1'. Spore print rather bright rusty-brown 2

   2(1') Pileus viscid, plicate-striate; cheilocystidia not abruptly capitate . . . . **Bolbitius**
   A small genus on dung and vegetable debris or wood, of wide distribution but seldom occurring in abundance. As Singer has pointed out, very little is actually known about a number of the species.

   2'(1') Pileus and cheilocystidia not as above 3

3(2') Pileus plicate-sulcate; cheilocystidia ampullaceous . . . . **Galerella**
   There is much to be said for including *Galerella*, *Pholiotina*, and *Conocybe* in 1 genus, as some investigators are doing at present.

3'(2') Not with both of the above features 4

   4(3') Mediostratum of lamellar trama greatly reduced; cheilocystidia mostly capitate
   . . . . **Conocybe**

   4'(3') Mediostratum well-developed; veil present or absent; cheilocystidia rarely capitate
   . . . . **Pholiotina**

J. Strophariaceae

The epicutis of the pileus is composed of appressed narrow tubular hyphae which are often gelatinous or (rarely), in some species, have some-what ascending hyphal ends. The hypodermium is often subcellular. The spores in water mounts are dark yellow-brown to rusty brown, or violaceous to violaceous-brown, but, if violaceous or purple-brown, they soon become dull yellow-brown in KOH either when fresh or revived. Typically, they have an apical germ pore or at least a minute apical discontinuity and their surface is smooth. The spore deposit is violaceous, fuscous, violaceous-brown, dark yellow-brown or dull fulvous. The stipe and pileus are confluent and the veils typically are thin to well-developed or absent. The type is *Stropharia*.

KEY TO GENERA OF STROPHARIACEAE

1. Spore deposit with a distinct violaceous to purple-brown tone 3
1'. Spore deposit dull-brown to rusty-brown 2

   2(1') Stipe excentric and about as long as pileus is wide . . . . **Pleuroflammula**
   Species of this genus approach those of *Crepidotus*. The genus is important by way of indicating a direction of evolution from *Pholiota*.

   2'(1') Stipe central and annulate or not; chrysocystidia present or absent; spores typically with at least a minute apical discontinuity . . . . **Pholiota**

A large cosmopolitan genus of wood decaying fungi with occasional species appearing terrestrial. It is an important genus from the standpoint of slash disposal. The concept presented here is broader than that of Singer (Smith and Hesler, 1968).

3(1) Stipe excentric and shorter than diameter of pileus . . . . **Melanotus**

3'(1) Stipe central, annulate or not; chrysocystidia present or not . . . . **Psilocybe**
    Stropharia and *Naematoloma* are included here. Quélet applied the name *Geophila* to this assemblage. The genus differs from *Pholiota* chiefly in the color of the spore deposit. See *Melanomphalia* also.

K.   Coprinaceae

The pileus cuticle is typically cellular or hymeniform, but the original cutis is sometimes broken up as the pileus expands. The spores (if thick-walled) typically have an apical pore and the spore deposit is black, dark brown, coffee-brown or various shades of chocolate brown, or (rarely) brick-red to pink. Brachybasidioles are often present (see *Coprinus* and *Pseudocoprinus*) but not constantly so in *Psathyrella*. The basidiocarps are typically very fragile. Coprinaceae live chiefly on trash or dung of herbivores. The type is *Coprinus*.

KEY TO GENERA OF COPRINACEAE

1. Hymenophore undergoing autodigestion; lamellae with practically parallel sides and very thin . . . . **Coprinus**
    A moderately large genus cosmopolitan in distribution. Fruiting occurs during spring, summer, or fall depending on the species. Many inhabit the dung of herbivores, and the remainder live on dead plant remains including dead wood. Many of the species fruit well in culture and are good subjects for laboratory studies.

1'. Hymenophore not undergoing autodigestion 2

    2(1') Base of stipe furnished with a membranous cup-shaped volva . . . . **Macrometula**
    The genus may be characterized as a *Psathyrella* with a membranous volva. Known from the type which was collected in a aroid house at Kew.

    2'(1') Volva if present more fibrillose than membranous 3

3(2') Pileus plicate-striate and hymenium containing brachybasidioles regularly
  . . . . **Pseudocoprinus**
    This genus is exactly intermediate between *Coprinus* and *Psathyrella*. It is important as an evolutionary step in the series of steps leading to *Coprinus*.

3'(2') Not with both of the above characters 4

    4(3') Lamellae obscurely mottled at maturity by patches of maturing spores 5

    4'(3') Lamellae not mottled, in section narrowly wedge-shaped; spore deposit black, purple-brown, gray-brown or reddish; typically with very fragile basidiocarps . . . .
  . . . . **Psathyrella**
    One of our truly large and cosmopolitan genera, over 400 species being known from North America alone. It is a trash-loving genus, but many appear to be secondary invaders in the process of rotting wood (Smith, 1972).

5(4) Pleurocystidia in most species (at least some on a single lamella) in fascicles of 2–4; spores typically ornamented and lamellae rusty brown and somewhat mottled at maturity

. . . . subgenus *Lacrymaria* of **Psathyrella**

5′(4) Pleurocystidia not in fascicles (cystidia may be absent); gills fuscous to black at maturity
. . . . 6

6(5′) Spores large as a rule; stipe slender; spore surface ornamented with irregular patch-like pieces of outer wall material . . . . subgenus *Panaeolina* of **Psathyrella**

6′(5′) Spores smooth, large, blackish brown under the microscope, in face view often obscurely angular . . . . **Panaeolus**
The genus is a satellite of *Psathyrella*. Many occur on dung or rotting wood but some are truly terrestrial. It is cosmopolitan, but not a large genus. Some authors add the feature that the spores do not disintegrate as readily in concentric sulphuric acid as do those of most species of *Psathyrella* (Ola'h, 1969).

L. Cortinariaceae

The epicutis of the pileus typically consists of narrow tubular to moderately broad somewhat inflated hyphae appressed to form a cuticle or, at times, a lax trichodermium which is rarely cellular. The stipe, if present, and the pileus are confluent and the hymenophoral trama is regular. The spore print is buffy tan, to tawny, to rusty brown, orange, ferruginous, clay-color, or date-brown. The spores are ornamented or smooth, but generally lack a distinct apical pore. The type is *Cortinarius*.

KEY TO GENERA OF CORTINARIACEAE

1. Stipe featuring a long, tapering, pseudorhiza; veil lacking; spores ornamented; pileus slimy to subviscid . . . . **Phaeocollybia**
Typically in conifer forests and especially with species of *Picea*. The center of distribution appears to be in the Pacific Northwest in North America (Smith, 1957).

1′. Not with above combination of features 2

2(1′) A membranous annulus (remains of partial veil) present on stipe; stipe 1–2 cm thick usually . . . . **Rozites**
Aptly characterized as a *Cortinarius* with a membranous partial veil. It is most abundant in the Southern Hemisphere. Our only common northern species, *R. caperata*, is a popular esculent.

2′(1′) Stipe (if annulate) with the ring composed of material of an outer veil (see *Cortinarius*), or the stipe only 1–3.5 mm thick 3

3(2′) Spore deposit yellowish; spores ellipsoid to globose, pilear epicutis a trichodermium tending to become an epithelium (Singer); Southern Hemisphere in distribution
. . . . **Descolea**
Forming mycorrhiza with broad leaved trees, especially *Nothofagus*.

3′(2′) Not as in above choice 4

4(3′) Spore deposit ochraceous tawny to rich cinnamon or fulvous; spores verrucose-rugulose, lacking an apical pore; inner veil cortinate and always present at first
. . . . **Cortinarius**
This is the largest genus of lamellate fungi in the world and is abundant in temperate areas of both hemispheres. In northern areas it is about as abundant, specieswise, in hardwoods as under conifers. *Leucocortinarius* (Lange) Singer, which I have seen in Belgium, appears to me to be a *Tricholoma* related to *T. terriferm* Peck.

4'(3') Not as above . . . .

5(4') Spore deposit typically bright rusty-orange or orange-ochraceous; spores verruculose; growing typically from rotting wood . . . . **Gymnopilus**
  Common in the tropics and rather frequent in northern conifer forests. It is a "satellite" of *Cortinarius* but typically is not a mycorrhiza former.

5'(4') Not with above combination of features 6

  6(5') Stipe excentric to lateral or basidiocarp sessile 7

  6'(5') Stipe typically central 9

7(6) Basidiocarps sessile or with a short "false" stipe visible only on the underside of the pileus
. . . . **Crepidotus**

7'(6) Basidiocarps with a distinct stipe (see *Crepidotus nyssicola* also) 8

  8(7') Spores ornamented . . . . **Pyrrhoglossum**
    This genus differs from *Gymnopilus* in pleurotoid habit and from *Crepidotus* in having microscopic features of *Gymnopilus* including the bright fulvous spores.

  8'(7') Spores smooth (see *Simocybe* also) . . . . **Phaeomarasmius**
    This genus connects to *Pholiota*. The spores have a faint apical discontinuity and are dull rusty brown. In the present restricted sense the genus contains only a few species.

9(6') Spores smooth, angular, nodulose, or coarsely echinate (rarely); lamprocystidia often present; growing on humus or very decayed wood; stipe typically 2 mm or more thick; pileus not riscid . . . . **Inocybe**
    This is one of the common genera throughout the world as well as one of the largest. Nearly all species are terrestrial and fruit from spring through to late fall and winter.

9'(6') Not as above 10

  10(9') Pileus usually viscid fresh; veil often present; spore-print clay-color (rarely porphyry brown); stipe fleshy and usually over 4 mm thick . . . . **Hebeloma**
    Easily distinguished from *Cortinarius* when species without a veil are compared; but the veiled species are easily confused with *Cortinarius*. The prominent cheilo-cystidia usually indicate a *Hebeloma*. It is fairly common in the range of *Cortinarius* and it is assumed that the species are mycorrhiza formers.

  10'(9') Not as above (stipe usually less than 4 mm thick, fragile) 11

11(10') Spores smooth (rarely appearing minutely punctate) 12

11'(10') Spores distinctly ornamented in some way 13

  12(11) Habit of fruit-body omphaloid or collybioid; spore deposit pale buff to pale cinnamon; spores typically collapsing readily . . . . **Tubaria**

    A widely dispersed genus occurring on soil especially where bark has been piled and allowed to rot.

  12'(11) Habit of basidiocarp naucorioid; spores not readily collapsing; pilear epicutis with scattered pileocystidia or these in groups (see *Galerina* also). . . . . . . **Simocybe**

13(11') Spores with a sharply defined smooth plage area (or spore appearing almost smooth except for a line delimiting the plage area) . . . . **Galerina**
    A common and cosmopolitan genus in mossy areas (tundra, sphagnum, along mossy streams, on mossy logs, summer and fall; cosmopolitan. (Smith and Singer 1964)

13'(11') Plage area of spore not delimited as in above choice 14

  14(13') Outer spore-sac loose, wrinkled, and sloughing off irregularly; spores with a minute apical pore; fruit bodies not associated with living alder plants

. . . . **Melanomphalina**
It has the aspect of *Omphalina* but with dull brown spores. It is also very close to *Galerina* in spore features since some of the latter have a loosely fitting outer layer readily separating around the plage area. However, in *Melanomphalina*, the spore print is said to be grayish olive to blackish.

14′(13′) Not as above 15

15(14′) Terminal cells of hyphae of the pilear epicutis often cystidioid; spores large (10 μm or more long); spore ornamentation warty-rugulose to punctate; pilear cuticle at times loosely cellular; associated with living *Alnus* and *Salix* . . . . **Alnicola**

15′(14′) Species of *Galerina* in which a plage is lacking on the spore or in which the spore has an apical pore or a perfectly smooth wall will key out here.

M. Paxillaceae

The pileus and stipe, when present, are confluent. The spore print is ochraceous to coffee-brown, to olive-brown or olivaceous and the spores lack an apical pore. The lamellae are well formed and thick to thin and close; at times they are intervenose. The hymenophore is typically readily separable from the pileus. The type is *Paxillus*.

KEY TO GENERA OF PAXILLACEAE

1. Stipe typically central; spores large and boletoid or boletelloid 2

1′. Stipe central to lacking; spores mostly less than 10 μm long 3

2(1) Spores ornamented with longitudinal wings . . . . **Phylloboletellus**
This genus may be better regarded as a subgenus of *Phylloporus* when more is learned about spore ornamentation in the latter.

2′(1) Spores smooth (under the light microscope); lamellae thick, bright-yellow when young . . . . **Phylloporus**

It is more abundant in the tropics than in temperate regions of the northern hemisphere (Corner, 1970).

3(1′) Spores smooth . . . . **Paxillus**
Cosmopolitan but only a few species in any one region.

3′(1′) Spores ornamented (under the light microscope) 4

4(3′) Spores verrucose; coscinocystidia present . . . . **Linderomyces**

4′(3′) Spores more or less echinulate; coscinoids none 5

5(4′) Pileus cuticle a trichodermium; lamellae subdistant . . . . **Neopaxillus**
Possibly better regarded as a section of *Linderomyces*, but I have not examined specimens.

5′(4′) Pileus cuticle of appressed interwoven hyphae; lamellae thin and close . . . **Ripartites**
It is also keyed in the Tricholomataceae (see Singer).

N. Gomphidiaceae

The spore deposit is fuscous to olive-brown. The lamellae are typically decurrent, close to distant and thickish. The spores are large and boletoid in shape (10 μm or more long) while the pleurocystidia are large (up to 100 μm or more long) and subcylindric to narrowly clavate. The species form mycorrhiza with conifers. The type is *Gomphidius*.

KEY TO GENERA OF GOMPHIDIACEAE

1. Outer veil composed of sphaerocysts . . . . **Cystogomphus**
    Except for the veil, the characters are those of *Chroogomphus*. Apparently monotypic.

1'. Outer veil fibrillose (dry or slimy) 2

    2(1') Lamellae and usually pileus and stipe at first dull to bright-ochraceous; some tissue of the basidiocarp distinctly amyloid under a microscope . . . . . . **Chroogomphus**

    2'(1') Lamellae typically pallid when young; stipe pallid except for lemon-yellow base in many species, often blackening; slime-veil present or absent; context of pileus whitish when young . . . . **Gomphidius**
       The Pacific Northwest in North America appears to be the center of distribution for both *Chroogomphus* and *Gomphidius*.

O. Russulaceae

This large family features spores varying from broadly elliptic to globose and having amyloid ornamentation. This is combined with the feature that the tissue of the stipe and pileus are composed of nests of sphaerocysts typically bounded by connective hyphae. The type is *Russula*.

KEY TO GENERA OF RUSSULACEAE

1. Latex absent from the basidiocarp . . . . **Russula**
    A large genus readily recognized at sight but with many species and these difficult to identify. Cosmopolitan and many very likely form mycorrhiza. Fruiting is characteristically midsummer and early fall in the North Temperate area. The genus connects to *Macowanites* in the Gasteromycetes.

1'. Latex present (cut apex of stipe with a sharp razor to demonstrate this character)
  . . . . **Lactarius**
    There are many species that can be recognized at sight and many are considered very desirable for the table. As for *Russula* the species are probably mycorrhiza formers. The distribution and fruiting times are also similar.

III. HYMENOGASTRALES

As defined here, the order contains Gasteromycetes in which the hymenium is relatively persistent to maturity whether they are tuberlike, or have a columella or stipe-columella, or a stipe. Many of the latter were formerly in the Secotiaceae. These fungi connect up to the families of the Agaricales, as has been shown in the last decade or two, and it is this approach to their systematics that is used here.

KEY TO FAMILIES OF HYMENOGASTRALES

1. Spores purple-brown to fuscous, dark-chocolate or blackish (in KOH) 2

1′. Spores paler to hyaline under the microscope 4

   2(1) Spores without an apical pore, boletoid in shape and large (10 μm long or more) . . . . **Gomphidiaceae** p. 442
      The basidiocarps remind one of an unexpanded button of a *Gomphidius* or *Chroogomphus*.

   2′(1) Spores not as in above choice 3

3(2′) Spores mostly under 10 μm long, fuscous to dark purple brown or blackish revived in KOH, no germ pore readily evident . . . . **Agaricaceae** p. 444
   The basidiocarps remind one of an unexpanded *Agaricus* button.

3′(2′) Spores colored as above but with a readily observed apical pore in most species . . . . **Coprinaceae** p. 444

   4(1′) Spores (in KOH) yellow-brown to rusty-brown; if ornamented the ornamentation not amyloid (if columella is absent and spores are smooth see *Rhizopogonaceae*). . 5

   4′(1′) Spores paler than in above choice 9

5(4) Spores with an apical germ pore (it may be very narrow) 6

5′(4) Spores typically lacking a distinct apical pore 7

   6(5) "Pileus" with a cellular to hymeniform epicutis (study young material) . . . . **Bolbitiaceae** p. 445

   6′(5) "Pileus" with an epicutis of appressed ± radial-interwoven hyphae . . . . **Strophariaceae** p. 445

7(5′) Spores longitudinally grooved to ridged, minutely warty to warty-rugulose or outer wall often loosely enveloping the spores . . . . **Cortinariaceae** p. 444

7′(5′) Spores smooth to punctate 8

   8(7′) Spores yellowish to rusty-brown and punctate-ornamented by pores through the thickened wall . . . . **Cribbiaceae** p. 443

   8′(7′) Spores smooth (see *Strophariaceae*) and hymenophore tubulose . . . . **Boletaceae** p. 442

9(4′) Spores typically globose to broadly elliptic (mostly over 8 μm wide) and either having a coating of amylaceous material over the surface or the ornamentation amyloid at least in part; basidiocarp typically with heteromerous trama in peridium and stipe (if the latter is present) . . . . The "**Astrogastraceae**" p. 442

9′(4′) Not as above 10

   10(9′) Spores angular as in *Entoloma* and nearly hyaline to tinged reddish in KOH . . . . **Entolomataceae** p. 442

   10′(9′) Not as above 11

11(10′) Spores typically dextrinoid with or without an apical pore (if lacking a pore the wall not ornamented) . . . . **Secotiaceae** p. 444

11′(10′) Spores not as above 12

   12(11′) Spores yellowish and with pores projecting through a thick inner wall . . . . **Cribbiaceae** p. 443

   12′(11′) Spores not ornamented as in above choice 13

13(12′) Spores subhyaline to pale ochraceous to brownish, smooth or (if amyloid) practically so, rarely slightly angular; basidiocarp tuberlike; columella absent to inconspicuous . . . . **Rhizopogonaceae** p. 442

13′(12′) Spores hyaline to yellowish or pale clay-color in KOH; ornamented or not; columella typically readily observed in sections . . . . **Genera of Uncertain Position** p. 445

A.   *The Gasteroid Members of the Gomphidiaceae*

KEY TO GENERA OF GASTEROID MEMBERS OF THE GOMPHIDIACEAE

1. Basidiocarp tissue amyloid in some part and, when fresh, ochraceous to reddish in color . . . . **Brauniellula**

1′. Basidiocarp tissue not amyloid in any part; context of pileus whitish at first (never ochraceous) . . . . **Gomphogaster**

B. *The Gasteroid Genera Connecting to the Entolomataceae*

KEY TO GASTEROID MEMBERS OF THE ENTOLOMATACEAE

1. Stipe-columella present . . . . **Rhodogaster**

1′. Stipe-columella absent . . . . **Richoniella**

C.   *The Gasteroid Genera with Probable Connections to the Boletaceae: The Rhizopogonaceae*

KEY TO GENERA OF RHIZOPOGONACEAE

1. Stipe-columella distinctly present; hymenophore tubulose . . . . **Gastroboletus**

1′. Not as above 2

   2(1′) Stipe-columella "rudimentary" (a basal stump of a stipe which sends narrow branches through the gleba as a dendroid columella); clamps regularly present on hyphae of the basidiocarp . . . . **Truncocolumella**

   2′(1′) Stipe-columella absent (or present rarely as a thin line up to 0.5 mm wide); typically lacking a basal point of attachment to the substratum and surface usually sparsely coated with rhizomorphs; spores amyloid or not (including *Nigropogon*) . . . . **Rhizopogon**

D.   *The Gasteroid Genera Connecting to the Russulaceae (The "Astrogastraceae")*

KEY TO GENERA OF ASTROGASTRACEAE

1. Stipe-columella absent as such, the sterile column "reduced" to a basal area of tissue with 1 or more branched extensions into the gleba, or no interior sterile tissue present other than

tramal plates 2

1′. Stipe-columella well defined; gleba usually exposed at the base next to the stipe and in age often separating from the columella portion 6

  2(1) Fresh specimens exuding a latex when cut open 3

  2′(1) Latex absent 4

3(2) Basidiocarps multipileate, up to 13.5 cm broad; columella present; lacking heteromerous structure in the trama; spores prominently ornamented (echinate) with amyloid spines; trama deep purple with KOH in fresh state; South American . . . . . . **Hybogaster**

3′(2) Basidiocarps tuberlike, 1–5 cm in diameter; columella greatly reduced; KOH not giving above reaction . . . . **Zelleromyces**

  4(2′) Amyloid ornamentation of spore a thin layer partly covering the spore surface and thickest around the sterigmal scar . . . . **Mycolevis**

  4′(2′) Not as above 5

5(4′) Mediostratum of tramal plates lacking sphaerocysts . . . . **Martellia**

5′(4′) Mediostratum of tramal plates containing some sphaerocysts . . . . . . **Gymnomyces**

  6(1′) Latex present in fresh specimens . . . . **Arcangeliella**

  6′(1′) Latex absent 7

7(6′) Mediostratum of tramal plates lacking sphaerocysts . . . . **Elasmomyces**

7′(6′) Mediostratum of tramal plates containing at least some sphaerocysts 8

  8(7′) Cuticle of peridium consisting of large pseudoparenchymalike cells; leptocystidia typically present in hymenium . . . . **Cystangium**

  8′(7′) Cuticle of peridium various but not as above; hymenial cystidia mostly as pseudo-cystidia . . . . **Macowanites**

E. Cribbiaceae

KEY TO GENERA OF CRIBBIACEAE

1. Consistency of basidiocarp decidedly gelatinous . . . . **Cribbea**

1′. Consistency of basidiocarp fleshy 2

  2(1′) Epicutis of peridium an epithelium . . . . **Setchelliogaster**

  2′(1′) Epicutis not an epithelium 3

3(2′) Peridium typically absent at maturity; forming a thin covering over the gleba at first . . . . **Singeromyces**

3′(2′) Peridium present on mature basidiocarps 4

  4(3′) Clamp connections absent (see *Hymenogaster* also) . . . . **Paxillogaster**

  4′(3′) Clamp connections present . . . . **Austrogaster**

F.   Gasteroid Fungi Connecting to the Agaricaceae and Coprinaceae

KEY TO GENERA OF GASTEROID MEMBERS OF AGARICACEAE AND COPRINACEAE

1. Gleba lamelloid and hanging free from the margin of the expanded apex of the stipe; volvate; spores black . . . . **Montagnea**

1′. Not as above 2

    2(1′) Stipe elongate and 1–3 mm thick (stature of *Galeropsis*) . . . . . . **Panaeolopsis**

    2′(1′) Stipe much thicker (1–2 cm) 3

3(2′) Stipe fleshy and typically short (extending 1–2 cm below the lower edge of the peridium) . . . . **Endoptychum**

3′(2′) With a distinct well-developed semiwoody stipe-columella 4

    4(3′) Stipe annulate but not volvate . . . . **Longula**

    4′(3′) Stipe volvate 5

5(4′) Spores with a distinct apical pore . . . . **Polyplocium**

5′(4′) Spores lacking a distinct apical pore . . . . **Gyrophragmium**

G.   Gasteroid Genera Connecting to the Cortinariaceae

KEY TO GENERA OF GASTEROID GENERA OF CORTINARIACEAE

1. Stipe-columella present; columella percurrent; spores finely warty to verrucose-rugose (as in *Cortinarius*) . . . . **Thaxterogaster**

1′. Not as above (hypogeous basidiocarps of *Thaxterogaster* may lack a stipe-columella but have a percurrent columella) 2

    2(1′) Tramal plates gelatinous-cartilaginous at maturity; basidiocarp typically firm and rubbery; spores longitudinally grooved or striate; peridium often absent
    . . . . **Gautieria**

    2′(1′) Not as above 3

3(2′) Spores nearly smooth to verrucose to warty rugulose or with a sacklike outer loosely fitting wrinkled layer, rarely somewhat longitudinally striate . . . . **Hymenogaster**

3′(2′) Spores distinctly longitudinally striate to grooved; tramal plates not gelatinous; peridium often staining blue if injured when fresh . . . . **Chamonixia**

H.   Secotiaceae

KEY TO GENERA OF SECOTIACEAE

1. Spores dextrinoid and with a germ pore (examine a number of spores since a pore is not readily visible on all) 2

1′. Spores not both dextrinoid and with a pore 5

    2(1) Stipe volvate . . . . **Secotium**

2′(1) Stipe not volvate 3

3(2′) Clamps present on some hyphae of basidiocarp; spores ornamented at maturity
. . . . **Neosecotium**

3′(2′) Clamp connections consistently absent 4

    4(3′) Spores smooth . . . . **Endolepiotula**

    4′(3′) Spores with exceptionally coarse ornamentation . . . . see *Octavianina*

5(1′) Spores smooth and dextrinoid but lacking a pore . . . .**Notholepiota**

5′(1′) Spores coarsely ornamented . . . . see *Octavianina*

I. Gasteroid Genera Connecting
to the Bolbitiaceae and the Strophariaceae

KEY TO GENERA OF GASTEROID MEMBERS
OF THE BOLBITIACEAE AND STROPHARIACEAE

1. Gleba grayish to grayish brown to purplish brown when fresh; peridium with an epicutis
of appressed hyphae; chrysocystidia present . . . . **Weraroa**

1′. Gleba yellow-brown to rusty-brown (fulvous) 2

    2(1′) Gleba ochraceous tawny to tawny or brighter fulvous when fresh or dried; "pileus"
    cuticle cellular (study young specimens because the layer soon collapses) . . . . . .
    . . . . **Gasterocybe**

    2′(1′) Not as above 3

3(2′) Cystidia absent; spores verrucose-reticulate . . . . **Hypogaea**

3′(2′) Spores smooth 4

    4(3′) Chrysocystidia present . . . . **Weraroa**

    4′(3′) Chrysocystidia absent 5

5(4′) "Pileus" and stipe distinctly squamulose . . . . **Tympanella**

5′(4′) "Pileus" more or less glabrous 6

    6(5′) Spores with a broad germ pore at apex; stipe slender (1–3 mm); leptocystidia absent
    . . . . **Galeropsis**

    6′(5′) Spores not truncate from an apical pore (pore very narrow—as in *Pholiota*); stipe
    short and thick; pleurocystidia present as leptocystidia . . . . **Nivatogastrium**
    See also *Neosecotium*.

J. Genera of Uncertain Position

KEY TO GENERA OF UNCERTAIN POSITION

1. Stipe-columella present 2

1′. Stipe-columella absent (stipe lacking and columella reduced or lacking) 4

    2(1) Volva present . . . . **Brauniella**

    2′(1) Volva absent 3

3(2') On hardwood logs; pileus with dark fibrils or squamules; spores hyaline under microscope
. . . . *Lentodium* stage of *Lentinus tigrinus* (American form)

3'(2') On soil; peridium absent; gleba morchelloid; spores yellow-brown; South American
. . . . **Gymnopaxillus**

    4(1') Basidiocarp lacking a peridium; cystidia vinaceous in KOH; spores smooth but exhibiting logitudinal dextrinoid bands as revived in Melzer's . . . . **Protogautieria**

    4'(1') Not as above 5

5(4') Spores coarsely echinulate (spines to 2 μm long), hyaline in KOH; clamps present; spore ornamentation inamyloid or rarely with a very faint suggestion of an amyloid reaction
. . . . **Hydnangium**

5'(4') Not as above 6

    6(5') Spores large (over 25 μm), smooth, ochraceous in KOH . . . .**Gigasperma**

    6'(5') Not as above 7

7(6') Latex present (section fresh specimens) . . . . **Leucogaster**

7'(6') Lacking a latex (but see *O. papyracea*) . . . . **Octavianina**

REFERENCES

The list serves both for the Agaricales and the secotioid fungi and their relatives, and it is not to be assumed that omission from the list of any particular work means that that work was considered unimportant. The list is an introduction to the literature mostly covering the period 1945–1970.

Bas, C. (1960). Notes on Agaricales. II. *Persoonia* **1**:303–314.

Bas, C. (1961). The genus *Gloiocephala* Massee in Europe. *Persoonia* **2**:77–89.

Bas, C. (1965). The genus *Squamanita*. *Persoonia* **3**:331–359.

Bas, C. (1969). Morphology and subdivision of *Amanita* and a monograph of its section Lepidella. *Persoonia* **5**:285–579.

Bigelow, H. E. (1965). The genus *Clitocybe* in North America. Section Clitocybe. *Lloydia* **28**:139–180.

Bigelow, H. E. (1968). The genus *Clitocybe* in North America. II. Section Infundibuliformes. *Lloydia* **31**:43–62.

Bigelow, H. E. (1970). *Omphalina* in North America. *Mycologia* **62**:1–32.

Bigelow, H. E., and A. H. Smith. (1969). The status of Lepista—a new section of *Clitocybe*. *Brittonia* **21**:144–177.

Boedijn, K. B. (1960). The Strobilomycetaceae of Indonesia. *Persoonia* **1**:315–318.

Bresinsky, A., and J. Huber. (1967). Schlüssel für die Gattung *Hygrophorus* (Agaricales) nach Exsikkatenmerkmalen. *Nova Hedwigia* **14**:143–185.

Corner, E. J. H. (1966). A monograph of Cantharelloid Fungi. *Ann. Bot. Mem.* **2**:1–255.

Corner, E. J. H. (1970). *Phylloporus* Quélet and *Paxillus* Fries in Malaya and Borneo. *Nova Hedwigia* **20**:793–822.

Corner, E. J. H., and C. Bas. (1962). The genus *Amanita* in Singapore and Malaya. *Persoonia* **2**:241–304.

Dennis, R. W. G. (1970). Fungus flora of Venezuela and adjacent countries. *Kew Bull.*, *Add. Ser.* **III**:1–531.

Dennis, R. W. G., P. D. Orton, and F. B. Hora. (1960). New Check List of British Agarics and Boleti. *Trans. Brit. Mycol. Soc.*, *Suppl.* **43**:1–225.

Donk, M. A. (1962). The generic names proposed for Agaricaceae. *Beih., Nova Hedwigia* **5**:1–320.

Essett, H. (1964). "Atlas mycologiques. I. Les Psalliotes." Paris.

Favre, J. (1955). Les Champignons Supérieurs de la zone Alpine du Parc National Suisse. *Beitr. Kryptogamenflora Schweiz* **10**:1–212.

Favre, J. (1960). "Catalogue descriptif des champignons supérieurs de la zone subalpine du Parc National Suisse," pp. 323–610. Liestal.

Groves, J. W. (1962). "Edible and Poisonous Mushrooms of Canada." Ottawa.

Harmaja, H. (1969). The genus *Clitocybe* in Fennoscandia. *Karstenia* **10**:5–120.

Heim, R. (1959). Les Champignons hallucinogènes du Mexique. Chap. IV. Étude descriptif et taxinomique des Agarics hallucinogènes du Mexique. *Arch. Mus. Hist. Natur. Paris.* [7] **5**:123–204.

Heim, R. (1966). Breves diagnoses latinae novitatum genericarum specificarumque nuper descriptarum. *Rev. Mycol.* [N.S.] **31**:150–159.

Heim, R. (1968). Deuxième mémoire sur les Cyttarophyllés. *Bull. Soc. Mycol. Fr.* **84**:103–116.

Heinemann, P. (1955). Champignons recoltes au Congo Belge par madame M. Goossens-Fontana. I. Boletineae (note complémentaire 2). *Bull. Jard. Bot. Belg.* **25**:169–181.

Heinemann, P. (1961). Les Boletinées. *Bull. Natura. Belg.* **42**:333–362.

Heinemann, P. (1964). Boletineae du Katanga. *Bull. Jard. Bot. Belg.* **34**:426–478.

Hesler, L. R. (1967). *Entoloma* in southeastern North America. *Beih., Nova Hedwigia* **23**:1–195.

Hesler, L. R. (1969). North American species of *Gymnopilus*. *Mycol. Mem.* **3**:1–117.

Hesler, L. R., and A. H. Smith (1960a). Studies on *Lactarius*. I. North American species of sect. Lactarius. *Brittonia* **12**:119–139.

Hesler, L. R., and A. H. Smith. (1960b). Studies on *Lactarius*. II. North American species of sections Scrobiculus, Crocei, Theiogali and Vellus. *Brittonia* **12**:306–350.

Hesler, L. R., and A. H. Smith. (1963). "North American species of *Hygrophorus*." Univ. of Tennessee Press, Knoxville.

Hesler, L. R., and A. H. Smith. (1965). "North American Species of *Crepidotus*." Hafner, New York.

Horak, E. (1964). Fungi austroamericani. I. *Tricholoma* (Fr.) Quélet. Sydowia **17**:153–167.

Horak, E. (1964). Fungi austroamericani. VI. *Sydowia* **17**:206–213.

Horak, E. (1964a). Fungi austroamericani. VII. *Sydowia* **17**:297–301.

Horak, E. (1964b). Fungi austroamericani. IX. *Sydowia* **17**:308–313.

Horak, E. (1968). Synopsis generum Agaricalium. *Beitr. Kryptogamenflora Schweiz* **13**:1–741.

Horak, E. (1971a). Studies on the genus *Descolea* Singer. *Persoonia* **6**:231–248.

Horak, E. (1971b). A contribution towards the revision of the Agaricales (Fungi) from New Zealand. *N.Z. J. Bot.* **9**:403–462.

Horak, E. (1971c). Contributions to the knowledge of the Agaricales s. 1. (Fungi) of New Zealand. *N.Z. J. Bot.* **9**:463–493.

Horak, E., and M. Moser. (1963). Fungi austroamericani. VIII. *Nova Hedwigia* **10**:329–338.

Huijsman, H. S. C. (1961a). Sur trois Psilocybes. *Persoonia* **2**:91–95.

Huijsman, H. S. C. (1961b). Observations sur le genre *Hohenbuehelia*. *Persoonia* **2**:101–107.

Imazeki, R., and T. Hongo. (1965). "Colored Illustrations of Fungi of Japan," Osaka, Japan.

Kühner, R., and H. Romagnesi. (1953). "Flore Analytique des Champignons Supérieurs." Paris.

Lange, M. (1952). Species concept in the genus *Coprinus. Dansk Bot. Ark.* **14**:1–164.

Lange, M., and A. H. Smith. (1953). The *Coprinus ephemerus* group. *Mycologia* **45**:747–780.

Leclair, A., and H. Essette. (1969). "Atlas mycologiques. II. Les Boletes." Paris.

McNabb, R. F. R. (1967). The Strobilomycetaceae of New Zealand. *N.Z. J. Bot.* **5**:532–547.

McNabb, R. F. R. (1968). The Boletaceae of New Zealand. *N.Z. J. Bot.* **6**:137–176.

Malençon, G., and R. Bertault. (1970). "Flore des Champignons Supérieurs du Maroc." Vol. 1. Rabat.

Mazzer, S., and A. H. Smith. (1967). New and interesting boletes from Michigan. *Mich. Bot.* 6:57–67.

Miller, O. K., Jr. (1964). Monograph of *Chroogomphus* (Gomphidiaceae). *Mycologia* 56: 526–549.

Miller, O. K., Jr. (1968). A revision of the genus *Xeromphalina*. *Mycologia* 60:156–188.

Miller, O. K., Jr. (1970). The genus *Panellus* in North America. *Mich. Bot.* 9:17–30.

Miller, O. K., Jr. (1971). The genus *Gomphidius* with a revised description of the Gomphidiaceae and a key to the genera. *Mycologia* 63:1129–1163.

Miller, O. K., Jr., and L. Stewart. (1971). The genus *Lentinellus*. *Mycologia* 63:33–369.

Møller, F. H. (1950). Danish *Psalliota* species. *Friesia* 4:1–60.

Møller, F. H. (1958). "Fungi of the Faeröes," Part II (appendix to Part 1), pp. 223–249. Copenhagen.

Moser, M. (1951). Cortinarien-Studien I-Phlegmacium. *Sydowia* 5:488–544.

Moser, M. (1952). Cortinarien-Studien. *Sydowia* 6:17–161.

Moser, M. (1960). "Die Gattung *Phlegmacium*. Die Pilze Mitteleuropas," Vol. 4, pp. 1–440.

Moser, M. (1967). Basidiomyceten. II. Röhrlinge and Blätterpilze (Agaricales). *In* "Kleine Kryptogmenflora" (H. Gams, ed.), 3rd ed., Vol. 2, Part b–2, pp. 1–443. Stuttgart.

Ola'h, G. M. (1969). Le genre *Panaeolus*: Essai taxinomique et physiologique. *Rev. Mycol., Mem. Ser.* 10:1–273.

Orton, P. D. (1960). New check list of British Agarics and Boleti. Part III. Notes on genera and species in the list. *Trans. Brit. Mycol. Soc.* 43:159–439.

Otieno, N. C. (1969). Further contributions to a knowledge of termite fungi in East Africa. The genus *Termitomyces* Heim. *Sydowia* 22:160–165.

Pantidou, M. E., and J. W. Groves. (1966). Cultural Studies of Boletaceae. *Can. J. Bot.* 44: 1371–1392.

Pegler, D. N. (1966). Tropical African Agaricales. *Persoonia* 4:73–124.

Pegler, D. N., and T. W. K. Young. (1971). Basidiospore morphology in the Agaricales. *Beih., Nova Hedwigia* 35:1–210.

Perreau, J. (1964). Complément à l'étude des ornementations sporales dans le genre *Boletellus*. *Ann. Sci. Natur.: Bot. Biol.* [12] *Veg.* 5:753–766.

Perreau-Bertrand, J. (1961). Recherches sur les ornementations sporales et la sporogenèse chez quelques espèces des genres *Boletellus* et *Strobilomyces* (Basidiomycetes). *Ann. Sci. Natur.: Bot. Biol.* [12] *Veg.* 2:399–489.

Pilát, A. (1951). Ceské druhy žampionu (*Agaricus*). (The Bohemian species of the genus *Agaricus*). *Acta Mus. Nat. Prag. Ser. B* 7:1–142.

Pilát, A. (1959). "Nase Houby II." Czech. Akad. Publ., Prague.

Pilát, A. (1969). "Houby Ceskoslovenska." Czech. Akad. Publ., Prague.

Romagnesi, H. (1967). "Les Russules d'Europe et d'Afrique du Nord." Bordas, France.

Romagnesi, H. (1956). "Nouvel Atlas des Champignons," Vol. I. Bordas.

Romagnesi, H. (1958). "Nouvel Atlas des Champignons," Vol. II. Bordas.

Romagnesi, H. (1961). "Nouvel Atlas des Champignons," Vol. III. Bordas.

Shaffer, R. L. (1957). *Volvariella* in North America. *Mycologia* 49:545–589.

Shaffer, R. L. (1962). The subsection Compactae of *Russula*. *Brittonia* 14:254–284.

Shaffer, R. L. (1964) The subsection Lactarioideae of *Russula*. *Mycologia* 56:202–231.

Shaffer, R. L. (1970). Notes on the subsection Crassotunicatinae and other species of *Russula*. *Lloydia* 33:49–96.

Singer, R. (1945). The Boletineae of Florida with notes on extralimital species. I. Strobilomycetaceae. *Farlowia* 2:97–141.

Singer, R. (1960). Three new species of Secotiaceae from Patagonia. *Persoonia* 1:385–391.

Singer, R. (1961a). Type studies on Basidiomycetes. *Persoonia* **2**:1–62.

Singer, R. (1961b). Monographs of South American Basidiomycetes. Pt. 4. *Inocybe* in Amazone region. Suppl. to Pt. 1. *Pluteus* in South America. *Sydowia* **15**:112–132.

Singer, R. (1962). New genera of Fungi VIII. *Persoonia* **2**:407–415.

Singer, R. (1962) [1963]. "The Agaricales in Modern Taxonomy," Cramer, Weinheim.

Singer, R. (1965). Monographic studies on South American Basidiomycetes, especially those of the east slope of the Andes and Brazil. 2) The genus *Marasmius* in South America. *Sydowia* **18**:106–358.

Singer, R. (1965). "Die Röhrlinge. Teil I. Die Pilze Mitteleuropas," Vol. 5, pp. 1–131. Klinkhardt, Bad Heilbrunn.

Singer, R. (1967). "Die Röhrlinge. Teil II. Die Pilze Mitteleuropas," Vol. 6, pp. 1–151. Klinkhardt, Bad Heilbrunn.

Singer, R. (1969). Mycoflora australis. *Beih., Nova Hedwigia* **29**:1–405.

Singer, R., and A. P. L. Digilio. (1952). Prodromo de la Flora Agaricina Argentina. *Lilloa* **25**:5–462.

Singer, R., and K. Grinling. (1967). Some Agaricales from the Congo. *Persoonia* **4**:355–377.

Singer, R., and A. H. Smith. (1958a). Mycological investigations on teonanácatl, the mexican hallucinogenic mushroom. Pt. II. A taxonomic monograph of *Psilocybe* sect. Caerulescentes. *Mycologia* **50**:262–303.

Singer, R., and A. H. Smith. (1958b). Studies on secotiaceous fungi. III. The genus *Weraroa*. *Bull. Torrey Bot. Club* **85**:324–334.

Singer, R. and A. H. Smith. (1958c). Studies on secotiaceous fungi. I. A monograph of the genus *Thaxterogaster*. *Brittonia* **10**:201–216.

Singer, R., and A. H. Smith. (1959a). Studies on secotiaceous fungi. VI. *Setchelliogaster* Pouzar. *Madrono, San Francisco* **15**:73–79.

Singer, R., and A. H. Smith. (1959b). Studies on secotiaceous fungi. V. *Nivatogastrium* gen. nov. *Brittonia* **11**:224–228.

Singer R., and A. H. Smith. (1960a). Studies on secotiaceous fungi. VII. *Secotium* and *Neosecotium*. *Madrono, San Francisco* **15**:152–158.

Singer, R., and A. H. Smith. (1960b). Studies on secotiaceous fungi. IX. The Astrogastraceous series. *Mem. Torrey Bot. Club* **21**:1–112.

Singer, R., J. E. Wright, and E. Horak. (1963). "Mesophilliaceae" and "Cribbeaceae" of Argentina and Brazil. Monographs of South American Basidiomycetes, especially those of the east slope of the Andes and Brazil. VI. *Darwinia* **12**:598–611.

Smith, A. H. (1947). "North American Species of Mycena," Univ. of Michigan Press, Ann Arbor.

Smith, A. H. (1951). North American species of *Naematoloma*. *Mycologia* **43**:467–521.

Smith, A. H. (1957). A contribution toward a monograph of *Phaeocollybia*. *Brittonia* **9**:195–217.

Smith, A. H. (1960). *Tricholomopsis* (Agaricales) in the Western Hemisphere. *Brittonia* **12**: 41–70.

Smith, A. H. (1963). New astrogastraceous fungi from the Pacific Northwest. *Mycologia* **55**:421–441.

Smith, A. H. (1965). New and unusual Basidiomycetes with comments on hyphal and spore wall reactions with Melzer's solution. *Mycopathol. Mycol. Appl.* **26**:385–402.

Smith, A. H. (1966). Notes on *Dendrogaster*, *Gymnoglossum*, *Protoglossum* and species of *Hymenogaster*. *Mycologia* **58**:100–124.

Smith, A. H. (1972). North American species of *Psathyrella*. *Mem. N. Y. Bot. Gard.* **24**:1–633.

Smith A. H., and L. R. Hesler. (1962). Studies in *Lactarius*. III. The North American species of section Plinthogali. *Brittonia* **14**:369–440.

Smith, A. H., and L. R. Hesler. (1968). "The North American Species of Pholiota." Hafner, New York.

Smith, A. H., and R. Singer. (1958). Studies on secotiaceous fungi. VIII. A new genus in the Secotiaceae related to *Gomphidius. Mycologia* **50**:927–938.

Smith, A. H., and R. Singer. (1959). Studies on secotiaceous fungi. IV. *Gastroboletus, Truncocolumella* and *Chamonixia. Brittonia* **11**:205–223.

Smith, A. H., and R. Singer. (1964). "A Monograph on the genus *Galerina*." Hafner, New York.

Smith, A. H., and H. D. Thiers. (1964). "A Contribution Toward a Monograph of North American Species of *Suillus*." Ann Arbor, Michigan (distributed by Stechert-Hafner Service Agency, Inc., Darien, Connecticut).

Smith, A. H., and H. D. Thiers (1971). "The Boletes of Michigan." Univ. of Michigan Press, Ann Arbor.

Smith, A. H., and S. M. Zeller. (1966). A preliminary account of the North American species of *Rhizopogon. Mem. N.Y. Bot. Gard.* **14**:1–177.

Smith, A. H., H. D. Thiers, and R. Watling. (1966). A preliminary account of the North American species of *Leccinum*, section Leccinum. *Mich. Bot.* **5**:131–167.

Smith, A. H., H. D. Thiers, and R. Watling. (1967). A preliminary account of the North American species of *Leccinum*, sections Luteoscabra and Scabra. *Mich. Bot.* **6**:107–153.

Smith, H. V. (1954). A revision of the Michigan species of *Lepiota. Lloydia* **17**:307–327.

Snell, W. and E. Dick. (1970). "The Boleti of Northeastern North America." Cramer, Weinheim.

Stevenson, G. (1961). The Agaricales of New Zealand. I. *Kew Bull.* **15**:381–385.

Thiers, H. D. (1957). The agaric flora of Texas. I. New species of agarics and boletes. *Mycologia* **49**:707–723.

Thiers, H. D. (1960). The agaric flora of Texas. III. New taxa of brown and black-spored agarics. *Mycologia* **51**:529–540.

Thiers, H. D. (1963). The bolete flora of the Gulf Coastal Plain. I. The Strobilomycetaceae. *J. Elisha Mitchell Sci. Soc.* **79**:32–41.

Thiers, H. D. (1964). The genus *Xerocomus* Quélet in northern California. *Madrono, San Francisco* **17**:237–249.

Thiers, H. D. (1965). California boletes. I. *Mycologia* **57**:524–534.

Thiers, H. D. (1966). California boletes. II. *Mycologia* **58**:815–826.

Thiers, H. D. (1967). California boletes, III. The genus *Suillus. Madrono, San Francisco* **19**: 148–160.

Thiers, H. D. (1971). California boletes. IV. The genus *Leccinum. Mycologia* **63**:261–276.

Thiers, H. D., and A. H. Smith. (1969). Hypogeous Cortinarii. *Mycologia* **61**:526–536.

Thiers H. D., and J. M. Trappe. (1969). Studies in the genus *Gastroboletus. Brittonia* **21**:244-254.

van Waveren, E. K. (1968), The "Stercorarius group" of the genus *Coprinus. Persoonia* **5**:131–176.

van Wavern, E. K. (1970). The genus *Conocybe* subgen. Pholiotina. 1. The European annulate species. *Persoonia* **6**:119–165.

van Waveren, E. K. (1971). Notes on the genus *Psathyrella*. I. *Psathyrella gracilis* and *P. microrrhiza. Persoonia* **6**:249–280.

Watling, R. (1968). Observations on the Bolbitiaceae. IV. A new genus of Gastromycetoid Fungi. *Mich. Bot.* **7**:19–24.

Watling, R. (1970). Boletaceae: Gomphidiaceae; Paxillaceae. *In* "British Fungus Flora" (D. M. Henderson, P. D. Orton, and R. Watling, eds.), Vol. 1, pp. 1–125. Roy. Bot. Gard., Edinburgh.

Watling, R. (1971). Observations on the Bolbitiaceae. IV. Developmental studies on *Conocybe* with particular reference to the annulate species. *Persoonia* **6**:281–289.

Basidiomycotina
Gasteromycetes
(See also Chapter 23)

CHAPTER 24

Gasteromycetes

D. M. DRING

The Herbarium
Royal Botanic Gardens
Kew, Surrey, England

I. GENERAL CHARACTERISTICS

Class Gasteromycetes

In this class, the hyphae are septate, with or without clamp connections. They are usually saprobic in soil or sometimes on dead wood or dung; occasionally they form mycorrihzal associations, and are seldom parasitic. The fruiting bodies range from being just visible to the naked eye to very large and complex. They have basidiospores which at maturity form a gleba or spore mass usually enclosed within a peridium. The basidiospores are not discharged violently from the basidium, and are usually symmetrical.

The gasteromycetes present the widest range of form and structure of fruiting body which can be found in any fungal group of comparable size. For this reason, they are difficult to define, but for most purposes they may be accepted as being those higher basidiomycetes in which the basidium is not involved in the discharge of the spore. Other means of discharge, some of them violent, have been developed in some cases, but these take place independently of the basidium. Thus, whereas most basidiomycetes are said to produce ballistospores, gasteromycetes produce *statismospores*.

The basidia of almost all gasteromycetes are, at least for the early part of their development, enclosed within the fruit-body. This habit of *angiocarpy* again serves to distinguish them from most basidiomycetes, which are gymnocarpous. The peridium which encloses the gleba may be a single-layered structure, or may consist of a number of more or less easily defined layers. In the Lycoperdales, the outer layer of the basically two-layered peridium is called the *exoperidium*; the inner is called the *endoperidium*. It is better not to use these terms for two-layered peridia in other groups since this might suggest homology. In a few cases there is never a peridium, or only a vestige.

The presence or absence of a hymenium is the most fundamental criterion for subdivision of the gasteromycetes. About half of the members of the

451

group have a well-developed hymenium at some stage in their ontogeny; the others are entirely without one. In the Lycoperdales and Phallales, for example, the developing basidia are arranged in a continuous hymenium, just as in other higher basidiomycetes. The hymenium is usually much convoluted but is still quite recognizable. It is bounded on one side by the trama, and on the others by an air space. In other groups, such as the Sclerodermatales, however, the basidia are produced either as isolated cells distributed throughout the gleba, or as knots or islands of fertile cells. Even if the basidia are loosely organized into sheets, as in some of the Nidulariaceae, the sheets are not continuous throughout the fruit-body. In addition, these *pseudohymenia* may be distinguished from the true hymenium in that the ends of the basidia do not project free into an air space. This distinction suggests a different origin for the two groups, and that the hymenial gasteromycetes have, in the course of evolution from a hymenomycete ancestor, abandoned the use of the ballistospore mechanism but nevertheless retained the arrangement whereby the basidium projects into an air space. The nonbasidial gasteromycetes, or *plectobasidial* fungi, have no air space into which a basidium could discharge, even if it were otherwise able to do so. Furthermore, in the plectobasidial fungi, several generations of basidia are usually developed in a single fruit-body, though not all of them function directly as spore-bearing organs. In a few genera, not all directly related one to another, the basidiospores are shed from their basidia before they are ripe, and their development continues with the help of a placenta of nurse-hyphae, which themselves are probably modified, second-generation basidia.

An origin for these plectobasidial gasteromycetes can be hypothesized from relatively lowly basidiomycete stock in which basidia were not geneti-

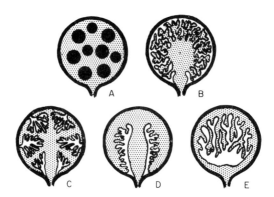

FIG. 1. Developmental diagrams of gasteromycete fruit-bodies. A, Lacunar; B, forate ("coralloid"); C, multipileate; D, pileate; E, aulaeate.

cally fixed simply in the role of spore guns, and were not marshalled into a hymenium. It is intended that this point will be elaborated elsewhere (Dring, 1974), but for the moment, it is necessary for the essential difference between the members of the two subclasses to be appreciated.

Lohwag (1926) drew up some diagrams illustrating different developmental types of fruit-body. These diagrams have achieved the status of classics in gasteromycetology, and can hardly be omitted from an introductory account. In Fig. 1 they are reproduced with some modification, although the reader is warned that the utmost caution is necessary in their interpretation, since they are two-dimensional representations of three-dimensional objects. The type of structure universally known as coralloid is formed by invagination of the primordial tissue, not by outgrowth of coralloid processes. It is, in fact the very opposite of coralloid, as is made clear in Fig. 2, and the term "forate" is substituted in the remainder of this study.

Equally important, there are several other types of development of the fruit-body in gasteromycetes which are not represented in Lohwag's diagrams. One useful addition is the aulaeate type (Fig. 1E) (Kreisel, 1969).

In some cases, the mode of development has a profound effect on the form of the mature gleba. Thus, in the Secotiaceae, the pileate development is betrayed by the agaricoid final form of the gleba, which consists simply of more or less radially arranged, dried tramal plates bearing the collapsed basidia and undischarged basidiospores. For this reason, the secotioid genera are dealt with in this volume under Agaricales (see p. 444).

In the Phallaceae, however, though the development is also of the pileate type, the hymenium and hymenophoral trama autolyse at maturity, and the gleba then consists of spores suspended in a glutenous and homogeneous matrix. The gleba of members of the Nidulariales consists of discrete seedlike pellets, each containing relatively few spores, but with many and varied vegetative elements. Each of the pellets, or peridioles, represents

Fig. 2. Developmental diagrams of gasteromycete fruit-bodies. A, 3-Dimensional view of multipileate type; B, 3-dimensional view of forate type.

one of the islands of fertile tissue formed in the process of the lacunar type of development.

In the groups with a powdery gleba, the latter often contains, in addition to spores, a capillitium. This is a more or less tangled mass of durable, specialized sterile hyphae. Such a capillitium is present in some rather distantly related groups, such as *Podaxon* and *Tulostoma*, but is more characteristic of the Lycoperdales. In the Lycoperdales it is convenient to distinguish two types of capillitial hyphae. First, there is the true capillitium, consisting of thick-walled, brown, aseptate, or infrequently septate, hyphae. Secondly, there is the paracapillitium (Kreisel, 1962) consisting of thin-walled, often collapsed, hyaline hyphae with many septa. These two types of hyphae may be regarded as corresponding, respectively, to skeletal (sometimes binding) and generative hyphae of Aphyllophorales. The two types differ in their reaction to cotton blue in lactophenol (the cyanophilous reaction of Kotlaba and Pouzar, 1964). Often both types of capillitium occur in the same gleba, and the capillitium is then *dimitic*.

Whatever the construction of the gleba, gasteromycete spores share a number of characteristics. Almost all, unlike those of hymenomycetes, are symmetrical about the axis of the sterigma which bears them. There is no hylar appendage. In many cases the sterigma, or part of it, remains attached to the discharged spore, which is then termed *pedicellate*.

Some gasteromycete spores are smooth but most are ornamented by warts, spines, striae, or reticulations. These are often caused by wrinkling of a more or less loose perisporial sac, but in some cases they consist of thickenings of an inner layer of the spore wall.

A few gasteromycetes have imperfect states belonging to the form-genus *Leucophlebs*. The conidia closely resemble the basidiospores of the corresponding perfect state (*Leucogaster*) in their ornamentation.

Among stalked gasteromycetes, it is useful to distinguish between those which possess a true stipe, and those having a *pseudostem*. In the case of *Lycoperdon* and *Phallus*, the pseudostem consists of a mass of spongy tissue in which the hyphae are not orientated parallel to the axis of the stipe. In *Lycoperdon* it is obviously modified glebal tissue, and is referred to as *subgleba*. A true stipe is to be found in *Tulostoma*, where it consists of more or less parallel hyphae, and is not merely an obvious modification of some preexisting organ. The true stipe of *Tulostoma* is the exact equivalent, though not perhaps the strict homologue, of the stipe of an agaric.

In the keys which follow, strong emphasis has been placed on developmental features because it is felt that they are of prime importance in obtaining a satisfactory arrangement of the genera. Unfortunately, information on its mode of development cannot always be deduced from a study of the mature fruit-body. Where possible, therefore, supplementary characters, not all of them so clear-cut, have been used.

II. IMPORTANT LITERATURE

Fischer, in Engler and Prantl, "Naturlichen Pflanzenfamilien," Aufl. II, **7a**, 1933; Lloyd, *Mycological Writings*, 7 vols., 1898–1925; Pilát, "Flora ČSR," **B1**, 1958; Zeller, Keys to the orders, families and genera of Gasteromycetes, *Mycologia* **41**:36–58, 1949; Demoulin, Les Gasteromycètes, *Naturalistes Belges* **50**:225–270, 1969.

III. KEY TO ORDERS

KEY TO ORDERS OF GASTEROMYCETES

1. Hymenium present. Basidia maturing more or less simultaneously. Spores often less than 10 μm, seldom reticulate 2

1′. Hymenium absent. Basidia borne singly or in groups scattered through the gleba, often maturing in several generations, not all of which are fertile. Spores often larger than 10 μm, often reticulate 6

    2(1) Development pileate or multipileate 3

    2′(1) Development otherwise 4

3(2) Gleba powdery at maturity. Fruit-body pileate, truly stipitate, of coprinoid facies, spores dark, distinctly agaricoid . . . . **Podaxales** p. 455

3′(2) Gleba fleshy or mucid at maturity. Fruit-body never with a true stipe. Spores hyaline and bacilloid or pale brown, ovoid . . . . **Phallales** p. 458

    4(2′) Development forate or a modification thereof 5

    4′(2′) Development aulaeate. Gleba fleshy (except in *Gastrosporium*). Spores ellipsoid, often ribbed . . . . **Hymenogastrales** p. 468

5(4) Gleba powdery. Spores usually globose, never ribbed. Peridium present
. . . . **Lycoperdales** p. 463

5′(4) Gleba cartilaginous. Spores ellipsoid, longitudinally ribbed. Peridium often absent
. . . . **Gautieriales** p. 468

    6(1′) Fruit-body sessile or occasionally with pseudostem of modified glebal tissue or rooting strands 7

    6′(1′) Fruit-body truely stipitate . . . . **Tulostomatales** p. 474

7(6) Gleba organized into peridioles (sometimes a single peridiole). Fruit body not exceeding 1 cm in diameter. Spores smooth, hyaline, often large . . . . **Nidulariales** p. 468

7′(6) Gleba not so organized. Fruit-body diameter usually greater than 1 cm 8

    8(7′) Gleba fleshy at maturity . . . . **Melanogastrales** p. 470

    8′(7′) Gleba powdery at maturity . . . . **Sclerodermatales** p. 471

IV. PODAXALES

The mycelium in this class is saprobic, often growing in termitaria. The fruit-body is stipitate and resembles *Coprinus* in outline; it is up to about 25 cm tall and 5 cm wide. The peridium is single, membranous, fragile, pale, and dehisces round the base where it joins the stipe-columella. The gleba is traversed by a percurrent columella; otherwise it is homogeneous at maturity. Development is pileate, and the hymenium is borne on irregular but recognizably lamellate tramal plates which break down following the development of the capillitium. The basidia persist in fascicles in the mature gleba. The spores are subglobose to ellipsoid, closely resembling those of *Coprinus* in their structural details. The capillitial threads are brown, straight, and aseptate, with a wall of spiral construction which unwinds to produce a ribbon. The stipe is continuous with the columella, and is tough, slender, and bulbous at the base.

There is a single family, Podaxaceae, and a genus, *Podaxis*, of which Morse (1933) considered that there was only one species, *P. pistillaris*. (Fig. 3A). Podaxales are widespread in the tropics and subtropics.

This is really an "order of convenience" since it is obvious that the fungus really belongs to the Agaricales in the sense in which they are interpreted here. The agaricoid appearance of the mature fruit-body and spores, and the pileate development, described by Fischer (1934), suggest that the presence of a capillitium is merely an example of parallel evolution and does not imply relationship with the Lycoperdales or Tulostomatales.

V. PHALLALES

The fruit-body is hypogeal, at least initially, and more or less globose, ranging from 0.5 to 5 cm in diameter. It consists of a usually smooth, membranous peridium covering a more or less well developed gelatinous layer. In this matrix, the gleba is suspended, either attached to the peridium at the base or by means of membranous peridial sutures passing directly through the gelatinous layer. The gleba is fleshy, labyrinthine, sweet-smelling, greenish, or brown.

In some families the peridium is indehiscent, and the fruit-body may be considered mature at the stage described above. The spores in these groups are typically brown, ovoid, of the order of 10×5 μm. Spore dispersal is by burrowing animals. In the remaining families, however, there is a receptacle of compressed spongy or tubular tissue in addition to the gleba and gelatinous tissue filling the peridium. When the peridium of these fruit-bodies breaks open, the receptacle expands, carrying the gleba out of the "egg" and raising it above the surface of the ground. Meanwhile, the hymenophoral trama autolyzes and is reduced to a dark, often putrid-smelling slime in which the spores are suspended. The spores are small, bacilloid, about $2-4 \times 1.5$ μm, almost hyaline, and obviously suited for ingestion by flies which feed upon the gleba.

There is no up-to-date account of the group but Lloyd (1909) and Fischer (1890, 1893, 1900) have described most of the receptacular species.

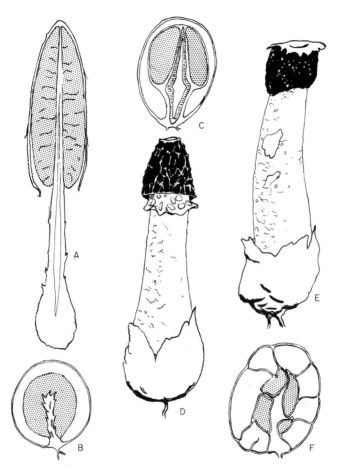

FIG. 3.   A. *Podaxis pistillaris*, vertical section of mature fruit-body showing basal dehiscence, percurrent stipe-columella and gleba (stippled); B, *Gelopellis* sp., vertical section of mature fruit-body (gleba stippled); C, *Mutinus caninus*, vertical section of unopened fruit-body (note unexpanded receptacle); D, *Phallus duplicatus*, mature fruit-body (note short indusium protruding from below cap, untenable as a generic character); E, *Itajahya rosea*, mature fruit-body (note apical calyptra); F, *Protubera clathroidea*, vertical section of mature fruit-body [note absence of receptacle (cf. Fig. 4B) and presence of peridial sutures].

Classically, the order contains only the well-known stinkhorns (Phallaceae) and Clathraceae, but as here treated, the hypogeal forms with indehiscent peridia are also included because of their developmental similarity to the epigeal forms.

Two different modes of development characterize the two lines into which the order is divided. On the one hand, the Hysterangiaceae, Protophallaceae, and Clathraceae are multipileate, and on the other, the Gelopellidaceae and Phallaceae are pileate. The relationship between these two modes of development is discussed by Dring (1966), but for the moment it may be accepted that they represent two parallel lines.

KEY TO FAMILIES AND GENERA OF PHALLALES

1. Peridium dehiscent. Receptacle present, epigeal. Gleba glutenous 2

1′. Peridium indehiscent. Receptacle absent. Usually hypogeal. Gleba soft but not glutenous
. . . . 4

    2(1) Receptacle sessile or stipitate, consisting of a spherical network or of several columns united at the top, or of spreading arms. Gleba usually borne on the inside of the receptacle. Peridial sutures present. Development of fruit-body multipileate
    . . . . **Clathraceae** 10

    2′. Receptacle otherwise. Peridial sutures absent 3

3(2′) Receptacle sessile, globose, with a nonmucid, sweet-smelling gleba borne on the inside. Development unknown . . . . **Claustulaceae**
    There is a single monotypic genus **Claustula** (*C. fischeri*), known only from 1 locality in New Zealand. A good description is given by Cunningham (1944), but developmental studies are needed before its anomalous position can be resolved.

3′(2′) Receptacle a simple, hollow column bearing the mucid, usually stinking gleba near the top, on the outside. Development pileate . . . . **Phallaceae** 6

    4(1′) Gelatinous layer below the peridium poorly developed, often cartilaginous, often interrupted . . . . **Hysterangiaceae** 22

    4′(1′) Gelatinous layer well-developed 5

5(4′) Gelatinous layer interrupted by peridial sutures. Gleba divided into many lobes. Development multipileate . . . . **Protophallaceae** 21

5′. Gelatinous layer continuous. Gleba a more or less globose mass with central columella. Development pileate . . . . **Gelopellidaceae**
    A single genus **Gelopellis**, with perhaps about 5 species. The mature fruit-body resembles exactly an unopened *Mutinus* in which the receptacle is wanting (Fig. 3B). About half the species have small hyaline spores of the phalloid type, the others have larger, ellipsoid, brown spores. Zeller (1939) gives a brief account.

    6(3′) Receptacle consisting of a simple, stalklike column bearing the gleba directly . . . 7

    6′(3′) Receptacle with separable, campanulate cap borne on apex of stipe and supporting the gleba 8

7(6) Gleba confined to a ring of tissue some distance below the apex of the stipe
. . . . **Staheliomyces**
    One species, *S. cinctus*; C. and S. America.

7′(6) Gleba borne subapically on the stipe . . . . **Mutinus**
    About 12 species (Fig. 3C), including segregates *Jansia* and *Xylophallus*. Widespread, usually associated with decaying wood.

    8(6′) Glebiferous cap reduced to a torn, scant remnant . . . . **Floccomutinus**
    One species, *F. zenkeri*. On decaying wood. Cameroons.

8'(6') Glebiferous cap covering the upper part of the stipe 9

9(8') Glebiferous cap cylindrical, surface covered with membranous and tubercular processes which support copious gleba. Apex of stipe covered initially by a calyptra. Stipe pink or white, hollow, with a thick wall . . . . **Itajahya**
One species, *I. rosea* (Fig. 3E); widespread in tropics.

9'(8') Glebiferous cap more or less campanulate, covering the apex of the stipe like a thimble over a finger-end. Calyptra absent. Surface of cap rugulose, papillate, or with a reticulum of ridges (this feature being the basis for separation of subgenera). Stipe hallow, its wall of spongy tissue or sometimes of a single or double layer of chambers, white, red, or orange
. . . . **Phallus**
About 18 species, widespread. In some species [which used to be segregated into the untenable genera *Dictyophora* and *Clautriavia*, see Dring (1964)] a netlike indusium hangs down under the cap, around the stipe like a skirt.

10(2) Receptacle sessile, consisting entirely of a fertile network 11

10'(2) Receptacle stipitate, with a fertile network or united or divergent arms 12

11(10) Receptacle white, pink, or red with obvious differentiation between base and apex, usually of spongy tissue retained within the cupulate volva at maturity . . . . . **Clathrus**
About 15 species (Fig. 4A,B); mainly tropical. Including *Clathrella*.

11'(10) Receptacle white, usually with no dorsiventral distinction, of obviously tubular construction, becoming free from the volva at maturity . . . . **Ileodictyon**
Two species, known best from Australasia. For descriptions see Cunningham (1944), where the genus is united with *Clathrus*, and Reid and Dring (1964).

12(10') Receptacular arms united above, free below 13

12'(10') Receptacle stipitate, or if of separate arms or columns, then these united below
. . . . 15

13(12) Gleba borne on special glebifer 14

13'(12) Gleba borne directly on underside of arch of receptacular arms of which there are 2–5. Occasional specimens have irregular lateral branches of arms . . . . **Linderia**
Two species of which *L. columnata* (Fig. 4D) is widespread outside Europe. *Linderiella* is a synonym.

14(13') Receptacular arch of 3–4 arms with gleba borne on lateral flaps of tissue
. . . . **Blumenavia**
Two species: Africa and S. America.

14'(13') Receptacular arch of 2–3 arms with gleba borne on a "lantern" hanging from the vault . . . . **Laternea**
Two species; C. and S. America. (Fig. 4C)

15(12') Receptacle red, clumsily constructed of irregularly spongy tissue, with short obconical stipe diverging into 4–6 pointed arms, initially united at their tips but soon becoming free. Spores rather large, 4–6 × 2–3 μm . . . . **Anthurus**
A single species (Fig. 4E); widespread outside Europe where, however, it is introduced in a few places.

15'(12') Receptacle more delicately chambered. Spores smaller 16

16(15') Receptacle a fertile network borne on a well-developed stipe 17

16'(15') Receptacle consisting of a stipe and arms either united above or divergent . . 19

17(16) Receptacle pink, red, or orange, consisting of stipe and fertile network only, meshes

rather small. Gleba often straying onto outside of receptacle . . . . **Simblum**
A single species *S. periphragmoides* (Fig. 4F) widespread in tropics and subtropics.

17'(16) Receptacle red, consisting of stipe, fertile network, and other structures 18

18(17') Fertile network bearing conspicuous dichotomous outgrowths at the intersections
of the arms. Gleba almost wholly external . . . . **Kalchbrennera**
A single species *K. corallocephala* (Fig. 5A); limited to Africa south of the Sahara.

18'(17') Receptacle consisting of a short stipe surmounted by about 6 columns, which
support a small fertile network at the top . . . . **Colus**
One species, *C. hirudinosus*, (Fig. 5D); chiefly Mediterranean in distribution.

19(16') Arms simple, relatively few, not or hardly divergent 20

19'(16') Arms bifid, or united into pairs, or if single, then more numerous, widely divergent.
Gleba borne on a disc between the arms or on the proximal part of the arms themselves
. . . . **Aseroë**
Two species, Australasia, S. America, S. E. Asia.

20(19) Stipe longer than arms, arms few (about 4), rather thick in proportion to their
length, at first united above, usually remaining upright, occasionally slightly divergent
in very ripe specimens. Gleba on adaxial and lateral faces of arms, sometimes straying
round to the abaxial face which always, however, bears a conspicuous, sterile groove
. . . . **Lysurus**
Three species (Fig. 5B); widespread.

20'(19) Stipe shorter than or equal to arms. Arms about 5, slender, normally remaining
united above, gleba borne on adaxial face only . . . . **Pseudocolus**
Perhaps 3 species, but not well-known. Resembles *Anthurus* in general outline, but
not closely related (Fig. 5E). Can be distinguished by the much more delicate construction, and smaller spores, about 2–3 × 1.5 μm.

21(5) Gleba powdery at maturity . . . . **Calverula**
One species, *C. excavata*, Florida.

21'(5) Gleba fleshy or cartilaginous . . . . **Protubera**
About 6 species; tropics and subtropics (Fig. 3F). Furtado and Dring (1967) and
references therein give accounts of the species.

22(4) Fruit-body more or less globose at maturity . . . . **Hysterangium**
About 40 species, cosmopolitan. Zeller and Dodge (1929) provide the best available
account.

22'(4) Fruit-body substipitate or pyriform 23

23(22) Columella not percurrent. Dehiscence by irregular splitting of peridium. Gleba in
irregular lobes which are attached by raised ridges to the peridium. Spores like those of
Phallaceae . . . . **Phallogaster**
A single species, *P. saccatus*, from eastern U.S.A. This species strongly reminds
one of *Claustula*.

23'(22) Columella percurrent and continuous with a short stipe. Dehiscence by sloughing off
of the whole, thin peridium. Development of an intermediate type, but suggesting that of
a primitive *Phallus*. Gleba becoming dry at maturity, spores brown, ellipsoid, pedicellate
. . . . **Rhopalogaster**
One species, *R. transversarius*, growing in damp places in association with rotten
wood; southeastern U.S.A.

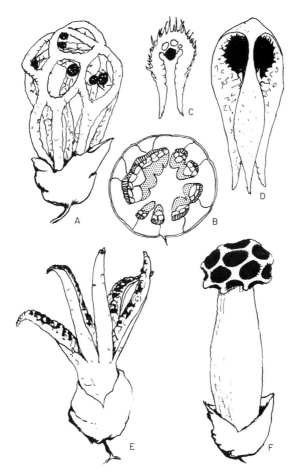

FIG. 4. Clathraceae. A, *Clathrus oahuensis*, mature fruit-body (note elongated lower meshes, a character of the untenable genus *Clathrella*); B, *Clathrus baumii*, vertical section of undehisced "egg" (note peridial sutures joining compressed receptacle to peridium); C, *Laternea pusilla*, mature fruit-body without volva (note gleba suspended on glebifer below fornix of receptacle, arms free below; D, *Linderia columnata*, mature fruit-body without volva (note gleba borne directly on arms, arms free below); E, *Anthurus archeri*, mature fruit-body; F, *Simblum periphragmoides*, mature fruit-body.

VI. LYCOPERDALES

The mature fruit-body is epigeal (the earlier stages are often hypogeal), more or less globose, ranging from a few mm to over 1 m in diameter; it is

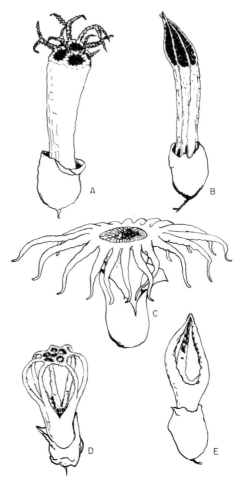

FIG. 5.   Clathraceae. A, *Kalchbrennera corallocephala*, mature fruit-body from which most
of the gleba has gone; B, *Lysurus mokusin*, mature fruit-body (the ribbed stipe is not a generic
character); C. *Aseroë rubra* var. *brasiliensis*, mature fruit-body; D, *Colus hirudinosus*, mature
fruit-body; E, *Pseudocolus javanicus*, mature fruit-body (note superficial resemblance to
Anthurus).

usually sessile, and a few have a pseudostem. The peridium is usually divisible into two major layers, the exo- and endoperidia; these layers are
often further subdivided. Glebal development is forate or a modification
thereof. At maturity they become powdery, they usually contain some sort
of capillitium, and they probably always have glebal membranes (Kreisel and
Dring, 1967). The glebal membranes and the capillitium are usually orga-

nized into a central tuft (pseudocolumella) and a peripheral part, there being a portion of the gleba between these two which is relatively free from capillitium. The spores are occasionally smooth, usually warted or spiny, and very seldom reticulate (at least under the light microscope). They are almost always globose, tinted to dark, and almost always well below 10 μm in diameter.

This order contains the puffballs and earthstars. *Scleroderma* shares the general shape and powdery gleba of the puffballs but can be distinguished on the basis of its simple peridium, often large reticulate spores, always without pedicels, homogeneous gleba betraying its lacunar development, and lack of a capillitium.

For similar reasons it is now usual to exclude *Astraeus*, in spite of its strong superficial resemblance to *Geastrum*. I would go further and place *Myriostoma* and *Broomeia* in the plectobasidial Gasteromycetes, though the adult fruit-bodies have many features in common with the Lycoperdales, and may have keyed out here on the basis of these characteristics.

KEY TO FAMILIES AND IMPORTANT GENERA OF LYCOPERDALES

1. Fruit-bodies hypogeal, more or less globose, about 1–2 cm in diameter. Peridium very fragile, falling away at maturity. Gleba of minute chambers, each with a hymenium, which separate at maturity like grains of sand. Capillitium and sterile base absent. Spores globose to ovoid. smooth, up to about 10 μm in diameter. Glebal membranes perhaps represented by the persistent peridiolar walls which are thin and acellular . . . . **Arachniaceae**

   The family comprises a single genus, **Arachnion**, with about 6 species, from N. and S. America, S. Africa, and Australasia; a very poorly known group of uncertain position. The presence of a distinct hymenium, in combination with other characters, make it difficult to place this genus outside the Lycoperdales. No doubt developmental studies will throw light on the true position.

1′. Fruit-bodies hypogeal or epigeal, peridium of more than 1 layer, not usually completely dehiscent. Gleba powdery at maturity, not peridiolar. Capillitium usually present . . . 2

   2(1′) Peridium usually indehiscent, rather thick, composed of 3–4 clearly defined layers. Gleba usually radially organized; spores elliptical or globose, warted
   . . . . **Mesophelliaceae** 4

   2′(1′) Peridium dehiscent, thick or thin. Gleba usually with pseudocolumella; capillitium branched or simple; spores brown, globose, occasionally elliptical, smooth, warted, or spiny, occasionally reticulate 3

3(2′) Exoperidium of 3 layers, opening by stellate dehiscence to reveal endoperidium which is apically dehiscent. Basidia capitate, sterigmata of equal lengths; capillitium aseptate, typically unbranched, always markedly organized into a pseudocolumella . . . . . **Geastraceae** 6

3′(2′) Exoperidium of 1 layer, often caducous. Spore sac dehiscing by apical pore or by attrition from above, occasionally by radial lobes. Basidia cylindrical, often with sterigmata of markedly varying lengths. Capillitium septate or not, typically branched
   . . . . **Lycoperdaceae** 7

   4(2) Epigeal. Exoperidium consisting of mycelial layer heavily encrusted with sand,

middle plectenchymatous layer, and inner fleshy layer. Endoperidium free, tough, and membranous; dehiscence apparently by attrition. Gleba of strongly radiating peripheral capillitium, the center occupied by a large firm core of undifferentiated fundamental tissue and held in place by tramal trabeculae; spores ellipsoid, smooth, with traces of gelatinous exospore; basidia ovoid . . . . **Mesophellia**
Four species known, (Fig. 6A); endemic to Australia, Tasmania, and New Zealand; see Cunningham (1944) for key.

4'(2) Without these characters in combination 5

5(4') Hypogeal. Exoperidium and endoperidium each of 2 layers. Gleba radiating from the columella which collapses to leave a central cavity in mature specimens; spores globose, brown, finely spiny about 3 μm in diameter . . . . **Radiigera**
A single species, *R. paulensis* from Brazil. The spores and organization of the gleba are strongly reminiscent of *Geastrum* of which members of the Mesophelliaceae are possible precursors.

5'(4') Epigeal. Exoperidium of two layers and endoperidium of a single layer; dehiscence by crudely radial splitting of peridia from the apex. No central core and gleba not obviously radially organized; spores and capillitium like those of *Mesophellia* . . . . . **Castoreum**
Three species, limited to Australia and Tasmania; see Cunningham (1944) for a key.

6(3) Endoperidium papyraceous, persistent, with apical ostiole. Gleba with a prominent pseudocolumella of highly compact capillitium . . . . **Geastrum**
Perhaps 50 species (Fig. 6B); widespread. The well-known earthstars, but see also *Astraeus* and *Myriostoma* which may have keyed out here. Many North Temperate species are dealt with by Staněk (1958). The others must be sought in regional floras; many species remain to be described.

6'(3) Endoperidium very thin, more or less adhering to the exoperidium when the latter splits and opens, laying bare the radially arranged gleba, the center of which is occupied by a large columella of undifferentiated sterile tissue . . . . **Trichaster**
Four species, including *Terrostella*, *Geasteropsis*, and *Geasteroides*, vicariously distributed in Europe, S. Africa, and N. America. The hard columella and the peridial characters suggest relationship with *Mesophellia*. See Zeller (1945) for a partial account.

7(3') Spores strongly reticulate. Exoperidium a sand case, dehiscing by falling away from above; endoperidium papyraceous, apically dehiscent. Gleba more or less homogeneous, capillitium septate usually breaking into short pieces . . . . **Abstoma**
Ten species; Australasia, Africa, America. Apparently quite rare, and because difficult to place, neglected. The large, reticulate spores suggest a place for these among the lacunar families, but on present knowledge they are placed here because of general resemblance to *Disciseda*.

7'(3') Spores smooth, spiny, or warted 8

8(7') Fruit-bodies gregarious, pulvinate, about 0.5–1 cm in diameter, borne on a subiculum growing on dead wood. Exoperidium scaly; endoperidium cream-colored, very thin, dehiscing from the apex and becoming reflexed. Gleba radially arranged, with small (to 4 μm), irregularly globose, hyaline, minutely spiny spores, the spines aggregated into an incomplete reticulum; capillitium hyaline with clamps, each thread surrounded by a gelatinous sheath made up of the remains of other, spirally arranged hyphae. Sterile base often zonate, of undifferentiated tissue . . . . **Lycogalopsis**
One species, *L. solmsii* (Fig. 6C); pantropical. Of uncertain position since the early developmental details are in dispute. However, it keys out best here, and the prominent

solid sterile base and radial gleba are reminiscent of Mesophelliaceae. See Martin (1939) for a good account of the species.

8'(7') Fruit-bodies otherwise 9

9(8') Fruit-body up to 20 cm in diameter, irregularly globose. Exoperidium thin and floccose; endoperidium 2–3 mm thick, hard, dehiscing by irregular lobes from the apex, opening in dry weather. Gleba homogeneous; capillitium aseptate, spiny; spores globose, up to 13 μm in diameter, coarsely wrinkled . . . . **Mycenastrum**
> Two species, of which *M. corium* is well known and widespread. Occupying a rather isolated position in the family; the capillitium suggests that of *Bovista* with suppressed branches having become spines. The large, coarsely reticulate spores suggest *Abstoma* and the thick peridium Mesophelliaceae.

9'(8') Fruit-body not as described 10

10(9') Fruit-body up to 3 cm in diameter, depressed globose, not attached to ground at maturity. Exoperidium a thick sand case, caducous except for a small basal part which persists as a disk; exoperidium tough and membranous, with definite apical stoma. Gleba powdery, often highly colored, homogeneous; capillitium of short hyphae; spores brown, globose, often extravagantly warted, often with long pedicels
. . . . **Disciseda**
> Probably about 20 known species, widely distributed, and many still to be described; 8 are given by Cunningham (1944), a few others by Bottomley (1948), some, mainly from N. America by Lloyd at various points in "Mycological Notes" (as *Catastoma*, a synonym) and from Europe by Moravec (1958). The ontogeny of the fruit-body is interesting in that it is said to turn over at maturity (Fig. 6D), the thick "basal" plate really being apical, and the "apical" ostiole marking the site of attachment of the basal rhizomorph.

10'(9') Fruit-body not as described 11

11(10') True capillitium absent or very sparse 12

11'(10') True capillitium abundant 13

12(11) Fruit-body 1–2 cm in diameter, pulvinate, highly colored, gregarious on dead wood. Exoperidium minutely granular, spiny or velutinous; endoperidium membranous, very thin above where it falls away in patches. Gleba with pseudocolumella, consisting of spores, paracapillitium, and persistent glebal membranes, not true capillitium; spores globose, about 4 μm in diameter, echinulate; paracapillitium septate, hyaline, collapsed; glebal membranes rugulose, gathered into a pseudocolumella. Subgleba small, chambered or compact . . . . **Morganella**
> Nine species (Fig. 7E) in all continents except Europe. The genus has been revised by Kreisel and Dring (1967) and 2 additional species added by Ponce de León (1971).

12'(11) Fruit-body 2–5 cm in diameter, cream-colored, at least when young, pulvinate, on soil, usually in turf. Exoperidium granular or strongly spiny, falling away; endoperidium membranous, often areolate, dehiscing irregularly from the apex. Gleba of spores, paracapillitium, and scanty, peripheral capillitium, barely heterogeneous, separated from subgleba by a diaphragm; spores smooth to minutely warted; paracapillitium sometimes slightly tinted, often not collapsed, septate. Subgleba chambered . . . . **Vascellum**
> Probably about 10 species (Fig. 7 F); worldwide. Partially revised by Ponce de León (1970); additional information tabulated by Dissing and Lange (1962) as *Lycoperdon*, in which genus they preferred to retain these species having a diaphragm.

13(11') Sterile base absent or virtually so or compact 14

13′(11′) Sterile base chambered 15

14(13) Fruit-bodies about 7 cm to 1 m in diameter, often detached from the ground at maturity. Exoperidium pale, falling away; endoperdium papery, brittle at maturity, and falling away in irregular patches. Gleba homogeneous; capillitium brown, branched, rigid, often constricted at the septa; spores globose, spiny . . . . . . **Langermannia**
    Probably about 6 species; worldwide. The giant puffballs, particularly the North Temperate *L. gigantea* belong here. Some species are dealt with by Kawamura (1937). *Lanpila* and *Lasiosphaera* are synonyms.

14′(13) Fruit-bodies about 1–5 cm in diameter, some detached from the ground at maturity. Exoperidium pale, membranous to warted; endoperidium membranous, dehiscing by an apical pore. Gleba homogeneous to slightly heterogeneous; spores globose to ovoid, smooth, spiny, or warted, often with a long pedicel; capillitium branched, often profusely so. Subgleba usually absent, if present of compact, undifferentiated tissue . . . . **Bovista**
    Fifty species; worldwide. The description corresponds to the broad concept of Kreisel (1967) who has monographed the genus (Fig. 7B,C). The narrower concept of the genus excludes all species except those with homogeneous gleba and profusely branched "bovistoid" capillitium, the fruit-bodies becoming detached from the

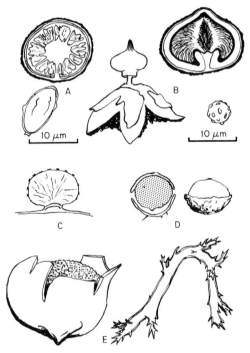

Fig. 6.   Lycoperdales. A, *Mesophellia arenaria*, vertical section of mature fruit-body and spore; B, *Geastrum striatum*, expanded fruit-body, vertical section of unexpanded fruit-body, spore; C, *Lycogalopsis solmsii*, vertical section of mature fruit-body; D, *Disciseda* sp., vertical section of dehiscing fruit-body, same fruit-body after inversion; E. *Mycenastrum corium*, mature fruit-body, capillitial hypha.

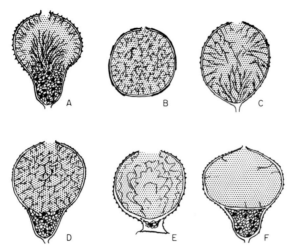

FIG. 7. Diagrammatic vertical sections of fruit-bodies of Lycoperdaceae. A, *Lycoperdon perlatum*; B, *Bovista plumbea*; C, *B. pusilla* (sometimes placed in *Lycoperdon*); D, *Bovistella paludosa*; E, *Morganella compacta*; F, *Vascellum pratense*.

ground at maturity. If the latter definition is accepted the remainder of the species are accommodated in *Lycoperdon*.

15(13') Dehiscence by apical pore, fruit-bodies medium-sized, pyriform 16

15'(13') Dehiscence by falling away of the upper part of the peridium, fruit-bodies medium-sized to large, usually pyriform 17

16(15) Fruit-bodies usually growing on the ground, occasionally on wood. Exoperidium scurfy to spiny; endoperidium opening by regular apical pore. Gleba heterogeneous with well-marked pseudocolumella; capillitium simple or slightly branched; spores globose, smooth to spiny, often with remains of a pedicel. Subgleba conspicuous, chambered not separated from gleba by a diaphragm . . . . **Lycoperdon**
Twenty-five species; widespread: the "typical" puffballs (Fig. 7A). This is the narrow generic concept of Kreisel (1967) for which see under *Bovista*.

16'(15) Fruit-bodies growing on the ground. Exoperidium scurfy-tomentose to slightly spiny; endoperidium with regular apical pore. Gleba hardly heterogeneous, pseudocolumella poorly developed; capillitial elements composed of well-marked main stem and many dichotomous lower-order branches ("bovistoid"); spores often long-pedicellate. Subgleba well developed, chambered, not separated from gleba by diaphragm
. . . . **Bovistella**
A single well-known species, *B. paludosa* (Fig. 7D); Europe and N. America. Kreisel (1962) gives a short account.

17(15') Capillitium sometimes branched but not spiny. Exoperidium membranous to very strongly warted; Endoperidium thin, fragile. Gleba homogeneous olivaceous or lilaceous; spores globose, occasionally ellipsoid, smooth, or spiny. Subgleba chambered, often well-developed, often tardily separated from gleba by compact zone which may be mistaken for a diaphragm . . . . **Calvatia**

About 20 species; worldwide including arctic-alpine. A recent monograph by Zeller and Smith (1964) of the N. American species takes a wider view of the genus and includes *Langermannia* and *Vascellum*.

17'(15') Capillitium with antlerlike branches ending in points. Exoperidium of large pyramidal warts; endoperidium fragile. Gleba homogeneous, brown; spores almost smooth, globose, 4 μm in diameter. Subgleba chambered, about one-quarter the size of the gleba

. . . . **Calbovista**

One species, *C. subsculpta*, from high altitudes in western N. America. See Morse (1935) for full description.

VII. GAUTIERIALES

The mycelium is saprobic in forest soils and (?) mycorrhizal. The fruit-body is hypogeal and subglobose to reniform and the peridium is thin, single, and often lacking at maturity. At maturity the gleba is cartilaginous and traversed by a branched columella. Development is forate with the glebal chambers becoming labyrinthine. The basidia are claviform and two- or four-spored. The spores are almost sessile, dark, ellipsoid, and often up to 20 μm long, with about eight longitudinal ribs. Cystidia are inconspicuous.

The order comprises the single family Gautieriaceae and genus *Gautieria* with about twenty species from Europe, North and South America, and Australasia. The habit, habitat, and spores suggest affinity with the Hymenogastrales from which they differ conspicuously in mode of development. The North American species have been revised by Zeller and Dodge (1918) and the development described by Fitzpatrick (1913).

VIII. HYMENOGASTRALES

The mycelium grows in the ground, often in (?) parasitic or (?) mycorrhizal association with green plants. The fruit-body is hypogeal and development is aulaeate (for a definition of aulaeothecium see Kreisel, 1969). The peridium is single or double. The gleba is fleshy to cartilaginous and occasionally powdery. The basidia are subfusoid and acrosporous with well-developed sterigmata. Spores are hyaline or pigmented, globose to ellipsoid and they usually have warts or longitudinal ribs. The capillitium is absent.

Some of the genera listed by Savile (1968) as having reasonably well-documented affinities with hymenomycetes have many of the characters listed above, particularly with regard to developmental features. These genera are, however, dealt with in the relevant hymenomycete keys.

KEY TO FAMILIES AND IMPORTANT GENERA OF HYMENOGASTRALES

1. Fruit-body globose up to 1.5 cm in diameter. Peridium double; outer peridium chalk white, flocculent, of collapsed hyphae mixed with oxalate crystals; inner peridium membranous,

cracking irregularly at maturity. Gleba homogeneous, olivaceous, powdery; basidia usually 8-spored; spores ochraceous, globose to subglobose, about 5 μm in diameter, ornamented with a few, small, irregular warts; cystidia and capillitia absent . . . . **Gastrosporiaceae**

> A single genus, *Gastrosporium*, and species, *G. simplex*, associated with grasses, etc., in xerothermic situations; Mediterranean region and central Europe. A good description is given by Pilát (1934).

1'. Peridium single; gleba not pulverulent; spores ellipsoid 2

> 2(1') Fruit-bodies minute to 500 μm in diameter, globose with a single globose glebal chamber. Hymenium not convoluted; basidia subfusiform, 1- to 4-spored; spores shortly ellipsoid, smooth, brown, filling the glebal cavity at maturity; capillitium and cystidia absent . . . . **Protogastraceae**
>
> > A single monotypic genus, *Protogaster* (*P. rhizophilus*), on living roots of *Viola,* Maine, U.S.A.
>
> 2'(1') Fruit-bodies usually more then 1 cm in diameter. Peridium of 1 or 2 indistinct layers. Gleba of a single labyrinthine cavity; basidia fusiform 1- to 4-spored; spores ellipsoid, colored, often longitudinally ribbed, irregularly warted or occasionally smooth; occasional capitate cystidia present . . . . **Hymenogastraceae**
>
> > A single genus, *Hymenogaster*, with about 70 species. There are treatments by Soehner (1962) and Dodge and Zeller (1934).

IX. NIDULARIALES

In Nidulariales, the fruit-body grows upon the ground or on an organic substrate. They are sessile and up to 1 cm in diameter, and often gregarious. The peridium is of one to many layers. The gleba consists of one to many, usually hard, seedlike peridioles and the spores are smooth, hyaline, thick-walled and often large. Clamp connections are present.

KEY TO FAMILIES AND IMPORTANT GENERA OF NIDULARIALES

1. Fruit-body up to 3 mm in diameter, initially globose, growing on wood or dung. Peridium of 6 layers, dehiscing by stellate apical splitting leaving the single globose peridiole free in the peridial cup; inner peridial layers becoming detatched from the outer except at the tips of the rays; differential absorption of water by the layers of the inner cup causing it to evert, shooting the peridiole away. Spores ellipsoid, smooth, hyaline, 10 μm in diameter; gemmae present. . . . . **Sphaerobolaceae**

> A single genus, **Sphaerobolus**, with 2 recognized species, but see Ingold (1971). Other accounts include those of Walker (1927) and Buller (1933).

1'. Fruit-body 1 mm to 1 cm in diameter, initially globose, usually becoming pulvinate or obconical. Peridium of 1 to 3 layers, dehiscing either irregularly or by a circumscissile epiphragm in which case the remains of the peridium form a cup. Peridioles occasionally 1, usually several to many per fruit-body. Basidia acrosporous, 4-spored with mucilaginous coat . . . . **Nidulariaceae** 2

> 2(1') Fruit-bodies pulvinate, dehiscing by irregular bursting or autolysis of peridium 3
>
> 2'(1') Fruit-bodies eventually cupulate following abscission of epiphragm 4
>
> 3(2) Peridium thick, cream to cinnamon, dimitic, of tinted, rigid, thick-walled, branched,

spiny hyphae, and hyaline, thin-walled hyphae with clamp connections. Peridioles brown, lenticular, numerous, covered in mucilage, free. Spores ellipsoid, hyaline, smooth
. . . . **Nidularia**
  About 3 species; widespread. Cejp and Palmer (1963) give a brief guide.

3'(2) Peridium thin, white, monomitic, of hyaline, thin-walled, clamped hyphae. Peridioles 1 to many, covered in mucilage, free. Spores ovoid, pip-shaped or ventricose, smooth; often mixed with metamorphosed basidia . . . . **Mycocalia**
  Five species, widely distributed. Cejp and Palmer (1963) give a good account.

  4(2') Peridioles not attached to wall of cup, lying free in a matrix of mucilage. Peridium cupulate, composed of 2 layers of hyphae. Spores ellipsoid . . . . **Nidula**
    About 4 species. Southern Asia, Australasia, the Americas.

  4'(2') Peridioles not noticeably mucilaginous, attached to wall of peridium by a funicle.
  5

5(4') Fruit-bodies growing on dead wood. Peridium 2-layered, dimitic. Peridioles with a thick, silvery coat (tunica), at least the majority attached to peridium by a simple elastic funicle. Spores very variable in size, ovoid . . . . **Crucibulum**
  A single widespread species, *C. vulgare.*

5'(4') Fruit-bodies growing on the ground, dead wood or dung. Peridium 5-layered. Peridioles with only a thin tunica, leaving the dark-colored inner layer visible, attached to peridium by a complicated funicle made up of sheath, middle-piece and purse, the last containing the coiled elastic thread. Spores smooth, ovoid sometimes to 40 μm long . . . . **Cyathus**
  The familiar birds' nest fungi. About 25 species; widely distributed, especially numerous in tropics. Brodie (1967) and Brodie and Dennis (1954) give an account of most of the species. For an account of methods of dispersal in this and *Crucibulum,* see Brodie (1951).

X. MELANOGASTRALES

Melanogastrales are saprobic or perhaps sometimes mycorrhizal. Their fruit-bodies are hypogeal, or sometimes they become epigeal at maturity; they are sessile or occasionally have a pseudostem. The peridium is single or two-layered and usually dehisces by attrition. The gleba is lacunar with infertile primary basidia and sporogenous secondary basidia (in some cases primary basidia become converted to "conidia"); the basidiospores are globose to long-ellipsoid, and smooth, with warted or reticulate, mucid exospore. They are hyaline or colored and the capillitium is absent. Clamp connections are absent.

KEY TO FAMILIES AND IMPORTANT GENERA OF MELANOGASTRALES

1. Pseudostem absent, fruit-body irregularly globose. Islands of primary basidia, gelatinizing to give way to secondary basidia (occasionally primary basidia produce "conidia"). Hypogeal, ? sometimes mycorrhizal . . . . **Melanogastraceae** 2

1'. Pseudostem present or not. Single primary basidia scattered through maturing gleba partially collapsing but persisting, forming foci towards which secondary, fertile basidia grow. Terrestrial or marine . . . . **Torrendiaceae** 4

2(1) Gleba pale, spores hyaline 3

2′(1) Gleba dark, strongly smelling, mucid, the lacunae clearly visible at maturity; "conidia" not produced; secondary basidia producing 4–8 ellipsoid basidiospores
. . . . **Melanogaster**
About 12 species widely distributed in N. Hemisphere often introduced in S. Hemisphere. Zeller and Dodge (1937) give an account.

3(2) Fruit-body cinnamon-colored, about 2 cm in diameter. "Conidia" not produced; secondary basidia with 8 ellipsoid, smooth hyaline spores about 4 × 2 μm. . . . . **Alpova**
One species, *A. cinnamomeus*, from Michigan, U.S.A.

3′(2) Fruit-bodies variously colored. Secondary basidia with 4 globose, reticulate or echinulate spores with conspicuous mucous sheath. In some cases development ceases at the stage of the primary basidia which produce "conidia" resembling the basidiospores (in these cases the fruit-bodies are referable to the imperfect genus *Leucophlebs*) . . . . **Leucogaster**
Ten species; N. Hemisphere. Most of the species revised by Zeller and Dodge (1924).

4(1′) Growing upon wood in the open sea. Fruit-body sessile subglobose, about 5 mm in diameter, orange. Peridium single, thin, falling away irregularly to release spore mass embedded in mucilage. Spores tetraradiate, sessile, hyaline; glebal islands consisting of an infertile primary basidium surrounded by 4-spored secondary basidia . . . **Nia**
A single species, *N. vibrissa*; Atlantic and Mediterranean. Details are given by Doguet (1967, 1968, 1969).

4′(1′) Terrestrial. Fruit-body initially hypogeal, consisting of developing gleba and compressed pseudostem, and surrounded by a simple universal veil. Glebal islands as in *Nia*; secondary basidia with 4 ovoid, hyaline spores borne on sterigmata. Pseudostem developed in same way as gleba but without spores (i.e., a subgleba), elongating at maturity to break the peridium which remains as a cupulate volva at the base and to expose the gleba . . . . **Torrendia**
A single species, *T. pulchella* (Fig. 8A); Portugal and Morocco. A description has been given by Malençon (1955).

XI. SCLERODERMATALES

The mature fruit-body is epigeal or hypogeal, more or less globose, or taking the form of an earth star, or consisting of many spore sacs seated on a stroma. The gleba is pulverulent, but betrays evidence of its lacunar development at least until just prior to maturity. Frequently, with several generations of basidia not all of them are sporogenous. Usually there is no capillitium. The spores are smooth, spiny, warted, or reticulate. Clamp connections are present or absent.

KEY TO FAMILIES AND IMPORTANT GENERA OF SCLERODERMATALES

1. Fruit-body sessile, initially subglobose but with thick and complex outer peridium which usually opens stellately from the apex to form an earth star, occasionally indehiscent; inner peridium of 1 or 2 layers. Gleba with abundant capillitium; spores irregularly warted or reticulated, globose or subglobose . . . . **Astraeaceae** 2

1′. Outer peridium relatively thin or peridium not obviously divided into layers 4

2(1) Fruit-body hypogeal, indehiscent, about 1 cm in diameter. Outer peridium cream,

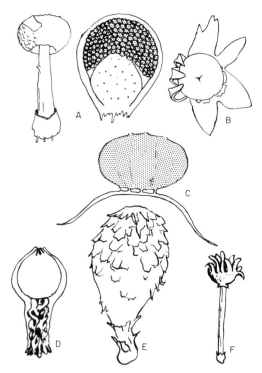

FIG. 8.   A, *Torrendia pulchella*, mature fruit-body and vertial section of undehisced fruit-body; B, *Astraeus hygrometricus*, expanded fruit-body (note remnants of outer layer of inner peridium); C, *Myriostoma coliformis*, diagrammatic vertical section of expanded fruit-body showing compound nature; D, *Calostoma lutescens*, mature fruit-body; E, *Phellorinia herculeana* ssp. *herculeana*, mature fruit-body; F, *Schizostoma laceratum*, mature fruit-body.

thick, made up of about 6 layers; inner peridium membranous, separable. Ripe gleba unknown, probably pulverulent; basidia clavate, pleurosporous; spores irregularly globose, about 4 μm in diameter with a small germ pore . . . . **Endogonopsis**
One species; eastern India. Developmental details are unknown but the thick, complex peridium and clavate pleurosporous basidia suggest a position near *Astraeus*. Some indehiscent forms of *Astraeus* may key out here.

2'(1) Fruit-body epigeal at maturity. Outer peridium stellately dehiscing from the apex consisting of an outer, floccose mycelial layer, a median fibrous layer, and an inner cartilaginous layer; endoperidium membranous of 2 separable layers, the outer ephemeral 3

3(2') Fruit-body up to 3 cm in diameter before dehiscence. Spore sac sessile; inner peridium of 2 layers, the outer layer ephemeral; dehiscence by apical slit. Glebal islands each formed of a group of infertile primary basidia which give place to fertile secondaries; secondary basidia clavate, pleurosporous spores discharged early and nourished by placental hyphae, at

maturity wrinkled, globose, about 10 µm in diameter; capillitium of branched, hyaline threads, clamped at the septa . . . . **Astraeus**
> One species, *A. hygrometricus* (Fig. 8B); among the most cosmopolitan of fungi (see Coker and Couch, 1928). Often placed with *Geastrum* with which it has strong superficial resemblance although true relationship is precluded by the very different development (see Malençon, 1955). Some forms are not dehiscent (see Heim and Wasson, 1970) and are apparently obligately mycorrhizal with dipterocarpous trees.

3′(2′) Fruit-body depressed, globose, up to about 6 cm in diameter before dehiscence. Spore sac raised on several pedicels, dehiscing by circular ostioles at the top, corresponding in number and relative position to the pedicels. Gleba divided into portions each corresponding to a pedicel and an ostiole, separated by sparse trabeculae of sterile tissue; spores brown, globose, about 4 µm in diameter with deep reticular ornament; capillitium hyaline with swellings at the septa . . . . **Myriostoma**
> One species, *M. coliformis* (Fig. 8C); widespread. Originally described from England but now apparently extinct there. Development unknown but by analogy with *Astraeus* probably lacunar. The spore sac and gleba apparently compound.

4(1′) Capillitium absent though flocci of collapsed trama may be present in gleba . . 6

4′(1′) Capillitium hyaline, branched, septate. Fruit-bodies pulvinate to subglobose frequently gregarious, often on a common subiculum. Peridium thin, usually tough, dehiscing apically. Basidia very large, branched, and lobed and bearing numerous spores on short sterigmata over the entire surface. Spores globose, reticulately wrinkled or warted . . . . **Glischrodermataceae** 5

5(4′) Dehiscence by irregular tearing of the apex . . . . **Lycoperdellon**
> Three described species; Afro-Iberian. Growing on the ground or on termitaria.

5′(4′) Dehiscence by regular ostiole with well-marked peristome . . . . . . **Glischroderma**
> One species, *G. cinctum*; Germany, England, and Morocco. Growing on charcoal. There is strong evidence, not only from the morphology but also from the cytology (Malençon, 1964), that the spore-bearing structures are indeed modified basidia.

6(4) Fruit-body compound, consisting of numerous spore sacs seated on a more or less massive stroma. Outer peridium forming a thin universal veil breaking away at maturity; inner peridia dehiscing by apical stomata. Capillitium present . . . **Broomeiaceae** 7

6′(4) Fruit-body simple, often strikingly yellow or orange. Peridium not divided into separable layers; dehiscence by attrition from the apex, occasionally irregularly stellate. Glebal islands consisting of groups of basidia which discharge their spores early, development being continued through the agency of placental hyphae; spores globose, brown, with well-developed spiny or reticulate ornament; capillitium absent but flocci persistent. Occasionally a pseudostem of well developed rooting strands
. . . . **Sclerodermataceae** 8

7(6) Compound fructification up to 15 cm in diameter, of 5–900 individual spore sacs seated in alveoli on the upper surface of a thick stroma. Peridium of spore sac covered with minute balls of hyphae, presumably sclerotial; ostiole with clearly defined conical fimbriate peristome. Spores brown, globose to ellipsoid about 6 µm in diameter; capillitium hyaline, branched, often sinuous . . . . **Broomeia**
> Two species, *B. congregata* and *B. ellipsospora* endemic to southern Africa. Development unknown, presumably lacunar. For complementary descriptions, see Bottomley (1948) and Dring and Rayner (1967). *Broomeia congregata* is apparently parasitic on *Acacia*, the other's substrate is unknown.

7′(6) Compound fructifications to 10 cm in diameter, of about 3–60 individual spore sacs each

seated in its own cup on the top of the stroma. Peridium of spore sac smooth; ostiole at first circular becoming irregular; peristome indefinite. Spores brown, about 7 μm in diameter, globose, irregularly warted; capillitium hyaline . . . . **Diplocystis**
One species, *D. wrightii*; W. Indies. An account is given by Lloyd (1920).

8(6′) Flocci fragile; gleba at maturity fully pulverulent. Dehiscence irregular, occasionally somewhat stellate. Spores spiny or reticulate or a combination of the two
. . . . **Scleroderma**
About 25 species; the common and widespread earth balls. There is a recent revision by Guzmán (1970).

8′(6′) Flocci persistent so that the gleba is divided into rather fragile, lenticular peridioles. Fruit-body variable in shape and size often with well-developed rooting base. Dehiscence by attrition from the apex. Spores brown, irregularly warted, globose, about 7 μm in diameter . . . . **Pisolithus**
One species is usually recognized, *P. arrhizus*, some forms of which are apparently saprobic and others mycorrhizal. Widespread particularly in subtropics. *Polysaccum* is a synonym.

XII. TULOSTOMATALES

The mature fruit-body is epigeal and the fertile portion is more or less globose and supported on a well-developed stipe. The peridium is simple or divisible into several layers. The gleba is powdery and a capillitium is present, at least in the early stages. Spores are globose to subglobose, smooth, spiny, warted, or reticulate. Clamp connections are present.

This order contains the stalked puffballs. Probably the two families are not very closely related. However, until further reliable developmental studies are forthcoming it seems wise to leave them together.

KEY TO FAMILIES AND IMPORTANT GENERA OF TULOSTOMATALES

1. Peridium complex consisting of 4 clearly defined layers, the outermost gelatinous or spiny, the second pigmented, the third very horny, the innermost membranous and remaining attached to the outer layers only at the top around the star-shaped apical pore and so hanging loose inside the horny layer. Stipe consisting of a continuation of the outer and the horny layers. Gleba pale, of claylike texture with annularly thickened capillitium initially which, however, disintegrates at maturity. Basidial islands formed from a primary infertile basidium surrounded by secondary fertile basidia; secondary basidia pleurosporous; spores large and often extravagantly ornamented either with a reticulum or long spines
. . . . **Calostomataceae**
A single genus, **Calostoma**, of which some 10 species are known (Fig. 8D); Australasia, southern Asia, the Americas. East Indian species are given in Boedijn (1938); some developmental details are given by Burnap (1897). *Mitremyces* is a commonly used synonym and aptly describes the configuration of the conspicuous peristome.

1′ Peridium relatively simple, often not divisible into well-defined layers, usually apically dehiscent. Stipe well-developed. Gleba powdery with well-developed capillitium; basidia pleurosporous . . . . **Tulostomataceae** 2

2(1′) True capillitium lacking though collapsed tramal hyphae or annularly thickened elaters present in abundance. Stipe massive, firmly attached to head 3

2'(1') True capillitium abundant. Stipe usually slender or, if massive, not firmly attached to head 6

3(2) Elaters not present in gleba; basidial fascicles persisting at maturity 4

3(2) Elaters present; gleba radially organized at least until just before maturity; basidial fascicles not persistent 5

4(3) Fruit-body up to about 15 cm tall; stipe and peridium confluent; volva usually lacking. Peridium pale, thick, ornamented when young by striking imbricate scales or large pyramidal warts, dehiscing widely from above. Stipe obconical. Gleba consisting of spores and collapsed tramal debris; spores brown, globose, warted, about 6 μm in diameter, tending to remain attached to the fascicles of collapsed basidia
. . . . **Phellorinia**
One species, *P. herculeana* (Fig. 8E), divisible into a number of superficially different subspecies. Widely distributed in warm arid regions. For a description, see Dring and Rayner (1967).

4'(3) Fruit-body up to about 50 cm tall. Stipe and peridium more or less confluent. Peridium pale, with deciduous horny scales or warts, dehiscing so as to expose the entire spore mass. Gleba held in place by fairly persistent tramal plates, foetid, brown; spores globose, about 6 μm in diameter. Stipe thick, slightly attenuated downward, expanding to a disk at the top. Volva initially with deliquescent inner layer . . . **Dictyocephalus**
One species, *D. attenuatus*; in warm arid regions of N. America and Africa. For detailed description, see Long and Plunkett (1940).

5(3') Fruit-body up to 20 cm tall; fertile portion carried on expanded disk of stipe. Outer peridium remaining only as a volva at maturity; inner peridium membranous, tough, dehiscing by several ostioles. Spores globose, slightly wrinkled, 6 μm in diameter; tramal debris and annular elaters present. Stipe at first smooth becoming scaly, slightly attenuated downward . . . . **Battarraeoides**
One species, *B. daguetii*, from southern N. America.

5'(3') Fruit-body as above but sometimes taller, and dehiscence circumscissile, the whole upper part of the peridium falling away in one piece . . . . **Battarrea**
Probably 2 species, *B. phalloides* endemic in southern England and often associated with decaying wood, and *B. stevenii* widespread in rather warmer more arid regions, growing in sandy soil.

6(2') Depressed globose fertile head firmly attached to discoid apex of obconical stipe. Outer peridium thin, fugaceous above, persisting below as a fibrillose cupulate volva; inner peridium pale, tough, and membranous smooth, dehiscing through an indefinite apical pore. Gleba ochraceous, more or less homogeneous at maturity but bearing traces of radial arrangement; spores ochraceous, globose, about 8 μm in diameter, moderately spiny; basidial fascicles persistent; capillitium hyaline, clamped. Stipe longitudinally grooved with a few scales. Volva ragged . . . . **Chlamydopus**
Two species; Australasia, Indian subcontinent, N. and S. America, in sandy soil. For a description of the better-known species, *C. meyeneanus*, see Cunningham (1944).

6'(2') Fertile head with socket below into which fits apex of usually slender, more or less cylindrical stipe. Peridia usually well-differentiated; outer peridium thin or thick, fugaceous or not; inner peridium tough and membranous. Gleba brown, homogeneous at all stages; spores globose to irregular, smooth or with irregular warts, occasionally with striae or reticula; capillitium abundant, hyaline or not. Volva usually reduced to a disk at base of stipe 7

7(6') Stipe slender, usually more or less cylindrical. Inner peridium not falling away in irregular patches from the apex. Fruit-body usually less than 10 cm tall 8

7'(6') Stipe not slender, but nevertheless well differentiated from head, attenuated upward. Fruit-body about 10 cm tall. Dehiscence by falling away in irregular patches from the apex. Gleba with hyaline capillitium; spores warted . . . . **Queletia**
> A single species, *Q. mirabilis,* growing upon spent tan in France, Britain, and U.S.A. (Penn).

8(7) Dehiscence by breaking of the peridium into irregular lobes which open outward. Outer peridium represented at most by small ragged volva at maturity. Spores pip-shaped, dark, smooth, about 4.5 μm in diameter. Capillitium of dark and hyaline hyphae, septate, fragmenting . . . . **Schizostoma**
> *Schizostoma laceratum* (Fig. 8F) is the only well-marked species but there are local variants. Africa, Persia, Khazakstan, W. Pakistan, N. and S. America. For a full description, see Dring and Rayner, 1967.

8'(7) Dehiscence by a well-defined pore, peristome definite or indefinite, plane or tubular. Outer peridium represented in the mature fruit-body by a collar, often in the form of a sand-case, around the base of the spore sac, or sometimes as a spinose covering like the exoperidium of *Lycoperdon*. Spores usually about 4 μm in diameter, globose to irregular, smooth, more often irregularly to regularly warted, occasionally striate or reticulate; capillitium hyaline to dark, thin-walled or thick, swollen at the septa or not. Stipe fitting closely or loosely into a socket in the fertile head. Volva occasionally present, usually obsolete . . . . **Tulostoma**
> Probably about 50 valid species; worldwide, usually growing on sandy soil, occasionally on rotten wood in forest. A useful introduction to the specific characters is given by Wright (1955).

REFERENCES

Boedijn, K. B. (1938). The genus *Calostoma* in the Netherlands Indies. *Bull. Jard. Bot. Buitenzorg* [3] **16**:64–75.

Bottomley, A. M. (1948). Gasteromycetes of South Africa. *Bothalia* **4**:473–810.

Brodie, H. J. (1951). The splash-cup mechanism in plants. *Can. J. Bot.* **29**:224–234.

Brodie, H. J. (1967). New records of Nidulariaceae from the West Indies. *Trans Brit. Mycol. Soc.* **50**:473–487.

Brodie, H. J., and R. W. G. Dennis. (1954). Nidulariaceae of the West Indies. *Trans. Brit. Mycol. Soc.* **37**:151–160.

Buller, A. H. R. (1933). "Researches on Fungi," Vol. 5. Longmans, Green, New York.

Burnap, E. (1897). Notes on the genus *Calostoma*. *Bot. Gaz.* **23**:180–196.

Cejp, K., and J. T. Palmer. (1963). The genera *Nidularia* and *Mycocalia* in Czechoslovakia. *Ceska Mykolo.* **17**:113–124.

Coker, W. C., and J. N. Couch. (1928). "Gasteromycetes of the Eastern United States and Canada." Univ. of North Carolina Press, Chapel Hill.

Cunningham, G. H. (1944). "Gasteromycetes of Australia and New Zealand." Dunedin.

Dissing, H., and M. Lange. (1962). Gasteromycetes of the Congo. *Bull. Jard. Bot. Brux.* **32**: 325–416.

Dodge, C. W., and S. M. Zeller. (1934). *Hymenogaster* and related genera. *Ann. Mo. Bot. Gard.* **21**:625–708.

Doguet, G. L. (1967). *Nia vibrissa*, remarquable basidiomycète marin. *C. R. Acad. Sci.,* Ser. D **265**:1750–1783.

Doguet, G. L. (1968). *Nia vibrissa*, gasteromycète marin. I. *Bull. Soc. Mycol. Fr.* **84**:343–351.

Doguet, G. L. (1969). *Nia vibrissa*, gasteromycète marin. II. *Bull. Soc. Mycol. Fr.* **85**:93–104.

Dring, D. M. (1964). Gasteromycetes of west tropical Africa. *Mycol. Pap.* **98**:1–60.

Dring, D. M. (1966). Morphological relationship between Phallaceae and Clathraceae. *Brit. Mycol. Soc. News Bull.* **25**:21–24.

Dring, D. M. (1974). "An Introduction to the Gasteromycetes." Academic Press, New York. In preparation.

Dring, D. M., and R. W. Rayner. (1967). Some gasteromycetes from Eastern Africa. *J. E. Afr. Uganda Natur. Hist. Soc.* **24**:5–46.

Fischer, E. (1890). Untersuchungen zur vergleichende Entwicklungsgeschichte und Systematik der Phalloideen. *Neue Denkschr. Schweiz. Ges. Naturw.* **23**:1–103.

Fischer, E. (1893). Untersuchungen. II. *Neue Denkschr. Schweiz. Ges. Naturw.* **33**:1–51.

Fischer, E. (1900). Untersuchungen. III. *Neue Denkschr. Schweiz. Ges. Naturw.* **63**:1–84.

Fischer, E. (1934). Zur Kenntnis der Fruchtkörperentwicklung von *Podaxis*. *Ber. Schweiz. Bot. Ges.* **43**:11–18.

Fitzpatrick, H. M. (1913). A comparative study of the development of the fruit-bodies of *Phallogaster, Hysterangium* and *Gautieria*. *Ann. Mycol.* **11**:119–147.

Furtado, J. S., and D. M. Dring. (1967). The rediscovery of *Protubera maracuja*, with additional descriptive notes. *Trans. Brit. Mycol. Soc.* **50**:500–502.

Guzmán, G. (1970). Monografia del genero *Scleroderma*. *Darwiniana* **16**:233–407.

Heim, R., and R. G. Wasson. (1970). Les putka des Santals, champignons doués d'une âme. *Cah. Pac.* **14**:1–86.

Ingold, C. T. (1971). The glebal mass in *Sphaerobolus*. *Trans. Brit. Mycol. Soc.* **56**:105–113.

Kawamura, S. (1937). *Lasiosphaera fenzlii*. *J. Jap. Bot.* **13**:748–751.

Kotlaba, F., and Z. Pouzar. (1964). Preliminary results on the staining of spores and other structures of Homobasidiomycetes in cotton blue, and its importance for taxonomy. *Feddes Repert.* **69**:131–142.

Kreisel, H. (1962). Die Lycoperdaceae der Deutschen Demokratischen Republik. *Feddes Repert.* **64**:89–201.

Kreisel, H. (1967). Taxonomisch-pflanzengeographiche Monographie der Gattung *Bovista*. *Beih., Nova Hedwigia* **25**:1–244.

Kreisel, H. (1969). "Grundzüge eines natürlichen Systems der Pilze." Fischer, Jena.

Kreisel, H., and D. M. Dring. (1967). An emendation of the genus *Morganella*. *Feddes. Repert.* **74**:109–122.

Lloyd, C. G. (1909). Synopsis of known phalloids. *Mycol. Writ.* **2**:1–95.

Lloyd, C. G. (1920). *Diplosytis*. *Mycol. Writ.* **6**:917.

Lohwag, H. (1926). Zur Entwicklungsgeschischte und Morphologie der Gasteromyceten. *Beitr. Bot. Zentralbl.* [2] **42**:177–234.

Long, W. H., and O. A. Plunkett. (1940). Studies in Gasteromycetes. I. The genus *Dictyocephalos*. *Mycologia* **32**:696–709.

Malençon, G. (1955). Le développement du *Torrendia pulchella*, et son importance morphogénetique. *Rev. Mycol.* **20**:81–130.

Malençon, G. (1964). Le *Glischroderma cinctum*, sa structure et ses affinités. *Bull. Soc. Mycol. Fr.* **80**: 197–211.

Martin, G. W. (1939). The genus *Lycogalopsis*. *Lilloa* **4**:69–73.

Moravec, Z. (1958). Disciseda. In "Flora ČSR" (A. Pilát, ed.), Vol. B–1, pp. 377–386. Czech Akad. Publ., Prague.

Morse, E. E. (1933). A study of the genus *Podaxis*. *Mycologia* **25**:1–33.

Morse, E. E. (1935). A new puffball. *Mycologia* **27**:96–101.

Pilát, A. (1934). Sur le genre *Gastrosporium*. *Bull. Soc. Mycol. Fr.* **50**:37–49.

Ponce de León, P. (1970). Revision of the genus *Vascellum*. *Fieldiana* **32**:109–125.

Ponce de León, P. (1971). Revision of the genus *Morganella*. *Fieldiana* **34**:27–44.

Reid, D. A., and D. M. Dring. (1964). British records. 71. *Trans. Brit. Mycol. Soc.* **47**:293–295.

Savile, D. B. O. (1968). Possible interrelationships between fungal groups. *In* "The Fungi" (G. C. Ainsworth and A. S. Sussman, eds.), Vol. 3, pp. 649–675. Academic Press, New York.

Soehner, E. (1962). Die Gattung *Hymenogaster*. *Nova Hedwigia, Beih.* **2**:1–113.

Staněk, V. J. (1958). *Geastrum*. *In* "Flora ČSR" (A. Pilát, ed.), Vol. B–1, pp. 402–526. Czech. Acad. Publ., Prague.

Walker, L. B. (1927). Development and mechanism of discharge in *Sphaerobolus iowensis* n. sp. and *S. stellatus*. *J. Elisha Mitchell Sci. Soc.* **42**:151–178.

Wright, J. E. (1955). Evaluation of specific characters in the genus *Tulostoma*. *Pap. Mich. Acad. Sci., Arts Lett.* **40**:79–87.

Zeller, S. M. (1939). New and noteworthy gasteromycetes. *Mycologia* **31**:1–32.

Zeller, S. M. (1945). Studies in the gasteromycetes, XI. The genera *Trichaster* and *Terrostella*. *Mycologia* **37**:601–608.

Zeller, S. M., and Dodge, C. W. (1918). *Gautieria* in North America. *Ann. Mo. Bot. Gard.* **5**:133–142.

Zeller, S. M., and C. W. Dodge. (1924). *Leucogaster* and *Leucophlebs* in North America. *Ann. Mo. Bot. Gard.* **11**:380–410.

Zeller, S. M., and C. W. Dodge. (1929). *Hysterangium* in North America. *Ann. Mo. Bot. Gard.* **16**:83–128.

Zeller, S. M., and C. W. Dodge. (1937). *Melanogaster*. *Ann. Mo. Bot. Gard.* **23**:639–655.

Zeller, S. M., and A. H. Smith. (1964). The genus *Calvatia* in North America. *Lloydia* **27**:148–186.

Author Index

Numbers in italics refer to the pages on which the complete references are listed.

Index to Genera, Families, Orders, and Higher Taxa

An asterisk (*) after a page number indicates an illustration. (†) after a page number indicates a key. *Note*: For new names see pp. 239, 382, 383, 384, 386, 387.

A

Abortiporus Murr., 404
Absidia van Tieghem, 212*, 213
Abstoma G. H. Cunn., 464
Acanthobasidium Oberw., 332
Acanthophysellum Parm., 332
Acanthophysium (Pilát) G. H. Cunn. em. Parm., 332
Acaulopage Drechsler, 232*, 234
Acephalis Badura & Badorowa = *Syncephalis*
Achlya Nees, 63*, 131, 132*, 134*, 135–138
Achlyogeton Schenk, 97
Achlyogetonaceae, 96
Achroomyces Bon., 313
Achrotelium Syd., 270
Ackermannia Pat. = *Sclerocystis*
Acrasiaceae, 18
Acrasidae, 11*, 17
Acrasiales, 18†
Acrasiomycetes, 2, 3, 5, 9–35, 10†
Acrasis van Tieghem, 18, 19, 20*
Actiniceps Berk. & Br., 356
Actinocephalum Saito = *Cunninghamella*
Actinomortierella Chalabuda = *Mortierella*
Actinomucor Schostakowitsch, 220
Actinostroma Klotzsch, 343
Acurtis Fr., 433
Acytosteliaceae, 26, 32
Acytostelium Raper, 10, 26, 32, 34*
Aecidium Pers. & Pers., 264
Aeruginospora Höhn., 426
Agaricaceae, 424, 434†, 441, 444†
Agaricales, 323, 407, 421–450, 423*
Agaricus Fr., 434
Agrocybe Fayod, 435
Alacrinella Manier & Ormières ex Manier, 243
Albatrellus Gray, 403, 410

Albuginaceae, 165, 168
Albugo Pers. ex Gray, 165, 168, 169*
Alectorolophoides Earle, 366
Aleurocorticium Lemke, 333
Aleurocystidiellum Lemke, 332
Aleurocystis Lloyd ex G. H. Cunn., 332
Aleurocystis Lloyd as McGinty, 333
Aleurodiscus Rabenh. em. Parm., 333
Allantula Corner, 356
Allochytrium Salkin, 104
Allomyces Butler, 94*, 105
Allopuccinia Jackson, 273
Allotelium Syd., 275
Alnicola Kühner, 439
Alpova Dodge, 471
Althornia Gareth-Jones & Alderm., 129, 130
Alveolaria Lagerh., 268
Amanita Pers. ex Hook., 421, 433
Amanitaceae, 423, 433
Amaurochaete Rost., 59
Amauroderma Murr., 417
Amaurodon Schroet., 339
Amoebidiaceae, 240*, 243†
Amoebidiales, 238, 243†
Amoebidium Cienk., 240*, 243
Amoebochytrium Zopf, 101
Amoebophilus Drechsler, 231, 234
Amporoina Singer, 432
Amphicypellus Ingold, 102
Amphinema Karst., 340
Amylaria Corner, 355, 417
Amylocorticium Pouz., 337
Amylocystis Bond. & Singer ex Singer, 411
Amylomyces Calmette = *Chlamydomucor*
Amyloporia Bond. & Singer ex Singer, 398*, 407
Amylostereum Boid., 345
Ancylistaceae, 222
Ancylistes Pfitzer, 220, 221, 227, 228*

487